APPLIED WEED SCIENCE

Second Edition

Merrill A. Ross
Carole A. Lembi
Purdue University
West Lafayette, Indiana

Prentice Hall
Upper Saddle River, New Jersey 07458

Library of Congress Cataloging-in-Publication Data

Ross, Merrill A.
　　Applied weed science / Merrill A. Ross, Carole A Lembi. — 2nd ed.
　　　　p.　　cm.
　　Includes bibliographical references and index.
　　ISBN 0-13-754003-5
　　1. Weeds—Control.　2. Weeds.　I. Lembi, Carole A.　II.　Title.
SB611.R67　1999
632'.5—dc21　　　　　　　　　　　　　　　　　　　98-22059
　　　　　　　　　　　　　　　　　　　　　　　　　　CIP

Acquisitions Editor: Charles Stewart
Editorial Assistant: Jennifer Stagman
Editorial Production Services: WordCrafters Editorial Services, Inc.
Managing Editor: Mary Carnis
Director of Production and Manufacturing: Bruce Johnson
Prepress Manufacturing Buyer: Marc Bove
Marketing Manager: Melissa Bruner
Cover Designer: Liz Nemeth
Printer/Binder: Courier Westford

 © 1999 by Prentice-Hall, Inc.
Simon & Schuster/A Viacom Company
Upper Saddle River, NJ 07458

Printed in the United States of America

10　9　8　7　6　5　4　3

ISBN 0-13-754003-5

Prentice-Hall International (UK) Limited, *London*
Prentice-Hall of Australia Pty. Limited, *Sydney*
Prentice-Hall Canada, Inc., *Toronto*
Prentice-Hall Hispanoamericana, *Mexico*
Prentice-Hall of India Private Limited, *New Delhi*
Prentice-Hall of Japan, *Tokyo*
Simon & Schuster Asia Pte. Ltd., *Singapore*
Editora Prentice-Hall do Brasil, Ltda., *Rio de Janeiro*

CONTENTS

PREFACE

The purpose of this book is to meet the need for a reference that relates the principles of weed science directly to weed management situations encountered in the field. Although the concept is simple, the objective is difficult to attain.

From many years of experience in extension and teaching, we have concluded that poor results from the application of weed management practices do not stem necessarily from a lack of knowledge but from an inability to understand how bits and pieces of weed science knowledge relate to crop management systems as a whole. To be successful in weed control, one must conceptualize weed and crop management as an integrated unit, recognizing that each individual management practice, whether designed for weed control or not, influences the outcome of a weed management program.

We have attempted to facilitate application of weed science principles to practical weed management in five ways. The first is to provide a body of knowledge about weeds and weed control methods that is organized, comprehensive, and as up-to-date as possible. The reader should recognize that modern weed management is a high technology endeavor. Therefore, learning terminology and technical information must be a high priority. Although all chapters provide technical information, the elements needed for a basic understanding of weeds and weed control are presented in the first half of the book. Important characteristic of weeds are covered in Chapters 1 (Characteristics, Biology, and Importance of Weeds), 2 (The Plant System), and 6

(Plant-Herbicide Interactions). Another important characteristic of weeds, their potential to develop herbicide resistance, was covered in just three paragraphs in the first edition of this book. In the 13 years since that time, the increased incidence of herbicide resistance in weeds and the introduction of herbicide resistant crops merit a new chapter on Herbicide Resistance (Chapter 8). The framework for understanding the principles of weed management is presented in Chapters 4 (Methods of Weed Control), 5 (Introduction to Herbicides), 6 (Plant-Herbicide Interactions), and 7 (Soil-Herbicide Interactions). Individual herbicides and herbicide groups are described in Chapters 9 through 12. The technology associated with herbicide formulations and packaging has changed and expanded so much in 13 years that what was once a single chapter on herbicide application methods has been expanded into two in order to adequately cover herbicide formulations and related information (Chapter 16) and application methods (Chapter 17).

The second feature of this book is based on the assumption that readers will have diverse backgrounds with respect to experience, familiarity with equipment, individual needs, and formal training in weed science, soil science, crop science, and plant biology. We intend for this book to be used by college students in introductory weed science courses; professional agriculturalists such as agribusiness marketing, service, and technical persons; pest management specialists; outstanding farmers and vocational agricultural teachers; and students in short courses. To compensate for the wide range of user understanding and knowledge, we have included background material that we believe is essential to the scientific use of weed control information. In particular, Chapters 2, 3, and 19 provide background on plant structure and function, soils, and tillage equipment and practices (an area in which college students from a non-farm environment have little or no background), respectively. Besides providing background information, these chapters describe how plant, soil, and tillage systems interact with weeds and weed control methods and vice versa.

In addition to classroom study, we hope this book will find its way into the libraries and offices of persons actually involved in solving weed control problems. We believe that a third important feature of this book is its reference value. Once the student has completed college or other training and is pursuing a career, he or she should be able to refer back to this book for weed science information not readily available elsewhere.

The fourth feature of this book is our attempt to illustrate an approach to solving weed problems. We describe the kinds of factual information that must be obtained to design a successful weed control program, emphasizing the need for integrating appropriate weed control practices with the crop management system throughout the growing season and even beyond. The framework for this approach is spelled out in the first part of Chapter 4. Placing this discussion toward the beginning of the book serves to emphasize its importance. It should not be ignored! In essence, the remainder of the book provides the information required to carry out the steps of this approach.

Finally, the last half of the book with chapters on Weed Life Cycles and Management,

Weed Management Situations, Aquatic Plant Management, and Troubleshooting demonstrate how one's approach and technical knowledge together can be employed to solve actual problems. We conclude with a new chapter entitled Weed Control in an Age of Rapidly Changing Technology. The technology of agriculture, indeed of all biology, is changing at a rapid pace, and we must recognize the potential benefits and pitfalls that new strategies and approaches provide in the area of weed control.

We sometimes restate the obvious in this text. Our experience convinces us that poor weed control results can be the consequence of not recognizing, using, or understanding some elementary concept or practice. We hope that such repetition will not detract from the overall usefulness of the book.

The major concepts and new terminology sections of each chapter are intended to aid the reader by providing special emphasis and a quick reference to the content of the individual chapters.

No text of reasonable length can cover all herbicides, application equipment, and every crop and weed found in North America. This diversity plus a rapidly changing technology results in the need for the practicing agriculturalist to acquire supplemental, up-to-date information. Federal and state agencies, university recommendations and publications, herbicide labels, equipment catalogues, etc., are logical sources. Similarly, no attempt has been made in this text to provide (with the exception of aquatic weeds) a guide to weed identification. A list of available weed identification guides is given in Appendix A.

Merrill A. Ross
Carole A. Lembi

Identify the Weed Leaves?
The leaf outline associated with each chapter number represents a different weed species. Can you identify them? The answers are given on page 440.

ACKNOWLEDGMENTS

Many people have helped in the preparation of the second edition of this book. We are deeply grateful for the technical assistance of Ruth Brown, Caroline Logan, Debra Lubelski, Jennifer Rankin, and Anita Eberle. We thank our colleagues, Tom Bauman, Mike Foley, and Tom Jordan, who have always been willing to share knowledge, insights, and reference books with us. We especially thank Tom Bauman and Dan Childs for teaching a semester of our weed science course so that one of us (M. A. R.) could devote more time to writing this book. We also thank Fred Warren and Steve Weller for allowing us to use the written materials from their Herbicide Mode of Action short course given at Purdue University each year. The materials provided by Dan Hess have been particularly helpful to us.

Over the years we have been inspired and helped by many individuals, including the undergraduate students in our classes, graduate students, past and current colleagues, family, and friends. We thank all of these people, without whose support, we could not have written this book. We also give a special thanks to Judith Goodrich, formerly of Burgess Publishing Company, wherever she may be, for convincing us that we could do it!

CHARACTERISTICS, BIOLOGY, AND IMPORTANCE OF WEEDS

Before we can intelligently approach the subject of weeds and their control, we must determine exactly what a weed is. Two frequently used definitions are *plants growing where they are not wanted* and *plants out of place*. Weeds also have been defined as plants whose virtues have yet to be discovered. Unfortunately, these definitions are limited and somewhat misleading.

The unwanted-plant or plant-out-of-place definitions do not distinguish plants that possess truly weedy characteristics from those that are only occasional nuisances. These definitions inaccurately imply that the physical location of a plant is the sole factor determining its potential as a weed. Volunteer corn in soybean fields or a petunia in a marigold patch are certainly out of place and unwanted, but neither has the potential to become the long-lasting, noxious weed problem that plants such as common cocklebur or field bindweed have become.

The undiscovered-virtues concept suggests that if we could only find a use for a weed, it would no longer be a weed. The fact that johnsongrass, one of the world's worst weeds, was introduced into this country as a forage crop did not prevent it from infesting millions of acres of valuable cropland and becoming an extremely noxious pest. Johnsongrass *still* would be a weed even if a use were found that would make it as valuable as corn, cotton, or tomatoes.

A weed is a weed because it possesses certain definable characteristics that set it apart from other plant species. These characteristics include one or more of the following features: abundant seed production (and thus potentially large populations), rapid population establishment, seed dormancy, long-term survival of buried seeds, adaptations for spread, the presence of vegetative reproductive structures, and the capacity to occupy sites disturbed by human activities. Because of these features, we think weeds are better described as *plants that interfere with the growth of desirable plants and that are unusually persistent and pernicious. They negatively impact human activities and as such are undesirable.*

Relatively few plants have the characteristics of true weeds. Of the total number of plants in the world (about 250,000 species), only a few thousand are thought to behave as weeds. Of these, about 200 species or 0.08% of the total are recognized as major problems in world agriculture (Holm et al. 1977, 1979, 1997). Holm et al. (1997) suggest that these 200 species account for 90% of the loss in world food crops. Only about 25 species or 0.01% of the total cause the major weed problems in any one crop. If you think in terms of a single crop, whether it be corn, tomatoes, turf, or Christmas trees, you probably will be able to list only a handful of weed species that cause continuing problems. Although a casual glance at woods, prairies, and other relatively undisturbed areas may reveal a variety of weed-like plants, most do not possess the characteristics that enable true weeds to survive and compete, under the environmental conditions imposed by modern crop management practices.

CHARACTERISTICS OF WEEDS

What are the specific features that set weeds apart from other plants, and how do these features permit weeds to interfere with the growth of desirable plants, be unusually persistent and pernicious, and negatively impact human activities?

Interference with Growth of Desirable Plants

Interference is the term used to describe the total impact of one plant on another. When managing crops, the concern is the ability of weeds to interfere with, or adversely affect, crop growth. Interference includes competition and allelopathy. *Competition* is the mutual struggle of two or more plants for some growth factor (water, light, or nutrients) that has become limiting (i.e., is in short supply). *Allelopathy* is the inhibition of the growth of nearby plants through the production of biological toxins.

Competition

All green plants require light, water, and mineral nutrients for growth. In crop production systems in which weeds are present (as they usually are), both crops and weeds compete for the same limited supplies of these essential growth substances. Since substances utilized by weeds are not available to support the growth of crop plants, crop vigor and yield often are drastically reduced. This effect is particularly noticeable when one or more of the important growth factors is in short supply. For example, when soil moisture is low, weed-infested corn shows leaf curling much earlier than weed-free corn does. Similarly, nutrient deficiencies in a crop are more evident when weeds are present (Figure 1.1).

The ability of weeds to compete successfully with crops for light, water, and nutrients depends on several interrelated factors. These include the timing of weed emergence in relation to crop emergence, the growth form of the weed, and the density of weeds present in the crop.

Timing of Weed Emergence. *The first plant that effectively obtains water, nutrients, and light from a site and becomes established at that site has a distinct competitive advantage over plants that develop later.* The first plant not only has the advantage of utilizing available resources but also, by virtue of its growth (for example, the development of a shading canopy or an extensive root system to better tap the water supply), can adversely affect the growth of plants that develop later. In practical terms, crops established before weeds

Figure 1.1 Under conditions of limiting nitrogen supply, giant foxtail-infested corn plants (left) are yellowed and stunted in contrast to weed-free corn (right).

emerge have a good chance of producing acceptable yields. If, on the other hand, weeds become established before the crop (for example, planting a crop into a no-till site in which the weeds have not been controlled, or planting a crop in a site with rapidly developing perennial weeds), crop yields will almost always be adversely affected. In fact, weeds that become established early reduce not only the development of crops but also the development of weeds that germinate later. For example, the successful control of early-germinating annual broadleaved weeds often gives late-germinating weeds such as annual grasses a much better opportunity to grow and compete with crop plants

The effect of weed competition is greatest when the crop is young; thus, crop yields are much more likely to be reduced by early-season weed competition than by late-season competition. During the early stages of development, crops are most susceptible to the adverse conditions caused by weeds, such as shading or competition for water and nutrients. In fact, if a crop with good competitive ability (rapid growth rate, dense canopy formation) can be kept free of annual weeds for 2 to several wk after planting (Table 1.1), weeds developing later in the season usually cause little or no loss of yield. For example, 1.5 common ragweed seedlings per meter of row emerging at the same time as white beans resulted in a 19 to 22% loss in seed yield (Chikoye et al. 1995). However, the same density of ragweed seedlings emerging when the white beans were in the second trifoliolate stage resulted in yield losses of only 4 to 9%. Similarly, downy brome caused two- to fivefold greater reductions in yield when it emerged within 3 wk after winter wheat than

Table 1.1 Weed-Free Period Required After Planting or Emergence to Prevent Crop Yield Loss

Crop	Weed-free Period (weeks)	Weed
Corn[a]	4	Giant foxtail
Sorghum[b]	4–5	Mixed annuals
Field beans[c]	4–5	Mixed annuals
Snapbeans[d]	2	Common purslane
Sunflowers[e]	4–6	Mixed annuals
Cotton[f]	6–8	Mixed annuals
Peanuts[g]	6	Smooth pigweed, large crabgrass
Sugarbeets[h]	10–12	Mixed annuals
Onions[i]	12	Redroot pigweed, kochia, grasses

Data from [a]Knake and Slife (1965), [b]Burnside and Wicks (1967), [c]Dawson (1965a), [d]Vengris and Stacewicz-Sapuncakis (1971), [e]Johnson (1971), [f]Buchanan and McLaughlin (1975), [g]Hill and Santelmann (1969), [h]Dawson (1965b), [i]Wicks et al. (1973).
Data for d, f, and i based on weed-free periods after crop emergence. All others based on weed-free periods after crop planting.

when it emerged six wk later (Blackshaw 1993). Crops that are not very competitive, such as onions and sugar beets, may require weed-free periods as long as 10 to 12 wk in order to obtain maximal yields (Table 1.1).

The sequence of events that occurs when crop seeds are planted and competition ensues due to the emergence of weeds has led to the development of a concept known as the *critical period* (Nieto et al. 1968, Weaver and Tan 1983, Zimdahl 1988, Hall et al. 1992). Knowing the critical period is useful for determining when weed control measures will be most effective in preventing crop yield loss.

The critical period refers to the time when weeds will have a negative impact on the crop. In general, if the crop and the weeds emerge together, crop losses will be minimal if weeds remain for the first few weeks after emergence. This initial period, however, is followed by a period over the next several weeks when the crop is adversely affected by the weeds. Crop losses during this period will continue to accumulate as weed pressure increases. This intermediate period represents the critical period. From the end of the critical period up to harvest, the presence of weeds will have little or no further negative impact on crop yield.

The critical period often is referred to as the *critical period of weed control* because it represents the period when weeds must be controlled to prevent significant crop yield losses. The critical period of weed control obviously must be timed for the period when the crop is being negatively impacted by weeds. The method for determining the critical period of weed control in relation to a particular crop and weed association is based on two types of experiments.

In the first set of experiments, usually termed weed-infested (or weed-removal) experiments,

weeds are allowed to emerge and grow with the crops for various lengths of time during the growing season. For example, weeds may be allowed to grow with the crop for 1 wk following emergence; at the end of 1 wk, the weeds are removed by hand weeding or herbicides. In another plot, weeds may be allowed to grow with the crop for 2 wk before being removed; in another plot, for 3 wk, and so on. The resulting crop yields at harvest establish the length of time that the crop can grow with the weeds without a significant loss of crop yield. This point marks the start of the critical period (Figure 1.2), when competition from weeds begins and weed control is necessary to prevent crop losses.

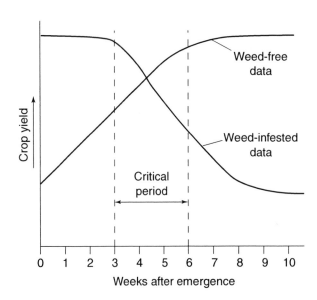

Figure 1.2 The critical period of weed control begins when crop yields in the weed-infested experiments begin to decline and ends when crop yields in the weed-free experiments are no longer negatively impacted by the emergence of weeds. In this example, the critical period is from 3-6 wk after crop and weed emergence.

The second set of experiments, usually termed weed-free experiments, is conducted on crops that are kept weed free for certain periods of time. The weeds are removed for various periods of time after crop emergence. For example, weeds may not be allowed to emerge in the crop until 1 wk after crop emergence, 2 wk after crop emergence, and so on. The resulting crop yield data determine the length of time that a crop must be kept weed free in order not to lose yield. This point marks the end of the critical period (Figure 1.2).

An ideal time for treatment is at the start of the critical period; in other words, treatment must be early enough so that weeds emerging with the crop do not reduce yield (3 wk in Figure 1.2). The treatment (or subsequent treatments) must last long enough to control weeds up to the time when newly emerging weeds no longer have an impact on yield (6 wk in Figure 1.2).

The timing of the critical period of weed control varies with the crop species and its ability to compete with weeds, the level of infestation, growing conditions, and even characteristics of the site. In some cases, the beginning of the critical period has not been easy to determine. For example, the beginning of the critical period for weed control of annual weeds in corn in southern Ontario was extremely variable, ranging from the 3- to the 14-leaf stages of corn development; however, the end of the critical period was relatively consistent at the 14-leaf stage (Hall et al. 1992). Although it was difficult to tell when control measures should be initiated, any weed-control efforts taken after the 14-leaf stage had little effect on crop yield. On the other hand, some experimental data appear to be relatively clear cut. For example, the critical period for johnsongrass control was determined to be between 3 and 6.5 wk after corn emergence in Texas (Ghosheh et al. 1996).

In some cases, the beginning of the critical period is at or even before crop emergence. This is particularly true of noncompetitive crops, in sites with heavy weed pressure, or in no-till systems in which weeds are present and established at planting. In cotton, which is not a very competitive crop, lint yield was reduced 11.2% for each week that removal of ivyleaf morningglory was delayed up to 9.5 wk. Only 0.2 % per week was lost for each week of competition after 9.5 wk (Rogers et al. 1996). Thus, weed removal starting at the time of emergence was essential to avoid significant yield losses. For a relatively competitive crop such as potatoes growing in quackgrass-infested fields in Québec, levels of weed pressure, as measured by weed density, af-

fected the start of the critical period. The critical period started at about 15 days after potato emergence at a low level of weed density, and at about 3 days after emergence at a medium level of weed density (Baziramakenga and Leroux 1994). At a high level of infestation, the critical period actually began before the potatoes emerged. Thus, early quackgrass control was needed in heavy infestations in order to reduce loss of crop yield. In no-till situations, weeds that are present must be controlled prior to crop emergence in order to prevent early-season competition and crop losses.

In some situations, late-germinating weeds can cause yield loss. Crops that are planted and develop during periods of cool weather (for example, sugar beets planted in March or winter grains planted in the fall) do not appreciably suffer from weed competition until warming of soil temperatures in late spring causes weeds to germinate and/or begin rapid growth. Onions, another crop that is slow to develop and never forms a shading canopy, is also a poor competitor with later-germinating weeds.

While late-developing weeds may not decrease yields in many crops, they can create serious problems by lowering crop quality, interfering with harvest operations, and producing seeds and vegetative structures that ensure continuing weed problems. Thus, yield loss is not the only factor to consider when deciding whether a weed control program is necessary.

Growth Form. The growth habits of plants (extent of root development, height, leaf area, amount of branching) and plant growth rates are determined by genetic and environmental factors. If environmental conditions (particularly those at the time of emergence) permit a plant to develop rapidly to its full height and canopy, it will have a distinct competitive advantage over plants that normally have slower growth rates, are shorter, or form less dense canopies. The resulting shade, in combination with the loss in water and nutrients, drastically reduces yields. For example, large weeds such as velvetleaf, jimsonweed, and giant ragweed, which germinate at the same time as soybeans, rapidly outgrow soybeans and form a shading canopy (Figure 1.3). Some large weeds not only produce a canopy but also are somewhat shade tolerant (cocklebur) or develop shade tolerance on later-developing leaves (giant ragweed) so that they produce lateral branches and leaves below the soybean canopy (Regnier and Stoller 1989, Webster et al. 1994), thus adding to their potential competitive ability. The ability of giant

Figure 1.3 Although these velvetleaf and soybean plants germinated at the same time, the more robust growth habit of the velvetleaf allows it to over-top the crop plants.

Figure 1.4 Low-growing weeds such as common purslane can compete with low-growing crops such as cucumber.

ragweed to use both strategies, rapid growth and shade tolerance within the canopy, may be the reason it causes extremely high yield losses in soybeans. For example, a density of only one giant ragweed plant per 15 ft of row has been reported to reduce soybean seed yield by 46 to 50% (Baysinger and Sims 1991; see Table 1.2).

Low-growing, rapidly spreading weeds such as common purslane or carpetweed can be extremely competitive in low-growing crops such as lettuce, onions, or carrots, which provide little or no canopy or shading throughout their growth period (Figure 1.4). These low-growing weeds, however, are seldom a problem in tall,

competitive crops such as soybeans or corn. In fact, a closed crop canopy inhibits the development of late-germinating weeds of nearly any potential size. Other weeds that normally are low growing but have a vining habit (e.g., annual morningglory or bur cucumber) can continue to develop through a closed crop canopy. These species then gain a canopy advantage by climbing over the taller crop plants.

The growth characteristics of serious weeds enable them to be so effective at obtaining light, water, and nutrients at the expense of crop plants that they can cause a significant loss in crop yield when only 1 or 2 plants per ft of row are present (Table 1.2). In one instance, a single

Table 1.2 Potential of a Single Weed to Reduce Crop Yields

Weed	Crop	Weeds per ft Row or Area	% Crop Yield Reduction
Giant foxtail[a]	Soybeans	1 weed/ft	13
Velvetleaf[b]	Soybeans	1 weed/3ft	34
Smooth pigweed[a]	Soybeans	1 weed/ft	25–30
Sicklepod[c]	Soybeans	1 weed/ft	30
Palmer amaranth[d]	Soybeans	1 weed/ft	64
Common cocklebur[e]	Soybeans	1 weed/ft	87
Giant ragweed[f]	Soybeans	1 weed/15 ft	46–52
Giant foxtail[g]	Corn	1 weed/ft	7
Hemp dogbane[h]	Corn	1 weed/ft	15
Fall panicum[i]	Corn	1 weed/4 ft	20
Hemp dogbane[h]	Sorghum	1 weed/ft	30
Tall morningglory[j]	Cotton	1 weed/3 ft	40–50
Sicklepod[k]	Cotton	1 weed/3 ft	40
Common cocklebur[l]	Cotton	1 weed/3 ft	40–60
Wild oat[m]	Sugarbeets	1 weed/ft	22
Fall panicum[n]	Peanuts	1 weed/16 ft	25
Wild oats[o]	Wheat	1 weed/ft^2	11
Canada thistle[p]	Wheat	1 weed/3 ft^2	60
Downy brome[q]	Wheat	13 weeds/ft^2	40

[a]Nave and Wax (1971), [b]Marwat and Nafziger (1990), [c]Bozsa et al. (1989), [d]Klingaman and Oliver (1994), [e]Waldrep and McLaughlin (1969), [f]Baysinger and Sims (1991), [g]Knake and Slife (1965), [h]Robison and Jeffery (1972), [i]Selleck (1980), [j]Buchanan and Burns (1971a), [k]Buchanan et al. (1980), [l]Buchanan and Burns (1971b), [m]Mesbah et al. (1995), [n]York and Coble (1977), [o]Sharma (1979), [p]Derscheid and Wallace (1959), [q]Rydrych (1974).

fall panicum plant per 16 ft of row reportedly reduced yields in peanuts by 25% (York and Coble 1977). Some of these observed reductions in yields also may be due to allelopathy, although the relative importance and extent of this phenomenon in crops has not yet been determined.

Weed Density. The numerical superiority that weeds exhibit greatly reduces the availability of water, nutrients, and light to crop plants and accounts for much of what we consider to be weed interference. As weed density increases, the severity of the effect on the crop also increases (Table 1.3, Figure 1.5). Weed densities in agricultural soils can be extremely high. In a typical midwestern agricultural soil 1 mo after secondary tillage, we counted an average of 168 annual weed seedlings per square foot (Figure 1.6). In johnsongrass-infested soils, we counted as many as 150 seedlings per square foot. If no further control measures were imposed, these high population densities would result in the total loss of noncompetitive crops such as onions and significant losses of up to 50% in competitive crops such as corn or soybeans. This kind of constant weed pressure on crops creates the tremendous need for weed control technology.

Allelopathy

Some weeds actively eliminate competition by producing toxins that enter the soil and prevent the normal growth of other plants. This phenomenon, known as allelopathy, reduces crop development more than is normally expected from competition for water, light, and nutrients alone. A familiar example, although not a weed, is the walnut tree, which produces a toxin that eliminates the growth of many broadleaved plants from around the base of the tree. Farmers in the northern states of the Midwest have long recognized the difficulty of establishing crops in land previously infested with quackgrass, and quackgrass rhizomes and residues have been shown to alter the growth of

Table 1.3 Effect of Increasing Weed Densities on Crop Yields

Velvetleaf in Soybeans[a]			Common Cocklebur in Seed Cotton[b]		
Velvetleaf Density (Plants/ft²)	Soybean Yield		Cocklebur Density (Plants/3 ft Row)	Seed Cotton Yield	
	lb/A	% Reduction		lb/A	% Reduction
0	3330	—	0	2472	—
0.5	2562	23	1	956	61
1	1990	43	1.5	517	79
2	1530	54	3	225	91
4	1120	66	6	190	93

White Mustard in Peas[c]			Common Cocklebur in Bell Peppers[d]		
Mustard Density (Plants/ft²)	% Yield Reduction		Cocklebur Density (Plants/25 ft Row)	Fruit Yield	
	Shelled Peas per acre	Shelled Peas per plant		lb/A	% Reduction
1	0	0	0	4240	—
3	58	47	2	3040	28
9	88	70	4	2800	34
27	94	94	6	2080	51

Canada Thistle in Alfalfa[e]				
Canada Thistle Density (Plants/ft²)	Total (1962–1965) Alfalfa Dry Matter (Tons/A)			
	Dry Wt. Available	% Reduction	Dry Wt. Consumed by Livestock	% Reduction
0	15.7	—	8.5	—
0.5	12.4	21	5.2	39
1	12.4	21	6.1	29
2	8.2	48	3.8	56

Data from: [a]Hagood et al. (1980), [b]Buchanan and Burns (1971b), [c]Nelson and Nylund (1962), [d]Mendt and Monaco (unpublished data), [e]Schreiber (1967).

Figure 1.5 Soybean plants grown at densities of (from left to right) 0, 0.25, 0.5, 1.0, and 4.0 velvetleaf plants per ft². Number of pods per soybean plant were 40, 25, 15, 12, and 4, respectively. (E. S. Hagood, Jr., Virginia Polytechnic University)

small-grain crops and corn. Other examples of weeds with the potential to express allelopathic effects on crops and thus improve conditions for their own growth and spread include giant foxtail, large crabgrass, johnsongrass, yellow and purple nutsedges, Canada thistle, marestail, common sunflower (Figure 1.7), and at least 80 other genera (Putnam and Weston 1986).

Unfortunately, the contribution of allelopathy to interference is not well documented. It is difficult to separate the adverse effects caused by allelopathy from the adverse effects caused by direct competition for water and nutrients. It is

one matter to show that weed residues and leachates can prevent crop seed germination or growth in the greenhouse and quite another to demonstrate it in the field when weeds and crops are growing side by side.

The end result of interference, whether it be from nutrient or water depletion, shading, toxin production, or a combination of these factors is to reduce the quantity and quality of the crop. A frequently quoted estimate of the effect of weeds on crops is that *for each pound of weed dry matter produced, a loss of a pound of crop dry matter can be expected.* Although this generality

Figure 1.6 A 1 ft² plot of agricultural soil 1 mo after secondary tillage. Plot contained seedlings of 125 giant foxtail, 27 common lambsquarters, 17 black nightshade, 13 carpetweed, 3 redroot pigweed, and 1 Pennsylvania smartweed.

Figure 1.7 Allelopathic effects of leachates of common sunflower (middle) and Jerusalem artichoke (right) on corn seedlings. Corn seedlings at the left are untreated. (T. N. Jordan, Purdue University)

may not apply to every situation, it gives an estimate of the crop losses that can be expected from interference.

Persistence

Weeds appear year after year in virtually every site disturbed by humans. Their persistence and ubiquity are due primarily to their ability to produce numerous, long-lived, and easily transportable seeds. For annual weeds, these seed-related characteristics are essential for survival and success. If a weed is perennial and produces vegetative reproductive structures in addition to seeds, its ability to persist and spread is even greater.

Seed-Related Characteristics

Successful weeds possess the following seed-related characteristics:

Weeds Produce Large Numbers of Seeds. Weeds can produce tens of thousands of seeds per plant (Table 1.4), whereas most crop plants produce only several hundred seeds per plant. In addition, crop seeds almost always are harvested, so relatively few are deposited on the soil. Weed seeds, on the other hand, are not harvested (except unintentionally) and, since they often mature before the crop, frequently are deposited back on the soil surface to germinate at a later time.

Some competitive crops, such as volunteer corn in soybeans, can act as weeds the next season if their seeds are not completely removed during harvest. Corn, however, does not exhibit either seed dormancy or long-term survival of

Table 1.4 Approximate Number of Weed Seeds Produced per Plant

Yellow nutsedge[a]	2,400
Pennsylvania smartweed[a]	3,000
Barnyardgrass[a]	7,000
Giant foxtail[b]	10,000
Common ragweed[c]	15,000
Velvetleaf[d]	17,000
Jimsonweed[e]	23,400
Shepherd's purse	38,500
Curly dock[f]	40,000
Common purslane[a]	52,000
Common lambsquarters[a]	72,000
Stinkgrass[a]	82,000
Redroot pigweed[a]	117,000
Black nightshade[e]	178,000
Russian thistle[g]	200,000
Witchweed[h]	500,000

Data from: [a]Stevens (1932), [b]Schreiber (1965), [c]Dickerson and Sweet (1971), [d]Chandler and Dale (1974), [e]Muenscher (1955), [f]Cavers and Harper (1964), [g]Evans and Young (1974), [h]Shaw et al. (1962).

buried seed, so the problem of volunteer corn as a weed is temporary.

The soil serves as a repository for seeds. Such accumulated deposits of seeds are called *seedbanks*. The number of viable weed seeds in seedbanks can be tremendously high, usually in the tens of millions per acre. In a study of seedbanks at eight midwestern sites, the majority of seeds (50 to 90%) were dead; however, the number of viable seeds was still very large, ranging from 600 to 162,000 seeds per m² (Forcella et al. 1992) or roughly 2.5 to 650 million seeds per acre. In the plots where we counted l68 seedlings per square foot (approximately 7 million per acre), soil samples revealed l3,000 weed seeds per cubic foot or approximately 280 million seeds per acre-furrow slice (an acre 6 in. deep). Khedir and Roeth (1981) reported 15 million seeds of just one species, velvetleaf, in an acre-furrow slice of soil. These huge reservoirs of seeds permit weeds to reestablish themselves quickly each time the soil is disturbed.

The tillage system can have an impact on the number of seeds available for germination in the seedbank. In conventional tillage systems, seeds that are produced by uncontrolled weeds are mixed through the plow layer. In general, seedbank populations under conventional tillage remain stable over the years as long as weed control is consistent. In contrast, uncontrolled weeds in no-till systems deposit their seeds directly on the soil surface where they remain in a position to germinate. The buildup of weed seeds in the upper soil layers of untilled sites is frequently observed. For example, under marginal levels of control, 15 million giant foxtail seeds per acre were found in the 0- to 1-in. layer of soil in no-till plots versus 7 million seeds under conventional tillage (Schreiber 1992). In a study on the vertical distribution of weed seeds under different tillage systems, Yenish et al. (1992) showed that moldboard plowing resulted in a uniform distribution of weed seeds through the plow layer, whereas most of the seeds remained on or near the soil surface in no-tilled sites (Figure 1.8). Chisel plowing resulted in an intermediate distribution of seed through the soil profile.

Viable seeds in the seedbank can be dormant or nondormant, and the number of seeds of any given weed species that actually germinate and emerge depends on dormancy factors as well as environmental factors. Thus, emergence varies with the species. For example, the percentage of viable seeds that emerged as seedlings in field plots in the midwestern study cited earlier (Forcella et al. 1992) ranged from <1% for yellow

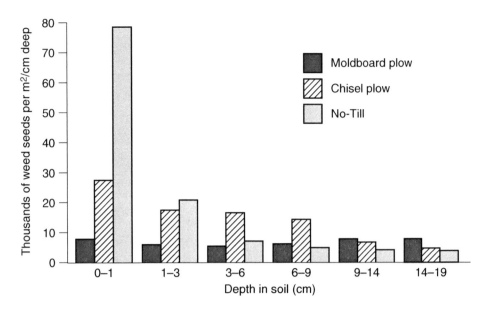

Figure 1.8 Effect of tillage on distribution of weed seeds in Wisconsin soils. Data from Yenish et al. (1992).

rocket to 30% for giant foxtail. Knowledge of the potential of seeds to germinate (i.e., the number of seeds per unit area and the percentage of seedlings that actually emerge) can be useful in developing weed management models that predict the density of weed seedlings in an area in any given year. Such information can be used to evaluate potential crop yield loss and the need for herbicides or other control methods. In a study of three weed species on five sites in west-central Minnesota, about one-half of the arable fields were predicted to harbor weed seedbanks so small that the resulting emergence and weed–crop competition did not justify the expense of controlling weeds with herbicides in a given year (Forcella et al. 1993).

Such low weed seed numbers are often due to the use of effective control measures that prevent weeds from producing seeds. The consistent use of herbicides and conventional tillage over a 6-yr period reduced seedbanks by as much as 97% (Schweizer and Zimdahl 1984). However, the 3% of seeds that remain, if allowed to germinate and set seed, can start a new infestation. Many weed species produce such large numbers of seeds that at least a small portion of the seeds will still be present, even after prolonged adverse conditions such as repeated tillage and herbicide applications that stress or kill most weeds before they set seed. A study conducted in Minnesota to evaluate the effects of various tillage and cropping systems on velvetleaf seed populations revealed that, even after 5 yr of continuous corn production with preemergence herbicides, hand

weeding, and fall plowing, 8% of the original seed number, or about 85 ungerminated seeds per square foot, still could be found (Lueschen and Andersen 1980). If only one of these 85 velvetleaf seeds germinated and was permitted to set seed, a new population of approximately 17,000 seeds (Table 1.4) would be produced to start a new infestation.

Weeds Produce Viable Seeds Under Adverse Conditions. Successful weeds can produce viable seeds even under conditions of poor fertility, low water supply, cool temperatures, a shortened growing season, and after mowing. Weeds such as foxtail, common cocklebur, common lambsquarters, and jimsonweed can germinate as late as mid-August in Indiana and still produce viable seeds before the first killing frost (Table 1.5; Figure 1.9). Mowed plants often sprout from the base of the cut stem and grow new prostrate spreading shoots that eventually will flower. In a study conducted at Purdue, giant foxtails planted in mid-May and then mowed as late as mid-August (after flowering) were able to set a second smaller crop of seed with viabilities as high as 78% (Schreiber 1965).

Weed Seeds Exhibit Periods of Dormancy. Seed *dormancy* is a condition in which seeds fail to germinate even when appropriate levels of moisture, air, temperature, and light are available. Weed control would be greatly simplified if all the seeds of a weed species were to germinate at one time. Application of control measures

Table 1.5 Effect of Planting Date on Panicle Length, Number of Seeds, and Germination of Giant Foxtail (West Lafayette, Indiana[a])

Date Planted (1961)	Panicle Length (cm)	Seeds/Panicle	% Germination
May 24	20	920	99
June 7	18	850	97
June 21	17	770	100
July 5	14	600	97
July 19	12	590	94
August 2	11	380	98
August 16	8	150	89
August 30	6	20	5
September 13	No panicles formed after this date.		

[a]Data from Schreiber (1965).

to prevent seedlings from developing into flowering plants would virtually eliminate a species from a field. Most weed seeds, however, undergo periods of dormancy that prevent this from occurring and ensure that they will continue to germinate over a period of many years. In contrast, the seeds of most crop plants do not become dormant, and their viability usually lasts less than a year under field conditions.

Seed dormancy is an important contributor to weed persistence because seeds are prevented from germinating during periods of adverse conditions. For instance, the weed seeds of summer annuals and most perennials will not germinate in the fall of the year in which they are produced. This feature prevents the seedlings from encountering lethal winter conditions. This dormancy at harvest is called *primary dormancy*, and it is in contrast to *secondary dormancy*, in which dormancy is induced over a period of time by environmental conditions such as burial in the soil or exposure to poor germination conditions. Another state of inactivity is termed quiescence. *Quiescence* is the lack of germination because some factor (e.g., oxygen, water, light) required for germination is missing.

Other terms commonly used to describe dormancy are *innate, enforced,* and *induced* (Harper 1957). *Innate dormancy* is the same as primary dormancy in that the seed will not germinate at harvest. Enforced and induced dormancies are equivalent to secondary dormancy in that they occur after exposure of the seed to certain adverse environmental conditions. Seeds with *enforced dormancy* will germinate once exposed to favorable conditions for germination, such as adequate moisture, light, temperature, and so on. Enforced dormancy is the same as quiescence. In contrast, seeds that have *induced dormancy* will not germinate even when given favorable germination conditions. Eventually, the periods of both innate and induced dormancy will end so that seeds that are still viable will germinate given good germination conditions. No matter what type of dormancy occurs, the end result is long-term survival of weed seeds.

The actual mechanism of dormancy varies with the species and is very complex. In many weeds, dormancy can be broken by one or a combination of several environmental conditions, for example, exposure to cool and moist conditions or to light and low temperatures. In

Figure 1.9 Late-season jimsonweed plant with mature flower. Plant stem is only 4 in. tall and yet can still produce viable seed.

some seeds that have primary or innate dormancy, the embryos are not fully mature when they are released from the plant. The period during which the seeds are exposed to environmental conditions that overcome dormancy or the period that they require to mature and become ready to germinate is termed *after-ripening*. The embryos of cow parsnip seeds, for example, initially constitute 0.4% of the dry weight of the seed but must continue to develop until reaching 30% of the dry weight to germinate (Barton 1965). Other seeds are physiologically immature when shed from the parent plant and require more time to become physiologically mature. In some cases, changes in hormonal content or a shift in storage products may have to occur.

Other seeds with innate dormancy have hard seed coats that restrict the emergence of the embryo. These seeds will not germinate until the seed coat has ruptured through some kind of mechanical activity (such as freezing and thawing or abrasion by cultivation) or has decomposed through microbial action. The coats of other seeds are impermeable to water and/or oxygen and must be broken, modified, or partially decomposed in order to give the embryo access to these essential growth factors. In some seeds chemical inhibitors must be leached away by rain for germination to occur.

In some cases, a cool, moist period is needed to break dormancy. This after-ripening period is termed *stratification* and usually involves exposure of the imbibed seeds to 33 to 50° F temperatures for periods varying from 2 wk to 8 mo. This cold requirement prevents the immediate germination of seeds newly produced in the late summer or fall. The actual stratification period may occur during the fall and late winter. Cold winter temperatures will prevent germination, but once temperatures in the spring start to warm, the seeds will be ready to germinate assuming moisture, light, and other growing conditions are appropriate. An after-ripening requirement for cool, moist conditions is common among the seeds of weed species found in the Midwest. Other species, however, require other temperature conditions. For example, wild oat seeds will not germinate unless given warm, dry conditions.

Many weed seeds require light to germinate. This type of dormancy may be present at seed maturation or may be induced by burial in the soil (secondary dormancy). It serves as a survival mechanism by preventing seeds from germinating at depths below which the shoots can emerge from the soil and initiate photosynthesis. This type of dormancy is regulated by the presence of the pigment *phytochrome*, which is present in the seed at very low concentrations. Phytochrome exposed to a high ratio of red light/far red light induces germination. The duration of the exposure to light required to break dormancy can be very short, on the order of milliseconds. Thus, one effect of tillage can be the exposure of dormant seeds to red light, even if for a brief period, and the initiation of germination. Researchers are exploring the possibility of conducting tillage operations at night to prevent the light-induced germination of some weed seeds.

Weed Seeds Buried in the Soil Remain Viable for Years. Whether truly dormant or quiescent, weed seeds can be stored by burial in the soil for 30 or 40 yr or more (Table 1.6). In 1980 seeds of moth mullein (*Verbascum blattaria*) removed from bottles buried by Dr. W. J. Beal[1] at Michigan State University 100 yr earlier were found to germinate and produce normal plants (Kivilaan and Bandurski 1981).

Long-term survival of weed seeds in the soil, coupled with the large numbers produced, ensures that once a weed becomes well established, it will continue to reestablish itself almost indefinitely. This means that a weed may be effectively controlled for several years and return as a major problem when control practices or production systems are altered or seeds are brought to the soil surface.

Table 1.6 Longevity of Weed Seeds Buried in the Soil[a]

Weed	Years[b]
Quackgrass	1–6
Common milkweed	3
Wild oat	4–7
Shattercane	10
Common cocklebur	16
Foxtail	20
Field bindweed	20+
Johnsongrass	20
Canada thistle	21
Jimsonweed	40
Common lambsquarters	40
Redroot pigweed	40
Velvetleaf	40

[a]Data from Holm et al. (1977), Klingman and Ashton (1982), Martin and Burnside (1980), Salisbury and Ross (1992).
[b]Viability may be anywhere from 1 to 100% at the end of the period. The fact that seeds can survive over these periods is more important than the actual percentages.

[1]Of the 20 pint-sized bottles buried by Dr. Beal, only 6 remain and these are not scheduled to be unearthed until the year 2040!

Weed Seeds Survive Adversity. Since seeds are natural resting bodies with a relatively low moisture content and frequently are well protected by seed coats resistant to breakdown or the passage of water or oxygen, they provide a stage in the life cycle that is extremely resistant to destruction. They resist freezing, drought, fire, passage through animal digestive tracts, submersion in water, and short periods of exposure to silage processes. Some seeds (or their fruits) repel grazing animals because they are covered with spines or have bad tastes or odors. Inactive seeds are not affected by herbicides.

Weed Seeds May Be Difficult to Detect in or Remove from Crop Seed. Many weed seeds are small and inconspicuous; an extreme example is the microscopic seed of the parasitic plant witchweed. Screening or removal of witchweed seeds from machinery by washing is an almost impossible task. Several of the most troublesome weeds produce seeds that are similar in size and shape to those of the crop plant. Examples include wild oat in grain crops, dodder in small-seeded legumes, and croton and black nightshade berries in soybeans (Figure 1.10). Similarity in size impedes the selective removal of the weed seed from the crop seed, thus ensuring that the weed will be reestablished again when contaminated crop seed is planted.

Many Weed Seeds and Fruits Have Adaptations That Aid in Dispersal. Weed seeds are excellent travelers. Hooks and spines adhere to feathers, fur, hair, and clothing; feathery hairs, parachutes, and wings aid in dispersal by wind; corky or inflated structures increase buoyancy in water (Figure 1.11). Even weed seeds without special adaptations can be moved easily by

Figure 1.10 Soybean seeds (upper left) and black nightshade berries (right). Note the similarity in size and the staining (lower left) of contaminated soybean seeds.

farm machinery, manure spreading, automobiles, airplanes, ships, and contaminated farm and forest products.

Moving water such as flood water and irrigation systems fed by open canals contribute significantly to the spread of weeds. Seeds from over 130 species of weeds were collected from irrigation water in Washington State (Kelley and Bruns 1975). The investigators concluded that if the water had been evenly distributed over the land served by the irrigation system, 35,000 seeds would have been deposited per acre. Any farmer growing crops on land subject to overflow from adjacent streams can attest to this continuing source of weed seed.

Vegetative Reproductive Structures

Many of our most troublesome and persistent weeds such as field bindweed, leafy spurge, johnsongrass, quackgrass, and yellow nutsedge are perennials that reproduce vegetatively. *Vegetative reproductive structures serve as major food storage organs and possess numerous buds capable of generating new plants.* The vegetative reproductive structures commonly found on perennial weeds are rhizomes, tubers, bulbs, stolons, and creeping roots (see Chapter 2 for a description of these structures).

Vegetative reproductive structures contribute in many ways to the overall success of perennial weeds.

Vegetative Reproductive Structures Allow the Plant to Have Another Form of Overwintering Structure in Addition to Seeds. The movement and storage of food materials (primarily carbohydrates) into these organs during the late summer and early fall permits the production of vigorous new shoots the following spring. In addition, buds on some of these vegetative structures can harden off and become dormant, providing an effective means of surviving adverse conditions in a manner similar to seeds. For example, Tumbleson and Kommedahl (1962) found only 12% germination in fall-harvested tubers of yellow nutsedge but 95% germination in spring-harvested tubers. Hardshell bulbs of wild garlic can remain viable in the soil for as long as 5 yr (Tinney 1942).

Vegetative Reproductive Structures Permit the Plant to Have Another Means of Propagation in Addition to Seeds. All vegetative reproductive structures can produce new shoots, including creeping roots that form adventitious

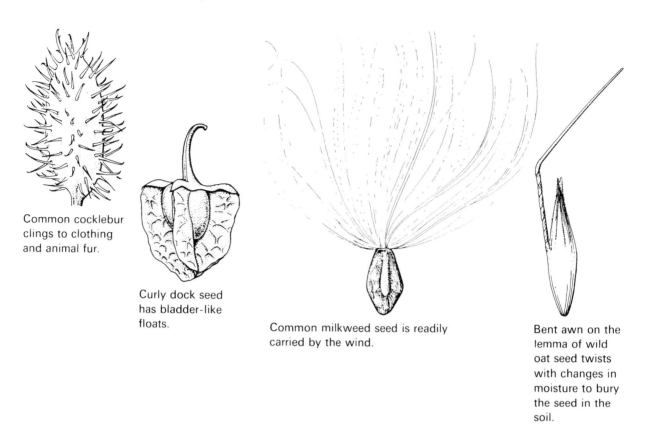

Common cocklebur clings to clothing and animal fur.

Curly dock seed has bladder-like floats.

Common milkweed seed is readily carried by the wind.

Bent awn on the lemma of wild oat seed twists with changes in moisture to bury the seed in the soil.

Figure 1.11   Seeds with adaptations for spread.

shoots. The number of new shoots can be huge. A plot of johnsongrass-infested soil in Europe measuring 10 ft² in area and 1 ft deep contained as many as 2.6 lb of rhizomes extending 91 ft in length and with over 2000 viable buds (Holm et al. 1977). Similar data were collected for johnsongrass-infested areas in southern Indiana (Figure 1.12). Yellow nutsedge also is

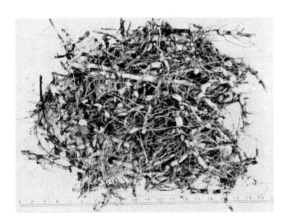

Figure 1.12 Johnsongrass rhizomes weighing 2.1 lb were collected from a plot of soil 10 ft² in area and 1 ft deep in southern Indiana. This quantity amounts to 4.5 tons of rhizomes per acre.

prolific vegetatively. In 1 full yr in Minnesota, the sprouting of one yellow nutsedge tuber yielded over 1900 new shoots and 6900 new tubers (Tumbleson and Kommedahl 1961). The number of yellow nutsedge tubers in some fields in Wisconsin has been estimated at 36 million per acre (Simkins and Doll 1981). These large numbers ensure the transplant and spread of weedy perennials. The dissemination of vegetative reproductive structures usually is restricted to localized areas; seeds are much better adapted for spread over long distances.

Vegetative Structures Allow the Plant to Extend Itself to New Sites for Water and Nutrients. The creeping roots of common milkweed have been reported to penetrate the soil to depths of 12.5 ft (Martin and Burnside 1980), and those of field bindweed have been found at depths of 27 ft with a spread of 18 ft (Holm et al. 1977). A single plant of johnsongrass can produce 180 to 270 ft of new rhizomes in 1 mo under ideal growing conditions (McWhorter 1973).

Plants Developing from Vegetative Structures Tend to Grow More Quickly Than Those Developing from Seeds. Plants that develop from vegetative structures have larger amounts of stored materials to draw on for growth, so they are larger and can begin to compete with crop plants sooner than seedling plants can (Figure 1.13).

Buds on Vegetative Reproductive Structures Serve as a Mechanism for the Survival and Rapid Reestablishment of a Perennial Weed After the Parent Plant Has Been Severely Disrupted or Partially Destroyed. Cutting vegetative reproductive structures into smaller segments stimulates the production of more plants by removing the apical dominance exerted by the growing tip over adjacent buds. Extremely small segments of many vegetative structures are capable of producing new plants. For example, a 1/2-in. piece of the creeping root of Canada thistle can produce a new plant. Both common milkweed and horsenettle can produce new plants from root segments less than 1 in. long.

Perniciousness

The dictionary definition of pernicious is "highly injurious or destructive in character; deadly." Weeds can indeed be destructive to people, crops, and animals.

The pernicious character of weeds is most frequently thought of in terms of the deleterious effects they have on crops. Crop losses reflect the end result of interference. Whether measured by loss in yield (quantity), loss in quality, or by the added costs of weed control, the economic consequences of crop losses are staggering.

The estimated average annual loss caused by weeds with current control strategies in 46 crops surveyed in the United States in 1991 was $4.1

Figure 1.13 Plants emerging from segments of johnsongrass rhizomes are larger and more vigorous than seedling plants (center). Both rhizomes and seedling plants were planted at the same time.

billion (Bridges and Anderson 1992). If herbicides had not been available for weed control, this loss would have amounted to $19.6 billion, almost a five-fold increase. The impact on weeds varied depending on the crop species and its ability to resist interference. For example, yield loss in peanuts, a very poor competitor, was estimated to be 70% (without the use of herbicides) in Georgia; in contrast, yield loss in wheat, a crop that is solid seeded and produces a dense, shading canopy, was only 17% without the use of herbicides in North Dakota. The use of herbicides reduced the loss in peanuts to 10% and in wheat to 3%. In a 1991 survey of 58 crops in Canada, the estimated average annual loss caused by weeds was $984 million, even when the best management practices were used.

Thus, crop losses occur even when herbicides and other management practices are used. In fact, weeds cause even greater economic hardships because the weed control practices must be paid for. In 1994 Bridges estimated that the total annual economic impact of weeds (crop losses plus weed control costs) to the U.S. economy was approximately $20 billion. Most of this economic impact was due to losses and costs in agricultural crops ($15.2 billion). The economic impact caused by weeds in noncrop situations such as golf courses, highway rights-of-way, animal health, and industrial and aquatic sites accounted for the other $4.8 billion (Bridges 1994).

Weed control costs vary with the crop. Purchased inputs for no-till soybeans, corn, and wheat from a southern Indiana farm gives an idea of the cost of herbicides relative to other costs (e.g., seed, fertilizer) of production. In 1996, herbicides for soybeans cost $30 per acre or approximately 47% of the $63.20 in total purchased input. For corn, that cost was $32 per acre or approximately 28% of the $113.60 in total purchased input, and for wheat it was $6 or approximately 6% of the $96.10 in total purchased input. Several factors help determine the relative costs of herbicides from one crop to another and include the competitive ability of the crop, the weeds present, the contribution of nonchemical control practices, the tillage method, management decisions, and the value of the crop.

Some economic losses are not initially obvious. For example, in irrigated systems, weeds use the water intended for the crops. Norris (1996) estimated the impact of barnyardgrass, at one plant per 6.5 m of row, on irrigated sugar beets and tomatoes. If barnyardgrass was not controlled, water use would increase by about

6 cm (approximately 2 1/2 in.) of water/ha. At an average water value of 16.5 cents /mm ($50.00/300 mm), the weeds would increase production costs by about $22/ha or almost $9/acre.

Weed losses are not limited to the United States and Canada. In 1977 Holm et al. estimated that an average of 30 to 35% and sometimes as much as 80% of the potential rice yields in Southeast Asia can be reduced by weeds. A single species of witchweed, *Striga hermonthica,* infests an estimated 21 million ha in Africa, causing an estimated grain yield loss of 4.1 billion kg/yr (Sauerborn 1991). Parker (1991) indicated that crop losses of 30 to 50% due to *Striga* were common under typical field conditions and that losses over a whole region could average 5 to 15%. The truly pernicious character of witchweed is that, because it is a plant that parasitizes crop roots, it causes most of its damage while still underground. Therefore, hand weeding efforts are virtually useless once the plant has emerged above the soil and can be seen.

Another example of a weed that causes serious crop losses in other countries is littleseed canary grass (*Phalaris minor*). This plant is a major weed in wheat fields in India and parts of the Mediterranean. Crop yields can be reduced by 30% by 150 plants/m² (Balyan and Malik 1989), and populations as high as 1000 to 2000 plants/m² have been observed in farmers' fields in India (Malik and Singh 1993).

Losses are not just limited to crops. A single species of poisonous weed in one Latin American country was responsible for the death of 15,000 head of livestock, and aquatic weeds severely reduce the potential for irrigation and fish production in many countries of Africa and Asia.

Thus, weed control is an essential part of crop and animal production throughout the world. In the United States and other developed countries, the energy used to control weeds is primarily mechanical and chemical. In developing countries, the energy used to control weeds is primarily human energy. In 1971 Holm suggested that the number one work task of the world's human population is hand weeding crops. There is no reason to assume that the situation has changed to any great degree over the years.

As impressive as weed associated losses are when expressed in terms of crop yields and costs, weeds can have an impact on crops in other ways.

Weeds Cause Loss of Crop Quality. Weeds can have a detrimental effect on crop quality as well as quantity, particularly in harvested crops in which the presence of weed contaminants can result in direct monetary loss due to dockage. Dockage can result from weeds that cause objectionable odors (wild garlic in wheat, mustards in milk), staining of crop or edible seeds (nightshade berries in soybeans or dry beans; Figure 1.10), and toxicity (jimsonweed seeds in soybeans). The penetration of quackgrass rhizomes into "root" crops such as potato tubers and carrot roots can cause perforation and distortion resulting in loss of quality. Crop spoilage can occur from excess moisture of contaminating weed parts; for example, wild sunflower heads can cause spoilage in harvested wheat.

Weeds Limit the Choices of Crop Rotation Sequences and Cultural Practices. A field seriously infested with a perennial weed is not an appropriate site for the establishment of a crop where economical selective control measures in that crop are not available. In addition, a serious infestation of weeds with a life cycle or seasonal growth period similar to that of a potential crop limits the alternatives for rotation. An example is wheat in a wild garlic-infested area.

Weeds Harbor Other Crop Pests Such as Plant Pathogens, Nematodes, and Insects. This problem is particularly serious in perennial weed infestations in which disease or insect pests can overwinter in the underground vegetative reproductive structures. Overwintering rhizomes of johnsongrass harbor viruses responsible for two devastative corn diseases, maize dwarf mosaic (MDM) and maize chlorotic dwarf virus (MCDV). During the next growing season, these viruses are transmitted to the corn by insects (Figure 1.14). Horsenettle is a host for several important diseases and insect pests of vegetable crops. These include the disease organisms that cause potato and tomato mosaic and leafspot of tomato, and insects such as potato flea beetles, onion thrips, and potato stalk borers.

Weeds Interfere with Crop Harvesting. Weeds in vegetables directly interfere with hand picking. For example, laborers tend to avoid areas that are heavily weed infested because the weeds slow the picking process, which reduces wages (Schroeder 1993). Mechanical harvesting is also affected by weeds, making it difficult to clean crops of soil and weeds. Grassy weeds and vines become wrapped around the rollers or cylinders of mechanical harvesters, necessitating frequent stops and cleaning of equipment

Figure 1.14 Stunting and chlorosis of corn plants are symptoms caused by maize dwarf mosaic virus disease carried over the winter in johnsongrass rhizomes. Plants in left background are uninfected. (H. L. Warren, Virginia Polytechnic University)

Weeds or Weed Seeds in Harvested Crops Necessitate Extra Cleaning and Processing Procedures. This is particularly true when the weed propagule is similar in size and shape to the harvested crop. Wheat harvested with wild garlic as a contaminant must be dried before the wild garlic bulbs can be removed from the grain by forced air separation. Since some of the garlic odor can be retained by the wheat, it must be blended with noninfested lots to eliminate the garlicky odor. Special techniques must be used to remove dodder seeds from the seeds of small-seeded legume crops such as red clover and alfalfa.

Weeds Lead to Increased Transportation Costs. In 1969 and 1970, for example, 16 million tons of grain from Canada were delivered by rail to West Coast ports. Transported along with this grain were 487,000 tons of wild oat seeds or the equivalent of 33 train car loads per day for every day of the year. Excess transportation costs were estimated at $2 million, and the dockage loss for cleaning the grain of wild oat contamination was estimated at approximately $8 million (Shuttleworth 1973).

In addition to crop losses, weeds cause many other serious problems.

Weeds	Examples
Reduce land values	Farm real estate; for example, loss of value/acre on land severely infested with perennial weeds; lakeshore property located in aquatic weed-infested areas (Figure 1.15).
Cause allergies and rashes	Allergies: ragweed in fall, grasses in

(Table 1.7). Mechanical harvesting of cotton was impossible at ivyleaf morningglory densities above 1 to 3 plants per 5 linear ft of row (Rogers et al. 1996). Harvesting of weeds along with the crop adds to the wear on expensive machinery. Desiccants often are used to dry weeds prior to harvest so they will not interfere with harvesting. The cost of the chemical and its application are added expenses.

Table 1.7 Effect of Annual Morningglories on Soybean Harvesting Ability[a]

No. Weeds per ft. Row	Soybean Yield (bu/A)	Soybean Yield % Reduction	Soybean Lodging[b]	Harvesting Ability[c]
0	26.9	—	1.2	1.0
0.5	23.6	12	2.1	2.8
1	21.4	20	3.0	4.0
2	19.8	26	3.0	4.3
4	19.6	27	3.3	4.3
8	15.0	44	3.9	4.8
10–12	15.7	42	3.8	4.9

[a]Data from Wilson and Cole (1966)
[b]1.0 = no lodging; 5.0 = severe lodging (100% of plants down)
[c]1.0 = easily harvested; 5.0 = harvest nearly impossible

Figure 1.15 Aquatic weeds reduce the appeal and value of this real estate.

	early summer; rashes: 2 million cases of skin poisoning are reported each year from poison ivy, poison oak, and poison sumac (Mitich 1995).
Injure and poison livestock	Greatest loss of livestock in open rangeland, primarily western states (halogeton, larkspur, locoweeds).
Reduce aesthetic values	Around homes, schools (Figure 1.16), office buildings; in gardens, turf, nursery plantings, parks, and along roadsides.
Interfere with transportation	Can come through asphalt and widen cracks in road pavement; aquatic weeds interfere with boat travel.
Interfere with recreation	Aquatic weeds interfere with fishing, swimming, boating, water skiing, and other recreational activities.
Create fire hazards	Around areas where flammable products such as lumber and oil products are stored; around electrical substations.
Obstruct powerlines	Prevent access for repairs; cause power outages (Figure 1.17).

Figure 1.16 Weeds emerging through cracks in playground asphalt.

Figure 1.17 Brush can interfere with access to utility lines.

Figure 1.18 Weeds obstruct visibility at road crossings.

Obstruct visibility   At crossroads and railroad crossings (Figure 1.18).

ECOLOGY AND ORIGIN OF WEEDS

The characteristics that permit weeds to compete and survive adversity enable them to thrive under conditions of environmental instability. Unfortunately, almost all human activity results in unstable or disturbed environments. Examples of such activity include farming cropland; building and maintaining rights-of-way; clearing land for homes and golf courses; constructing ponds, reservoirs, and lake channels; and overgrazing pastures and rangeland. In stable *climax communities* (e.g., eastern woodlands, prairies, or cold deserts), plants we call weeds have little opportunity to develop and rarely are present in significant numbers.

Although we think of weeds as invaders, they are a part of natural plant succession which, if allowed to proceed unimpeded, results in the eventual conversion of the disturbed site to a climax community. Plant *succession* typically proceeds from a condition of instability to a condition of stability (e.g., from weedy annuals to herbaceous perennials to woody shrubs to short-lived trees to the stable climax community such as eastern woodlands). The planting of crops to provide maximum yields of seed, forage, and fiber maintains the disturbed environment and is a direct contradiction to plant succession. For crops to be maintained, energy must be put into the system. Part of this energy is required to control weeds and prevent the first step in plant succession.

Weeds probably have always been a feature of disturbed sites. Before human life, natural selection presumably resulted in plants that were adapted to unstable or disturbed areas such as eroding stream banks, volcanic deposits, floodplains, and burned areas. Eventually, these plants also moved to areas disturbed by human activities, probably around human habitations first and then to gardens and fields as plants were domesticated as crops. The easiest weeds to control probably were eliminated as serious competitors early, but those particularly well adapted to disturbed crop sites spread and are still with us today.

In terms of geographic origin, some of our most serious weeds, such as the ragweeds, common cocklebur, common sunflower, and common milkweed, are native plants (Table 1.8). Several of the mallow weeds (family Malvaceae) such as prickly sida, Venice mallow, and spurred anoda are from South America. South America is also the source of some of our most serious aquatic weeds such as water hyacinth. Of 500 weed species whose origins have been investigated (Muenscher 1980), approximately 39% are native to North America; another 3% originated in South and Central America. Fully 35% were introduced from Europe during the period of the westward migrations. These plants quickly spread throughout North America, partly because their natural enemies (or biocontrol agents) were not brought with them. Other weeds, which apparently were cosmopolitan in distribution, were found in both the western and

Table 1.8   Geographic Origins of Some Common Weeds[a]

Native to North America	Wild onion, three-seeded mercury, common milkweed, field dodder, bigroot morningglory, ragweed, fall panicum, common cocklebur, poison ivy, white snakeroot, common sunflower
Native to South and Central America	Redroot pigweed, Venice mallow, spurred anoda, prickly sida, water hyacinth
Native to Europe (and Eurasia)	Wild garlic, leafy spurge, buckhorn plantain, chickweed, curly dock, quackgrass, bull thistle, Canada thistle, wild oat, yellow foxtail, common purslane, common lambsquarters
Native to Asia or Africa	Johnsongrass, wild carrot, giant foxtail, velvetleaf, kudzu
Cosmopolitan	Knotweed, yellow rocket, jimsonweed, carpetweed, black nightshade

[a]Data from King (1966) and Muenscher (1980).

eastern hemispheres at the time observations were first made.

Some alien plant species are especially invasive and even have displaced relatively stable native vegetation. Leafy spurge (*Euphorbia esula*) and yellow starthistle (*Centaurea solstitialis*) in western and prairie grasslands, kudzu (*Pueraria lobata*) along forest edges in the southeastern United States, garlic mustard (*Alliaria petiolata*) on forest floors of the eastern United States, purple loosestrife (*Lythrum salicaria*) in wetlands, melaleuca (*Melaleuca quinquenervia*) in the Florida Everglades, and hydrilla (*Hydrilla verticillata*) in lakes and rivers of the southeastern United States are examples of exotic species that have displaced native species and are continuing to spread. Many new species of weeds and other pests are introduced first into Florida (e.g., tropical soda apple, Chinese tallow tree, melaleuca, water hyacinth, and hydrilla), where, because of a mild climate, they can gain a foothold and then spread to other parts of the southeastern United States. Hydrilla, for example, has been found as far west as California and has spread north along the eastern seaboard. The invasion of exotic species into Hawaii has altered the habitat of native birds and other animals, and the rapid spread of non-native grasses has increased the incidences of fire, which kill many native Hawaiian plant species.

Invasions, even by species present in this country for a long time, continue. When United States National Park superintendents were surveyed in 1994 about conditions in their parks, 61% of the 246 respondents indicated that nonnative plants were a moderate or major problem, and 59% of Nature Conservancy stewards in 1992 ranked pest plants among their top-ten conservation challenges (see Randall 1996). It is clear that invasive plant species will be a serious problem for the natural areas of the United States for many years to come. The rate of spread of noxious invading species into the rangeland and forested areas of western federal lands in the United States has been estimated to be 4600 acres per day (U.S. Department of Interior 1995).

CLASSIFICATION OF WEEDS

Weeds can be described in several different ways: by habitat, life cycle, morphology or structure, and physiology. Each of these classification categories is defined here because the terminology involved is commonly used to describe weeds and control practices for them.

I. HABITAT
   A. TERRESTRIAL: Plants that live on land.
   B. AQUATIC: Plants that are structurally modified to live in water (watermilfoil, hydrilla), on water (duckweed, water hyacinth), or around water (cattails, bulrushes).
II. LIFE CYCLE (for detailed descriptions, see Chapter 2)
   A. HERBACEOUS PLANTS: Plants with nonwoody aerial stems that die each year in temperate climates.
      1. ANNUALS: Plants that live for one growing season only.
      2. BIENNIALS: Plants that require parts of two growing seasons to complete their life cycle.
      3. PERENNIALS: Plants that live indefinitely.
   B. WOODY PLANTS: Plants with woody aerial stems that persist from year to year.

   Thus, all woody plants are *perennial*.

      1. TREES: Woody perennials with a single main stem or trunk.
      2. SHRUBS: Woody perennials with more than one principal stem arising from the ground. Usually shorter than trees.
III. MORPHOLOGY
   A. DICOTS: Plants whose seedlings produce two cotyledons (seed leaves). Usually typified by net leaf venation and flower parts in fours, fives, or multiples thereof. Typically called broadleaves. Examples include jimsonweed, morningglory, and ragweed.
   B. MONOCOTS: Plants whose seedlings bear only one cotyledon. Usually typified by parallel leaf venation and flower parts in threes or multiples of three. Typically called grasses or grasslike plants. Actually, there are many groups of monocots (for example, orchids and lilies), but the major weeds are found in two groups, the true grasses and the sedges.
      1. GRASSES: Leaves are two-ranked (attached in two rows along the stem) and usually have a ligule and sometimes an auricle. The leaf sheaths are split around the stem with the stem being round or flattened in cross section with hollow internodes. Examples include crabgrass, foxtails, johnsongrass, and quackgrass.

2. SEDGES: Leaves lack ligules and auricles and the leaf sheaths are continuous around the stem. The stem in many species is triangular in cross section with solid internodes. Examples include yellow nutsedge and purple nutsedge.

IV. PHYSIOLOGY

Plants in which photosynthesis occurs via the Calvin-Benson cycle are termed C_3 *plants*. The first stable product of photosynthesis, phosphoglyceric acid, has three carbon atoms. In other plants, the first stable photosynthetic product is a four-carbon acid. These C_4 *plants* tend to be more efficient at photosynthesis than C_3 plants and are better competitors, particularly at higher temperatures. Of the world's 18 worst weeds, 14 have a C_4 pathway (McWhorter and Patterson 1980).

A. C_3 PLANTS: Crops: soybeans, peanuts, carrots, cotton, wheat; weeds: curly dock, common lambsquarters, cocklebur, jimsonweed.

B. C_4 PLANTS: Crops: sugarcane, corn; weeds: yellow and purple nutsedge, johnsongrass, barnyardgrass, bermudagrass, crabgrass, redroot pigweed, Russian thistle.

PARASITIC AND POISONOUS WEEDS

Parasitic Weeds

Parasitic plants form protoplasmic connections with their host plant, deriving food, water, or both from the host tissues. Three economically important parasitic weeds found in the United States are witchweed, dodder, and the mistletoes.

Witchweed is a serious problem in Africa and parts of Asia and was introduced into the United States in the 1950s (also see Chapter 3). The genus of witchweed, *Striga*, consists of 35 species, of which at least 11 attack various cereal crops (Raynal-Roques 1991). The only species found in the United States, *Striga asiatica*, is a parasite of corn, sorghum, sugarcane, and other grass crops. Seeds germinate in response to strigol, a compound secreted by the host plants. The witchweed seedlings depend on the host plant for both water and nutrients during the first 30 days of development, which takes place underground. After emergence, witchweed develops its own chlorophyll- and food-synthesizing capabilities but continues to draw water from the host. The seed can survive at least 20 years in the soil, and a single plant can produce up to 500,000 seeds. The losses in yield of corn, millet, and sorghum to witchweed have been particularly devastating to the health and economies of the peoples of Africa. Estimates of crop losses range from 15% to 40% of Africa's total cereal harvest. Current efforts to reduce the severity of witchweed infestations include recent introductions of witchweed-resistant sorghum varieties into ten African countries (Vogler et al. 1995). These varieties do not produce strigol; thus, the witchweed seeds cannot germinate and parasitize the crop.

Other parasitic plants include the dodders and mistletoes. Dodder (*Cuscuta* spp.) is a nonphotosynthetic plant throughout its life cycle and thus derives both food and water from the host plant. It produces long yellow stems that twine around and penetrate the foliage of the host. It infests many different plants but is most serious as a weed in small-seeded legumes (e.g., alfalfa) grown for seed and in some horticultural and floricultural crops. The mistletoes are parasites of trees. The common green mistletoe (*Viscum album*) of Christmas fame is a pest in fruit and nut orchards, where it weakens the host and can lower yields. Dwarf mistletoe (*Arceuthobium* spp.), a nonphotosynthetic parasite, is a major pest in coniferous forests such as ponderosa pine and Douglas fir harvested for timber. Over 50% of the forests in some parts of the Rocky Mountain states are thought to be infested with dwarf mistletoe. It causes a decline in growth, increased branch mortality, and severely deformed trunks, thus greatly lowering the yield and quality of the wood.

A parasitic plant that has appeared infrequently in the United States is broomrape (*Orobanche* spp.). Like witchweed, this plant infects plant roots and the seed is stimulated to germinate by exudates from the host plant, but the host range is quite different. It affects broadleaved crops, such as tomatoes, tobacco, carrots, and beans, rather than grass crops. The broomrapes are important weeds in the Mediterranean countries including northern Africa; in the United States it has been reported in California, Texas, and Georgia. *Orobanche minor* was probably introduced into southwest Georgia in 1992 and had spread from a one-county area to five counties by 1995.

Current distribution information on parasitic species listed as Federal Noxious Weeds can be obtained by contacting the NAPIS database on the World Wide Web at http://ceris.purdue.edu/napis/.

Poisonous Plants

Poisonous plants can affect both humans and livestock. Livestock problems are more common in the western states and provinces, where there are large acreages of rangeland. Poisonous plants may be grazed when desirable forages have been eaten or when they appear before the desirable forages in the spring. Examples of poisonous plants of western ranges include tall larkspur, locoweed, lupine, and halogeton. Other plants toxic to livestock include wild black cherry and chokecherry, cockleburs in the seedling stage, bracken fern, and whorled milkweed. Symptoms to animals can include mild sickness, loss of product quality and quantity (for example, milk production), abortion of the fetus, or death depending on the weed species and quantity consumed.

Poison hemlock and water hemlock cause problems to humans as well as livestock because these plants can be confused easily with edible, nontoxic members (e.g., wild carrot, wild parsnip) of the same family (Apiaceae). Persons using jimsonweed as a medicine or as a substitute for marijuana have been poisoned. As already noted, poison ivy and poison oak are well known for their ability to cause severe dermal rashes.

MAJOR CONCEPTS IN THIS CHAPTER

Weeds differ from other plants: they are plants that interfere with the growth of desirable plants and are unusually persistent and pernicious. They negatively impact human activities and as such are undesirable.

Of 250,000 plant species, only a few thousand can be documented as weeds somewhere in the world. Only 200 or 0.08% are major problems in world agriculture.

The ability of weeds to negatively affect crop growth is termed interference. Interference includes competition and allelopathy.

Weeds interfere with crops and persist because:

- Weeds have the ability to compete successfully for light, water, and nutrients.
- Weeds reproduce profusely by seeds and sometimes by vegetative structures.
- Weed seeds survive burial in the soil (usually for decades).
- The high number of seeds produced, coupled with long-term survival in the seed bank, results in high weed seed populations in soils.

- Seed dormancy ensures periodic germination and prevents seedlings from encountering unfavorable or lethal conditions.
- Weeds can reproduce under unfavorable conditions.
- Weed seeds are adapted for spread with crop seeds, by natural agents, and by humans.
- Vegetative reproductive structures contribute to the success of some of our most serious creeping perennial weeds by providing more vigorous, faster growing plants and by providing another method of reproduction in addition to seeds. These structures resist destruction and provide a mechanism for rapid reestablishment after severe disturbance.
- Besides competing for light, nutrients, and water, some plants are allelopathic, that is, they secrete biological toxins that suppress the growth of other plants.

Weeds are pernicious in a number of ways:

- Crop-related problems include loss of yield and quality and interference with harvesting and processing. Weeds also limit crop management options and harbor other crop pests.
- Additional problems caused by weeds include injury and discomfort to humans and to livestock. They create fire hazards and unsightly areas; contribute to unsanitary conditions; obstruct visibility; and interfere with recreational activities, transportation, and the delivery of public utilities.

Some important ecological principles regarding weeds are:

- Weeds are invaders of disturbed sites (they appear early in the natural successional sequence).
- High-yield cropping systems result in disturbed sites. The maintenance of the high-yield system is in direct conflict with natural succession.
- Some exotic weed species can invade stable climax communities and displace native species.
- The first plant to become established has the competitive advantage.
- Weeds that compete with the crop early in its life (during establishment) reduce yields the most.
- Later-germinating weeds can cause other problems such as loss in crop quality, difficulty in harvesting crops, etc.

- Each pound of weed dry matter produced results in an equal loss in crop dry matter.

TERMS INTRODUCED IN THIS CHAPTER

After-ripening
Allelopathy
Annual plants
Aquatic plants
Biennial plants
C_3 plants
C_4 plants
Climax community
Competition
Critical period
Critical period of weed control
Dicots
Dormancy
Enforced dormancy
Grasses
Herbaceous plants
Induced dormancy
Innate dormancy
Interference
Monocots
Parasitic plants
Perennial plants
Phytochrome
Poisonous plants
Primary dormancy
Quiescence
Secondary dormancy
Sedges
Seedbank
Shrubs
Stratification
Succession
Terrestrial plants
Trees
Vegetative reproductive structures
Weed
Woody plants

SELECTED REFERENCES ON THE BIOLOGY AND ECOLOGY OF WEEDS

Anderson, R. N. 1968. Germination and Establishment of Weeds for Experimental Purposes. Weed Science Society of America, Lawrence, Kansas.

Bewley, J. D. and M. Black. 1994. Seeds, Physiology of Development and Germination, 2nd ed. Plenum Press, New York.

Bridges, D. C., Ed. 1992. Crop Losses Due to Weeds in the United States. Weed Science Society of America, Lawrence, Kansas.

Bridges, D. C. 1994. Impact of weeds on human endeavors. Weed Tech. 8:392–395.

Cousens, R. and A. M. Mortimer. 1995. Dynamics of Weed Populations. Cambridge University Press., Cambridge, UK.

Cronk, Q. C. B. and J. L. Fuller. 1995. Plant Invaders: The Threat to Natural Ecosystems. Chapman & Hall, London.

D'Antonio, C. M. and P. M. Vitousek. 1992. Biological invasions by exotic grasses, the grass/fire cycle and global change. Ann. Rev. Ecol. Syst. 23:63–87.

Egley, G. H. and S. O. Duke. 1985. Physiology of weed seed dormancy and germination. Pages 28–64 in Weed Physiology, Vol. I. Reproduction and Ecphysiology, S. O. Duke, Ed. CRC Press, Boca Raton, Florida.

Forcella, F., R. G. Wilson, K. A. Renner, J. Dekker, R. G. Harvey, D. A. Alm, D. D. Buhler, and J. Cardina. 1992. Weed seedbanks of the U.S. Corn Belt: magnitude, variation, emergence, and application. Weed Sci. 40:636–644.

Harper, J. L., Ed. 1960. The Biology of Weeds. Blackwell Scientific Publications, Oxford, England.

Hill, T. A. 1977. The Biology of Weeds. In Studies in Biology, Vol. 79, Edward Arnold Ltd., London.

Holm, L., J. Doll, E. Holm, J. Pancho, and J. Herberger. 1997. World Weeds, Natural Histories and Distribution. John Wiley and Sons, New York.

Holm, L. G., J. V. Pancho, J. P. Herberger, and D. L. Plucknett. 1979. A Geographical Atlas of World Weeds. John Wiley and Sons, New York.

Holm, L. G., D. L. Plucknett, J. V. Pancho, and J. P. Herberger. 1977. The World's Worst Weeds—Distribution and Biology. The University Press of Hawaii, Honolulu, Hawaii.

Holzner, W. and M. Numata, Eds. 1982. Biology and Ecology of Weeds. Dr. W. Junk Publishers, The Hague, The Netherlands.

King, L. J. 1966. Weeds of the World—Biology and Control. Interscience Publishers, New York.

Muenscher, W. C. 1955. Weeds. Macmillan, New York.

Mulligan, G. A., Ed. 1979. The Biology of Canadian Weeds—Contributions 1–32. Pub. 1693. Information Services, Agriculture Canada, Ottawa.

Musselman, L. J., Ed. 1987. Parasitic Weeds in Agriculture. Vol. I, Striga. CRC Press, Boca Raton, Florida.

Parker, C. and C. R. Riches. 1993. Parasitic Weeds of the World: Biology and Control. CAB International, London.

Press, M. and J. Graves, Eds. 1996. Parasitic Plants. Chapman & Hall, New York.

Putnam, A. R. and W. B. Duke. 1978. Allelopathy in agroecosystems. Annu. Rev. Phytopath. 16:431–451.

Putnam, A. R. and C-S. Tang, Eds. 1986. The Science of Allelopathy. John Wiley and Sons, New York.

Radosevich, S., J. Holt, and C. Ghersa. 1997. Weed Ecology, 2nd ed. John Wiley and Sons, New York.

Randall, J. M. 1996. Weed control for the preservation of biological diversity. Weed Technol. 10:370–383.

Rice, E. L. 1984. Allelopathy, 2nd ed. Academic Press, Orlando, Florida.

Salisbury, F. B. and C. W. Ross. 1992. Plant Physiology, 4th ed. Wadsworth Publishing Co., Belmont, California.

Taylorson, R. B. 1987. Environmental and chemical manipulation of weed seed dormancy. Rev. Weed Sci. 3:135–154.

Westbrooks, R. G. and J. W. Preacher. 1986. Poisonous Plants of Eastern North America. University of South Carolina Press, Columbia.

Williams, M. C. 1994. Impact of poisonous weeds on livestock and humans in North America. Rev. Weed Sci. 6:1–27.

Wilson, R. G. 1988. Biology of weed seeds in the soil. Pages 25–39 in Weed Management in Agroecosystems: Ecological Approaches, M. A. Altieri and M. Liebman, Eds. CRC Press, Boca Raton, Florida.

THE PLANT SYSTEM

Plants are complex, dynamic organisms. They are made up of well-defined structures in which thousands of vital processes occur in well-ordered and integrated sequences. *To control a weed, one or more of these intact structures or processes must be disrupted sufficiently to render the weed harmless.* In this chapter we first review the fundamentals of plant structure and function and then provide a description of processes and life cycles that relate to weed control practices. Specific weed control practices and plant–herbicide interactions are discussed in Chapters 4 and 6, respectively.

THE PLANT SYSTEM

Plant Structures

Plant structures frequently are identified according to location (Figure 2.1). The above-ground portions of the plant are referred to collectively as the *shoot*. The major underground structure is the *root*.

Plants are made up of components of decreasing size and complexity. Progressing from the largest to the smallest level of organization, the components of a plant are the whole plant, organs, tissues, cells, organelles, and molecules.

Plant Organs

The largest structural units of a plant are the organs. Organs are composed of tissues and tissues are composed of cells. There are four plant organs, and their functions are as follows:

Stem	Support, some photosynthesis, food storage
Leaves	Photosynthesis, gas exchange
Roots	Water and nutrient uptake, anchorage, food storage
Flowers	Sexual reproduction, resulting in the formation of seeds

The stem, leaves, and roots are the *vegetative structures* of a plant. The flowers are the *floral structures*.

All stems, including horizontal stems such as rhizomes, stolons, and tubers, are divided into *nodes* (sites where the leaves are attached) and *internodes* (Figure 2.1). Nodal areas can be recognized by the presence of a leaf (or leaf scar if the leaf has fallen off the plant), a bud located in the axil of the leaf, or in some instances, by the presence of roots (particularly on horizontal stems). Even underground stems (rhizomes, tubers) frequently possess obvious evidence of nodes (equidistantly spaced leaf scars with buds and rudimentary leaves).

Buds in the leaf axils are responsible for the production of lateral shoots (branches) from erect stems (Figure 2.2) and the production of new erect shoots from rhizomes and tubers.

Leaves are the primary photosynthetic organs. They consist of a blade (the flattened photosynthetic surface), a petiole, and sometimes stipules (flaplike pieces of tissue, usually paired, at the base of the petiole). In broadleaved plants, the leaf petiole (or blade where the petiole is absent) emerges directly from the point where it is

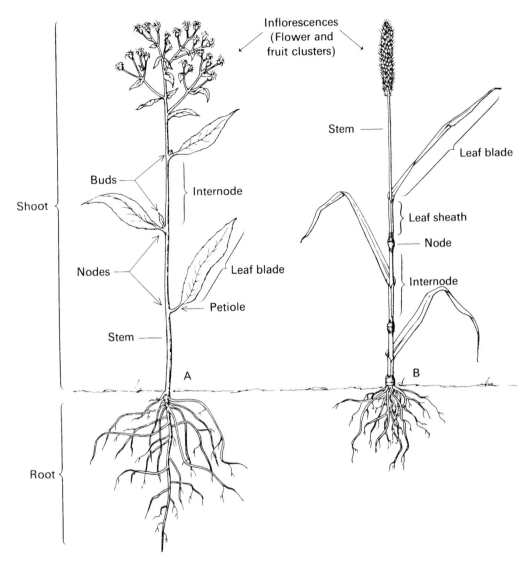

Figure 2.1 Organs and major parts of a mature (A) broadleaved plant and (B) grass plant.

attached at the node. The leaf blade of a grass plant also is attached at the node, but the blade ensheaths the stem (leaf sheath) for a distance above the node before emerging freely from the plant (Figure 2.1B).

Roots can be distinguished from stems because they lack regularly spaced nodes and internodes. Another difference between roots and stems is that lateral roots originate from an internal tissue in the root called the pericycle (Figure 2.2) rather than from buds. Lateral roots emerge at more or less random points along the length of the main root. The pericycle also is the origin of stems that develop from creeping perennial roots.

Roots are absorptive structures that can penetrate deep into the soil. In fact, perennial weeds with a creeping root system tend to penetrate much deeper into the soil profile than do peren-

nial weeds with underground stems (rhizomes, tubers). For this reason, perennials with creeping roots such as field bindweed and leafy spurge usually are more difficult to disrupt or control with tillage and other methods than are rhizomatous perennials such as quackgrass and johnsongrass.

Flowers are the sites of sexual reproduction and seed production. Flowers are clustered into *inflorescences* (Figure 2.1). Examples of inflorescences include the "head," or cluster of flowers typical of the thistles and other composites, and the "spike," typical of some grasses. Flowers range in size from extremely reduced and simple structures (e.g., in grasses, ragweed, cocklebur) that are wind pollinated to insect-pollinated flowers that tend to have larger, more colorful petals (e.g., nightshades, morningglories, thistles). *Pollination* is the process by which pollen, which contains the sperm, is delivered to

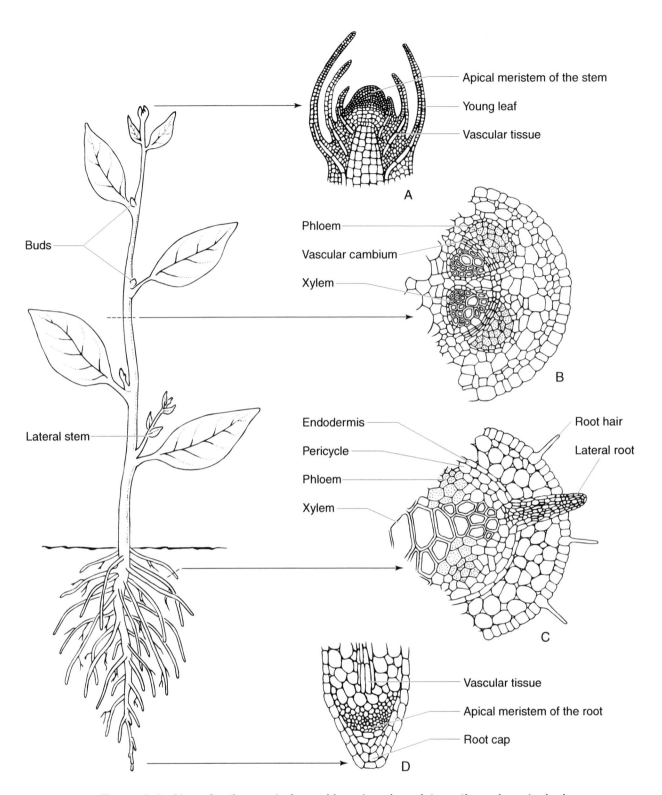

Figure 2.2 Vascular tissues (xylem, phloem) and meristematic regions (apical meristems of the stems and roots, vascular cambium, pericycle) of a broadleaved plant: (A) longitudinal section of a stem tip, (B) cross section of a stem, (C) cross section of a root, (D) longitudinal section of a root tip.

the pistil, which contains the egg(s). *Fertilization,* or the union of sperm (delivered to the ovary portion of the pistil by the pollen tube) and eggs (present in the ovary), results in the formation of seeds. As the seeds mature, fruit tissue develops around the seeds. Fruits serve to protect and disseminate seeds.

Plant Tissues

Tissues are groups of cells of similar structure that perform special functions. The major groups of plant tissues are the *vascular* or *conducting tissues* (xylem, phloem), the *structural tissues* (parenchyma, sclerenchyma, and collenchyma) that make up the bulk of the plant, and the *epidermal tissue* that forms a single layer of cells that completely surrounds the plant.

The vascular tissues in the stems of broad-leaved plants are located in bundles or as a cylinder around the periphery of the stem; the vascular tissues in the roots of broadleaved plants are located at the center of the root (Figure 2.2). The vascular tissues are scattered throughout the stems of grasses and other monocots (not shown).

A very important function of the structural tissues (besides as filler tissue) is photosynthesis, which takes place in the leaves in specialized parenchyma cells called the spongy mesophyll and palisade parenchyma (see Figure 2.6B). The

epidermal cells of the stems and leaves produce an outer waxy layer called the *cuticle* that protects the plant from minor mechanical damage and prevents excessive water loss (see Figure 2.6B). Certain cells of the root epidermis elongate to become the root hairs (Figure 2.2) that provide increased surface area for water and mineral uptake.

New cells and tissues are produced by the *meristematic regions* of the plant (e.g., the apical meristem, vascular cambium). The functions of the meristematic regions and vascular tissues are discussed more fully in the sections on plant growth and transport pathways, respectively.

Plant Cells

The basic living unit of a plant is the cell. Without the proper functioning of cells, a plant cannot survive. When we talk about the ways in which a weed can be killed (desiccated, broken or fractured, poisoned, smothered, or starved) we are talking about these events occurring to cells. Plant cells are surrounded by a cell wall and plasma membrane and contain a viscous fluid in which various membrane systems (e.g., the endoplasmic reticulum) and organelles (e.g., mitochondria, chloroplasts) are suspended. Briefly, the parts of a cell (Figure 2.3) and their importance to plant survival are as follows:

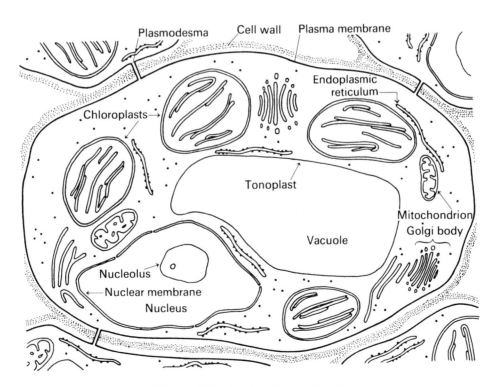

Figure 2.3 Parts of a plant cell.

Cell wall: Surrounds the protoplast and is composed primarily of cellulose strands or fibrils (microfibrils). The cell wall is a "non-living" structure. Its major function is to provide structure and rigidity to the cells. The microfibrils of the cell wall are hydrated (saturated with water) and loosely woven; they present no barrier to most molecules moving toward the cytoplasm.

Plasma membrane: Completely encloses the cytoplasm and is the major barrier to the penetration of some solutes. Nonpolar (uncharged) or lipophilic (fat-soluble) molecules penetrate the plasma membrane more readily than do polar, charged molecules (ions) or hydrophilic (water-soluble) molecules. Water and gases (e.g., carbon dioxide, oxygen), however, can readily penetrate the plasma membrane. The plasma membrane maintains the integrity of the cell; if disrupted or punctured, the cell contents leak out. This membrane *must* be intact for normal metabolism to occur.

Plasmodesma (pl.-desmata): Protoplasmic channels that pass through the wall and connect protoplasts of adjacent cells.

Nucleus: The site of the genetic material (DNA) that contains the information code for the plant. Its proper functioning is essential for the synthesis of ribonucleic acids and proteins. The DNA is localized in chromosomes. Linear segments of the chromosomal DNA, known as genes, code for hereditary traits. Nuclear division is called mitosis and results in the formation of two daughter cells, each of which contains chromosomes identical to those of the mother cell.

Mitochondrion (pl.-ria): The site of two of the three stages of respiration (Krebs cycle, electron transport chain). During respiration, sugars are metabolized to produce carbon dioxide, water, and chemical energy in the form of adenosine triphosphate (ATP). ATP is then utilized in numerous synthetic and metabolic processes in the cell, including the active pumping of minerals into the cell.

Chloroplast: The site of photosynthesis; water and carbon dioxide are converted to sugars with light acting as the energy source and chlorophyll pigments acting as the energy-trapping agents.

Ribosomes, endoplasmic reticulum, and Golgi bodies: Ribosomes are the sites of protein synthesis. The endoplasmic reticulum and Golgi bodies act as an interconnected membrane system to produce new plasma membrane and the noncellulosic components of cell walls. Cellulose itself is produced at the plasma membrane.

Vacuole: Water-filled area surrounded by a membrane known as the tonoplast; involved in the water economy of the cell and storage of waste materials.

Cytoplasm: The cell sap; the living nonmembranous portion of the protoplast.

Protoplast: The living parts of the cell; excludes nonliving parts such as the cell wall and vacuole.

Plant Growth

Growth of a plant involves two basic cellular processes: cell division and cell enlargement. These processes occur at meristematic regions located at the tips (*apical meristems*) of stems, buds, and roots. Certain internal tissues such as the *pericycle, vascular cambium,* and *intercalary meristem* (in grasses) are also meristematic and produce new cells (Figures 2.2 and 2.4).

The functions of the various meristematic regions are as follows:

Apical meristem of the stem	Increases length of the stem; formation of leaves, buds, and flowers
Buds	Formation of lateral shoots (branches) or flowers
Apical meristem of the root	Increases length of the root
Pericycle (found in roots only)	Formation of lateral roots
Leaf margins	Increase width of leaves
Vascular cambium	Increases both stem and root diameter in broadleaved species
Intercalary meristem	Increases length of stem in grasses

Intercalary meristems (Figure 2.4) are located at the base of each internode of a grass stem and account in part for the rapid elongation of the grass stem just prior to flowering.

The differentiation of newly formed cells into tissues occurs near the meristematic regions. *Therefore, control methods that affect meristems not only cause normal growth to cease but also may have some type of formative effect at the growing points.* Such an effect is most readily visible on stem tips, root tips, developing leaves, or flower buds.

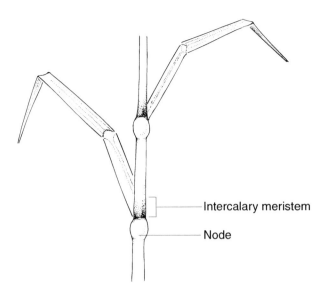

Figure 2.4 Location of intercalary meristems on a grass plant.

Plant Development

The pattern in which cells differentiate and become incorporated into tissues and tissues into organs results in the final form of the plant and accounts for the variation in structure among species. This overall process is known as plant development. For weed control purposes, it is important to understand the basic differences in seedling development between broadleaves and grasses (Figure 2.5).

In both broadleaves (dicots) and grasses (monocots) the first structure to emerge from the seed during germination is the root tip (radicle). Emergence of the stem and stem tip follows. In broadleaved seedlings the internodal cells of the stem (hypocotyl, epicotyl, and first internode) expand rapidly, thus increasing the overall height of the plant. Within a week or so after germination, the stem tip is elevated completely above the soil surface (Figure 2.5A, middle). Food material for the elongation and development of both the root and stem tips is provided by the two seed leaves (cotyledons) until the first foliage leaves are expanded and can begin photosynthesizing.

Some broadleaves (e.g., biennials during their first year of growth, perennials with crowns) form low-growing *basal rosettes* of leaves. The stem tip is located close to or below ground level and does not elongate until flowering (bolting) occurs during the second year of growth (e.g., in biennials).

In grasses the first structure to emerge from the soil is the protective sheath-like coleoptile. The developing leaves break through the coleop-

tile and unfold. Unlike the broadleaved plant, however, the internodal regions do not elongate immediately, and the stem tip of the grass seedling remains close to the ground (Figure 2.5B, middle). Not until several weeks after seed germination do the internodal regions expand, mainly by cell elongation and by the formation of new cells from the intercalary meristems. The stem greatly increases in length just prior to the production of flowering structures.

At the seedling stage, the growing points of most broadleaves are above the soil surface and are exposed, whereas the growing points of grasses are located at the base of the plant and are relatively well protected.

Transport Pathways

Materials essential for the normal growth and development of plants must be transported from their origin or point of entry to a site of utilization. These materials include water (a major component of all cells and a starting material for photosynthesis), minerals such as nitrogen and phosphorus (components of proteins and nucleic acids, respectively), and organic compounds such as sugars (the structural components of cells and sources of energy). There are two major transport systems in plants (Figure 2.6):

Apoplast: The major pathway for water and minerals: accounts for movement of materials from the soil upward through the plant and into the leaves.

Symplast: The major pathway for sugars: accounts for sugar movement from the site of production (leaves) to a site of storage or utilization (e.g., underground plant structures or developing stem tips).

The importance of these two pathways in the movement of externally applied materials such as herbicides warrants a more detailed description of each.

Apoplast

The apoplast is a more or less continuous system of cell walls, intercellular spaces, and interconnecting cells of xylem tissue. The components of this system are *nonliving.* Transpiration, the driving force for the movement of water and minerals in the apoplast, is a purely *physical* process. Transpiration is initiated by the evaporation of water from the leaf surface. As the water evaporates, more water moves to the leaf surface to fill the deficit. In this way, water is "pulled" up through the plant from the soil system through the roots and upward through the stem to the

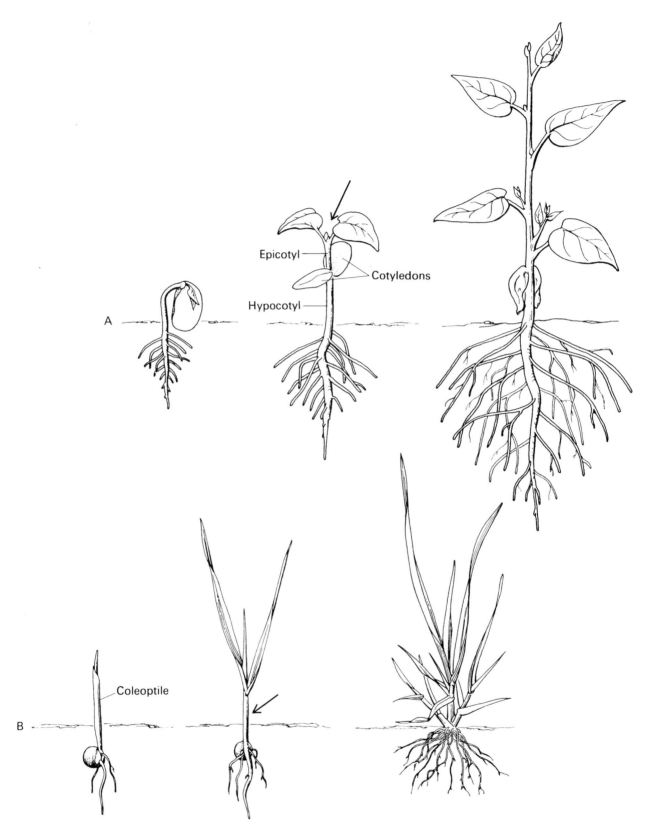

Epicotyl

Cotyledons

Hypocotyl

A

Coleoptile

B

Figure 2.5 Comparison of the germination and early development of (A) a broadleaved plant and (B) a grass plant. Left: Germination. Middle: 1 wk after emergence. Right: A well-established young plant. Arrows point to location of shoot tips.

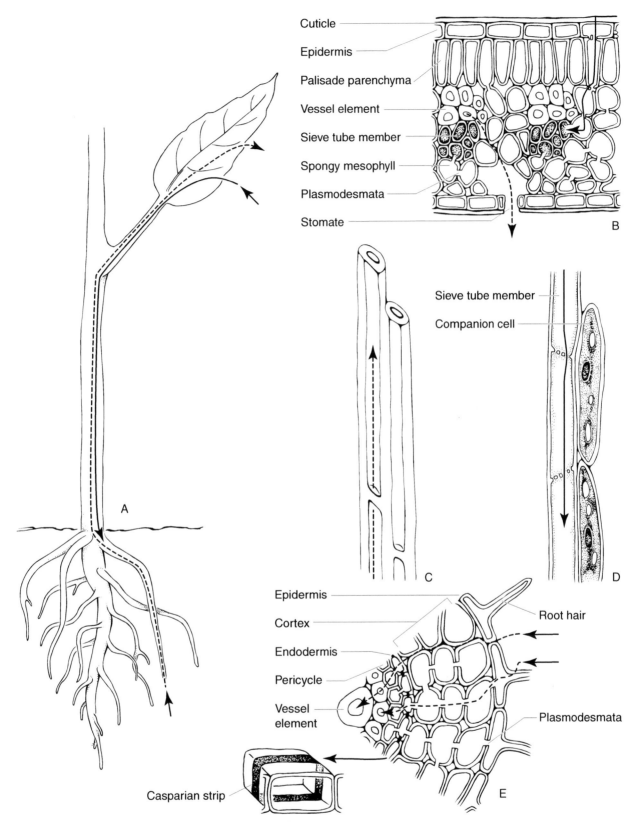

Figure 2.6 Transport pathways in plants (dashed line indicates pathway for apoplastic transport; solid line for symplastic transport): (A) overall pathway of symplastic and apoplastic transport in a plant, (B) cross section of a leaf, (C) vessel elements of xylem tissue (note thick cell walls), (D) sieve tube members and companion cells of phloem tissue, (E) cross section of root showing two pathways for movement of water and ions through the cortex.

leaves. This upward movement of water provides the mechanism for moving minerals to various parts of the plant.

Apoplastic transport begins with the movement of water into the root through the epidermal surface and root hairs (Figure 2.6E). Unlike the epidermal cells of stems and leaves, which are covered with a water-repellent cuticle, the surfaces of root hair cells have little or no cuticle. Water can enter the root by diffusing through the cells walls. In addition, since the plasma membrane of cells is not a barrier to its movement, water can penetrate easily into the cytoplasm of the root hair cells. From the root hairs, the water moves into the root cortex. Once inside the cortex, water can move freely through the cell walls and intercellular spaces or can move from one cell to another through plasmodesmata. Movement of water is toward the central core of vascular conducting tissue (the stele), which contains the xylem (xylem is made of up vessel elements, see Figure 2.6E), to replace the water that is lost to the transpiration stream.

In contrast to water, charged or ionic particles are repelled by the plasma membrane (minerals such as nitrogen, phosphorus, potassium, etc., are usually taken up in the ionic form). Although ions can move through the cell walls and intercellular spaces with water, at some point they must actually be pumped across the plasma membrane, using ATP, into the cytoplasm. The presumed reason for this requirement is the presence of the endodermis, a specialized layer of cells that encircles the stele. The cell walls of endodermal cells are impregnated with a waxy substance called suberin. This layer of suberin, called the Casparian strip, repels charged molecules that try to move through the cell wall portions of the endodermis. Therefore, in order to enter the stele, ions must be pumped into the cytoplasm of the endodermis. It also is thought that ions can be pumped into the cytoplasm of root hair cells and cortex cells, moving from cell to cell to the endodermis via plasmodesmata. Once ions move through the cells of the endodermis into the stele region, they are pumped, again using ATP, across the plasma membrane of living cells and move into the nonliving vessel elements and tracheids that make up the xylem. The Casparian strip prevents the backward movement of ions through the cell walls of the endodermis into the cells of the root cortex. In addition, charged molecules cannot reenter the cytoplasm of endodermal cells because of the presence of the plasma membrane and the lack of an appropriate ATP pump for backward movement. No matter whether ions move

through the root via a living system of cells and plasmodesmata or through the cell wall system (until they get to the endodermis), they eventually end up in the stream of water that is flowing through the xylem.

The vessel elements and tracheids of the xylem form a nonliving series of cylinders (Figure 2.6C) through which water and dissolved substances move upward through the root and stem and into the leaves. Although a portion of the water is used for cell enlargement and photosynthesis, the majority passes from the xylem of the leaf to leaf cell walls and then to openings (stomata) in the leaf surface (Figure 2.6B). At the cell wall–air interface, water is converted to the vapor phase and is removed from the leaf by diffusion.

Symplast

The symplast is a more or less continuous system of protoplasts of plant cells (connected by plasmodesmata) and includes the interconnected cells of the phloem tissue. The symplast is a *living* system; in order for substances to move through this system, the component cells *must be alive.* The driving force for phloem movement is the mass flow of solutes dissolved in water from regions of high solute concentration to regions of low solute concentration. This gradient is created by the synthesis and loading of sugars (solutes) at the "source" (e.g., photosynthetically active leaves) and the utilization of sugars at the "sinks." Sinks include actively growing parts of the plant such as stem tips and buds, actively respiring parts such as roots, or parts in which sugars are being converted to starch and stored, for example, in underground reproductive structures such as rhizomes, creeping roots, or tubers. This movement of organic materials frequently is referred to as the "source to sink" concept.

The symplast is the pathway by which substances applied to the foliage of plants move to underground structures. Passage into the leaf is not as easy as passage into the root, mainly because the waxy cuticle covering leaf and stem surfaces is an effective barrier to water and water-soluble substances. Substances that do move through the cuticle enter the living system of the plant by penetrating the plasma membrane (see Chapter 6 for a discussion of the mechanisms of penetration) and passing into the cytoplasm of cells (Figure 2.6B). Movement from cell to cell is through plasmodesmata. Materials eventually move into the components of the phloem known as the sieve tube members (Figure 2.6D). The end walls of sieve tube members contain pores that allow materials to pass from one cell to another. Although devoid of nuclei, sieve tube members are living cells. Closely associated with the sieve tube members are companion cells. Although the role of companion cells is unknown, these cells do have nuclei and also remain alive as long as the sieve tube cells remain viable.

Remember that the cells of the leaves and phloem must be intact and living for movement through the symplast to occur.

WEED LIFE CYCLES

Herbaceous (nonwoody) weeds can be divided into five life cycle categories (Table 2.1): summer annual, winter annual, biennial, simple perennial, and creeping perennial. Woody plants that are weedy are typically shrubs and vines. This classification system allows us to generalize about the common characteristics of life cycle groups before dealing with more specific differences, as described in Chapter 13. Life cycle also determines the relative susceptibility of established plants to control practices.

In this description, the production of new stems, leaves, and roots on a plant is called *vegetative growth.* The term *vegetative reproduction* refers to the generation of new plants from vegetative structures. This is in contrast to the production of new plants from flowers and seeds, which is termed sexual reproduction.

Annuals

Annual plants live for one growing season only. Seeds germinate and plants grow vegetatively, flower, produce seed, and die in one growing season or less. Individual established plants are relatively easily controlled. Season-long pressure on crops can be attributed to multiple flushes of new plants from seed over the growing season. Annuals survive in nearly all crop management systems where soil is unshaded. The two types of annual plants are: summer annuals and winter annuals.

Summer annuals germinate in spring and summer and mature before winter. Summer annual broadleaves include the pigweeds, lambsquarters, jimsonweed, velvetleaf, cocklebur, several smartweeds, common and giant ragweeds, ivyleaf and tall morningglories, knotweed, and common sunflower. Summer annual grasses include the foxtails, crabgrasses, barnyardgrass, fall panicum, and shattercane. Summer annuals are problems in crops planted in the spring and early summer and other open sites including bare spots in perennial crops (forages, turf) and winter cereals.

Winter annuals germinate in late summer, fall, and winter. They mature in spring or early summer. Winter annual broadleaf weeds include most mustards, henbit, and common chickweed. Winter annual grasses include downy brome, cheat, and false timothy. Winter annuals are problems in crops planted in the fall and late summer (winter cereals), in open sites such as harvested fields of summer annual crops, and in bare spots in dormant perennial crops including pastures, alfalfa, and turf grasses.

In climates where the summers are cool (e.g., the northern portions of the United States, Canada), winter annuals may become summer annuals. For example, wild oats are winter annuals in warm climates (e.g., California) but summer annuals in cooler climates (e.g., North Dakota and southern Canada).

Biennials

Biennial plants live for two growing seasons. Seeds germinate, and plants grow vegetatively as rosettes the first year. Exposure to cold during the winter induces the rosette to bolt (send up a flowering stalk) when the weather warms. The biennial then flowers, matures its seed, and dies during the summer or fall of the second growing season. Completion of the life cycle requires portions of two growing seasons and the winter between them. Musk thistle, bull thistle, wild carrot, common burdock, evening primrose, common mullein, and wild parsnip are biennials. Sites severely disturbed in late fall, winter, or early spring are poor habitats for biennials because the plants cannot overwinter successfully. Biennials are mostly associated with perennial crops and undisturbed sites.

Perennials

Perennial plants survive indefinitely (three and usually more years). In other words, flowering does not trigger senescence and death of the entire plant; underground structures continue to live.

Simple perennials start from seed, grow vegetatively, and form a crown of tissue at or below the soil surface on the upper end of a taproot. The root and crown survive indefinitely, generating new leaves and shoots. Shoots periodically produce flowers and seeds. Seeds provide the mechanism for spread. Dandelion, curly dock, chicory, plantains, and pokeweed are simple perennials. Simple perennials need habitats relatively free

Table 2.1 Some Examples of Weeds with Different Life Cycles

Annuals
  Summer annuals
  Abutilon theophrasti (velvetleaf), *Amaranthus blitoides* (prostrate pigweed), *Amaranthus retroflexus* (redroot pigweed), *Ambrosia artemisiifolia* (common ragweed), *Ambrosia trifida* (giant ragweed), *Anoda cristata* (spurred anoda), *Avena fatua* (wild oats), *Cassia obtusifola* (sicklepod), *Cenchrus* (sandbur), *Chenopodium album* (common lambsquarters), *Cycosis angulatus* (burcucumber), *Datura stramonium* (jimsonweed), *Digitaria ischaemum* (smooth crabgrass), *Digitaria sanguinalis* (large crabgrass), *Echinochloa crusgalli* (barnyardgrass), *Eleusine indica* (goosegrass), *Eragrostis cilianensis* (stinkgrass), *Eriochloa villosa* (woolly cupgrass), *Helianthus annuus* (common sunflower), *Ipomea hederacea* (ivyleaf morningglory), *Ipomea lacunosa* (pitted morningglory), *Ipomea purpurea* (tall morningglory), *Kochia scoparia* (kochia), *Mollugo verticillata* (carpetweed), *Panicum capillare* (witchgrass), *Panicum dicotomiflorum* (fall panicum), *Polygonum* (knotweed), *Polygonum pensylvanicum* (Pennsylvania smartweed), *Polygonum persicaria* (ladysthumb smartweed), *Portulaca oleracea* (common purslane), *Salsola iberica* (Russian thistle), *Setaria faberi* (giant foxtail), *Setaria leutescens* (yellow foxtail), *Setaria viridis* (green foxtail), *Sida spinosa* (prickly sida), *Solanum ptycanthum* (black nightshade), *Sorghum bicolor* (shattercane), *Tribulus terrestris* (puncturevine), *Xanthium strumarium* (common cocklebur)

  Winter annuals
  Brassica nigra (black mustard), *Bromus secalinus* (cheat), *Bromus tectorum* (downy brome), *Capsella bursa-pastoris* (shepherd's purse), *Conyza canadensis* (horseweed), *Descurainia* (tansy mustard), *Lamium amplexicaule* (henbit), *Lepidium* spp. (pepperweed), *Raphanus raphanistrum* (wild radish), *Senecio vulgaris* (common groundsel), *Stellaria media* (common chickweed), *Thlaspi arvense* (field pennycress)

  Biennials
  Arctium minus (common burdock), *Carduus nutans* (musk thistle), *Cirsium altissimum* (tall thistle), *Cirsium vulgare* (bull thistle), *Daucus carota* (wild carrot), *Dipsacus sylvestris* (teasel), *Oenothera biennis* (evening primrose), *Onopordum acanthium* (Scotch thistle), *Pastinaca sativa* (wild parsnip), *Verbascum thapsus* (common mullein)

  Simple perennials
  Cichorium intybus (chicory), *Phytolacca americana* (pokeweed), *Plantago lanceolata* (buckhorn plantain), *Plantago major* (broadleaf plantain), *Rumex crispus* (curly dock), *Taraxacum officinale* (dandelion)

  Creeping perennials
    Rhizomes
    Cynodon dactylon (bermudagrass), *Ellytrigia repens* (quackgrass), *Equisetum* spp. (horsetail), *Lythrum salicaria* (purple loosestrife), *Muhlenbergia frondosa* (wirestem muhly), *Polygonum coccineum* (swamp smartweed), *Solidago* (goldenrod), *Sorghum halepense* (johnsongrass), *Typha* spp. (cattail)

    Tubers
    Cyperus esculentus (yellow nutsedge), *Cyperus rotundus* (purple nutsedge), *Helianthus tuberosus* (Jerusalem artichoke)

    Bulbs
    Allium vineale (wild garlic)

    Stolons
    Cynodon dactylon (bermudagrass), *Glechoma hederacea* (ground ivy), *Muhlenbergia schreberi* (nimblewill)

    Creeping roots
    Acroptilon repens (Russian knapweed), *Ampelamus asbidus* (honeyvine milkweed), *Apocynum cannabinum* (hemp dogbane), *Asclepias syriaca* (common milkweed), *Cardaria draba* (hoary cress), *Cirsium arvense* (Canada thistle), *Convolvulus arvensis* (field bindweed), *Convolvulus sepium* (hedge bindweed), *Euphorbia esula* (leafy spurge), *Ipomea pandurata* (bigroot morningglory), *Physalis subglabrata* (smooth groundcherry), *Pueraria lobata* (kudzu), *Rumex acetosella* (red sorrel), *Solanum carolinense* (Carolina horsenettle), *Sonchus arvensis* (perennial sowthistle)

  Woody perennial shrubs and vines
  Artemesia tridentata (big sagebrush), *Campsis radicans* (trumpet creeper), *Larrea tridentata* (creosotebush), *Lonicera japonica* (Japanese honeysuckle), *Quercus* sp. (Gambel's oak and other weedy oak species), *Rubus* (wild blackberry), *Tamarix ramosissima* (saltcedar), *Toxicodendron radicans* (poison ivy), *Vitis* (wild grape)

from frequent below-ground disturbance for survival. For example, they are found in perennial crops, undisturbed sites, and no-till sites.

Creeping perennials differ from the other life cycle groups by reproducing from vegetative structures in addition to seed. When creeping perennials start from seed, they grow vegetatively for a while and then produce vegetative reproductive structures (creeping roots, rhizomes, tubers, stolons, or bulbs) in less than one growing season. Vegetative reproductive structures are the major means of localized spread, competition, and survival. Creeping perennials survive in most crop management systems because of their rapid and vigorous recovery capabilities. Creeping roots, rhizomes, tubers, etc., have large amounts of stored food and numerous buds capable of generating new shoots. These vegetative buds vary from those actively growing to those that are inactive. Some may be deeply dormant. Severe disturbance of creeping perennial plants stimulates sprouting of buds and rapid generation of new plants. Substantial portions of underground vegetative reproductive structures lie below the plow layer and are protected from the effects of tillage. Individual vegetative reproductive structures allow the regeneration of perennial plants over one to several years while their buried seeds provide a mechanism for long-term survival.

Quackgrass, johnsongrass, wirestem muhly, nimblewill, yellow nutsedge, leafy spurge, cattail, wild garlic, Canada thistle, field bindweed, hedge bindweed, common milkweed, honeyvine milkweed, Jerusalem artichoke, smooth groundcherry, horsenettle, hemp dogbane, big root morningglory, perennial sowthistle, and swamp smartweed are creeping perennials. Most noxious weed lists are dominated by creeping perennials.

Woody perennial shrubs and vines produce woody stems that survive overwinter and produce new stems and leaves the following spring. Examples of shrubs that pose weed problems include multiflora rose, sagebrush, buckbrush, and Gambel's oak. Woody vines such as wild grape, poison ivy, trumpetcreeper, and Japanese honeysuckle produce new plants from creeping roots.

Vegetative Reproductive Structures

The structures that permit creeping perennial weeds to spread, produce new independent plants, and overwinter are described here (Figure 2.7).

Rhizomes are elongated horizontal underground stems (with nodes, internodes, and modified leaves). Leaves on rhizomes usually are reduced to papery scales that are white to dark brown in color. Root and new shoot growth from rhizomes always originate from buds at the nodes. Rhizomes also are modified for food storage.

Tubers are thickened underground stems borne on rhizomes. They also have nodes and internodes and may have leaf remnants, which are scale-like and white to dark brown. The internodes are greatly reduced in length compared to those on a rhizome. Root and new shoot growth from tubers always originates from buds at the nodes. Tubers also store food.

Bulbs are leaf tissue modified for food storage and borne on a small stem plate. Roots and new bulbs develop from this stem plate.

Stolons are horizontal aboveground stems. Because of their exposure to light, stolons and the leaves that are produced at the nodes are green. Roots and new plants originate at the nodes. Food storage is not a major function of stolons as it is in the other vegetative reproductive structures.

Creeping roots do not have leaves, nodes, and internodes. These roots can grow both horizontally and downward. They develop buds (from the pericycle) capable of generating new shoots at more or less random intervals along their length. Creeping roots also are modified for food storage. Roots tend to penetrate deeper into the soil than some of the other vegetative structures and appear to be more resistant to control measures.

SHOOT REGENERATION

Shoot regeneration is essential for continued survival of an established weed. Thus, any control practice that destroys a weed below the lowest point capable of regenerating a new shoot results in the death of the established plant. The position of shoot regeneration on a plant and the resistance of a plant to destruction vary dramatically with the weed life cycle (Figure 2.8).

Annuals

The cotyledons on the shoots of annual broadleaf plants are above the soil surface. The bud in the axil of the cotyledons is the lowest point from which an annual broadleaf can regenerate a new shoot. Any process that destroys an annual broadleaf weed below the cotyledons should kill it.

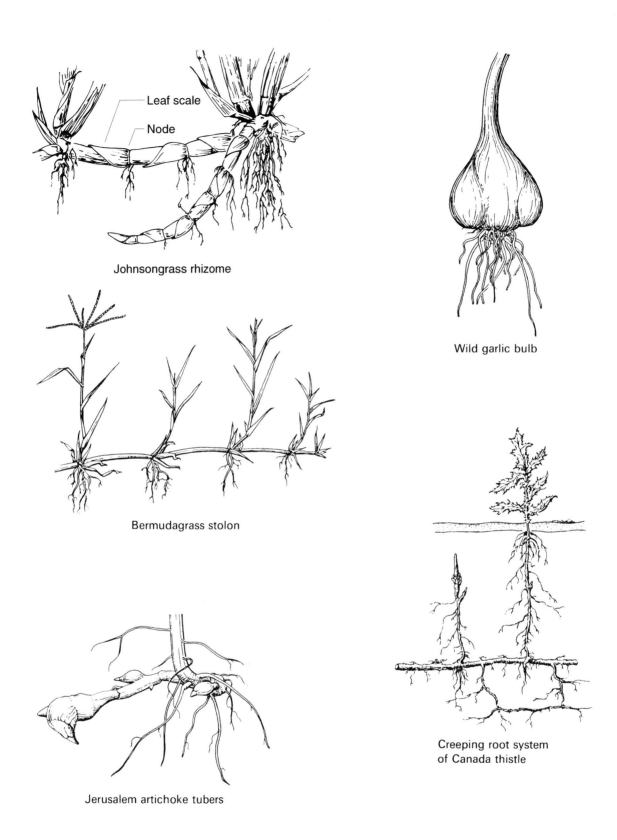

Leaf scale

Node

Johnsongrass rhizome

Wild garlic bulb

Bermudagrass stolon

Jerusalem artichoke tubers

Creeping root system
of Canada thistle

Figure 2.7 **Vegetative reproductive structures of creeping perennial plants.**

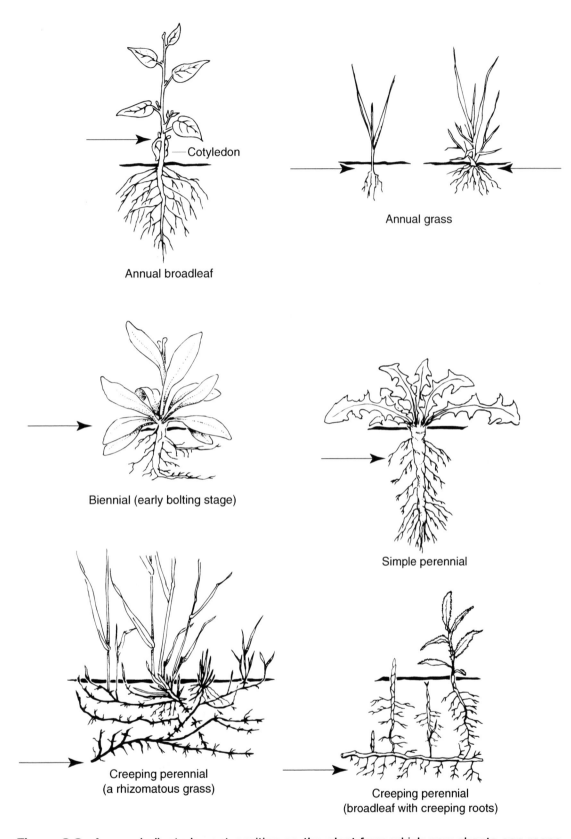

Cotyledon

Annual grass

Annual broadleaf

Biennial (early bolting stage)

Simple perennial

Creeping perennial
(a rhizomatous grass)

Creeping perennial
(broadleaf with creeping roots)

Figure 2.8 Arrows indicate lowest position on the plant from which new shoots can regenerate.

Annual grass shoots establish a growing point at or just below the ground. Since the growing point is somewhat protected, young annual grass plants need to be damaged under the soil surface for complete kill. Young grasses usually survive close mowing and light frost. Once an annual grass joints (develops internodes between leaves) and starts to send up seedheads, it becomes more susceptible to damage.

Biennials

The rosette of leaves produced during the first year is at the ground line; the apical meristem and all of the axillary buds are close together. Because of the low position, the growing points are somewhat protected from damage so that destruction is needed just below the soil surface. As a biennial bolts (sends up a seed stalk), the internodes elongate raising the meristems above the ground line so that they become more susceptible to destruction.

Herbaceous Simple Perennials

Herbaceous simple perennials grow from a crown of tissue one to several inches below the soil surface. The crown and taproot survive and generate new shoots even after substantial damage. A single close mowing, scraping at the soil suface, application of a contact herbicide, or cutting just below the soil surface usually will not kill an established simple perennial. Destruction two or more inches deep is required. If crowns cut off by tillage are not dried out they can be transplanted and grow again.

Creeping Perennials

Creeping perennials have numerous well-protected buds capable of generating new shoots. The vegetative reproductive structures have large amounts of stored food and numerous buds. They almost always penetrate 6 in. or more into the soil and sometimes extend several feet downward. These underground vegetative reproductive structures are capable of generating new shoots at any depth. Severe disturbance stimulates sprouting of buds and rapid generation of new plants that are far more aggressive than plants from seed. Rarely does a single treatment of any kind result in substantial control. Repeated treatments to exhaust buds and stored food are required to subdue established plants with these growth characteristics.

A perennial weed in the seedling and early vegetative stages behaves as an annual. During this period it can be treated as an annual. Once perennial characteristics are present, special control measures are needed.

MAJOR CONCEPTS IN THIS CHAPTER

Plants are complex organisms with well-defined structures in which a multitude of vital processes occur in well-ordered and integrated sequences. One of more of these structures or processes must be disrupted to control a weed.

Plants are made up of organs (stem, leaf, root, flower); organs consist of tissues (vascular, structural, epidermal) and regions (meristematic); and tissues are made up of cells. At the cellular level the plasma membrane is the major living barrier to the penetration of substances into the cell.

Meristems (growing points) exposed aboveground (e.g., stem tips of broadleaved weeds) are more susceptible to control practices than are meristems located at the soil surface (e.g., stem tips of annual grasses when young, perennials with crowns) or beneath the soil surface (e.g., buds on creeping roots, rhizomes, tubers).

The two major transport systems in plants for materials essential for normal growth and development are the apoplast and the symplast.

- The *apoplast* is the major pathway for water and minerals from the soil upward through the plant and into the leaves. The apoplast includes the xylem as the major conducting tissue and is a more or less continuous system of cell walls and intercellular spaces. The system is mostly nonliving, and transpiration is the driving force. The pathway from the soil to the shoot starts with water movement into the root hairs, through the cortex, into cell cytoplasms before or at the endodermis (the plasma membrane at the site of entry into and exit from the cytoplasms is the only living site through which apoplastically translocated substances must pass), into the vessel members of the xylem, and upward into the leaves.
- The *symplast* is the pathway by which substances applied to foliage move to underground structures. The symplast includes the phloem as the main conducting tissue and is a more or less continuous system of

cytoplasm throughout the plant. This system is living and must remain so for translocation to occur. Movement is with sugars from actively photosynthesizing leaves (sources) to sites of utilization (sinks). Sinks include actively growing stem tips, root tips, and buds and storage organs such as roots, tubers, and rhizomes, in which sugar is being converted to starch. A foreign substance applied to the foliage must first pass the cuticle (a waxy barrier on the shoot surface), penetrate cell plasma membranes, and move into the cytoplasm. Once inside the cytoplasm the substance can move throughout the plant.

Weeds are classified according to life cycle: annuals, biennials, and perennials.

Annual plants live only one year and reproduce by seed. They include summer annuals and winter annuals.

Biennial plants live two years. They overwinter in the basal rosette stage the first year, respond to cold, and then bolt and flower the second year.

Perennial plants live more than two years. They include simple perennials and creeping perennials.

Creeping perennial plants are characterized by the presence of vegetative reproductive structures such as rhizomes, tubers, bulbs, stolons, and creeping roots.

The ability of an established plant to resist control measures and regenerate itself depends on the lowest point on the plant capable of shoot regeneration. The plant's life cycle determines where this point is located. The lowest point at which annual broadleaves can regrow is generally above the soil surface; for annual grasses and biennial rosettes it is just below the ground level. Simple perennials can regenerate from several inches below ground, and creeping perennials can regenerate from any depth at which tissue is present (1 to several ft).

TERMS INTRODUCED IN THIS CHAPTER

Apical meristem
Apoplast
Basal rosette
Buds
Bulb
Casparian strip
Cell wall
Chloroplast
Coleoptile
Companion cell
Cotyledons
Creeping perennials
Creeping root
Cuticle
Cytoplasm
Endodermis
Endoplasmic reticulum
Epicotyl
Epidermal tissues (epidermis)
Fertilization
Floral structures
Flowers
Golgi bodies
Hypocotyl
Inflorescence
Intercalary meristem
Internode
Lateral root
Lateral stem
Leaves
Meristematic regions
Mitochondria
Node
Nucleus
Pericycle
Phloem
Plasma membrane
Plasmodesmata
Pollination
Protoplast
Radicle
Rhizome
Ribosome
Root
Shoot
Sieve tube member
Simple perennials
Source to sink
Stem
Stolon
Structural tissues
Summer annuals
Symplast
Transpiration
Tuber
Vacuole
Vascular cambium
Vascular tissues
Vegetative growth
Vegetative reproduction
Vegetative structure
Vessel element
Winter annuals
Xylem

SELECTED REFERENCES ON PLANT STRUCTURES AND PROCESSES

Esau, K. 1977. Anatomy of Seed Plants, 2nd ed. John Wiley and Sons, New York.

Harper, J. L., Ed. 1960. The Biology of Weeds. Blackwell Scientific Publications, Oxford, England.

Kigel, J. and D. Koller. 1985. Asexual reproduction of weeds. Pages 65–100 in Weed Physiology, Vol. I. Reproduction and Ecophysiology, S. O. Duke, Ed. CRC Press, Boca Raton, Florida.

Mauseth, J. D. 1995. An Introduction to Plant Biology, 2nd ed., Saunders College Publishing, Philadelphia.

Salisbury, F. B. and C. W. Ross. 1992. Plant Physiology, 4th ed. Wadsworth Publishing, Belmont, California.

Stern, K. R. 1994. Introductory Plant Biology, 6th ed. Wm. C. Brown Publishers, Dubuque, Iowa.

THE SOIL SYSTEM

Soil affects and is affected by weed control practices in many important ways. It is the anchoring medium and supplier of minerals and water for both weeds and crops, and it is a reservoir for large numbers of weed seeds and vegetative reproductive structures. Weed control measures such as tillage and applying herbicides to the soil actively involve the soil. Most herbicides, even those applied to the foliage, end up in or on the soil. It is impossible to approach the control of terrestrial weeds intelligently without considering the soil in which they grow.

In this chapter we briefly review some fundamentals of soil science and then relate these principles to weed growth and weed control measures exclusive of herbicides. Soil herbicide interactions are discussed in Chapter 7.

THE SOIL SYSTEM

For our purposes a paraphrased summary from the *Yearbook of Agriculture on Soil* (USDA 1957) is an appropriate introduction to soil: " . . . soil is a dynamic three dimensional natural body on the surface of the earth in which plants grow. It is composed of mineral and organic materials and contains living organisms. Soils develop their properties over time due to the integrated effect of climate, topography, and living organisms acting upon parent materials."

Individual soils differ from one another in their properties, which is significant because several soil types are likely to be encountered on the same farm, and two and often more may be present in the same field. Each individual soil is definable with respect to such features as texture, depth, pH, organic matter, porosity, slope, exchange capacity, potential to erode, and waterholding capacity. Although these categories appear to be relatively simple and straightforward, soil must be recognized as a *complex and dynamic system,* one in which the various components interact with one another and with outside elements.

In this section we discuss the four major characteristics of soil: (1) soil is composed of three "phases" (solids, liquids, and gases); (2) soil contains a colloidal component; (3) soil possesses living constituents; and (4) soil is open to outside forces.

The Soil Phases

Solid Phase

Soil solids (soil particles) consist of mineral components and organic matter. Mineral components constitute the major portion of the solid phase in mineral soils; they occupy approximately 50% of the soil volume. These components are classified according to size. Sand measures 2.0 to 0.2 mm in diameter; silt, 0.2 to 0.002 mm; and clay, less than 0.002 mm. Soil texture is determined by the percentages of sand, silt, and clay fractions. Soil textural classes are illustrated by the standard texture triangle (Figure 3.1).

Soil organic matter is the relatively stable fraction of decaying plant and animal residues.

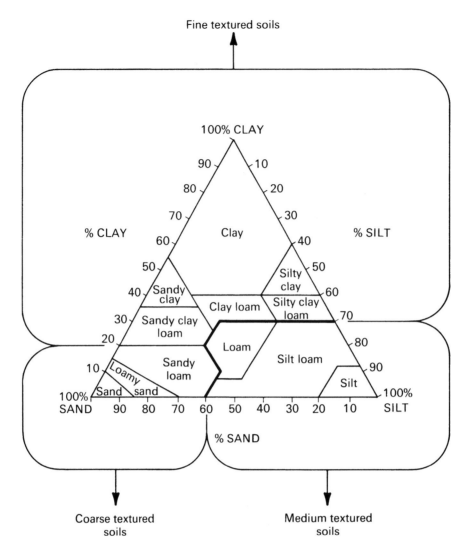

Figure 3.1 Standard soil texture triangle.

These residues eventually degrade to their inorganic constituents of carbon dioxide, water, and inorganic elements, so unless fresh residues are supplied continually, the organic matter content in agricultural soils gradually will decrease.

The rate of organic matter breakdown depends on climate, local topography, the amount of plant residue returned to the soil, and the relative amount of oxygen present in the soil environment. Under aerobic (oxygenated) conditions, microbes can oxidize organic matter into carbon dioxide and water. *Mineral soils* are formed on well-drained, aerobic sites and typically contain 0.5 to 6% organic matter. Under anaerobic (nonoxygenated) conditions, organic matter breakdown is much slower, and the continuous input of new plant residues results in a gradual buildup of organic matter in the soil. *Organic soils* are formed under wet, anaerobic conditions (in lakes, marshes, and swamps) that

protect the organic fraction from oxidation. The organic matter content of such soils may range from 10 to 80%.

Gas and Liquid Phases

The pore spaces (voids between the solid components) contain the gas (air) and liquid phases of soil. Air in soil contains somewhat less oxygen and many times more carbon dioxide than air above the soil surface; it is saturated with water vapor. The liquid phase or soil solution consists of water containing dissolved molecules and ions. The soil solution surrounds the soil particles and supplies nutrients and other materials to plant roots.

The amounts of air and water in the pore spaces fluctuate inversely with each other depending on the amount of water present. Soils that are too dry are unsuitable for plant growth because water is limiting. On the other hand, waterlogged soils lack sufficient oxygen for plant roots to function properly.

Colloidal Properties

Many soil properties are determined by the amount of *clay and organic matter.* These colloidal-sized particles (0.000001 to 0.001 mm diameter) have a high reaction potential because of their high surface-to-volume ratio and charge.

The extremely small size of clay and organic matter particles greatly increases the amount of reactive area. For example, a 1-cm cube has a surface area of 6 cm² or 0.0006 m². Division of the cube into smaller cubes with diameters the same as the largest colloidal diameter (0.001 mm) provides a surface area of 6 m². The same cube divided into cubes of the smallest colloidal diameter (0.000001 mm) provides a surface area of 6000 m² or nearly l 1/2 acres! In contrast to a gram of silt, which has a surface area of l m², one gram of montmorillonite clay has a surface area of 600 to 800 m².

Soil is negatively charged because the clay and organic fractions carry overall negative surface charges. This means that an anion (a molecule or ion with a negative charge) will be repelled and is likely to remain in the soil solution rather than be adsorbed on the colloid. On the other hand, cations (positively charged molecules and ions) will be attracted and adsorbed by the soil colloids. Most anions tend to be leached out of the soil, whereas cations tend to be retained. The capability of a soil to retain and exchange cations is termed the cation exchange capacity and depends primarily on the amount and kind of colloid present.

The cation exchange capacities of several different types of clays and stable organic matter are presented in Table 3.1. Substantial differences in adsorptive capacity occur among clay minerals. Most soils in the United States contain a mixture of clays, so clays in a typical medium-textured soil (from the Corn Belt, for example)

might have cation exchange capacities ranging from 20 to 65 meq/l00 g. The presence of organic matter (200 to 400 meq/100g) in a soil has an even greater impact on adsorptive capacity. A l% difference in organic matter results in a four- to twentyfold greater difference in adsorption than a 1% difference in clay.

Influence of Colloids on Soil Properties

The amount and kind of clay and organic matter present influence the potential a soil has for aggregation, tilth, water- and nutrient-holding capacities, water percolation rates, plasticity, crusting, and potential for compaction. The soil organic matter has an even greater influence on soil properties than clay does. The continuing decay of plant residues provides for the storage and release of plant nutrients, particularly nitrogen, phosphorus, sulfur, and potassium, and helps provide substances such as iron in forms available for uptake by plants. Organic matter also serves to stabilize the aggregation of soil particles. The improved aggregation of soil particles in a soil high in clay improves water penetration, porosity, tilth, and workability and reduces the susceptibility of the soil to compaction, crusting, and erosion. On coarse-textured soils (those with low amounts of clay), organic matter may be the major colloidal constituent.

Living Components

Soil is a living system populated by bacteria, fungi, algae, invertebrates, and vertebrates. Of these, the heterotrophic organisms (primarily bacteria, fungi, and earthworms) contribute directly to the degradation (breakdown and decay) of organic residues derived from plants and animals and to the recycling of important chemical elements. These organisms ultimately are responsible for the benefits derived from soil organic matter. In addition, bacteria and fungi are responsible for degrading herbicides to nonphytotoxic components and then to simple, basic elements and compounds such as carbon dioxide, water, ammonia, and sulfur.

The soil biota should be viewed as a dynamic living community with the capability of degrading nearly any biologically derived organic molecule, as well as most of those synthesized in the laboratory. Substances that contain carbon and can serve as food or energy sources will be biodegraded when added to the soil.

Table 3.1 Cation Exchange Capacities of Organic Matter and Several Types of Clays[a]

Soil Constituent	Cation Exchange Capacity (meq/100 g)
Organic matter	200 to 400
Vermiculite	100 to 150
Montmorillonite	80 to 150
Illite	10 to 40
Kaolinite	3 to 15

[a]Data from Bailey and White (1964).

Conditions that favor the functioning of biological systems enhance the activity of the soil biota. Reasonable temperatures above freezing and below boiling; adequate but not excessive moisture; and adequate sources of carbohydrates, major and minor elements, and oxygen are required by the heterotrophic organisms listed earlier. Extremes of temperature, pH, or moisture, as well as limiting essential nutrients or food sources, can be expected to reduce the activity of this component of the soil.

Soil as an Open System

Soil is an open system continuously subjected to outside influences. Soil comes in contact and interacts with plants, animals, wind, sunlight, precipitation, temperature changes, plant and animal residues, seeds, spores, chemicals from the atmosphere, humans and their manipulations, and various other substances and agents. Even though the soil resists changes, it is responsive to the actions impinging upon it. Thus, one seldom deals with precisely the same conditions every time weed control practices are used on a given soil. A good example is the frequent variation in soil moisture. A good weed control program must be flexible enough to meet changing soil conditions.

Soil as a Buffered System

Even though soil is open, it is *buffered* against permanent change. For example, soil is a good insulator. With the exception of the soil surface, it resists extreme changes in temperature. This is one reason seeds can lie dormant and remain viable in the soil for many years. The organic matter content of a soil, even under cropping conditions, remains relatively stable over several years. Only under drastic changes in management do organic matter percentages change rapidly. For example, if a prairie sod is converted to a clean tillage cropping system, organic matter content may drop rapidly due to the loss of plant residues and exposure of the soil to oxidizing conditions. Organic matter content then will become relatively stable. Even soil pH, a parameter that is manipulated frequently, requires the addition of 0.5 to 4 tons of finely divided calcium carbonate (lime) per acre to cause a change of one unit in the plow layer. Although the addition of foreign substances such as crop residues or pesticides may alter temporarily the characteristics of a soil bi-

otic community, permanent major alterations in the biota seldom occur, and these substances are degraded relatively soon. From a practical standpoint, unless it is subject to excessive erosion, the soil system remains a relatively constant and predictable entity.

WEEDS AND SOIL

In general, soils that support crops also support weeds. Occasionally, soils that provide poor growing conditions for crops will support an abundance of weeds. High or low pH, coarse-textured droughty soils, low fertility, soils high in salts, or overly wet or arid conditions favor certain weed species (Table 3.2). The success of such weeds, however, may be due not to their enhanced growth under these conditions but rather to the unsuitability of these sites for the growth of competitive plants. For example, weeds such as yellow nutsedge, smartweed, and cocklebur, which appear to have a special affinity for wet areas, may be successful in these sites because of the absence of competitive crop plants that have been flooded out.

Soil is the major repository (seed bank) for weed seeds and vegetative propagules. Seeds, which can survive in the soil for decades due to dormancy or limiting oxygen and light conditions, become buried to the depth of stirring by tillage, usually 8 to 10 in. Vegetative structures, on the other hand, can grow deep into the soil and frequently extend well below plow depth. The roots of creeping perennial broadleaved weeds in particular can be found in large masses several feet deep. The depth of root penetration depends on the weed species, soil texture, moisture regime, and the depth of the water table or soil layers impervious to root penetration. In arid climates, deep vegetative structures provide creeping perennial species with a distinct competitive advantage over other plants.

Table 3.2   Weeds Common to Specific Soil Conditions

Saline soils	Greasewood or chico (*Sarcobatus* sp.) Alkali sacaton (*Sporobolus airoides*) Desert saltgrass (*Distichlis stricta*) Alkali grass (*Puccinellia* sp.) Kochia (*Kochia scoparia*)
Dry soils	Halogeton (*Halogeton glomeratus*) Russian thistle (*Salsola kali*)
Wet soils	Yellow nutsedge (*Cyperus esculentus*) Barnyardgrass (*Echinochloa crusgalli*)
Acid soils	Red sorrel (*Rumex acetosella*)

WEED CONTROL PRACTICES AND THE SOIL

General Considerations

Weed control and other cultural practices can have a significant impact on the integrity and structure of the soil system. For example, weed control practices that remove protective plant cover expose the soil to the action of wind or water and can lead to serious soil losses. Such practices include repeated mechanical top kill, deep tillage, or the repeated use of foliar-applied herbicides and persistent soil herbicides. The impact of these practices, along with climatic forces (rainfall amount and intensity, wind velocity, etc.), slope, soil type, and cropping practices (rotations, etc.), are factors that must be considered in determining the susceptibility of an area to erosion.

The major problems arising from soil erosion are the loss of topsoil and the siltation of bodies of water. Associated with the latter is the movement of adsorbed pollutants such as phosphorus and pesticides into water. Weed control and cultural practices that minimize soil erosion always should be used in areas adjacent to or that drain into streams, ponds, and lakes.

Tillage

Much recent emphasis has been placed on the potential of tillage to cause physical damage to the soil (compaction, loss of structure, and loss of organic matter). Damage to the soil frequently is associated with tillage operations, but operations such as planting and harvesting also can contribute. All soils are susceptible to injury, although those high in clay and the sandy loams are perhaps more susceptible. Much of the compaction problem in soils could be minimized if the working of excessively moist soils were avoided.

Conventional tillage mixes the litter from the soil surface into the plow layer, which leaves the soil surface unprotected and increases organic matter breakdown. A century of continuous tillage in many soils has resulted in decreases in soil organic matter to 50% of the original levels. Tillage also negatively alters the microbial and invertebrate (for example, insect and earthworm) populations of the soil, components that normally process and incorporate organic residues into the soil, aerate soil, and improve soil structure and water infiltration. Some insects are effective predators of crop pests.

In natural systems plant litter on the soil surface buffers against extremes in temperature and provides a more stable habitat for micro-organisms and invertebrates. No-till cropping also leaves the litter on the surface, thus providing conditions more analogous to natural systems than does conventional tillage. In fact, organic matter has increased by as much as 1100 lbs (approximately 1/2 ton) per acre per year under no-till. Assuming this rate of addition of new organic matter and relatively little breakdown of the residues, it would take 3 yr, 10 yr, and 20 yr to increase the organic matter content by 1% in the upper inch, upper three inches, and upper six inches of soil, respectively. Obviously the greatest amount of accumulation in the short term is in the upper inch of soil, but the organic matter can move over longer periods of time to greater depths of the soil profile (for example, by leaching, movement by earthworms, or disturbance by planting equipment). In addition, the more carbon that is stored in the soil (about 1/2 of organic matter consists of carbon), the less carbon dioxide will be released into the atmosphere where it can combine with other greenhouse gases to contribute to global warming.

The surface litter left by no-till cropping also provides a more suitable habitat for increased numbers and a diversity of microbe, insect, earthworm, bird, and mammal populations. In fact, rodents can become a major pest problem in no-till systems. The herbicides on which no-till cropping is dependent appear to have little negative impact on these microbial and animal populations nor do they affect microbial processes such as organic matter formation, respiration, and nitrogen fixation (Edwards 1989). The loss of soil and habitat and the increase of suspended sediment in water that occurs with conventional tillage present a much greater negative impact to the environment than the use of herbicides on no-till systems.

Not only do tillage practices have an impact on soil, but soil characteristics such as texture, organic matter, and moisture have profound effects on tillage results. Tilling wet soils high in clay or silt results in extremely cloddy conditions and a less than desirable seedbed. Sandy soils, on the other hand, are much more friable and less prone to clod formation and compaction. They too, however, can be seriously damaged by soil compaction (some of the most serious compaction problems in the United States occur on the sandy loam soils of the southeastern coastal plains). The presence of organic matter improves tilth and aggregation in all soils and tends to ameliorate the degree of cloddiness under wet conditions. In addition, rocks, stumps, growing plants, and plant residues determine the suitability and effectiveness of various tillage machines and operations. For example, chisel plows are more suitable for use in rocky soils than other primary tillage tools.

Surface Conditions

Soil surface conditions also interact with weeds and weed control practices. Operations that result in firm seedbeds favor seed germination, particularly that of small-seeded weeds. This germination response can be observed readily in areas compacted by tractor or implement tires. Loose, relatively coarse seedbeds are less favorable for weed seed germination than finely worked compact ones.

Uneven, cloddy, or trashy soil surfaces can present obstacles to the proper performance of soil-applied herbicides. The importance of using soil-applied herbicides on a relatively smooth, dry surface free from trash and clods and the crop injury problems associated with movement of herbicides on uneven soil are discussed further in Chapter 7.

Another cultural consideration is the load-bearing ability of soil when equipment must be supported under varying weather conditions. For example, in orchards, sod strips help support spraying, pruning, and harvesting equipment. When wet, the order of increasing ability to support weight is tilled bare soil, undisturbed bare soil, and sod-covered soil.

MAJOR CONCEPTS IN THIS CHAPTER

Soils interact with weed management practices in several different ways. The weed itself, tillage practices, crop management practices, and herbicides all interact with the soil.

Soil consists of a solid phase surrounded by gas and liquid phases. The solid phase consists of sand, silt, and clay (the relative proportions of which determine texture). Stable organic matter constitutes an additional solid fraction.

The reactive portion of soil (soil colloids) consists of clay and organic matter particles which provide large amounts of surface with an overall negative charge.

Soil is a living system populated by numerous organisms. Moderate temperatures and adequate moisture, nutrients, energy sources, and oxygen enhance the activity of the soil biota.

Soil is an open system subject to weather, other natural agents, and to humans and their manipulations.

Soil is a buffered system that resists most changes.

Soil is the major repository for weed seeds and vegetative propagules. It serves as a medium for the storage of weed seeds and vegetative structures.

Some weeds are adapted for survival on saline, dry, wet, or acid soils.

Weed control practices can have adverse effects on soils. Soil erosion can be the result of vegetation removal or tillage practices and can, in turn, cause siltation and other pollution of water bodies. Tillage also can result in soil compaction.

No-till systems more nearly resemble natural systems by leaving plant residues on the soil surface. The presence of plant residues on the soil surface reduces erosion, and the lack of soil disturbance encourages the development of organisms that process and incorporate organic matter into the soil.

TERMS INTRODUCED IN THIS CHAPTER

Aerobic conditions
Aggregation
Anaerobic conditions
Anion
Buffering
Carbon dioxide
Cation
Cation exchange capacity
Clay
Coarse-textured soil
Colloid
Compaction
Degradation
Erosion
Gas phase
Heterotrophic organisms
Illite
Kaolinite
Liquid phase
Microbial populations
Mineral soil
Montmorillonite
Organic matter
Organic soil (peat, muck)
Plant residues
Porosity
Sand
Silt
Soil biota
Soil moisture
Soil pH

Soil solution
Soil texture
Solid phase
Tilth
Topsoil
Vermiculite
Water-holding capacity

SELECTED REFERENCES ON SOILS

Alexander, M. 1977. Introduction to Soil Microbiology, 2nd ed. John Wiley and Sons, New York.

Brady, N. C. 1996. The Nature and Properties of Soils. Prentice Hall, Upper Saddle River, New Jersey.

Follett, R. F. and B. A. Stewart. 1985. Soil Erosion and Crop Productivity. American Society of Agronomy, Madison, Wisconsin.

Foth, H. D. 1990. Fundamentals of Soil Science, 7th ed. John Wiley and Sons, New York.

Rendig, V. V. and H. M. Taylor. 1989. Principles of Soil-Plant Relationships. McGraw-Hill, New York.

Soil, The 1957 Yearbook of Agriculture. U.S. Dept. of Agriculture, Washington, D.C.

Tan, K. H. 1994. Environmental Soil Science. Marcel Dekker, New York.

METHODS OF WEED CONTROL

The contemporary applied weed scientist must function in an era of rapid changes that include shifting crop management practices, ever changing weed control technology, increasing concern for the environment, and the rapid adoption of computer and communication technology.

NEW DIRECTIONS IN WEED MANAGEMENT

A number of factors, primarily associated with environmental concerns, have resulted in major shifts in crop and weed management practices over the past two decades. For example, soil erosion from conventional tillage results in the contamination of surface water by soil sediments and accompanying agricultural chemicals. No-till crop management systems have replaced conventional tillage systems to keep soils in place. No-till, however, eliminates the ability to control weeds with tillage operations. To date, herbicides have replaced the weed control formerly provided by tillage operations.

Weed control technology itself is changing, also in part because of environmental concerns. The appearance of low levels of herbicides in surface and ground waters is responsible for the restricted uses of some herbicides and the development of new herbicides less likely to persist in the environment. In an effort to reduce herbicide use, efforts are being made to use myco-herbicides, allelopathic cover crops, and other biologically derived agents to broaden the spectrum of weed control practices. However, tillage and herbicides have carried the burden of weed control in recent decades and will continue to do so for at least another decade or more.

Space age technology is being adapted for use in weed management. Predictive models for weed and crop growth and development, weed–crop interactions, and the behavior of herbicides in plant and soil systems are being developed. Weed mapping is being facilitated with site-specific technology made possible by remote sensing and global positioning systems (GPS), which in turn are used to apply herbicides precisely(for more detailed information on these types of technologies, see Chapter 20).

Knowledgeable Practitioners Are Required

All of these changes in weed control technologies require competent practitioners in the field. Technology should be regarded as providing the management tools; it alone does not compensate for a knowledgeable field practitioner who can diagnose, make decisions, and interpret results.

Modern weed science is a relatively sophisticated discipline. The technology used is complex, and its implementation requires a complete and thorough understanding of plants, soils, crop management systems, and the many other environmental parameters that interact with weed control practices. Solutions for today's weed problems must be based on factual information and scientific principles and philosophies. The majority of this chapter presents specific information on weed control methods. Practitioners, however, must recognize that there is more to solving weed problems than simply being able to list methods. The willingness to analyze each component of the total cropping system and to plan a well-ordered, integrated program can mean the difference between success and failure. Indeed, the *approach* to weed control problems is just as important as the actual application of the methods. Accordingly we have devoted the first part of this chapter to what we call the scientific approach to solving weed problems. We urge you to become familiar with this approach and to use it whenever you encounter a weed control situation.

Modern weed control practices can be implemented most effectively by well-informed practitioners who provide accurate on-site diagnosis and prescribe efficient and appropriate control measures. At the same time, they must strive for improved control of weeds without causing detrimental effects on the environment. Such an approach is consistent with the goals of integrated pest management (IPM).

To be readily adopted, weed control practices must be easy to execute by the practitioner, provide consistent results, be safe to the environment, and be profitable to the provider and the grower.

Weed control and other crop management practices are dictated by weather over which the agriculturalist has no control. Planting, tillage, mowing, spraying, and harvest operations are delayed or prevented by precipitation and wet soils. Wind dictates periods suitable for spraying. Excess rainfall following tillage or herbicide application can cause substandard weed control. Consequently, the crop manager frequently encounters fields with undesired stages of weed and crop development, unplanted crops, or an uncontrolled weed situation. Thus, the ultimate test for your using the weed science principles covered in this textbook is the actual field situation.

A SCIENTIFIC APPROACH TO SOLVING WEED PROBLEMS

The scientific approach to weed control involves the following major components: (1) a thorough understanding of the overall objectives of weed management, (2) a step-by-step procedure for dealing with an individual weed problem, and (3) a realistic appraisal of the limitations and potentials of individual available technologies.

Objectives of Weed Management

The objectives of weed management programs are prevention, control, and eradication.

Prevention is keeping a weed from being introduced into an uninfested area. Successful implementation depends on sanitation, the prevention of seed production, and the prevention of seed and other propagule spread. Weed and seed certification laws and regulations are enacted to meet these objectives (see the section on preventive weed control for details on implementation).

Control is the suppression of a weed to the point that its economic (or harmful) impact is minimized. It is the practice most frequently conducted once a weed is established. Control methods do not prevent all plants in an area from reproducing. A reservoir of propagules usually remains so that control practices must be continued year after year.

Eradication is the elimination of all plants and plant parts of a weed species from an area. Eradication includes the destruction of seeds as well as any vegetative propagules such as rhizomes, creeping roots, and tubers. Elimination of a weed usually can be achieved only in the case of new, small-scale infestations. Once a large area becomes infested, almost no practical methods exist to eliminate long-lived buried seeds. Two exceptions are quackgrass and witchweed, which have been eliminated despite large but localized infestations. In the case of quackgrass, success can be attributed to short-lived seed. Witchweed is being eliminated gradually by the use of herbicides, fumigants, and ethylene gas (Sand and Manley 1990). When injected into weed-infested soil, ethylene stimulates the majority of seeds to germinate, and the seedlings can then be killed.

Although any or all of these goals may be attempted, emphasis usually is placed on control. Remember, however, that prevention and even eradication (e.g., treatment of spot infestations) can and should play important roles in the successful implementation of most weed control programs.

A Five-Step Procedure for Solving Individual Weed Problems

A well-planned, properly executed program consists of appropriate practices coordinated in a sequence from the initiation of the first crop management operation until the growing season is over (sometimes efforts even before or after the cropping season are justified). Such a program contains provisions for adjustments that might be needed to compensate for the impact of weather and management practices on crop and weed development. The impact of current practices on subsequent crops and weed development must be a component of the decision process.

We suggest a five-step procedure consisting of *(1) a diagnosis of the problem, (2) a survey of available control methods, (3) the selection of a program or plan of action, (4) the execution of the planned program, and (5) follow-up.* This approach may appear to be elementary, but we think it is vital to any successful weed management program. You should have these steps well in mind whenever you plan a program. Indeed, all of the material presented in this book either details information needed to achieve these individual steps or explains the rationale behind them.

The major components of each step are outlined below. Once you understand the importance of each component, using a checklist like the one on pages 51 and 52 will help you plan a weed control program for each situation.

I. PROBLEM DIAGNOSIS
   The first logical step in designing a program is to itemize and evaluate weed problems and other important factors that interact with weed control practices. Although the factors listed here do not exhaust the possibilities that should be considered during diagnosis, they do include many of the important ones, and they illustrate the scope of items you must consider when attacking weed problems.
   A. Weeds
      1. Weed Identification
         Species identification almost always is required. Because weeds differ in their life cycles, growth habits, reproductive methods, types of problems caused, and responses to individual control methods, different practices may have to be used for different species. Unless all problem species surviving the current cropping system are identified and characterized, results usually will be unacceptable. A list of weed identification guides is provided in Appendix 1.
      2. Weed Abundance and Distribution
         The relative abundance and economic importance of each species in the field should be determined and a priority

list of weeds to receive treatment should be established. Treatment plans should include not only those species causing the present economic damage but also those present in small numbers that may become economically important at a later time. The distribution of weeds over an area to be treated may influence the type and amount of weed control used. Weeds uniformly distributed over the entire area would justify broadcast herbicide applications either before or after weed emergence as contrasted to localized (spotty) infestations which might be most ecomonically controlled with localized (spot) treatments after weed emergence.
      3. New Weeds
         Serious weed populations invading a new area must be recognized and prevented from becoming established. A preventive program may provide large future dividends when a serious weed such as johnsongrass, Canada thistle, field bindweed, leafy spurge, purple loosestrife, bermuda-grass, or a herbicide-resistant population is kept from invading an uninfested area. The commonly held assumption that extremely serious weeds such as these can be readily controlled with modern weed control practices and need not be feared is a false one.
   B. Soil
      Understanding soils is most important when working with soil-applied herbicides. See Chapter 7 for a detailed discussion and examples of soil–herbicide interactions.
      1. Soil Conditions
         Soil conditions at the time of planting, tilling, or herbicide application are important. Excessive surface moisture, crop and weed residues, surface unevenness, and cloddiness can all impair the effectiveness of soil-applied herbicides. Different soil types within the same field can limit the choice of suitable herbicides; this limitation is particularly true in the case of strongly bound herbicides that have limited crop safety. Soil moisture can affect foliar-applied herbicide performance because plants under water stress tend to be more resistant.

CHECKLIST FOR A FIVE-STEP PROCEDURE FOR SOLVING INDIVIDUAL WEED PROBLEMS

PROBLEM DIAGNOSIS (Scouting and mapping)

Weed species

   Weeds present: annual weeds _____

                            perennial weeds _____

   Weed abundance: _____

Map of localized weed infestations

   Problem weeds along fencerows, roadsides, etc. _____

Soil factors

   Texture: Sand _____ Silt _____ Clay _____

   Organic matter: _____ pH _____ Fertility _____

   Crop residues present _____

   Soil conditions (e.g., different soil types in same field) _____

_____

Environmental considerations

   Erosion potential _____

   Nearby bodies of water receiving runoff from treated area _____

_____

   High value vegetation susceptible to normally used herbicides _____

_____

   Nearby residences, playgrounds, recreational areas _____

_____

Management practices

   Crop this year _____

   Possible crop next year _____

   Management sequences (tillage, mowing, irrigation, etc.) _____

_____

Past history

   Major weed problems encountered in previous seasons _____

   Herbicides applied in previous years and results _____

_____

Previous crops and tillage systems _____

SURVEY OF AVAILABLE CONTROL METHODS

Probable contribution of method to control of each weed species _____

_____

_____

Consistency when correctly executed _____

Adaptability to total cropping system _____

Flexibility of time for application (how long a period is required to get it done) _____

_____

PROGRAM SELECTION

Cost effectiveness _____

Available equipment _____

Available time _____

Managerial competence and experience to conduct program _____

Availability of custom applicators _____

Adaptability to total cropping system _____

Critical period for weed removal _____

Additional follow-up practices required for success _____

_____

Alternatives in case substandard results are obtained _____

_____

PROGRAM EXECUTION

Operations must be done at the right time.

Proper equipment must be used.

Equipment must be correctly maintained, adjusted, and operated (see Chapter 17 for
  checklist on herbicide sprayer calibration).

_____

APPROPRIATE FOLLOW-UP

Scouting for results _____

Mapping of problem areas _____

Remedial action as needed _____

Suggestions for future changes _____

2. Texture
   Soil texture (sand, silt, clay) determines such properties as tilth, water-holding capacity, rate of water *and* herbicide movement, cation exchange, and workability with various tillage equipment types and systems.
3. Colloidal Components
   Clay and organic matter are the colloidal fractions in which most of the surface reactions characteristic of soils take place. Soils high in organic matter bind herbicides so tightly that they may become ineffective. On the other hand, soils having both low organic matter and clay content may permit excessive herbicide leaching or lack sufficient adsorptive surfaces to prevent crop injury.
4. pH
   The pH of a soil determines the degree of ionization of some herbicides and thus their binding affinities, availability for weed control, and persistence.
C. Environmental Factors
   1. Soil Erosion Potential
      The potential of a soil to erode determines the amount of surface residue that should be left on the soil surface. This factor affects the types of acceptable tillage and other weed management practices and, in turn, the number and types of chemical weed control options.
   2. Nontarget Species
      The presence of herbicide-sensitive crops in an area may necessitate adjustments in the choice of herbicide and application techniques or timing in order to minimize the chances of injury. Drift during the application or movement of vaporized herbicides following application continues to be a major problem.
   3. Nontarget Sites
      The proximity of bodies of water, the position of water tables, or the proximity of housing or other urban developments may affect the choice of weed control techniques, particularly when herbicides are to be used. Restrictions regarding aerial and ground applications exist for some herbicides and other pesticides.

D. Crop Management Systems
   1. Management Practices
      The type and frequency of tillage, mowing, grazing, and crop rotation all determine the proper management and weed control practices. In many instances, crop vigor combined with the management practices in the system provide the *majority* of weed control obtained. This is true for turf, forage crops, rangeland, and pastures, all situations in which mismanagement often leads to increased weed problems. Economics and environmental concerns have resulted in management systems with fewer tillage and mowing operations. The loss of such operations greatly limits the number of weed control options available. Weed control that would have been obtained with those methods also is lost.
   2. Cropping Sequences
      The previous cropping sequence and desired future sequence must be taken into account, particularly when long-lived herbicides could be used in a program but a sensitive crop is involved. Cropping sequence also must be considered when a problem weed appears and effective control of it can be better obtained in some crops than in others.
E. Past Weed Control Programs and Results
   Past experience should serve as a basis for current choices. Practices that have worked consistently and well should be retained. Those that have given poor results should be carefully scrutinized and replaced if their ineffectiveness cannot be overcome.

II. SURVEY OF AVAILABLE CONTROL PRACTICES
   The evaluation of the currently available control practices is the second logical step in devising a program. You will need to draw on your own experiences, read pertinent educational and sales literature, and talk with people such as agricultural extension specialists and consultants. Evaluate all practices that will provide weed control, including not only individual herbicides but also the benefits that individual tillage operations, mowing operations, row spacings, planting dates, and other crop management practices can bring to the program. Evaluate each practice for the following information:

A. Effectiveness
Determining the effectiveness of each method to control *each* weed species present is essential.

B. Consistency
How consistent is the outcome? Can good results be expected each time the method is used or does it sometimes give incomplete results?

C. Fit within the Individual Program
Ascertaining the potential for integrating the practice into both crop management and weed development sequences is important.

D. Flexibility
How flexible is the practice with regard to timing? Methods that must be conducted within a short time (e.g., herbicide treatments or tillage operations restricted to a few days during the early stages of crop development) are not as dependable as those that can be used over long periods or that can be accomplished along with an essential crop management practice (e.g., treatments at planting time).

III. PROGRAM SELECTION
Select those practices that provide a weed control program that is effective, economical, and flexible. The goal is to provide a multicomponent program that will fit the specific weed situation rather than depend on a single practice that may or may not be effective. In designing an overall program, take the following aspects into consideration:

A. Economics
The cost of the individual components of the program must be weighed against probable economic returns. This includes not only the costs of the control methods but also the costs of scouting, mapping, and modeling.

B. Management System
The program must fit both the individual management system and the capabilities of the user.

1. Equipment
Is sufficient equipment available when it is needed to accomplish the various steps?

2. Custom Services
Can custom applicators be relied upon to conduct the essential parts of the program?

3. Time
Will sufficient time be available during critical peak work periods? This requirement is particularly important when a practice is effective only if conducted within a short interval.

4. Operational Capability
Does the managerial and worker staff have the capability and experience to carry out the steps of the program?

5. Crop and Management System
Do the steps fit into the cropping and management system?

C. Value of Early Season Programs
Remember that any practice that can be accomplished early in the season increases the chances for success and economic return. Maximum yields are obtained when weeds are removed during the first few weeks of crop development. Furthermore, early season programs provide more alternatives for controlling weeds when a method fails or when weeds escape normally successful primary control measures. A planned program should include alternative practices that can be used when early season operations fail or when they cannot be conducted because of poor weather conditions or other circumstances.

IV. PROGRAM EXECUTION
A program must be properly and efficiently conducted or the diagnosis, method evaluation, and program selection efforts will be wasted. Proper execution depends on the following:

A. Operations must be done at the right time, stage of weed development, and stage of crop development. They must be properly integrated into the sequence of management operations and done under suitable weather conditions.

B. Proper equipment must be employed. It must be correctly maintained, adjusted, and operated.

V. APPROPRIATE FOLLOW-UP
A. Follow-up means that once the program is under way, you should have the flexibility to do those extra evaluations and operations that will ensure program success. Monitoring of the results to observe initial weed control, later developing weeds, and crop vigor is the *minimum* effort that should be expended beyond the first planned control practice. Once the program is under way, fields should be inspected (scouted) pe-

riodically and remedial weed control methods should be applied when and where appropriate. Later treaments may provide limited yield benefit but can prevent weed buildup. Weeds escaping control measures should be indicated on maps of the field in question, since they are likely to increase as problems in subsequent years. Mapping may allow the sites where weeds are located to be correlated with remote sensing data.

B. Evaluation of results and documentation are needed for diagnosis and planning in subsequent seasons. Factors that should be recorded include problem spots in the field; scattered new weeds; weeds generally escaping current season control practices; soil and weather conditions before, during, and after use of a control method; size and stage of development of the crop and weed at time of treatment; and general crop vigor.

The critical steps outlined in this five-step approach ensure maximum weed control effectiveness and minimal crop damage. It is essential for the practitioner to realize that achieving an effective weed control program requires proper operation of equipment and practical crop management skills in addition to an understanding of weed control practices.

The Need for Realistic Expectations

Expectations concerning the degree of control likely to be achieved from a weed control program or practice sometimes are unrealistic. Although it may seem obvious that developing an effective weed control program depends on knowing the true potential of each practice, the tendency is for people to expect more from technology than it can deliver. A lack of understanding of weeds and their characteristics, a failure to recognize the sophistication of weed control technology and the complexity of interacting systems in modern crop production, the over-promotion of products and technologies, and the eternal belief that miracle cures abound all contribute to overly optimistic expectations and eventually to dissatisfaction with the results.

Inherent in the scientific approach is a sensible and realistic attitude about weeds and the degree of weed control that a program or a new or existing technology can provide. To achieve this attitude, keep the following generalizations in mind:

1. In most situations crop growth (to provide profitable yield, attractive turf, landscaped grounds, etc.) rather than weed control per se is the primary goal.
2. Once a weed has become established, it will be a problem for an extended time even when a good control program is in effect.
3. A single application of a weed control method (e.g., one cultivation, one hoeing, or one herbicide application in a season) will not control all weeds present.
4. A weed that is susceptible to a control practice at one stage in its life cycle is not necessarily susceptible at other stages.
5. A program consisting of several steps or methods rather than a single treatment usually is the proper approach to solving weed control problems.
6. Appropriate weed control practices are limited by and must be matched to the crop management system.
7. The elimination of a given practice such as tillage, mowing, or crop competition from a management system will remove an increment of weed control that must be replaced with alternate control methods.
8. Some weeds will adapt to *any* management system.
9. We are not on the verge of some new and wonderful technology that will negate the need for older technology such as sanitation, herbicides, tillage, and hand removal.
10. Rarely is a practice as good as its proponents claim nor as poor as its detractors insist.

WEED CONTROL METHODS

The categories of weed control that will be discussed in the remainder of this chapter are as follows: preventive, mechanical and physical, cultural (crop management), biological, and chemical. In actual practice, most of the emphasis and effort spent on controlling weeds is on mechanical and chemical methods. Crop shading is an important but frequently unrecognized component of weed control and preventive methods have been deemphasized in recent decades. On lands where weeds have been controlled successfully, however, implementing preventive methods is as essential a component of the crop management program as are the other types of technology.

PREVENTIVE WEED CONTROL

Preventive weed control is aimed at preventing weed spread by seed or vegetative propagules. Other terms that encompass the concept of prevention and that are frequently used with insect and disease control are *sanitation* and *containment*. Unfortunately, the overall concept of prevention has been greatly overshadowed in recent decades by the development of modern herbicides and mechanical weeding equipment. The availability of these tools has led to the mistaken impression that effective control always can be obtained *after* weeds have invaded a new area.

Another possible reason for the deemphasis on prevention has been the promotion of the pest management concept of an economic threshold. This concept holds that a control practice should not be initiated until the point at which damage to the crop equals the cost of control. The concept was developed for highly mobile and somewhat unpredictable insect populations that can fluctuate greatly during the season or between years. Theoretically and in practice, relatively high populations of insects can be present without causing sufficient damage to justify treatment. The insects are only treated when the populations cause enough damage to exceed the economic threshold.

This concept does not necessarily work well for weeds that spread slowly and persist indefinitely once established. If a newly introduced weed requires control methods other than those currently being used in a crop management system, or if it has the potential of causing more damage than those weeds already present, the economic threshold is reached *when the first plant goes to seed or produces vegetative propagules*. Although the damage caused by a single plant may not have much of an impact on crop yields, the potential for loss from the progeny of that plant plus the cost of control in future years make it imperative to try to eliminate the plant before it reproduces, *not* after its population has reached the level of a traditional economic threshold. Prevention is so important that it should be considered an integral component of the overall program for weed control in any crop management situation.

As discussed in Chapter l, true weeds are ideally adapted for spread. Various seed and fruit coat modifications such as barbs, hooks, wings, or corky tissues allow for the dissemination of seeds by animals, wind, and water. Most reproductive structures are resistant to desiccation, which means they can be uprooted from the soil and remain viable when carried to other sites. The major contributor to weed spread over long distances is human activity, primarily through the movement of machinery and contaminated crop seed. Efforts to reduce weed spread have led to legislation in the form of quarantines, seed laws, and weed laws.

Quarantines

Quarantines are used to isolate and prevent the spread of noxious weeds. The best known example of a weed quarantine is that established for witchweed (*Striga asiatica*), a parasitic plant that infests grass crops such as corn and sorghum (Figure 4.1). This annual plant was introduced into North and South Carolina in the l950s. The potential for damage from witchweed to U.S. agriculture was considered so great that in l957 it was the first weed to be placed under federal and state quarantines to prevent its spread (Figure 4.2). These quarantines regulate interstate and intrastate movements of soil, plants, plant parts, produce, mulching materials, farm product containers, and farm and construction equipment from witchweed-infested areas. The regulations are strictly enforced and apply to anything able to move seed. In addition, a control and eradication program has been under way since the l950s. This program, along with the quarantine, has resulted in the reduction of the area of infestation over the years, from a high of 430,000 acres to about 60,000 acres in 1995. (For access to current distribution information on species listed as Federal Noxious Weeds, contact the National Agricultural Pest Information System [NAPIS] database on the World Wide Web at http://ceris.purdue.edu/napis/.)

The Federal Noxious Weed Act (signed into law in l975 and funded for the first time in fiscal year l979) now provides for more adequate jurisdiction to prevent new weeds from becoming established in the United States. It was enacted to prevent the entry of weeds of foreign origin by providing for inspection at ports of entry; establish and enforce appropriate quarantines; and control or eradicate weeds that are new or

Figure 4.1 Witchweed-infested corn in North Carolina (R. Eplee, Whiteville Plant Methods Center, Whiteville, North Carolina).

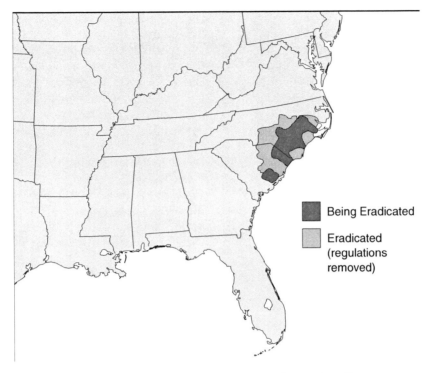

Figure 4.2 Areas of North and South Carolina under quarantine for witchweed as of August 5, 1996 (Retrieved from National Agricultural Pest Information System at http://ceris.purdue.edu/napis/.)

not widely distributed and that can cause damage to agriculture, livestock, fish and wildlife, and public health. Until this law was passed, inspectors at ports of entry had the authority to turn back only imports with insects, nematodes, or plant diseases. In fact, the quarantine for witchweed in the 1950s was set up because the plant was considered a plant disease (parasite) rather than a weed problem.

Seed Laws

Seed laws are enacted to ensure the purity of crop seeds used for planting and to prevent the spread of weed seeds. All states have seed laws designed to regulate the quality of agricultural seeds sold within their borders. These laws generally conform with the Federal Seed Act, although they vary from state to state as far as weed species are concerned. The Indiana Seed Law, for example, designates two types of noxious weed seeds: prohibited noxious and restricted noxious. In other states, these two categories often are called primary noxious and secondary noxious weed species. In Indiana, it is unlawful for crop seed to be sold with any prohibited noxious weed seeds, with more than 0.25% of restricted noxious weed seeds, and with 2.5% of all other weed seeds.

Seed laws are rigidly enforced. The inclusion of creeping perennial species such as johnsongrass and Canada thistle on the prohibited noxious lists of many states has been effective in hindering the spread of these weeds through crop seed.

Weed Laws

Weed laws are enacted to compel local officials and property owners to carry out weed sanitation measures on lands under their jurisdiction. Such lands include farms, roadsides, railroad tracks, noncrop areas, utility rights-of-way, and ditch banks. The enabling legislation for a weed law rests with the state legislature. In many states, the weed law is similar to the model uniform state weed law that was drafted in 1963 by the U.S. Department of Agriculture, state departments of agriculture, and the Weed Science Society of America. The model weed law permits the establishment of weed control districts at the local level under the direction of a superintendent responsible for supervising control and eradication programs. The superintendent often is the chief executive officer of a county, city, or township district (for example, a county commissioner or township trustee). Funding for the program usually comes from local taxes levied for this purpose.

Weed laws can be effective only if adequate funding is available, if the superintendent is both knowledgeable about weed control methods and persistent enough to see that the program is carried out thoroughly and effectively, and if a concerned public is willing to cooperate with authorities. Input concerning the weed species to be controlled should be generated from the local level as well as from higher levels. Unfortunately, state legislatures rather than local governments frequently designate the weeds to be controlled. In addition, funding is poor, and not enough pressure is exerted on local officials to enforce most weed laws.

Although some states impose fines on landowners who allow certain weeds (such as johnsongrass) to set seed, weed laws have not stopped the spread of noxious weeds. Their major positive effect has been to slow their spread by increasing public awareness of the noxiousness of certain problem weeds.

Preventing Weed Spread

Aside from working for legislation and abiding by regulations, anyone interested in controlling weeds should implement the following precautions:

1. Do *not* let plants reproduce. Plants going to seed or producing vegetative reproductive structures are the causes of continued weed buildup and spread.
2. Do *not* bury seed in uninfested areas. Once the seeds are in the soil they are protected and preserved for years or even decades.
3. Do *not* spread reproductive structures. Weeds normally do not travel long distances without human aid.

4. Do *not* plant weed-infested crop seed and transplants.
5. Do *not* bring weed-infested soil, bedding, hay, straw, or manure to clean areas.
6. Do *not* feed weed screenings to livestock unless they have been cooked.
7. *Do* clean harvesting, mowing, tillage, and earth moving equipment before moving to uninfested areas.
8. *Do* kill new plants prior to seed set (hand hoe if necessary).
9. *Do* kill weeds growing in such places as ditch banks, roadsides, fencerows, uncropped areas, and rights-of-way. They are easier to fight there than in a cropping situation (Figure 4.3).
10. *Do* spot-treat small infestations or isolated individual plants with herbicides. Remember, one plant may produce thousands of seeds and eventually thousands of new plants.

MECHANICAL AND PHYSICAL WEED CONTROL

Mechanical methods range in complexity from hand hoeing to tillage operations with multi-component machines such as cultivators and bed conditioners. The most commonly used mechanical methods are hand hoeing and pulling, tillage, and mowing. Manipulation of the environment by physical methods includes mulching, flooding, and fire and heat.

Hand Hoeing and Pulling

These methods are the earliest and most primitive types of weed control. They are still the ma-

Figure 4.3 Left: Keeping roadsides mowed is an important method of preventing weed spread into adjacent fields. Right: Unmowed roadside.

jor methods of control in underdeveloped countries. Of the 350 million farmers in the world, approximately 250 million still rely on hoes and wooden plows to weed and cultivate their crops (Hill 1982).

Because weeds are anchored in one place, they are more adapted to control by hand than are mobile pests such as insects or disease organisms. Hand hoeing, however, is more time consuming and expensive than other methods of weed control when conducted on large areas with heavy infestations of weeds. It is more practical for small areas or for high value crops. It is extremely valuable on scattered weeds that have appeared for the first time in an area and that are particularly noxious. The use of herbicides and close cultivation with tillage equipment reduces the amount of hoeing needed.

When herbicides are not or cannot be used, hand removal, along with tillage, is a major alternative for the control of weeds.

Tillage

A major benefit of tillage to modern cropping systems is weed control. Biennials and simple perennial weeds disappear from fields tilled annually because the underground roots that enable these plants to overwinter are killed before they can produce seed. Only annuals and creeping perennials are able to reestablish under an annual tillage regime. Annuals are successful because they can complete their life cycle and produce seed in the period between annual primary tillage operations. Creeping perennials not only can produce seed in the same time period but also can generate new plants from vegetative structures (including those that have been cut up by tillage).

The types of equipment and sequences used in tillage operations are described in detail in Chapter 18.

How Tillage Kills Weeds

Tillage destroys weeds by breaking them apart, cutting or tearing them loose from the soil, causing them to dry out (desiccate), smothering tender tissues by cutting off light and air, exhausting stored food reserves, or depleting reserves of seeds and vegetative propagules from the soil.

Cutting, breaking, or tearing the weed apart or loose from the soil frequently results in desiccation and death, particularly if the weed is a seedling, an annual, a biennial, or a simple perennial. Tops of creeping perennials also are

killed. Best results are obtained when the soil surface is dry so the disrupted weed is subjected to drying conditions. The soil beneath the surface also should be dry enough so that tillage equipment can be used without causing detrimental effects to the soil. When soils are sufficiently dry, substantial kill of vegetative reproductive structures of creeping perennial weeds also can be obtained. Viable weed seeds near the soil surface are not affected adversely by tillage if they have not started to germinate.

When soils are too wet or when rainfall occurs shortly after the tillage operation, weeds are more likely to survive. Rhizomes, stolons, and creeping roots from creeping perennials almost certainly will be transplanted. Crowns of simple perennials will continue to grow, as will annuals with intact root systems or annuals capable of rooting from the stem nodes. Working soils when they are wet not only results in poor weed control but also causes clods, heavy crusts, and compaction zones that interfere with normal crop emergence and growth.

Covering weeds with soil can smother tender meristems and other tissues, reduce light reception, and inhibit gas exchange to the rest of the plant. Thus, covering plant shoots with soil during tillage can kill growing points and effectively shut off photosynthesis. Small weeds emerging in crop rows can be partially controlled during selective cultivation by carefully covering them with soil under the taller crops. Coverage can be accomplished by regulating the speed of travel of the tillage implement and by adjusting cultivator tools and shields.

Tillage can help control perennial weeds. Generally speaking, tillage is more detrimental to creeping perennials when the shoots have been growing above the soil surface for several days than when they first emerge. The first developing shoot growth depletes food reserves from underground structures until leaf surfaces are adequate to export sugars. If tillage is performed often enough, carbohydrate replenishment and the formation of new underground structures can be prevented and the plant eventually will starve to death. Tillage also breaks up the vegetative reproductive structures and removes the apical dominance effect of terminal buds. Axillary (nonterminal) buds thus are able to sprout and add to the depletion of the food reserves within the vegetative structure. Obviously, repeated tillage operations or other additional subsequent control methods are required to eliminate the individual surviving plants that inevitably appear and reestablish themselves.

One or more growing seasons of intensive effort using clean tillage are required to deplete the vegetative reproductive structures of most creeping perennial weeds. When clean tillage (fallow) alone is used, the operations should be done at 2- or 3-wk intervals starting before the weeds reach the late bud stage and continuing through the end of the growing season. Obviously, such intensive clean tillage eliminates the possibility of growing a crop during this period and is appropriate only when soil erosion can be avoided.

The sequence of events that will lead to the most successful weed control using tillage is as follows: a period of warmth and moisture to encourage weed emergence, weed emergence, tillage, a dry period, more warmth and moisture, additional weed emergence, tillage, a dry period, etc. Total dependence on tillage for weed control reduces the chances for early planting.

If properly done, repeated tillage also reduces populations of weed seeds in the soil. Most soils contain a reservoir of seeds to the depth of plowing; however, only those close to the surface are able to germinate. When weed seeds are brought nearer to the soil surface by tillage, they are exposed to light, oxygen, and other factors needed for germination. They also may be exposed to insects, microbes, rodents, and alternate wetting and drying, any of which could result in their destruction.

Effect of Crop Management and Tillage Systems on Weed Control

The crop management system and the tillage equipment used for it dictate the weed control techniques available for that system. The selection or deletion of primary and secondary tillage operations determines the implements that are suitable for subsequent operations. For example, in reduced tillage systems in which the soil is not loosened or in which heavy crop residues are left on the surface, many implements normally used for secondary tillage, such as light harrows, do not work well because they get clogged with trash or cannot penetrate the soil. Implements for selective weeding, such as the rotary hoe and many cultivators, also do not work well under such conditions. Soil manipulation and weed control benefits from these implements are lost in such systems, so they must be replaced with effective alternative methods. This has been accomplished with herbicides and, in some instances, with specially designed tillage implements for secondary tillage, such as the rod weeder or chisels and cultivators with sweeps (see Chapter 18).

Disadvantages of Tillage

The conventional tillage system described in Chapter 18 exposes soil to the elements so that coarse-textured (sandy) and peat and muck soils are subject to wind erosion and sloping soils to water erosion. Recent emphasis on the reduction of nonpoint sources of stream pollution, such as eroding soil, is having an impact on the selection of tillage practices. As mentioned in the previous section, the conversion from conventional to minimum, conservation, or no-till systems usually requires the adoption of alternative weed control measures such as herbicides or specialized tillage equipment.

The stirring of seeds from the surface and back again may or may not be helpful, depending on the situation. Burial of weed seeds often preserves them. With new infestations, every effort should be made to prevent seed burial. This need becomes less important once a weed is well established, although continual burial of seeds does increase the potential population pressure from the seedbank. Frequently, a problem develops when initial control methods have killed most or all of the weeds that were close enough to the surface to germinate, and then a deep-running tillage tool brings weed seeds to the surface to reinfest the clean soil. Shallow tillage minimizes this problem. Tillage equipment also can spread both weed seed and perennial vegetative reproductive structures to uninfested areas.

Untimely rainfall interferes with tillage operations (particularly with selective weeding techniques such as rotary hoeing and row cultivation). When wet soils prevent the use of these techniques, alternate methods of control are required. Selective herbicide applications are most often used to accomplish this.

Tillage can damage crop roots, but such damage can be reduced if deep or late-season cultivations are avoided. Although it is common to associate root injury with the mechanical cultivation of herbaceous crops grown in rows, this type of injury also occurs from tillage between rows of woody plants such as orchard plantings, vineyards, ornamental tree or shrub nurseries, or forest plantations.

Much has been said about the energy requirements needed for conventional tillage operations, but to date the cost of fuels has been relatively inexpensive so the weed control benefits derived from tillage still are economical. When used judiciously, current conventional tillage operations are efficient, cost-effective, and dependable. At a l997 price of $1.25/gal for diesel fuel (the same price it was in 1982 when the first edition of this book was written), the cost of fuel for a relatively complete tillage program consisting of one plowing, two secondary tillage operations, and two cultivations is $6.25/acre (Table 4.1).

The deleterious effect of tillage on soil is receiving extensive emphasis at the present time. The breakdown of soil aggregates and the formation of compacted soil layers are two of the problems frequently attributed to tillage. We think many of these problems stem from the use of tillage equipment on wet soils. Even if this is only partially true, the valuable contributions of tillage to weed control must be carefully considered before an essential tillage operation is dropped from a cropping system. The working of soils at proper moisture content not only enhances the chances of successful weed control but also minimizes most of the adverse effects attributed to tillage.

Mowing

Mowing can be an effective weed control practice. For the most part, it is done as part of an overall crop management program.

Mowing controls weeds by removing tops prior to seed formation, by depleting underground food reserves, and by favoring the growth habits of competitive crops. Unfortunately, mowing is relatively ineffective on species that can grow and produce seed below the normal height of cutting.

The prevention of seed formation generally is considered the minimum effective effort that can be employed to stop the spread of weeds. Mowing often enough to prevent seed formation is the minimum requirement established by most weed laws. For mowing to prevent seed formation, the plant must have a relatively tall growth habit (most grasses do not), and the mowing must be done prior to pollination and fertilization. Since mowed plants generally survive and regenerate new tops, repeated mowing is required to prevent seed formation. For most species, three to six mowings per growing season are needed.

Mowing after flowers appear is more cosmetic than useful because many weed species produce viable seeds within a week of fertilization. For example, Canada thistle seed is viable within 8 days after pollination. Shattercane (Burnside 1973) and leafy spurge (Wicks and Derscheid 1964) produce viable seed 10 to 15 days after pollination. These intervals are so short that many of the plants in a population may be flowering while others are maturing their seed. At least one study has shown that seeds of plants such as tansy ragwort, annual sowthistle, and common groundsel are viable when the plants are cut at flowering (Gill 1938).

Mowing sometimes will kill tall annual weeds with one or two passes, provided the buds on the stem are above the cutting height. A frequent response to mowing, however, is for a plant that has had a single central shoot cut off to send up several new shoots from buds below the cut. Repeated mowing thus can change a normally upright, single-stemmed plant to a many-branched, prostrate one. When this happens, seeds still can be produced.

When repeated frequently, mowing can result in the depletion of food reserves and the gradual elimination of tall perennial plants (Figure 4.4). As with tillage, plant shoots should be allowed to grow and draw on the underground food reserves prior to mowing. In most cases, the first mowing should be done before mid- to late bud stage. Additional mowings probably will be required at 2- to 3-wk intervals. The effect of frequent mowings on rhizome production of johnsongrass, a tall growing grass, is illustrated in Table 4.2.

Another advantage of mowing is that it can stimulate the germination of underground buds

Table 4.1 Diesel Fuel Requirements for Field Operations in Conventional and Chisel Systems for Corn on Moderate Soils[a]

Tillage System and Field Operations	Fuel Required (gal/acre)
Conventional system	
Disk stalks	0.45
Moldboard plow	1.85
Disk	0.55
Field cultivate	0.60
Apply anhydrous	0.70
Plant	0.50
Cultivate	0.35
	5.00
Chisel system	
Chisel plow (coulter-chisel)	1.25
Disk	0.55
Field cultivate	0.60
Apply anhydrous	0.70
Plant	0.50
Cultivate	0.35
	3.95

[a]Data from Griffith and Parsons (1980).

Table 4.2 Effect on Rhizome Production of Repeated Mowings of Established Johnsongrass at Various Stages of Growth[a]

Stage of Growth at First Cutting	Number of Cuttings	Rhizome Production at End of Season (lb/A Dry Wt)
(End of previous season)	—	5616
1 ft high	9	214
2 ft high	6	523
Boot	4	1823
Bloom	3	2280
Late milk	3	3803
Mature seed	2	3818
End of growing season	1	6098

[a]Data from Sturkie, 1930.

on creeping perennials, thus depleting their numbers. Close mowing also can cause the underground system to develop closer to the soil surface than it normally would. This may result in better control when tillage is subsequently used or if the underground systems are sensitive to kill by freezing.

Mowing frequently is used in conjunction with the culture of hay, pasture, turf, or cover crops. Generally the crop itself is managed by mowing so that mowing coupled with other proper management practices such as grazing and good fertilization results in a competitive advantage for the crop and aids in controlling weeds. The combination of mowing and a competitive crop is more effective than mowing alone. Mowing not only weakens the weed but causes some competitive crops, such as turf, to thicken, thereby providing more shade and competition.

Roadsides, wastelands, ditch banks, and levees are areas where weeds can produce abundant seed to serve as new sources of weeds in adjacent production fields. If such wayside areas are covered with a frequently mowed turf, the production of noxious weed seeds can be minimized or even eliminated.

Mulches

Mulches control weeds by excluding light. They are used on small acreage sites such as home gardens and on relatively high value crops and are limited mostly to emerged, established, or transplanted crops.

Mulching material can be any substance that is opaque or thick enough to exclude light, is relatively cheap, and is reasonably easy to work with. Plant materials used for mulches include straw, sawdust, wood chips, bark chips, and grass clippings (Figure 4.5). Continuous strips or sheets of black plastic or paper also are used. Paper mulches have been used extensively in Hawaii for weed control in pineapple. Materials that can be applied in solid strips or sheets are more effective than particulate materials that are porous and may allow some weeds to emerge.

Figure 4.4 Repeated mowings can result in a shift from tall herbaceous perennial broadleaves such as goldenrod and tall ironweed (background) to lower-growing perennial grasses (foreground).

Figure 4.5 Straw mulch used for weed control in a cucurbit crop.

Perennial weeds with vigorous underground reproductive systems usually are not controlled well with mulches. The food reserves in vegetative reproductive structures permit the growth of vigorous plants that can push through particulate mulches and occasionally through black plastic mulches. We have observed bigroot morningglory shoots emerging through 6 in. of rolled asphalt.

Black or other opaque solid plastic sheets or plastic sheets with fine perforations are commonly used for continuous strip mulching. A combination of mulching materials can be used in ornamental plantings. Sheets of plastic are put down first and then covered and protected by a more decorative mulch such as rock, gravel, wood chips, or peat moss (Figure 4.6). The penetration of air, water, and nutrients to the growing plants are claimed to be benefits of using perforated sheets of plastic rather than solid sheets, although the benefits may not offset the extra cost.

Some problems inherent in using mulches include the following:

- The mulching material itself may be fairly costly or it may be costly to haul and apply.
- Not all weeds in an area are likely to be controlled.
- Thin spots in particulate mulches or tears in strip mulches will allow weeds to penetrate and become established.
- Organically derived particulate mulches such as bark, sawdust, wood chips, and sphagnum may decay in place and serve as suitable substrates for weed seed germination and growth.
- The presence of mulch limits the number of usable alternative weeding techniques and sometimes interferes with the operation of equipment used for other management practices such as mowing on adjacent areas.
- Mulches may alter temperature and moisture relationships.
- Uncovered black plastic tends to increase soil temperatures; many of the other materials tend to keep the soil cooler.
- Once the mulch has served its purpose, it must be disposed of.

Residues from previous crops may serve as a mulch and in some cases also confer an allelopathic effect on weed growth (see the section on allelopathy, page 67).

Flooding

Flooding is used for controlling weeds in rice and sometimes for controlling creeping perennial weeds. In the case of rice, the crop is able to grow under flooded conditions while many common weeds cannot. Some weeds such as barnyardgrass and arrowhead do survive and create problems.

Flooding can kill the vegetative reproductive structures of some creeping perennial weed species. To be successful, the soil or subsoil must be sufficiently impermeable to keep water from leaching. Six to 10 in. of standing water must be maintained for 3 to 8 wk during the summer so the weeds cannot get their tops above water. An adequate supply of water is thus a necessity.

Not all weeds with vegetative reproductive systems can be controlled with flooding because many vegetative buds can go dormant and survive. A short duration flooding of even 3 to 8 wk cannot be expected to have an impact on ungerminated weed seeds.

Fire and Heat

Flaming devices such as flame throwers are used to remove weeds from ditch banks, roadsides, and noncrop areas. Fire also is frequently

Figure 4.6 Rock mulch over plastic in an ornamental landscape.

used to remove undesirable brush and hardwood species from conifer forests. The major use of fire in row crops was in cotton in the 1950s and 1960s. Small weeds were killed with flamers directed beneath the tough-stemmed cotton plants. A first pass was usually made to sear and dry the plants. This was followed several days later by a second pass to burn off the dead plants. The cotton plants could further be protected by the use of a water shield sprayed above the burner to prevent excessive temperatures in the cotton canopy. The use of flaming eventually declined because of the rising costs of liquid propane gas; however, recent pressure to decrease the amount of pesticides used in cotton has led to a renewed interest in flaming, both for weed and insect control (Seifert and Snipes 1996).

Flaming generally is nonselective, although seedling grasses, which have growing points close to the ground, may survive. By removing shoot material, flaming provides basically the same level of weed control as a cultivation or an application of a postemergence contact herbicide. Therefore, perennial weeds are only temporarily set back, and seeds protected by soil are not affected at all. Flaming frequently is used for a cosmetic effect; that is, to eliminate large unsightly weeds. As with any method that provides shoot kill only, weed control benefits are nil if the weeds are burned after they have matured and set seed. As indicated with cotton, some insect and disease control can be achieved with burning.

The potential fire hazard of flaming must be taken into account. Muck soils (Figure 4.7), dry grasslands, wooden fence posts, and telephone poles often fall victim to burning.

Heat is used effectively to steam sterilize greenhouse and potting soils. In this case, weed seeds usually are eliminated along with other plant parts, insects, and disease organisms. Steam and hot water also have been investigated for killing weed shoots on railroad rights-of-way.

Research on devices that produce concentrated forms of energy to kill weeds includes that on solar-radiation collectors (Johnson et al. 1990) as the source of heat and machines equipped with high voltage electrodes that pass an electrical current through weeds they come into contact with (Vigneault et al. 1990). The electrical current is converted to heat, which kills the weed.

CULTURAL WEED CONTROL (CROP COMPETITION AND MANAGEMENT)

A vigorously growing crop often is the most effective and economical weed control practice available. Unfortunately, the contribution that crop competition makes to weed control often is overlooked.

Smother Crops

When maintained in dense stands, some crops are vigorous enough by themselves to keep weeds in check. Such crops are called smother crops. They generally are solid seeded or planted in closely spaced rows. Examples of traditional smother crops are alfalfa (Figure 4.8), foxtail millet, buckwheat, rye, sorghum, Sudan grass, sweet clover, sunflower, barley, soybeans for hay, cowpeas, clovers, and silage corn. Alfalfa grown as a smother crop in South Dakota reduced field bindweed stands by 31, 74, 91, and 95% when grown for 1, 2, 3, and 4 yr, respectively (Derscheid 1978).

Competitive Crops

Smother crops are not the only ones with sufficient competitive ability to yield substantial weed control. All the other forage and small-grain crops, perennial grass sods (both unimproved and turfgrass), most agronomic row crops, and some horticultural crops can provide heavy competition if managed correctly.

Shading is a major means by which crop plants suppress weeds. The crop must develop rapidly enough to get ahead and stay ahead of the weed. Any plant-related factor or management practice that hastens canopy closure by the

Figure 4.7 Using fire as a weed control agent on peat or muck soils is not advisable.

Figure 4.8 Well-maintained competitive crops such as alfalfa (left) and wheat (right) prevent the establishment of weeds. In the right-hand photo, a poorly managed stand of wheat (background) permits weeds such as common lambsquarter to take over.

crop favors the crop and enhances its competitive ability. Emerging from the soil first, having a height advantage, or growing at a faster rate than the weed are three factors that provide a competitive advantage. Conversely, anything that favors the weed and puts the crop at a disadvantage hinders weed control.

There are numerous examples in which lack of crop development has led to increased weed growth. Poor crop stands can be caused by an improper choice of crop variety, bad weather, low soil fertility, and herbicide injury. A poor stand of corn with skips in planting or flooded end rows is a prime candidate for an increased growth of yellow nutsedge, crabgrass, fall panicum, and numerous other weeds. These same weeds are problems in fields where corn hybrids are produced. In this case, a combination of slow growth rates, short inbreds, and wide row spacing to allow movement of hand laborers results in sunlight penetrating the crop canopy for most or all of the growing season (Figure 4.9).

In many annual crops, weeds are held in check by shading until the crop matures and the leaves dry or drop off. Light can then penetrate so that weeds develop. Even though these weeds may not affect yields directly, they can interfere with harvesting operations, and they serve to maintain or increase the weed population.

Weed problems frequently develop in perennial sod crops such as turf, pastures, and hay fields because of poor stand establishment or because stands have been thinned or spots killed by disease, drought, winterkill, insect damage, or improper management of grazing

Figure 4.9 Left: Poor crop establishment in end rows and wide (38-in.) spacing of corn inbreds result in late-season weed growth. Right: Taller hybrid corn plants grown in 30-in. rows and with fewer tillage and herbicide treatments are almost weed free.

animals. In these cases, weeds move readily into the newly opened and disturbed areas (Figure 4.10).

Management Considerations

Crops can be manipulated and managed to optimize shading ability. Plant population densities and row spacings that maximize shading by hastening crop canopy closure should be considered when planning a crop management program (Figure 4.11). Such considerations are especially critical when growing seasons are shortened (e.g., in double cropping) or with small crop plants. Unfortunately, narrow row spacing not only improves shading ability but also limits the ease with which a row cultivator can be used. On the other hand, if the canopy closes rapidly, fewer cultivations should be needed. Row spacing and canopy closure time should be considered not only for agronomic crops but for any applicable cropping situation.

Selecting well-adapted, fast-growing varieties favors the crop. In contrast, choosing slow-developing, short, small, or upright-leaved plants that allow deep penetration of sunlight into the crop canopy can be expected to favor weed development unless the spacing between plants and rows is reduced to speed canopy closure. Other conditions that favor the crop are proper planting, fertilization, mowing, irrigation, adequate drainage, and control of other plant pests such as diseases, insects, nematodes, and rodents.

In any cropping system, the crop canopy is open and susceptible to weed invasion at some point. Supplemental weed control practices such as tillage and herbicides should be aimed specifically at these particular stages or periods in crop development.

Crop Rotation

Another aspect of crop management is crop rotation. This option obviously is not available on established perennial crops such as permanent pastures, rangelands, orchards, vineyards, lawns, forests, and woody ornamental plantings. Rotation possibilities also are limited for crops such as forages, perennial pastures, asparagus, strawberries, cane fruits, and Christmas

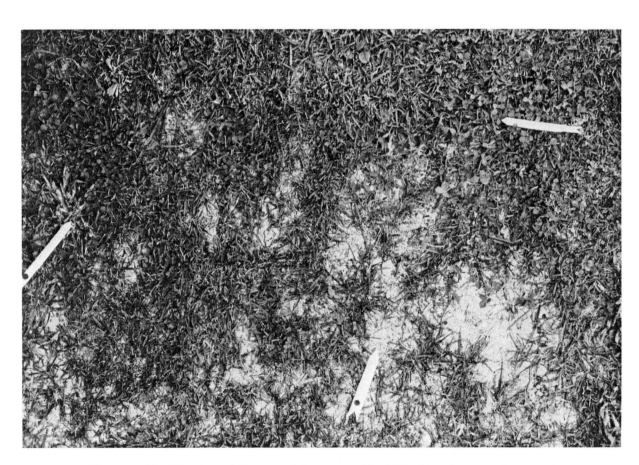

Figure 4.10 Weeds rapidly move into areas of turf killed by disease. Arrows at right and left indicate weeds. (W. H. Daniel, Purdue University)

Figure 4.11 Soybeans planted in narrow, 10-in. rows (left) develop a shading canopy much more rapidly than soybeans planted in 30-in. rows (right).

tree plantations. Rotation is most adapted for annual, biennial, or short-term perennial crops.

Rotation is useful because continuous and uniform management practices select for those weeds that are best adapted to the particular management system. Each crop and crop management system tends to develop its own characteristic weeds. In some instances, management practices that rely on the continuous use of certain herbicides year after year on the same crop have led to the appearance of herbicide-resistant weeds. Crop rotation, on the other hand, leads to the diversification of individual cropping practices, thereby causing changes in weed populations and species composition. The more dissimilar the crops and their management practices are in a rotation system, the less opportunity an individual weed has to become dominant or develop herbicide resistance. This principle also applies to insects, diseases, and other crop pests.

Crops are rotated much less frequently today than they were before World War II. Present-day economics favor the repeated use of one or two well-adapted, highly profitable crops tended with less equipment. Diversified cropping systems require large capital investments in a variety of equipment, a diverse technological capability, and in some instances, the inclusion of livestock, which requires further expense as well as a knowledge of husbandry. Crop rotation may prove more practical in the future should the buildup of difficult-to-control weeds, herbicide-resistant weeds, and other pests alter the economic picture.

To date, pest problems have been adequately met through the development of varietal resistance, modern pesticides, and other management practices. Rotation of weed control techniques (e.g., rotation of herbicides) appears to be a more practical approach to preventing the rapid buildup of weeds in modern circumstances than does rotating crops.

Allelopathy

Plants sometimes produce chemical substances that inhibit (or in some cases stimulate) the growth of nearby competing plants (allelopathy). Allelopathic properties have been reported for both weeds and crops. Allelopathic effects probably account for at least some of the interference that weeds exert on crops (see Chapter 1). The concept that crops could be allelopathic to weeds was demonstrated by Putnam and Duke (1974), who screened 526 accessions of cucumber for allelopathic activity against a mustard (Figure 4.12) and a grass weed. One accession inhibited weed growth by 87% and 25 inhibited growth by 50% or more. Other crop plants that have been shown to exhibit allelopathic properties to weeds include sorghum, domesticated sunflowers, celery, and sweet potato. Although readily demonstrated under greenhouse conditions, the actual impact that allelopathic crops have on weeds is difficult to assess in the field situation.

The most effective way in which allelopathy has been incorporated into crop management systems is by using residues (decomposing plant material) from allelopathic plants as mulches prior to crop planting. For example, residues of wheat and rye have successfully suppressed weed germination in field situations. The wheat residue can be from the previous year's crop that was left in the field to break down over the winter. Rye, and also wheat, can be grown as cover crops over the fall and winter and are killed

Figure 4.12 Left: Cucumber cultivar 169391 and mustard planted at the same time. The cucumber has inhibited germination and growth of the mustard. Right: Control in which the mustard was planted without the cucumber. (A. R. Putnam, Michigan State University).

before or at crop planting in the spring; the allelopathic residues result from the breakdown of herbicide-killed plant material. In order for allelopathic mulches to work, a management system that leaves the residues on or very near the soil surface, such as no-till or stale seedbed, must be employed. Sorghums and mustard crops also have potential as allelopathic mulches.

One constraint to the use of allelopathic mulches is that the crop must be planted in such a way that it is not adversely affected by the mulch. This can be accomplished by planting large-seeded crops with a no-till planter that cuts through the mulch and allows the seed to be placed below the mulch layer. Allelopathic mulches also are less likely to have an adverse impact on larger, transplanted crops such as tomatoes. Placing seeds deeper or transplanting will not ensure adequate crop establishment in every case, and conditions for planting must be worked out for each situation and crop.

The failure to consider and utilize the competitive abilities of crop plants results in an incomplete weed control program. Every effort should be made to utilize the competitive contribution from the crop in planning and executing a weed control program. Such an effort will enhance and complement other control practices and will result in the most economical and effective management program.

BIOLOGICAL WEED CONTROL

Biological weed control is the use of natural enemies to reduce weed populations to economically acceptable levels. The concept of biological control has been so widely publicized in recent years that the general public views it as a viable and readily available alternative to the use of pesticides. Although there have been some outstanding successes, particularly in noncropland situations such as rangelands and aquatic sites, it will be some time before biological controls are used for weed management with the frequency of either tillage equipment or herbicides. By their very nature, biological control organisms are selective in their food preferences and cannot provide the broad spectrum control that is achievable with herbicides and tillage.

Biological control is not new. Records dating from the 1860s show that the cochineal insect *Dactylopius ceylonicus* (Greene) was introduced into parts of India and Ceylon for the control of prickly pear infestations.

The effectiveness of biological controls has been limited for the most part to perennial weeds on low value land (rangelands, pastures, forests, roadsides, and recreation lands) and aquatic sites. Perennial weeds growing in these relatively undisturbed areas are ideal targets because the biocontrol agent has a chance to build up its populations over an extended period of time. Additionally, using biological controls on these large areas often is more economical and feasible than using either mechanical or chemical control methods.

Biocontrols have not been effective for rapidly developing annual weeds in annual crops. Environments in which intensive cultivation, crop rotation, or heavy pesticide treatment is utilized are extremely unstable and lessen the survival chances of the biocontrol agent. More important, control organisms tend not to build up their populations rapidly early in the season, when competition from weeds causes the greatest yield losses.

Biocontrols are chosen specifically for their selectivity, so they cannot be expected to control a complex of weeds effectively. Their greatest value has been and probably will continue to be in stable environmental situations in which a single aggressive weed is the major problem.

Even with these limitations, research to find new biological controls continues. At the current time, three approaches to biological control are being used or developed. The first is the classical approach. In this case the biocontrol organism, usually an insect, is introduced once and is allowed to reproduce or spread on its own through the weed-infested area. The second approach is the introduction of grazing birds, mammals, or fish. Bird and mammalian grazers, unlike insects, require special husbandry. The third approach is that of a manipulated agent. In this case the biocontrol organism, usually a disease-causing (pathogenic) bacterium or fungus, is grown, concentrated, packaged, and then applied to the target weeds using techniques that are similar to those used with herbicides. Like herbicides, they must be applied each time a flush of the target weed appears. This area has received much research interest in recent years.

The Classical Approach

The classical approach provides an economical and relatively permanent solution when a single weed species is the major problem and a biological control agent that can propagate and spread itself is available for its control. Inherent in the method is acceptance of a low level of the weed population so the biological control agent can survive.

To be successful, a biological control agent must possess certain characteristics other than an ability to reduce a weed population to nondestructive levels: it must not harm nontarget desirable plants, it must be able to reproduce quickly enough to prevent the weed from competing with desirable plants or from reestablishing major infestations, it must survive and maintain a population equilibrium between itself and noncompetitive levels of the weed, and it must be adapted to the environment of the host. It also should be free of its own predators or pathogens.

A potential problem in the use of biological controls is the possible change in food preference from weeds to desirable plants. However, almost none of the several hundred insects introduced into this country as biocontrol agents have shifted food preference to other species. It is also significant that the most successful instances of biological control have been on weeds introduced from foreign countries. These weeds often arrive without the natural enemies that may have reduced their abundance in their countries of origin. This explains why some introduced weeds spread so rapidly in their new environments and why much of the search for biological controls takes place in the native home of the weed. For example, Goeden (1975) reported that three-fourths of the 80 weeds investigated at the time by U.S. federal and state agencies for their susceptibility to biocontrols were alien (two-thirds also were perennial).

Examples of Classical Biological Control Agents

Insects have provided the majority of biocontrol agents for terrestrial weeds. Large numbers of plant-feeding species exist among the insects, they frequently have a high degree of host specificity, and they have the capacity to destroy both vegetative and reproductive parts of the host plants. Although much research has been directed toward finding pathogenic organisms such as bacteria and fungi to control weeds, few have proved successful. Weeds are indeed susceptible to pathogenic organisms, but in many cases, the organism causes localized damage rather than destroying the whole plant or population of plants. The most effective use of pathogenic organisms is as bioherbicides, which are described in the section on manipulated agents.

Prickly pear was the target for perhaps the most successful biological control effort when the eggs of the imported cactus moth (*Cactoblastis cactorum* [Berg.]) were introduced into Australia in 1925. In that year, prickly pear claimed over 60 million acres of rangeland. The larvae of the insect thrived on the weed. Bacterial soft rots, introduced as secondary parasites by the boring larvae, added to the damage, and most of the rangeland eventually was reclaimed for its intended use. In the United States, spectacular results were obtained on St. Johnswort, a perennial plant that had claimed about 2 million acres of rangeland in California and Oregon. Introduction of the St. Johnswort beetle (*Chrysolina quadrigemina* [Suffrian]) in 1944 has reduced that infestation to about 1% of its original abundance.

In some cases, several insects, each feeding on a different part of the plant, are introduced. For example, tansy ragwort, a poisonous plant, has been successfully controlled in California and Oregon using three insects (Pemberton and Turner 1990; McEvoy et al. 1991). The larvae of the cinnabar moth feed on the foliage and flowers of the ragwort. The larvae of the seedhead fly feed on developing ragwort seeds, while the larvae of the flea beetle feed on the root crowns and stems of the plant. A similar approach using five insects is currently being tested for control of purple loosestrife in wetland areas.

Other insects that have been tested for their weed control potential include the thistle-head weevil (*Rhinocyllus conicus*) for musk thistle (Figure 4.13), the root-boring flea beetle (*Aphthona* spp.) for leafy spurge, the seedhead gall fly (*Urophora* spp.) for diffuse and spotted knapweed, and a moth (*Coleophora* spp.) for Russian thistle. All of these weeds are either perennials or are problems in noncropland situations. For consistent results, these biocontrol agents may need to be reapplied once or twice a year, in which case they should probably be classified as manipulated agents.

When the thistle-head weevil was introduced in 1968 to control musk thistle, its ability to feed on native thistle species was known. The potential for damage to native species was thought to be minimal because the weevil clearly preferred feeding on musk thistle. However, in 1997, the first report appeared that showed the weevil was reducing seed production significantly in five native thistle species in the field (Louda et al. 1997). A case can now be made that the weevil never should have been introduced in the first place. Clearly, present day attitudes about maintaining healthy and diverse native plant communities will have to be taken into account before new biological control agents can be introduced into the environment.

Successful biocontrol agents for aquatic weeds are described in Chapter 15.

Grazing Animals

Grazing animals such as geese, goats, and cows have been used effectively for certain weed con-

Figure 4.13 Left: Musk thistle infestation in Montana rangeland before introduction of the *Rhinocyllus* weevil; photo taken in 1975. Right: By 1977, musk thistle plants had been eliminated from the site. (N. E. Rees, USDA/ARS, Montana State University)

trol situations. It is essential that the animals display a certain amount of selectivity; that is, that they find undesirable plants more palatable than the other vegetation present. Use of such animals requires good husbandry. They cannot simply be turned out into the crop or pasture and left to fend for themselves. Essential management practices include supplemental feeding to provide nutrients not supplied by the weeds; fencing, herding, and sheltering of the animals; protection from predators; sales or disposal of excess animals; and general health care.

Geese, particularly the White Chinese breed, have been used for grass control in orchards, vineyards, nursery crops, mint, strawberries, and in row crops such as beans, hops, asparagus, cotton, potatoes, and onions. Geese will consume almost all immature grasses and nutsedges and reportedly prefer bermudagrass and johnsongrass even when plants are large. They nibble emerging perennial grass shoots until the plants eventually starve and die. Obviously, geese would not be suitable for grass crops such as corn, sorghum, and other grains. Three to five geese per acre reportedly are adequate for severely infested grassy fields. At an estimated annual cost of $15.00/acre, they seem to be ideally suited for such situations. They must, however, be supplementally fed with grain; they should have access to a pond or ditch; they must be fenced from neighboring grass crop fields; and they are extremely susceptible to predation from dogs, coyotes, weasels, and foxes. Also, alternate control methods for the broadleaved weeds that always are present in crop situations still would be needed.

Other examples of weed control with grazing animals include using goats and sheep on ditch banks and using goats for controlling brush in pastures (although such a use may not be the ideal husbandry for goats); spiny weeds such as blackberry and sweet briar; and allergenic plants such as poison ivy, poison oak, and poison sumac. Cattle can be used for the control of johnsongrass in fescue pastures. Cattle will eat johnsongrass preferentially and can provide excellent control when properly managed. Sheep and cattle have been used for weed control in tree crops such as under pine, fir, and oil palm plantations.

Examples of grazers used in aquatic sites are given in Chapter 15.

Manipulated Agents

Manipulated agents are disease organisms or insects that are grown on artificial media in large quantities. The organisms are concentrated, packaged, and introduced into the environment when the conditions are favorable for infection or feeding. The advantage of applying concentrated organisms directly to the target vegetation, much as is done with herbicides, is to increase the level of contact between the foliage and the biocontrol agent. This approach should result in a more rapid and higher degree of control than can be obtained from the natural spread of the pathogen. Either the reproductive propagules (e.g., spores of fungi) or the actual organism is applied to the target.

Like herbicides, manipulated agents must be reapplied to obtain continuing control; however, the negative environmental impacts of using natural agents, particularly agents that are native to the area of weed infestation, should be fewer than with synthetic herbicides.

At this time, the organisms with the greatest potential as manipulated agents are bacterial and fungal pathogens. Fungal pathogens are referred to as *mycoherbicides*. A more general term, to include other manipulated agents, is *bioherbicide*.

Two bioherbicides have been sold in recent years, and one is registered but not yet sold. The fungus *Colletotrichum gloeosporioides* f. sp. *aeschynomene* is packaged as Collego for the control of northern jointvetch, an important weed in rice. Three components are packaged separately and added to water just before application: the dried spores, a rehydrating agent, and a surfactant. The rehydrating agent improves the wetting of the spores and the plant surface, and the surfactant presumably improves spore retention and hyphal penetration into the leaf. Another fungus, *Phytophthora palmivora*, has been sold as DeVine for the control of strangler vine in Florida citrus groves. The spores of this soil-borne organism are not stable, and the product has a shelf life of only 6 wk. However, because the marketing area is small, the product can be delivered from the formulation site to the field as needed. The third product that has not yet been sold in the United States is BioMal for the control of round-leaved mallow. It consists of the fungus *Colletotrichum gloeosporioides* f. sp. *malvae* packaged with a silica gel carrier.

Some of the characteristics of these products point out the potential problems in the development of bioherbicides as a successful weed control strategy. Although a bacterium or fungus may show potential for severely injuring a weed in laboratory or greenhouse tests (over 100 have been identified as possible candidates), the organism, once released to the field, is subjected to many adverse environmental conditions. These conditions include temperatures outside of its growth range, lack of adequate moisture, interference from competing pathogens, and injurious ultraviolet radiation. Many foliar pathogens require several hours of free moisture on the plant surface for spore germination, formation of infection structures, and actual infection. This requirement necessitates the development of additives such as the rehydrating agent sold with Collego. The loss of viability when stored for long periods of time is another hindrance to development. A shelf life of at least 6 to 18 mo will be required if bioherbicides are to be used over larger marketing areas than is DeVine. As Zorner et al. (1993; as cited in Jackson et al. 1996) pointed out, "research efforts must be shifted from the discovery of bioherbicides to solving the production, storage, and efficacy problems that plague all bioherbicides."

The selectivity of bioherbicides to single weed species can be both a benefit and a detriment. Selectivity is beneficial because the organism will not affect nontarget species; however, the market potential for a bioherbicide that controls only one species may be limited unless the weed is not easily controlled using other methods. The gradual shift away from chemical herbicides may encourage the continued development of bioherbicides, and, just as is done with chemical herbicides, mixing several bioherbicides together can broaden the spectrum of weeds that is controlled. The costs for developing bioherbicides are less than for chemical herbicides. Collego cost approximately $2 million in research and development in the late 1970s and early 1980s compared with $15 to $20 million to discover and develop a chemical herbicide at that time (Templeton 1986).

Probably the greatest potential for bioherbicides is to complement a total weed management program. For example, DiTomaso et al. (1996) showed that a mycoherbicide (*Colletotrichum coccodes*) developed for velvetleaf control produced better results when it was used in a mix of the weed and crop rather than on the weed alone. The pathogen stunted the velvetleaf to such an extent that the soybean crop was able to overtop and shade out the velvetleaf. Combinations of management practices, such as the use of bioherbicides and crop competition, need to be further explored.

Another option in the use of bioherbicides is the isolation of compounds from bacteria and fungi that are toxic to the target weed. Examples include the toxins fumonisin B_1 from the fungus *Fusarium moniliforme* (Abbas and Boyette 1992)

and the AAL-toxin from the fungus *Alternaria alternata* f. sp. *lycopersici* (Abbas et al. 1995). However, the toxicology of these compounds to humans and other nontarget species will have to be established before they can be used on a wide scale.

CHEMICAL WEED CONTROL

The major area of expansion in weed control technology since World War II has been the development of herbicides. Herbicides lead other individual pesticide groups from the standpoint of total acreage treated, total tonnage of pesticides produced, and total dollar value from pesticide sales. In 1990, herbicides accounted for 65% of total pesticides used in the United States and 85% of pesticides used on cropland (Delvo 1990). Herbicides have been accepted to this degree because of the unparalleled success associated with their use and because of the many contributions they have made to vegetation management and crop production systems. Those contributions include the following:

1. Herbicides permit the control of weeds where row cultivation is impossible; for example, within the crop row, in crops planted in rows less than 20 in. apart, in solid-seeded or sod crops, and in no-till systems.

2. Herbicides reduce the number of tillage operations as well as the critical timing needed for such operations, particularly at planting time. Controlling weeds with herbicides permits earlier planting dates since one or more of the shallow tillage operations needed to reduce weed populations before planting can be replaced with a herbicide treatment. The fact that herbicides can be applied at or before planting lessens dependence on the rotary hoe, early-season cultivation, and the dry weather conditions necessary for conducting these operations. This is one of the major reasons for the increase in the past 40 yr in the total number of acres that one person can farm successfully.

3. Chemicals have reduced the amount of human effort expended in hand weeding. Where effective herbicides are available, weeding costs can be greatly reduced. Except where it is done as a hobby, hand weeding is hard work and for many people is an almost unbearable task. Effective use of herbicides permits a more complete mechanization of crop production and harvest. When hand labor is reduced, so are the management problems associated with hiring, overseeing, and housing workers; keeping records; and dealing with governmental agen-

cies and unions that accompany the hiring of farmworkers.

4. Weeds not economically controlled by other methods frequently can be controlled effectively and at relatively low cost with herbicides. Even some of the serious creeping perennial weeds have yielded to chemical control. Nearly maximum crop production can now be attained in fields in which, just three decades ago, one or more seasons of clean fallow had to be employed to reduce perennial weed populations and obtain profitable yield. Other examples include the elimination of weeds that develop after layby (the last cultivation time) in row crops treated with directed sprays and the control of weeds in turf, rangelands, and aquatic sites.

5. The use of herbicides has permitted a greater flexibility in the choice of a management system. Today, farmers can use fewer crops in rotation, have the option of removing one or more tillage or mowing operations, and eliminate periodic fallow every three or four seasons. Recent advances in the utilization of conservation tillage systems (e.g., no-till) can be attributed directly to the replacement of tillage operations with herbicides. Besides conserving energy and soil, other advantages of fewer tillage operations include less mechanical damage to crops caused by root pruning and stem injury and the reduction of structural damage to the soil.

Although herbicide use is not without problems, the benefits have far exceeded the risks. The most common problems associated with herbicides are related to their inherent ability to kill plants. Problems include injury to nontarget vegetation, crop injury, and residues in the soil from the previous season. Inconsistent weed control also is a frequent problem. The continued use of some herbicides has resulted in the occurrence of herbicide resistance in some weed species. Fish kills can occur when herbicides enter ponds, lakes, or streams. Usually such a problem is due not to the toxicity of these compounds but to the fact that herbicides kill algae and other aquatic plants which, as they decompose, deplete oxygen from the water.

In general, the public has a negative perception of any synthetic chemical applied to the environment, including herbicides. The inability of the public to distinguish among insecticides, fungicides, and herbicides has led herbicides to be categorized with chemicals that have been detrimental to the environment. For example, the organochlorine insecticides such as DDT bioaccumulated in the fatty tissues of wildlife causing reproductive failure.

Bioaccumulation has not been attributed to herbicides; however, the very fact that herbicides may be present, even at extremely minute concentrations, in the environment causes substantial public concern. These concerns have resulted in stringent restrictions and regulations on the use of herbicides and have greatly increased the costs of development of new herbicides. A negative aspect has been the loss of certain herbicides for limited acreage crops (e.g., tobacco, broccoli, asparagus, garlic, red beets, blueberries). The low acreage, and therefore low financial return, does not justify the cost of development or registration of herbicides for the crop. Many limited-acreage crops are high value, and the resultant liability to manufacturers should crop injury or poor performance occur makes the profitability of the herbicide questionable relative to the income generated.

A problem that can be attributed to the beneficial rather than the negative aspects of herbicide performance is the tendency of many practicing agriculturalists to expect that any and all weed problems can be solved effectively with herbicides. When this attitude prevails, other valuable weed control practices such as sanitation, tillage, and crop competition tend to be ignored.

Although infrequent, problems of injury to applicators, bystanders, and livestock have occurred. The removal of extremely toxic compounds (primarily sodium arsenite, which caused injury to livestock that consumed treated vegetation) from general use has largely eliminated this problem.

All of the potential problems listed here can be minimized by the proper selection, storage, handling, and application of herbicides. Education is the key to using herbicides wisely and safely.

MAJOR CONCEPTS IN THIS CHAPTER

A planned approach to solving weed-related problems is necessary for the most effective application of the sophisticated technology available. Such an approach includes an understanding of the interacting systems (plants, soils, crop management, environment).

Components of the suggested approach include:

- A clear understanding of overall weed management objectives (prevention, control, eradication).
- A step-by-step procedure for solving weed problems (problem diagnosis, survey of

available practices, program selection, program execution, follow-up).

- A realistic appraisal (understanding) of the capabilities of individual control practices.

Control methods include preventive practices, mechanical and physical removal (hand removal, tillage, mowing, mulches, flaming), crop management (competition, rotation, allelopathy), biological control methods, and chemicals. All are applicable to certain situations. Older weed control technology should be added to, not disregarded.

Prevention is aimed at not spreading weeds to uninfested areas with seeds and vegetative propagules. Sanitation, seed laws, weeds laws, preventing seed formation, not burying seeds in the soil, and spot removal of isolated plants or patches are components of prevention. The benefits of prevention warrant more effort than is currently expended.

Mechanical and physical weed control has long been the cornerstone of modern weed control technology and basically constituted most of the weed control accomplished until the l950s.

- Hand pulling and hoeing are the most primitive mechanical weed control methods. Although costly per unit area, they are appropriate for high value situations, mixed species (ornamental plantings), home gardens, and scattered weeds.
- Tillage with modern equipment is synonomous with advances in agriculture and provides effective weed control efficiently and economically. Soil erosion and the need for timely operations are the two major limitations. Use of tillage equipment on wet soils can result in compacted soil layers. The current emphasis on reduced tillage operations to prevent soil erosion has been made possible through the use of modern herbicides. Removal of a tillage operation from a managment system removes some increment of weed control that must be replaced.
- Mowing can provide control of tall growing weeds and at least prevent them from producing seed. Mowing is most effective when used in conjunction with a competitive sod or forage crop. Species with growing points near the soil surface are not controlled effectively by mowing. Mowing after pollination has limited to no adverse effect on seed production.
- Mulches control weeds by excluding light. They are appropriate on small areas and on high value crops and are best suited to emerged, established, or transplanted crops. Black (or other opaque) plastic sheeting is a commonly used mulch, although a variety of materials can be used. In permanent plantings plastic mulch frequently is used under a decorative particulate mulch.

Crop competition frequently is the most effective and economical weed control practice available, yet its contribution usually goes unrecognized and seldom is considered when a weed control program is planned and developed.

- Shading is the major factor by which crops suppress weeds. Any factor that hastens canopy closure and shading by the crop will enhance its competitive ability. Supplemental weed control is needed most at those times when the crop does not produce full shading. Crops can be managed to provide maximum shading.
- Crop rotation is feasible for annual, biennial, or short-term perennial crops. It is not available for long-term, established perennials crops. The more dissimilar the crops and their management practices are in the rotation system, the less opportunity an individual weed has to become dominant. A practical substitute for crop rotation is to employ varying weed control practices as much as possible.
- The allelopathic properties of cover crops such as rye, and plant residues such as wheat stubble, can provide effective weed control in some crops.

Biological weed control is the use of natural enemies to reduce weed populations to noneconomical levels. Success has been limited primarily to perennial weeds on low value land and aquatic sites.

- Classical biocontrol agents such as insects are released once and are allowed to spread on their own and maintain a long-term equilibrium with a nonharmful level of the host weed.
- Grazing animals such as geese and mammals require a certain amount of husbandry in order to survive and thrive.
- Manipulated agents mostly consist of packaged bacteria, fungi, or insects that are applied directly to the target vegetation. Since these organisms cannot perpetuate themselves long term in the host's environment, they need to be reapplied as needed. A few manipulated agents (e.g., Collego, DeVine) have been marketed.

Chemical weed control is widely accepted because it is effective at reasonable cost. Herbicides have provided the most spectacular advances in weed control during the past four decades and likely will continue to do so in the near future.

TERMS INTRODUCED IN THIS CHAPTER

Bioherbicide
Biological control agent
Biological weed control
Canopy closure
Chemical weed control
Classical approach (biological weed control)
Clean tillage (fallow)
Competitive crop
Control
Crop rotation
Crop shading
Cultural weed control
Desiccation
Economic threshold
Eradication
Flaming
Flooding
Grazers
Hand hoeing
Herbicide
Manipulated agents
Mechanical weed control
Mowing
Mulch
Mycoherbicide
Narrow crop rows
Noxious weed
Particulate mulch
Physical weed control
Prevention
Preventive weed control
Quarantine
Sanitation
Seed laws
Sheet mulch
Smother crop
Tillage
Weed control districts
Weed laws
Witchweed

SELECTED REFERENCES ON WEED CONTROL METHODS

Anderson, W. P. 1996. Weed Science: Principles and Applications, 3rd ed. West Publishing Co., Minneapolis/St. Paul, Minnesota.

Ashton, F. M. and T. J. Monaco. 1991. Weed Science Principles and Practices, 3rd ed., John Wiley and Sons, New York.

Auld, B. A. and L. Morin. 1995. Constraints in the development of bioherbicides. Weed Technol. 9:638–652.

Charudattan, R. and H. L. Walker, Eds. 1982. Biological Control of Weeds with Plant Pathogens. John Wiley and Sons, New York.

Delvo, H. W., compiler, 1990. Agricultural Resources—Inputs Situation and Outlook Report. Resourc. Technol. Div., Econ. Res. Ser., U. S. Dept. Agric., Washington, D.C. AR-17.

Duke, S. O., J. J. Menn, and J. R. Plimmer, Eds. 1993. Pest Control with Enhanced Environmental Safety. American Chemical Society, Washington, D.C.

Harley, K. L. S. and I. W. Forno. 1992. Biological Control of Weeds: A Handbook for Practitioners and Students. Inkata Press, Melbourne and Sydney, Australia.

Hoagland, R. E., Ed. 1990. Microbes and Microbial Products as Herbicides. Am. Chem. Soc., Washington, D.C.

Klingman, G. C. and F. M. Ashton. 1982. Weed Science. John Wiley and Sons, New York.

Popay, I. and R. Field. 1996. Grazing animals as weed control agents. Weed Technol. 10:217–231.

Putnam, A. R. and W. B. Duke. 1978. Allelopathy in agroecosystems. Ann. Rev. Phytopathol. 16:431–451.

Rice, E. L. 1984. Allelopathy, 2nd ed. Academic Press, Orlando, Florida.

Sand, P. F., R. E. Eplee, and R. G. Westbrooks. 1990. Witchweed Research and Control in the United States. Monograph Series of the Weed Science Society of America, Number 5, Lawrence, Kansas.

TeBeest, D. O., Ed. 1991. Microbial Control of Weeds. Chapman and Hall, New York.

Watson, A. K. 1989. Current advances in bioherbicide research. Brighton Crop Protection Conference—Weeds. 8B-3:987–996.

Watson, A. K., Ed. 1993. Biological Control of Weeds Handbook. Monograph Series of the Weed Science Society of America, Number 7, Lawrence, Kansas.

Wiese, A. F., Ed. 1985. Weed Control in Limited-Tillage Systems. Monograph Series of the Weed Science Society of America, Number 2, Lawrence, Kansas.

INTRODUCTION TO HERBICIDES

Almost 300 herbicides, as listed in the journal *Weed Science* (1996, Vol. 44:964–969), have been developed over the years. Of these, approximately 140 herbicides are currently available for use (*Herbicide Handbook* of the Weed Science Society of America, 1994; *Weed Technology,* 1996, Vol. 10:1002–1004). This large number of chemistries, most of them available in the United States, is a testament to the widespread adoption and success of chemical weed control.

This chapter will detail the history of herbicide development and the characteristics that make herbicides so amenable to weed control in diverse sites and circumstances. As with any tool, weed control with herbicides presents certain advantages and disadvantages. The chapter will conclude with a description of problems inherent with herbicide use and the governmental regulation of herbicides.

HISTORY

The Early Years (1900–1941)

Herbicide technology had its beginnings about 1900. Although salt, ashes, smelter wastes, and other industrial by-products had been applied for centuries at high rates (and near their source of supply) to control vegetation, it was the discovery of the fungicidal properties of Bordeaux mixture that led to the first serious attempts at chemical weed control. The discovery (near Bordeaux, France) in 1896 that a lime-copper-sulfur mixture would control downy mildew on grapes led to the extensive testing of this fungicide and other copper and inorganic salts for the control of diseases on a variety of crops. Researchers also observed that selective control of broadleaf weeds in cereal crops could be obtained with several of these inorganic salts (Smith and Secoy 1976).

Thus, the principle of selective removal of weeds from crops with chemicals was established. A major developmental effort occurred from 1900 to 1915 in France, Germany, and the United States with most of the emphasis placed on weed control in cereal crops. Solutions of copper nitrate, ammonium salts, sulfuric acid, iron sulfate, and potassium salts were shown to be selective herbicides (National Research Council 1968).

The practice of spraying with these compounds for the selective control of broadleaf weeds in small grains was most extensively used in Europe and the British Isles, where the prevalence of small, intensively cultivated farms ensured that the chemicals would be applied in a timely and careful manner. Furthermore, the penetration and effectiveness of these water-soluble salts were favored by the high humidity conditions prevalent in these areas. In the United States, however, the development and extensive use of selective chemical controls lapsed after 1915. Inadequate spraying equipment, large acreages, lower humidity conditions, and the introduction of cleaner, weed-free seeds and effective fallow systems have been suggested as reasons (Crafts 1975).

Between 1900 and 1940 a limited number of other herbicides were introduced. These included compounds such as the arsenicals, chlorates, and borates as highly persistent herbicides; carbon bisulfide and the thiocyanates as fumigants; ammonium sulfamate for the control of woody plants; herbicidal oils; and the phenols, the first specific group of organic compounds to be used for weed control. At the start of World War II, then, the list of available herbicides consisted of about a dozen chemicals of somewhat limited utility.

2,4-D to Glyphosate (1941–1980)

The discovery of the herbicidal properties of 2,4-D (2,4-dichlorophenoxyacetic acid), first synthesized in 1941, triggered the development of modern herbicide technology. Evaluation of the compound for insecticidal and fungicidal properties proved negative. Concurrent investigations of the role of auxin hormones and synthetic compounds with growth-regulating effects led to the discovery that 2,4-D had similar growth-altering properties on plants. These investigations led to the discovery of the herbicidal properties of 2,4-D. Much of the early work was carried out under wartime regulations and secrecy; however, soon after the war, the compound was released for general investigation and use.

2,4-D proved to be an outstanding herbicide. It was effective at low rates (1/4 to 4 lb/acre), it was cheap to produce, and it had a broad spectrum of uses. It controlled most broadleaved weeds selectively in most economically important grass crops. It also proved effective for controlling serious creeping perennial broadleaved weeds. The material is symplastically translocated, so it could be applied to foliage and would move to the underground portion of the weed. This characteristic provided much longer lasting and more severe damage to perennial weeds than that achieved with mechanical top removal alone. Controlling weeds with 2,4-D was found to be reasonably predictable and consistent. The compound proved to be easy and safe to handle and apply, and large areas could be sprayed quickly. Also, at normal use rates, long-term herbicide residues were not a problem. All factors considered, 2,4-D was and still is an outstanding herbicide.

As successful as 2,4-D was, it could not by itself solve all weed problems. The commercial success of 2,4-D led to the development in the 1950s of other herbicides, including other phenoxy herbicides such as MCPA, silvex, and

2,4,5-T; amitrole; and the phenylurea herbicides such as monuron and diuron. These plus other herbicides provided a critical mass of usable chemistry such that by the late 1950s chemical control was a success on a broad range of crops and weeds. In the late 1950s and the 1960s, some truly outstanding herbicides were developed and introduced. Herbicides such as atrazine, simazine, EPTC, dicamba, linuron, trifluralin, alachlor, and DCPA provided excellent weed control and greatly facilitated the continued adoption of chemical weed control to replace tillage and hand labor. The introduction of paraquat, along with advances in planting equipment, made no-till planting a viable option.

The first hints of regulatory pressure on the herbicide industry also appeared during these years. The 1960s saw the first major withdrawal of a herbicide from the market when sodium arsenite was removed because of toxicity to livestock that consumed treated vegetation, and the food uses of amitrole were suspended after a highly publicized scare over its use on cranberries just before the 1962 Thanksgiving season.

In the 1970s, tank mixes of soil-applied herbicides became the standard practice for the control of annual weeds in corn and soybeans. Many herbicides with chemistries similar to those already in existence were introduced, filling gaps and niches in crop or weed selectivity, increasing the options and ease of application, and reducing environmental persistence. For example, linuron, a standard herbicide for annual broadleaf weed control in soybeans, was effective but not particularly easy to apply. Because of its tendency to bind tightly to soil organic matter, it could only be applied to the soil surface, and its dosage had to be changed according to changing organic matter within an individual field. The introduction in 1973 of metribuzin, which was much less sensitive to soil organic matter, provided a herbicide that could be used at a single dose on a much broader range of soil types and could be applied both as a surface treatment and soil-incorporated treatment. Because it was convenient to apply, metribuzin largely replaced linuron as the treatment of choice for the next decade. Cyanazine, with a shorter persistence time in soil, was introduced as an alternative to the more persistent atrazine.

The introduction of glyphosate, a nonselective herbicide, in the late 1970s provided outstanding control of most perennial grass and many perennial broadleaf weeds. The lack of soil residual meant that crops could be seeded immediately into glyphosate-treated areas, and adverse effects to the environment and the user were minimal. This era also marked the removal of 2,4,5-T and silvex from the U.S. market, mainly due to the presence of a highly carcinogenic form of dioxin as a contaminant in these herbicides and negative publicity associated with Agent Orange, a herbicide mixture containing 2,4-D and 2,4,5-T, used as a defoliant in Vietnam.

Selective Postemergence Herbicides and Herbicide-Resistant Crops (1980 to the Present)

Herbicide development in the 1980s was marked by the introduction of selective postemergence treatments in major crops. Herbicide groups such as the sulfonylureas, imidizolinones, and aryloxyphenoxy propionates not only provide excellent selectivity but are used at extremely low dosages of less than a third of a pound per acre. Chlorimuron, primisulfuron, and other sulfonylurea herbicides, for example, are used at fractions of an ounce per acre. In fact, since 1982 when herbicide production peaked at 500 million pounds, the total pounds of herbicides used on crops in the United States decreased by 10% over the next ten years (Gianessi 1992), in part due to the adoption of herbicides that were active at lower dosages as well as the reduction in use rates of older herbicides. With the high costs of reregistration required by the federal government, many older herbicides and minor crop uses for other herbicides were dropped by manufacturers.

The decade of the 1980s was also the first decade in which the rate of soil loss due to erosion decreased rather than increased. No-till and other conservation tillage practices were widely adopted, thanks to the availability of the new selective postemergence herbicides for weed control. Probably the greatest contributions these herbicides provided in no-till was the effective control of perennial weeds and escapes from initial burn-down/residual treatments at planting.

The 1990s were marked by the first introductions of herbicide-resistant crop cultivars. The development of glyphosate-resistant (Roundup Ready) soybeans, for example, allows the use of glyphosate, a herbicide that combines broad-spectrum weed control with minimal adverse environmental impact, over large acreages of crops. The development of other herbicide-resistant crops is continuing at a rapid rate.

Another important trend during this period was the establishment of herbicide restrictions, including those on the use of atrazine and others because of possible surface and groundwater contamination.

In summary, the history of herbicide development has been marked by an increase in the number of herbicides for major crops, an increase in the total number of weed species controlled in those crops, the opportunity to achieve good weed control in restricted tillage management systems such as no-till, and improved human and environmental safety. The introduction of herbicide-resistant crops promises significant increases in crop safety. Future developments in herbicide chemistry will likely continue along these lines.

CHARACTERISTICS OF HERBICIDES

An outstanding herbicide is effective at low rates, economical to manufacture, has a broad spectrum of uses, is safe and easy to handle and apply, and is relatively benign in the environment.

In order to achieve the benefits of herbicides, professional agriculturalists must have a working knowledge of the nature and characteristics of the product. Familiarity with individual herbicides allows one to understand the principles involved in their use and to predict their effects on plants and their reactions in soils.

This section begins with a description of herbicide nomenclature and classification. Nomenclature (i.e., how herbicides are named) provides the basis for communicating about herbicides and permits the immediate recognition of an ingredient when multiple labels and formulations are available. Knowing the precise meaning of terms related to herbicide classification is essential in describing herbicide use and provides a basis for correct application.

These sections are followed by a general discussion of herbicide chemistry. Chemistry not only affects the herbicidal activity of a compound but also determines its toxicity to nontarget organisms, persistence, and fate in the environment.

Herbicide Nomenclature

Three types of names are listed on the containers of commercially available herbicides. The product marketed as Roundup Ultra has the names Roundup Ultra, glyphosate, and N-(phosphonomethyl) glycine listed at the top of the label.

N-(phosphonomethyl) glycine, the *chemical name*, describes the chemistry of the compound.

This chemical entity is the biologically active ingredient in the package, and it is referred to as the *active ingredient.*

Glyphosate is the *common name* given specifically to N-(phosphonomethyl) glycine. Each herbicidal chemical has one common name assigned to it. In many cases, the common name is a simplified version of the chemical name. Common names must be approved by an appropriate authority before being accepted by working weed scientists. In the United States, the key authority is the American National Standards Institute (ANSI), which receives input from the Terminology Committee of the Weed Science Society of America. A list of approved chemical and common names is published in the last issue each year of the journal *Weed Science.* All the herbicide names used in this book (unless otherwise indicated) are common names.

The third type of name, given to a compound for marketing purposes, is the *trade name.* Roundup was the first registered trade name for glyphosate. A newly developed compound usually has only one trade name since the company that develops the chemical obtains proprietary use of that product for a period of 17 years after the product patent is obtained. Once the company's patent expires, however, other companies can market the product under different trade names.

Sometimes a chemical sold by an individual company has more than one trade name. For example, different trade names may be used for different uses or formulations. Monsanto Company originally marketed glyphosate as Roundup and more recently as Roundup Ultra (41% isopropylamine salt of glyphosate with surfactant) for nonaquatic uses, as Rodeo (53.8% isopropylamine salt of glyphosate without surfactant) for aquatic uses, and as Bronco (14.8% isopropylamine salt of glyphosate and 27.6% alachlor) for reduced tillage systems. Sethoxydim is packaged by BASF Corporation for agronomic use as Poast and Poast Plus and for ornamental use as Vantage. Occasionally, two companies will market the same proprietary chemical, resulting in two trade names for a single chemical. An example is the herbicide metribuzin, which is marketed as Lexone by DuPont and as Sencor by Bayer Corporation.

Herbicide Classification

How herbicides are described or categorized is a matter of convenience. It depends in large part on those features or aspects that the describer wants to emphasize. In fact, we can think of at least eight ways in which herbicides commonly and frequently are classified. Each of these categories and terms must be understood precisely in order to interpret information relating to herbicides and their proper use. Herbicides are classified as follows:

Degree of Response Differences among Plant Species. A *selective* herbicide is a chemical that is more toxic to some plant species than to others. In order to be of practical use, the difference in toxicity must be large enough so that one plant population can be removed without significant injury to the remaining plant population. In fact, one of the challenges in herbicide development is finding a chemical that can consistently achieve a difference of this magnitude between weeds and crops. *Nonselective* herbicides are toxic to all species present.

Coverage of the Plant or Soil. *Foliar applications* are made directly to the leaves or foliage of plants. *Soil applications* are made to the soil rather than to the vegetation.

Length of Soil Persistence. According to the *WSSA Herbicide Handbook, residual* herbicides are those that injure or kill germinating seedlings for relatively short periods of time (less than one season) when applied at recommended use rates. *Persistent* herbicides can harm susceptible crops planted in normal rotation after the treated crop is harvested or interfere with the regrowth of native vegetation in noncrop sites over extended periods of time. We prefer to divide persistent herbicides into two categories according to the length of time that the herbicide persists and causes plant injury. *Persistent* herbicides harm susceptible plants during the first and sometimes into the second season after application; *highly persistent* herbicides harm susceptible plants during the second season and sometimes for longer periods of time. In many cases, the highly persistent herbicides are used for long-term, total vegetation control.

Coverage of the Target Area. *Broadcast* treatments are applied over the entire area. *Band* treatments are applied to continuous, somewhat restricted areas such as on or along a crop row rather than over the entire field. *Spot* treatments are applied to localized areas such as scattered individual plants or clumps and small patches of weeds. A relatively small fraction of the total area is treated.

Application in Relation to Crop and Weed Development. Herbicides applied prior to plowing are called *preplow* herbicides. *Preplant* herbicides are applied to the soil surface before seeding or transplanting. *Early preplant* describes herbicides applied two or more weeks before planting. For example, they are frequently soil-active compounds that are applied in no-till situations before the annual weeds germinate. *Preemergence* herbicides are applied prior to the emergence of either the crop or a specific weed but not necessarily to both. When referring to the crop, *preemergence* generally means after planting and prior to crop emergence. In established crops (e.g., turf, orchards) preemergence usually means prior to weed emergence. *Cracking* (a term used to describe germinating legume crops) treatments are applied at the time when the soil above the emerging crop seedlings is beginning to crack. These treatments also are referred to as *at emergence* treatments. *Postemergence* herbicides are applied after the crop or specified weed has emerged. Unless otherwise indicated, it is assumed that these applications are applied above the plants. Postemergence treatments are further subdivided into early, late, and directed applications. *Early postemergence* follows emergence but is conducted during the initial growth phase of the crop or weed seedlings. *Late postemergence* treatments occur after the crop or weeds are well established. *Directed postemergence* refers to the specific placement of the herbicide to cover the weeds but miss as much of the crop as possible. *Layby applications* are made at the time of the last cultivation.

Methods of Soil Application. Soil-applied herbicides may be *surface applied, incorporated* (mixed or blended into the soil), *layered,* or *injected.* Layering and injecting involve the placement of the herbicide beneath the soil surface with an injection blade, knife, or tine, and are rarely used.

Plant-Related Characteristics. The pathway of herbicide movement in plants and the mechanisms by which herbicides kill plants are additional features used for classification. When referring to pathways of movement in the plant, herbicides tend to be either *symplastic* (phloem translocated), *apoplastic* (xylem translocated), or *contact* (no translocation). The types of herbicides according to their mode of action include *auxin-type growth regulators; aromatic amino acid inhibitors; branched-chain amino acid inhibitors; carotenoid pigment inhibitors; lipid biosynthesis inhibitors; "classical" photosynthetic inhibitors; "rapidly acting" photosynthetic inhibitors; photosystem I energized cell membrane destroyers; protoporphyrinogen oxidase inhibitors; glutamine synthesis inhibitors; microtubule/spindle apparatus inhibitors (root inhibitors); shoot inhibitors; cell division inhibitors;* and *cell wall formation inhibitors* (see Chapter 6 for descriptions of these modes of action).

Herbicide Chemistry. Herbicides often are classified into groups according to their common chemistry. Examples include the *phenoxy acids, benzoic acids, imidazolinones, sulfonylureas, aryloxyphenoxy propionates, triazines, diphenylethers, dinitroanilines,* and *thiocarbamates.* An elementary knowledge of basic herbicide chemistry and chemical nomenclature can be useful in explaining many of the characteristics of herbicides and herbicide groups.

Herbicide Chemistry

Most herbicides are organic compounds; that is, they contain carbon. The major elements contained in organic herbicides, in order of decreasing occurrence, are carbon, hydrogen, oxygen, nitrogen, chlorine, phosphorus, sulfur, and fluorine.

The basic structural components of most organic herbicides are carbon chains (aliphatic groups) and rings (aromatic groups).

The simplest aliphatic compound consists of one carbon and four hydrogens (CH_4) and is a gas named methane. The aliphatic compound with a chain of two carbons is called ethane (C_2H_6); three carbons, propane; four carbons, butane, etc. (Figure 5.1). When the radicals of these compounds are included as a portion of a larger molecule, they are similarly named but with -yl as the ending: methyl, ethyl, propyl, butyl, etc. An example is ethyl alchohol.

An aliphatic group with —COOH (a carboxyl group) at the end of the chain is called a carboxylic acid. A carbon chain in which one or more of the hydrogens is replaced by —OH (a hydroxyl group) is called an alcohol. A carboxylic acid with two carbons in the chain is called ethanoic (or acetic) acid; the two-carbon alcohol chain is called ethanol (ethyl alcohol). Names of the other carbon chains are given and the numbering system for the carbons is illustrated in Figure 5.1.

Ring compounds or aromatics are derivatives of benzene (C_6H_6). A benzene ring minus a hydrogen is termed a phenyl group and in this form can accept an appropriate substitution. For example, the replacement of the hydrogen atom with a hydroxy (OH) group results in the formation of phenol. Other substituted ring struc-

Number of carbons:

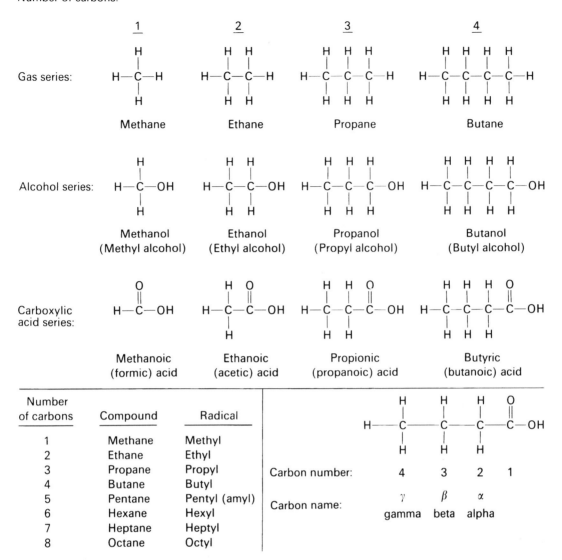

Figure 5.1 Chemistry and nomenclature of simple aliphatic compounds.

tures are shown in Figure 5.2. They include benzoic acid, toluene, and aniline. The positions on the benzene ring are numbered from 1 to 6 and also may be named. Positions 2 and 6 are the ortho positions; 3 and 5, the meta positions; and 4, the para position (Figure 5.2).

Derivatives of organic compounds are named according to the functional groups and substitutions present. Some of the most frequently encountered are shown in Figure 5.3.

Some examples of how standard chemical terminology is used are given in Figure 5.4.

How Chemistry Affects Herbicidal Properties

The chemistry of a compound determines how the herbicide will act in biological and physical systems such as plants, animals, soils, and water.

Compounds made up of simple chains of carbon, hydrogen, nitrogen, sulfur, and oxygen tend to be degraded readily by microbes. Examples include glyphosate, glufosinate, fosamine, and several thiocarbamate herbicides. Compounds with aromatic structures and halogen (chlorine, bromine, fluorine, iodine) substitutions tend to be longer lived than straight chains. Examples include the chloracetamide, triazine, sulfonylurea, and dinitroaniline herbicides. The more chlorinated a compound is in a series of analogous compounds, the longer it will persist in the soil.

The substitutions and alterations that can be made to an organic acid such as 2,4-D illustrate the effects that chemical structure can have on important herbicidal and mixing properties.

The acid form of 2,4-D is only slightly soluble in water and oil, so it has limited use in

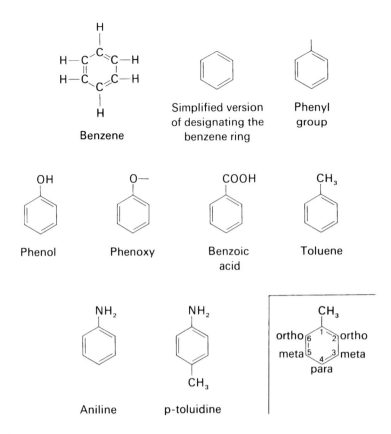

Figure 5.2 Chemistry and nomenclature of simple aromatic compounds.

commercial formulations. The acid can be reacted with bases to form salts and with alcohols to form esters. The properties of salts and esters are quite different.

Several of the common salts that can be formed include the sodium, potassium, ammonium, lithium, and several amine salts. These compounds ionize (dissociate) in water to form charged particles (Figure 5.5). All the salts just listed are soluble in water, with the amine salt being the most soluble. The amine salt is the

most commonly used salt formulation. Salts are not soluble in oil and when used as foliage treatments must be applied with wetting agents to increase wetting and penetration into the tissue.

The esters of 2,4-D are nonpolar molecules and do not ionize (Figure 5.5). They are insoluble in water but highly soluble in oil. During formulation, they usually are diluted in oil-based solvents and mixed with an emulsifying agent. The emulsifier acts to keep the oil–ester droplets

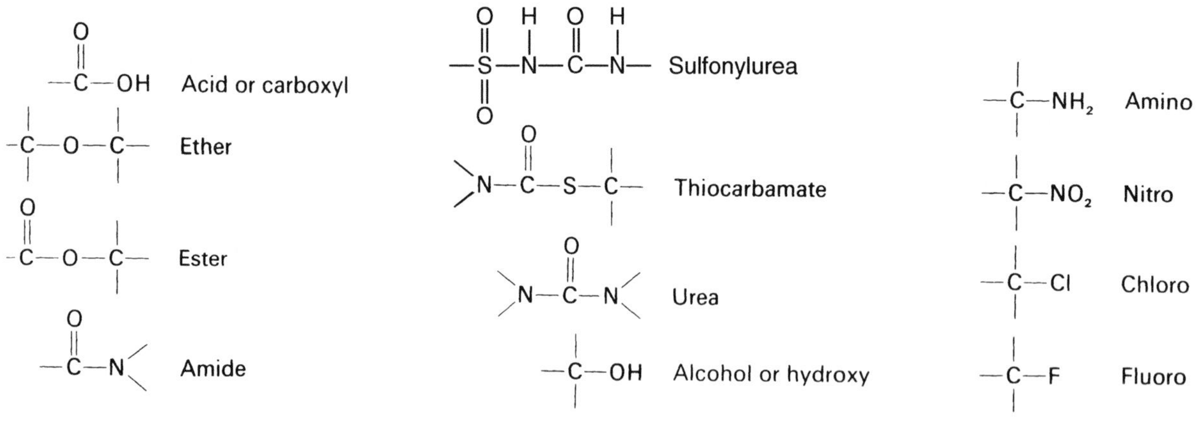

Figure 5.3 Major functional groups, linkages, and substitutions common to many herbicides.

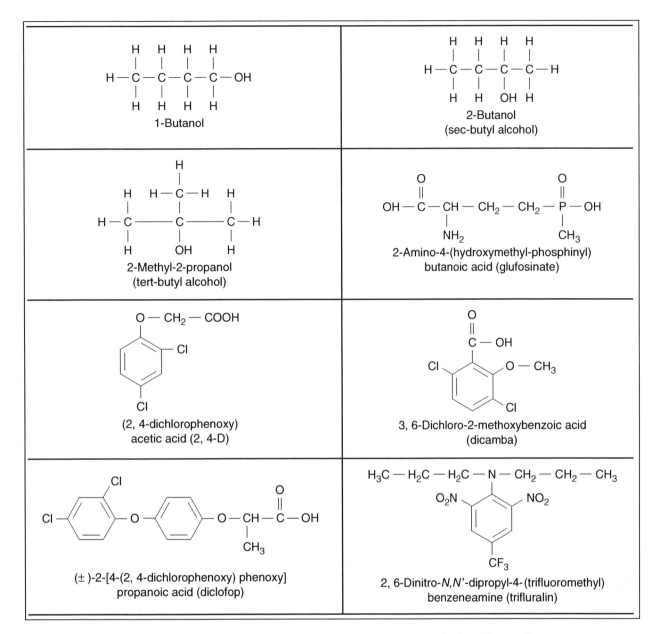

Figure 5.4 Examples of how chemical terminology is used.

suspended in water when water is used as the spray carrier. At other times, oils may be used directly as carriers for the application of esters, particularly for the treatment of woody species. The ester formulations more readily wet and penetrate plant cuticles and are more toxic to plants than the salt formulations.

The salts of phenoxy herbicides are nonvolatile, whereas the ester formulations tend to have varying degrees of volatility. Volatility of an ester is determined by the molecular weight of the alcohol. Short-chain, low-molecular-weight alcohols of five or fewer carbons lead to the formation of highly volatile esters. Low volatile esters, on the other hand, are heavier and consist of longer-chained alcohols with an ether link-

age. Although called low volatile esters, they still are somewhat volatile. High volatile esters are rarely or no longer used because of the injury problems that can occur when they volatilize into the air and move to nontarget plant species.

Another property affected by chemistry is that of stability in hard water; that is, water having a high calcium or magnesium content. Since 2,4-D salts dissociate in water, the sodium and potassium molecules are released leaving the 2,4-D anion. Calcium and magnesium, however, can rebind to the 2,4-D anion causing the formation of precipitates that clog filters and nozzles. The amine salts are less susceptible than the alkali salts to forming precipitates; the esters have excellent stability in hard water.

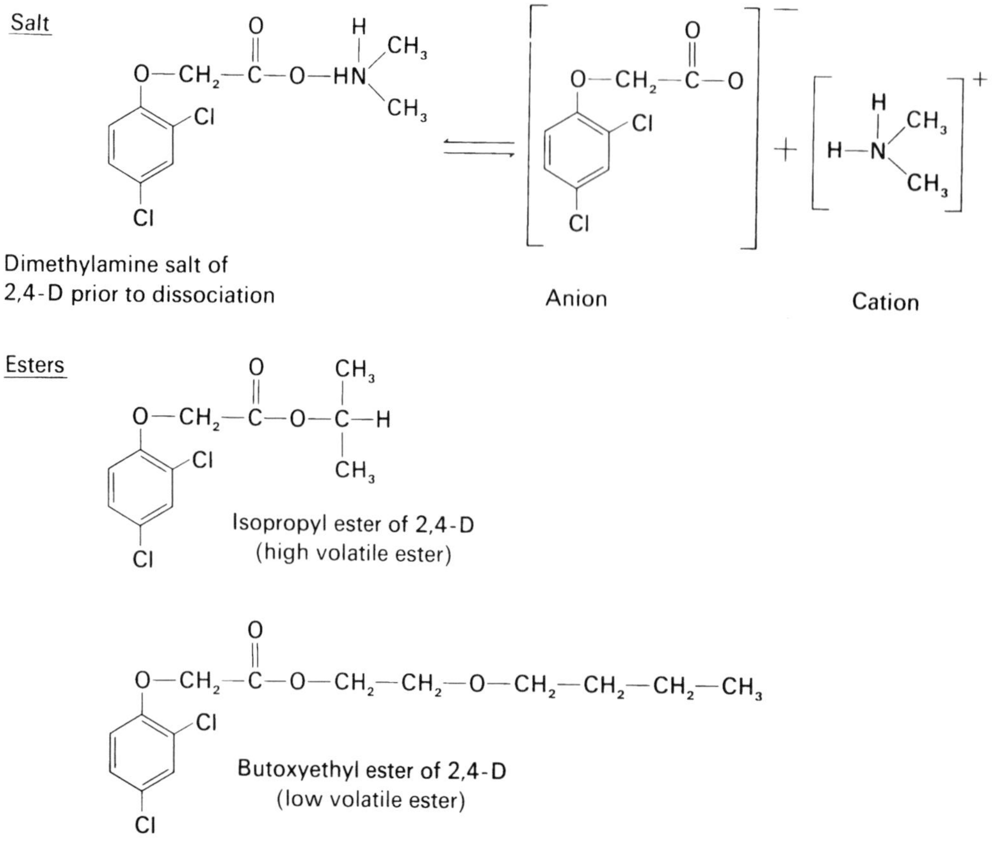

Figure 5.5 Major forms of 2,4-D.

As might be expected, salts of 2,4-D are formulated as water-soluble liquids, whereas esters are formulated as emulsifiable concentrates.

A summary of the properties of the various forms of 2,4-D is presented in Table 5.1. All auxin-type growth regulator herbicides are similarly affected and include the other phenoxy herbicides, the benzoic acid herbicides, and the picolinic acid herbicides. In fact, any herbicide with a reactive carboxyl group (—COOH) has the potential to be formulated as an acid, monovalent metallic salt, amine, or es-ter. For example, the aryloxyphenoxy herbicides are formulated as esters. Acifluorfen, a diphenylether, is formulated as a sodium salt.

In some cases, salts are formed on reactive groups other than carboxyl groups. Examples include glyphosate, which is formulated as an amine salt on the —P—OH group, and benta-zon, which is formulated as a sodium salt on an —S=O group.

Other examples in which chemistry affects the characteristics of the herbicide are paraquat and diquat. These herbicides are cationic but

Table 5.1 Comparison of Properties of Different Forms of 2,4-D[a]

| Form of 2,4-D | Solubility | | Volatility | Typical Formulation | Stability in Hard Water |
	Water	Oil			
Acid	Very low	Low	Very low	—	—
Standard amines	High	None	None	Soluble liquids	Fair
K, Na, Li salts[b]	High	None	None	Soluble liquids	Poor
				Soluble powders	Poor
High vol ester	None	High	High	Emulsifiable concentrates	Good
Low vol ester	None	High	Moderate	Emulsifiable concentrates	Good

[a]Modified from Klingman and Ashton (1982).
[b]Not used in 2,4-D formulations currently. Na and K salts are, however, used for other carboxy acid herbicides (e.g., dicamba).

are packaged as the chlorine (paraquat) or bromine (diquat) salt in order to maintain stability in the package. Once they are dispensed to water, the salt readily dissociates to release the cationic herbicide, which is highly soluble in water.

CONCERNS REGARDING HERBICIDE USE

A survey of the public conducted by King (1991) found that 89% of all respondents were more concerned about pesticides than all other food issues, and 63% said that "the dangers to human health of pesticides outweighed their benefits in protecting crops." These fears are gross overreactions, and yet, because pesticides, including herbicides, are toxic at some dose, caution dictates that they be subject to the most critical and rigorous testing for negative human and environmental impact.

Concerns about herbicide use generally fall into three categories: potential for movement off target, toxicity to humans, and negative effects on the environment.

Movement Off Target

Herbicides are chemicals that injure or kill weedy plants. Since the environment contains many different kinds of plants, off-target vegetation provides a natural bioassay system in which straying herbicides can be detected easily. Each time injurious quantities of herbicide move or drift off target, the result is a *highly visible* symptom expression on nearby sensitive plants and an advertisement to the whole community that the professional applicator or homeowner has made a mistake.

Prevention of off-target injury from herbicides is essential if their full potential as weed control agents is to be realized without complaints, attempts at governmental restrictions, and open hostility toward herbicide use from the general population.

Herbicide injury to off-target vegetation most frequently occurs from volatilization and vapor drift and from spray drift. Horizontal movement of herbicides through the soil or on the soil surface away from the target site also can injure plants; downward movement into groundwater is another environmental concern.

Volatilization and Vapor Drift

Volatilization and *vapor drift* occur when the herbicide is placed on target and then evaporates (volatilizes). The herbicide vapors move away from the target site and damage nearby plants. The vapor pressure (tendency to evaporate) of the herbicide is a major factor in determining potential volatility problems.

Vapor pressures vary greatly among herbicides and herbicide groups. Herbicides with acid or phenol groups as part of their chemistry (for example, phenoxy acids) can be formulated with varying degrees of volatility. These herbicides can be synthesized as the parent acid or phenol compound or as inorganic salts (e.g., sodium, potassium, lithium, or ammonium salts), amine salts, low volatile esters, or high volatile esters of the parent compounds. The order of increasing volatility of these compounds is: salts (least volatile), acids or phenols, low volatile esters, and high volatile esters. Obviously, *one of the best ways to reduce vapor movement is to use the salt formulations* of these compounds.

The minimum dosage needed to elicit a visible response in a plant is an important factor in damage to off-target vegetation. For example, the auxin-type growth regulator herbicides that include the phenoxies, benzoics, and picolinic acid derivatives cause plant responses at dosages well below those required for kill. Therefore, the volatilization of small amounts of these herbicides can result in appreciably more damage to off-target vegetation than the volatilization of compounds that elicit responses only as they approach herbicidal levels.

Wet surfaces of plants and soils can result in increased herbicide evaporation and movement off target because the moisture tends to compete for adsorptive sites that normally would hold the herbicide in place. Increasing temperatures and air movement also speed up volatilization. A classic example of off-target injury due to volatilization is that of clomazone, particularly when applied to wet soil surfaces. Vapor drift of clomazone to nontarget species results in a highly visible whitening of the vegetation. Such injury has been greatly reduced by the microencapsulation of clomazone so that the herbicide is surrounded by a polymer shell that prevents evaporation. Incorporation of clomazone into the soil rather than surface application also reduces off-target movement.

Spray Drift

Spray drift is the movement of herbicides off target following their release from application equipment. In contrast to volatilization, the chemical never reaches the intended target. Spray drift is caused by the presence of small-diameter spray droplets that fall slowly and are easily moved off target by air currents. In addition, water can evaporate from the spray droplets, leaving solid herbicide particles suspended in the air (aerosols) and susceptible to air movement.

Herbicide drift can be effectively reduced by using appropriate equipment or spray additives that eliminate small droplets (see Chapter 17 on Herbicide Application).

Movement On or Through Soil

The lateral movement of herbicides in surface water away from the target site can occur when applications are followed immediately by heavy rains. Injury to nontarget plants is most obvious with highly persistent herbicides. Injury to plants usually occurs over relatively short distances, generally less than 100 feet and frequently only a few feet. The problem can be avoided by applying the herbicide when torrential rains are not predicted. Incorporation of the herbicide into the soil will reduce its lateral movement; however, incorporation may not be an option on sites where highly persistent herbicides frequently are used, such as industrial sites and along railroad tracks. If problems persist, the applicator should consider switching to a safer chemistry. Excess doses also add to lateral movement.

Groundwater Issues

Another environmental concern is not due to injury to nontarget vegetation but rather to the presence of herbicides in groundwater. The downward leaching of herbicides is influenced by many factors (see Chapter 7 on Soil–Herbicide Interactions) and is not a problem for the vast majority of herbicides. However, atrazine and other triazine herbicides have been detected in groundwater. Most detections seldom exceed health standards established by the Environmental Protection Agency (EPA); however, at times the levels detected do exceed federal safety standards.

As the most widely used herbicides in the midwestern corn-growing region, triazines account for more than 12% of the total pesticide use in the United States; in 1991, atrazine was used on approximately 66% of the corn acreage (Gianessi and Puffer 1991). Atrazine has been detected in ground and surface waters including the river basins of major midwestern rivers, such as the Mississippi River. Groundwater monitoring by the EPA in the 1980s detected atrazine in approximately 4% of nearly 15,000 wells.

Atrazine labels were revised in 1992 to reduce the overall amount of atrazine used, reduce the conditions that increase the potential for atrazine runoff from treated fields, and protect surface water sources of drinking water. Restrictions were placed on the total amount of atrazine that could be applied to an individual field within a calendar year (amounts were reduced even further for application to highly erodible soils), and various setbacks were established for mixing or applying atrazine near wells, sinkholes, surface bodies of water, and areas of potential runoff. For example, atrazine cannot be mixed, loaded, or applied within 50 ft of any wells, including abandoned wells, drainage wells, and sinkholes. In addition, all products containing atrazine are designated restricted-use pesticides, which means that purchasers must show proof that they currently hold a valid pesticide applicator certification permit.

Another triazine, cyanazine, is being gradually removed from the market and will not be sold after 1999; however, existing supplies can be used through 2002. The status of atrazine, other triazine herbicides, and other herbicides (specifically the chloroacetamide herbicides alachlor, metalochlor, and acetochlor) with the potential to be found in groundwater is continuously reviewed by the EPA and subject to additional regulation as the need arises.

The removal of atrazine use would have an adverse economic impact on the agricultural community and cause the loss of a very effective herbicide. Although other herbicides could be substituted for atrazine, concerns about the most likely alternatives, the branched-chain amino acid inhibitors (for example, the sufonylurea and imidazolinone herbicides), include environmental persistence and the potential for weed resistance.

Toxicity and Exposure

One of the basic principles integral to all chemicals is that organisms (or biological systems) exposed to them follow typical dose-response curves. No response is observed until some minimum dosage is reached. Above this level, the organism shows a response proportional to the amount of chemical present, so that responses increase with increasing dosage until the maximum possible response is attained (frequently death). Increasing dosages above this maximum response level should yield no additional effect. The fact that there is a concentration below which no response will be elicited should always be kept in mind when interpreting statements related to the effects of herbicides on nontarget organisms and human health. This principle (the dose makes the poison) is true for any herbicide, whether present in the soil, water, or plant or animal systems.

The Environmental Protection Agency (EPA) is assigned the task of overseeing pesticide testing procedures for human and environmental effects. Before a pesticide can be sold in the United States, it must be registered by the EPA, which may require over 100 safety and environmental tests for each potential pesticide. Data submitted to the EPA must include the chemical and physical properties of the pesticide, its fate in the environment, a determination of herbicides and herbicide metabolites in feed and food crops, toxicological information, and ecological studies to determine the effects on nontarget plants and animals (Table 5.2). These data usually are required not only for the herbicide itself but also for its major breakdown products. This allows the EPA to evaluate the relative benefits and risks of herbicide use.

Human Health

A large portion of the development of a herbicide is devoted to determining its effect on human health. These effects usually are expressed in terms of toxicity and exposure. *Toxicity* is a measure of the amount of compound that is harmful or lethal. *Exposure* is the probability of encountering a harmful dose of a compound. A compound that is extremely toxic but is used in such a way that exposure is low may be less hazardous than a compound that is less toxic but whose application requires greater exposure of the applicator.

Potential adverse effects on humans are derived from toxicology studies on experimental animals such as rats, rabbits, and dogs. Compounds also are evaluated for their potential to cause reproductive problems and for possible carcinogenic (cancer-producing), teratogenic (birth defect-producing), and mutagenic (genetic abnormality-producing) properties. These studies are conducted by the company or by independent laboratories on behalf of the company seeking registration of a specific herbicide, and all data are submitted to the EPA for review. Products that meet the evaluation criteria are given an EPA registration number. Only products with an EPA registration number can be offered for sale to the public.

Data submitted to EPA must show the amounts of herbicide and herbicide breakdown products present in food, feed crops, and animals grazing on treated areas or feed crops in order to establish tolerances. A *tolerance level* is designated as the maximum amount of herbicide allowed on the crop. The tolerance level is then multiplied by the amount of the crop or food product an individual might consume over a period of time (the exposure). This value is compared with data from toxicology tests to determine if there is a human health hazard (see the following discussion of RFD).

From a historical standpoint, one of the most significant impacts on human health regulations was the 1958 Delaney clause found in the Federal Food, Drug, and Cosmetic Act. This clause prohibited the government from setting pesticide residue tolerances for processed foods when the pesticide was found to accumulate in processed foods and cause cancer in humans or experimental animals. Thus, no pesticide could be sold in the United States if it was shown to accumulate in processed foods and have any potential cancer-

Table 5.2 Partial List of Data Required by EPA for Pesticide Registration

Formulation data	Aerobic soil metabolism	Lifetime mouse feeding
Analytical methods	Anaerobic aquatic metabolism	Mutagenicity: dominant lethal gene test
Boiling point	Anaerobic soil metabolism	Mutagenicity: heritable translocation
Chemical nomenclature	Ecosystem study	Teratology: rat & rabbit
Color	Hydrolysis	Reproduction
Corrosiveness	Leaching, absorption, decomposition	Subchronic rat & dog feeding
Explosiveness	Off-target movement	**Hazards to nontarget plants & animals**
Flammability	Photodegradation: water, soil, & air	Aquatic ecosystem
Manufacturing process	Rotational crop residue & response	Avian oral studies
Melting point	Soil dissipation	Bee studies
Odor	Water dispersal	Beneficial insects
pH	**Crop residue & exposure data**	Bluegill accumulation
Purity of materials	**Hazards to human & domestic animals**	Bluegill & trout toxicity
Solubility	Acute delayed neurotoxicity	Catfish accumulation
Specific gravity	Acute oral toxicity	Crab toxicity
Stability	Acute inhalation toxicity	Daphnia toxicity
Structural formula	Eye irritation	Fish & invertebrate life cycle
Storage stability	Dermal irritation	Mallard & quail dietary toxicity
Vapor pressure	Dermal sensitization	Mallard & quail reproduction
Environmental chemistry data	Excretion, metabolism, accumulation	Marine fish toxicity
Absorption	First aid & antidote	Micropods
Aerobic aquatic metabolism	Lifetime rat feeding	Oyster, shrimp, and worm toxicity

causing activity, no matter how low the cancer-causing dosage might be or how low the chances were of a person coming into contact with or consuming that dosage. Interestingly, the clause was only established for "processed" foods (e.g., tomato paste, tomato sauce), not unprocessed or "raw" crops (e.g., tomatoes from a farm field).

This inconsistency plus other problems led to the elimination of the Delaney clause from federal law in 1996. At that time a new set of governmental regulations under the Food Quality Protection Act (FQPA) was established. This act regulates both raw and processed foods similarly under the wording "a reasonable certainty that no harm will result from aggregate exposure to the pesticide chemical residue, including all anticipated dietary exposures and all other exposures for which there is reliable information." Thus, the likelihood of exposure to a deleterious quantity of the pesticide became an important factor in determining the health risks of the pesticide.

Other provisions of the act include more complete evaluations of the effects of pesticides on infants and children and screening for estrogen-mimicking pesticides. In addition, the EPA must produce and distribute to grocery stores a pamphlet that describes what steps consumers can take to lessen their dietary risks from pesticides.

Human Toxicology. Potential toxicity to the user is generally expressed in terms of the acute toxicity and chronic toxicity. *Acute toxicity* is the potential of a substance to cause injury or illness shortly after exposure; *chronic toxicity* is the potential of a substance to cause injury or illness after repeated exposure over an extended period of time. Acute toxicity is generally expressed as an oral LD_{50}. The oral LD_{50} is the single dose in mg/kg (mg of compound per kg of body weight) taken by mouth that will kill 50% of the population of test animals. The test animal often used to develop the LD_{50} is the albino rat. LD_{50}s also are established to determine dermal toxicity (toxicity upon exposure to the skin). The categories of acute toxicity

for herbicide labeling are shown in Table 5.3. A list of the oral LD_{50}s of herbicides is provided in Table 5.4. The majority of herbicides are less toxic than aspirin, and many are less toxic than table salt. In some cases the ingredients used to formulate the product are more toxic than the herbicide itself. For example, trifluralin formulated as an emulsifiable concentrate has an LD_{50} of 3700 whereas the LD_{50} of technical grade trifluralin is greater than 5000 mg/kg.

Subchronic (90-day exposures) and chronic toxicities (12-, 18-, and/or 24-mo exposures), reproductive effects, teratogenicity and several other categories of effects are expressed as NOAELs. A NOAEL is the No-Observable Adverse Effect Level; in other words, it is the highest pesticide dose that does not cause observable harm or side effects to experimental animals. The NOAEL is expressed as mg/kg/d (mg of compound per kg of body weight per day). In general, the safety level for human exposure to a pesticide is set 100 to 1000 times below the NOAEL in order to provide a significant safety margin. This new value is called the Reference Dose (RFD). The RFD serves as the lowest toxicological value such that exposure below the RFD is not expected to cause harm to humans

For example, the NOAEL for chronic toxicity from a herbicide when fed to rats over a 24-mo period might be 400 mg/kg/d. With the hundred fold safety factor, the RFD for the herbicide is 4 mg/kg/d. Therefore, if the tolerance level for that herbicide on a food crop is 6 mg/kg, and an individual with a body weight of 70 kg has the potential to eat 0.5 kg of the crop per day, then the calculation (6 mg/kg times 0.5 kg/day divided by 70 kg) results in a value of 0.43 mg/kg/d, well below the RFD.

Carcinogenicity. The EPA has established an alphabetic lettering system for categorizing the potential of pesticides to cause cancer (Table 5.5). Pesticides with the greatest potential to cause cancer are classified as Category A compounds; pesticides in which no evidence of car-

Table 5.3 Toxicity Categories for Herbicide Labeling[a]

Signal Word[b]	Toxicity	Oral LD_{50} (mg/kg)	Dermal LD_{50} (mg/kg)	Approximate Amount Needed to Kill Average Adult
DANGER	Very high toxicity	50 or less	200 or less	A taste to a teaspoonful
WARNING	High toxicity	51 through 500	201 through 2000	A teaspoonful to an ounce
CAUTION	Moderate to low toxicity	Greater than 500	Greater than 2000	Greater than an ounce

[a]Adapted from EPA toxicology guidelines.
[b]All labels must state "Keep Out of Reach of Children."

Table 5.4 Mammalian Toxicity Groupings of Commonly Used Herbicides[a] (and Some Commonly Used Substances)

	Herbicide[b]	Oral LD$_{50}$ mg/kg to Rats[c]
Oral LD$_{50}$ = approximately 50 mg/kg or less	Acrolein	29
Oral LD$_{50}$ = approximately 50 to 500	Bromoxynil	440
	(Caffeine)	192
	Copper sulfate	470
	Cyanazine	182
	Diquat	230
	Endothall (amine salt)	206
	Haloxyfop acid	337
	(Nicotine)	53
	Paraquat	112
Oral LD$_{50}$ = approximately 500 to 5000	(Aspirin)	1240
	(Table salt)	3320
	Acetochlor	2148
	Aciflurofen	1540
	Alachlor	930
	Atrazine	3090
	Bensulide	770
	Bentazon	1100
	Butachlor	2000
	Butylate	4659
	Cacodylic acid	2756
	Chlorimuron	4102
	Clethodim	1630
	Clomazone	2077
	Clopyralid	4300
	Cycloate	3200
	2,4-D	764
	2,4-DB	1960
	Dazomet	650
	Diallate	1050
	Dicamba	1707
	Dichlobenil	4460
	Dichlorprop	800
	Diclofop methyl ester	557
	Difenzoquat methyl sulfate salt	617
	Diuron	3400
	DSMA	1935
	EPTC	1652
	Fenoxaprop ethyl ester	3310
	Fluazifop-P butyl ester	4096
	Fomesafen acid	1250
	Glufosinate	2170
	Hexazinone	1690
	Linuron	1254
	MCPA acid	1600
	MCPB acid	680
	Mecoprop	650
	Mefluidide	>4000
	Metolachlor	2877
	Metribuzin	1090
	Molinate	720
	MSMA 51% w/v aqueous	2833
	Naptalam (Na salt)	1770
	Pebulate	1675
	Prometon	4345
	Prometryn	4550
	Propachlor	1800
	Propanil	1080
	Pyrazon	2200
	Pyridate	4690
	Quinclorac	>2610
	Quizalofop ethyl ester	1670
	Sethoxydim	2676
	Tebuthiuron	644
	Terbacil	1255
	Thiobencarb	1033
	Triallate	1100
	Triclopyr	713

Table 5.4 (Continued)

	Herbicide[b]	Oral LD$_{50}$ mg/kg to Rats[c]
Oral LD$_{50}$ = approximately 5000 or greater[d]	Amitrole	>5000
	Asulam	>5000
	Benefin	>5000
	Bensulfuron	>5000
	Bifenox	>5000
	Bromacil	5175
	Chlorsulfuron	5545
	DCPA	>10,000
	Desmedipham	>10,250
	Dithiopyr	>5000
	Ethalfluralin	>10,000
	Ethametsulfuron	>5000
	Ethofumesate	>6400
	Flumetsulam	>5000
	Flumiclorac pentyl ester	>5000
	Fluometuron	6416
	Fluridone	>10,000
	Fosamine NH$_4$ salt	24,400
	Glyphosate	5600
	Imazamethabenz methyl ester	>5000
	Imazapyr acid	>5000
	Imazaquin	>5000
	Imazethapyr	>5000
	Isoxaben	>10,000
	Lactofen	5960
	Metsulfuron	>5000
	Napropamide	>5000
	Nicosulfuron	>5000
	Norflurazon	9000
	Oryzalin	>5000
	Oxadiazon	>5000
	Oxyfluorfen	>5000
	Pendimethalin	>5000
	Phenmedipham	>8000
	Picloram	>5000
	Primisulfuron	>5050
	Prodiamine	>5000
	Pronamide	>16,000
	Siduron	>7500
	Simazine	>5000
	Sodium chlorate	5000
	Sulfometuron	>5000
	Thiazopyr	>5000
	Thifensulfuron	>5000
	Triasulfuron	>5000
	Tribenuron	>5000
	Trifluralin	>5000
	Triflusulfuron	>5000

[a]Data taken from *Herbicide Handbook* of the Weed Science Society of America, Lawrence, Kansas (1994).
[b]All data are for the technical grade compound.
[c]LD$_{50}$ may vary according to sex of rat. Check the *Herbicide Handbook* of the WSSA.
[d]For herbicides with an oral LD$_{50}$ above 5000, the amount needed to kill an average adult is approximately a pint to a quart.

cinogenicity has been detected in at least two animal studies are classified as Category E compounds (however, this classification system is currently under review and is subject to change). The majority of herbicides are in the C to E categories (Table 5.5).

Fate and Impact on the Environment

Toxicology studies are required on certain wildlife species of birds, fish, and invertebrates (Table 5.2). In addition, the exposure of wildlife to pesticide residues in the environment must be estimated. The Estimated Environmental Concentration (EEC) is determined by measurements of pesticide residues in the foods that birds and mammals might consume and the concentrations of the pesticide in water caused by runoff or spray drift that an aquatic organism might come in contact with. These studies may be conducted in the field and require periodic sampling and analysis;

Table 5.5 Lettering System Used by the EPA to Categorize Chemicals for Their Carcinogenic Potential and the Number of Herbicides Under Each Category as of July 15, 1996[a]

Category	Carcinogenic Description	Evidence	Number of Herbicides
A	Human	Sufficient evidence of cancer/ causality from human studies	0
B[1]	Probable human	Limited evidence of cancer from human studies	0
B[2]	Probable human	Sufficient evidence of cancer from animal studies	6
C	Possible human	Limited evidence of cancer from animal studies	33
D	Not classifiable	Absence of data or inadequate evidence of carcinogenicity	16
E	Noncarcinogenic	No evidence of carcinogenicity in at least two animal studies	21

[a]System of nomenclature currently under review.

however, many of these fate studies also can be done with computer models. The EEC and toxicity data are used in calculations to determine the potential for harm to nontarget species.

Toxicity data for aquatic species are expressed in LC_{50} (concentration of the compound in water in mg/l that causes mortality to 50% of the population after exposure). A list of herbicide toxicities to fish is given in Table 5.6. Compounds with an LC_{50} of less than l mg/L are highly toxic to fish and should not be permitted to enter directly into water. The majority of herbicides pose little risk to wildlife and fish near or in water; however, the dinitroaniline herbicides (trifluralin, benefin, ethalfluralin, oryzalin, pendimethalin, prodiamine) are quite toxic to fish. These products are not a problem when used correctly. Label instructions alert the applicator to potential problems with these herbicides, and generally fish toxicity occurs when these herbicides are misused, for example, when they are sprayed directly over water or are allowed to contaminate water when filling the spray tank.

In addition to impacts on wildlife, the general question of environmental fate must be provided for all pesticides that are approved by the EPA. Data required include the rate of pesticide degradation under various environmental conditions, the identity of the breakdown products, and the likelihood that a pesticide or its metabolites will migrate and/or accumulate in the environment. Specific tests or computer simulations are conducted to determine hydrolysis (breakdown in water), photodegradation in water and soil, aerobic soil and anaerobic aquatic metabolism, mobility via runoff and leaching, and terrestrial field dissipation. These data also are used in various ways to calculate the EEC.

General Considerations

Most herbicides used according to label directions are relatively safe to the user, wildlife, and the environment. The reasons are as follows:

1. The toxic effects of herbicides are relatively specific for plant processes (for example, photosynthesis). Therefore, they tend to be low in mammalian toxicity.
2. Most herbicides are susceptible to breakdown by microbes, plant tissues, and physical processes. Therefore, there are few problems associated with long-term persistence.
3. Herbicides tend to be either water soluble or, as noted earlier, susceptible to breakdown. Therefore, little or no buildup of herbicides or their residues occurs in the fatty tissue of animals.
4. Most herbicides are of relatively low toxicity to fish and other aquatic organisms. Those compounds with high toxicity tend to be soil adsorbed and seldom move into water systems. Fish toxicity problems can occur when a herbicide accidentally is spilled into a stream. Sometimes applications to ditch banks and subsequent runoff can move significant quantities of a herbicide into the water; such applications must be avoided.

All herbicides are poisonous to some degree and must be handled with care. In all cases, the applicator should read and consult the label for specific directions and precautions. Data

Table 5.6 Toxicity of Herbicides to Fish[a]

Herbicide[b]	96 h LC_{50} (mg/l)	
	Bluegill[c]	Rainbow trout[c]
Actochlor	1.3	0.45
Acifluorfen	62	17
Acrolein	0.024	0.024
Alachlor	2.8	5.3
Atrazine	42	9.9
Bensulfuron	>150	>150
Bentazon	616	190
Chlorimuron	>100	>1000
Chlorsulfuron	>250	>250
Clethodim	33	18
Clomazone	34	19
Copper sulfate	44 (0.884 in soft water)	0.135 (in soft water)
Cyanazine	23	9
2,4-D dimethylamine salt	524	250
2,4-D isooctyl ester	>5	>5
Dicamba	135	135
Diclofop methyl ester	—	0.35
Diquat dibromide salt	175	21
Endothall potassium salt	308	187
Endothall amine salt	0.94	0.12
EPTC	27	19
Ethalfluralin	0.012	0.0075
Fenoxaprop	3.34	—
Fluazifop-P butyl ester	0.53	1.37
Fluridone	14.3	11.7
Glufosinate	—	>320
Glyphosate acid	120	86
Haloxyfop acid	548	0.4
Imazapyr acid	>100	>100
Imazaquin acid	>100	>100
Imazethapyr acid	420	340
Lactofen	>560	>0.1
Linuron	16	16
Mefluidide	>100	>100
Metolachlor	10	3.9
Metribuzin	80	76
Metsulfuron	>150	>150
Nicosulfuron	>1000	>1000
Oryzalin	2.88	3.26
Oxyfluorfen	0.2	0.4
Paraquat	—	32
Pendimethalin	0.199	0.138
Primisulfuron	>180	210
Prodiamine	>0.552	>0.829
Quizalofop	0.46–2.8	10.7
Sethoxydim	100	32
Simazine	>32	70.5
Sulfometuron	>12.5	>12.5
Thifensulfuron	>100	>250
Triclopyr	891	552
Trifluralin	0.058	0.041

[a]Data taken from *Herbicide Handbook* of the Weed Science Society of America, Lawrence, Kansas (1994).
[b]All data are for technical grade compound.
[c]Bluegills and trout were chosen as representative of warm-water and cold-water species, respectively.

developed for registration by the EPA is used to develop the directions on the label. In addition to the label, regulations from the Occupational Safety and Health Administration (OSHA) mandate that a Material Safety Data Sheet (MSDS) be available for all people who handle, transport, or work with the product. The MSDS provides information on hazardous ingredients, physical and chemical properties, accompanying hazards, primary routes of entry into the body, exposure limits, emergency first aid procedures, and responsible party contacts.

In addition to information provided by the label, the MSDS, and training programs, good common sense always is required when herbicides are being used.

Herbicide Registration

Pesticides cannot be sold legally until registered by the EPA. Potential herbicidal compounds undergo extensive and rigorous testing for efficacy, crop and weed susceptibility, environmental hazard, and effects on animal systems. Most of the developmental work (discovery, toxicology, efficacy) is conducted by the manufacturer, with additional research (consistency of performance, efficacy in specific management systems) contributed by universities and private research groups. The cost of developing a new herbicide may be as much as $50 million to $100 million. Approximately 7 to 10 yr of testing usually are required before a compound can be marketed. Since patents protect the use of a herbicide for 17 yr, only 10 yr or less of exclusive marketing can be expected. The financial risks are indeed great for a chemical company, since few compounds that are tested prove to be efficacious with low risks to humans and the environment. On average, only one in 35,000 chemicals survives from first discovery to the marketplace.

A substantial portion of the data obtained during the research and development stage is used to obtain registration for the product. The Federal Insecticide, Fungicide and Rodenticide Act (FIFRA) requires that all pesticides be registered before they can be sold to the public; offering for sale any unregistered product is a serious breech of the law. The responsibility of the EPA is to ensure that the pesticide poses no undue environmental and health hazards when used as instructed. Registration also is contingent on the development of a label that contains sufficient information and warnings to provide safe and correct handling and use of the product.

Once registration is granted, the label becomes a legal document that gives the manufacturer the right to sell the product. Under the law, any person misusing or recommending use of a pesticide contrary to label instructions is liable to prosecution and possible fines, a jail sentence, or both. Some flexibility in label interpretation is permissible, however. For example, it is legal to use herbicides at rates lower than those recommended on the label. Herbicides also can be used to control weeds not listed on the label as long as the application is to the crop site specified on the label. However, herbicide labels may state that the company will not guarantee the results of the product when these options are used.

Besides the federal registration for each use, the herbicide also must be registered with the appropriate individual state regulatory agency.

The state lead agencies, such as the state Departments of Agriculture, are given the task of enforcing both state and federal pesticide laws and providing pesticide applicator training and certification. States can impose additional restrictions and regulatory oversight on the use of pesticides. For example, in California, county commissioners must be notified at least 24 h in advance of the intention to use a restricted-use pesticide. The application for a permit, which is issued for each site and time that the pesticide is used, allows the commissioner to determine whether an adverse environmental or human health impact could result from the use of the pesticide.

Reregistration

The EPA is required by law to reregister existing pesticides that were originally registered prior to November 1, 1984, when the standards for approval and test data requirements were much different than they are today. The goal is to generate a complete scientific database for each pesticide. A comparison of the information in the *Herbicide Handbook,* published in 1994 by the Weed Science Society of America, with earlier editions from the 1970s and 1980s shows that gaps in many data sets regarding behavior in plants, behavior in the soil, and particularly toxicological properties have been filled in.

One of the results of reregistration was that the uses of some herbicides on certain crops were dropped or restricted in some way. In some cases the data indicated potential persistence or toxicological problems with the use of the herbicide on certain crops; in other cases, the company did not pursue reregistration for certain crop uses because development of the data was more expensive than the financial return to be gained from selling the herbicide for that crop. Examples of herbicide uses that were eliminated include simazine use on artichoke, alfalfa, pineapple, and sugarcane and alachlor use on cotton and potatoes. Alachlor also was reclassified from a general-use pesticide to a restricted-use pesticide.

In a few cases, the new data required for reregistration resulted in the removal or reduction of label restrictions on herbicides. For example, water treated with diquat for underwater weed control could not be used for 14 days for drinking, livestock consumption, and irrigation of food crops. Swimming was restricted for one day. New data showed that diquat residues in natural waters were so short lived that the restrictions were reduced to 0 days for swimming, 1–3 days for drinking, 1 day for livestock consumption, and 5 days for irrigation of food crops.

Experimental Use Permits

During the last stages of pesticide development, a company usually requests a permit from the EPA to use a limited amount of the compound. The amounts to be used and the geographic location for distribution must be designated. Once distributed, detailed records are kept of the amounts used by each applicator and the results. Depending on the compound and its progress through registration, the crop may be destroyed. An experimental use permit (EUP) usually is sought 1 to 2 yr prior to full registration and is helpful to the company in obtaining performance information based on actual user conditions.

General- and Restricted-Use Pesticides

FIFRA also authorizes the EPA to classify all pesticides into general- and restricted-use categories. *General-use pesticides* are those products that are not expected to cause unreasonable adverse effects on the environment and to humans. Any person can purchase and use general-use pesticides. *Restricted-use pesticides* are those that can be sold only to applicators trained and certified in the handling and use of these pesticides to avoid possible adverse health or environmental effects. EPA-designated state agencies are responsible for administering the certification test.

The EPA requires commercial applicators who use restricted-use pesticides to record pertinent information such as crops and sites treated and the amount of restricted-use pesticide applied. The U.S. Department of Agriculture (USDA) requires private applicators (e.g., farmers) to keep such records. States may require additional record keeping, including records on general-use pesticides.

Many herbicides are classified as general-use pesticides. Some restricted-use herbicides include atrazine, alachlor, diclofop, paraquat, and picloram. Most fumigants are restricted-use pesticides because of the hazards involved with their handling.

Special Registration Procedures

Because of special needs or requirements, the registration process can be facilitated or hastened in several ways.

Section 24(c) of FIFRA. The 24(c) section of the federal regulation provides the designated state regulatory agency with the means of registering a pesticide for a special local need. The need for the product usually is ascertained by state experts through contact with growers, industry groups, and researchers. The manufacturer of the product must seek approval from the state; once state approval has been granted, the EPA has 90 days to respond. If not challenged by the EPA, the use becomes a legal part of the chemical label for *that state only.* The user must have a copy of the new label at the time of application.

Section 18 of FIFRA. Section 18 of the federal regulation provides for the approval of a product for an emergency situation when no other adequate control measure is available. An emergency local need is requested of the EPA by the designated state regulatory agency, usually for a product not labeled for the specific situation or for a new product that has not been labeled federally. A section 18 clearance is given only for a specific length of time because its purpose is for an emergency pest outbreak.

The IR-4. The IR-4 (Interregional Research Project No. 4) serves as a mechanism for generating and compiling data for herbicides with potential uses in low-volume markets such as limited acreage crops. The cost of conducting the research and developing a compound for a minor acreage crop often is greater than the potential profit, so manufacturers are hesitant to commit the necessary funding for label development and registration. If a need for a particular herbicide use for a minor acreage crop can be ascertained by state and national experts, federal assistance in obtaining data for efficacy and crop residue analysis can be provided to state researchers. The IR-4 petition and supporting data are then submitted to the EPA, and if approved, full federal registration of the product for the minor crop will be obtained. The company prints the new uses on the label and assumes all liabilities associated with the new uses.

MAJOR CONCEPTS IN THIS CHAPTER

Nonselective chemical weed control has been obtained for centuries with high rates of salt, ashes, and smelter wastes. Selective removal of weeds from crops was demonstrated at the turn of the century.

The discovery of the herbicidal properties of 2,4-D during the 1940s triggered the development of several hundred compounds currently available for weed control.

Key developments in the history of herbicides include the introduction of effective soil-applied herbicides such as atrazine; herbicides such as paraquat that control weeds in no-till systems; glyphosate, a broad-spectrum herbicide that provides effective perennial weed control; selective postemergence herbicides that are effective at extremely low dosages; and herbicide-resistant crops.

Herbicides considered outstanding have a broad spectrum of uses and are effective at low rates, economical to manufacture, safe and easy to handle and apply, and relatively benign in the environment.

The informed herbicide user should be familiar enough with individual herbicides to understand the principles involved in their use and to predict their interactions with plants and soils.

Herbicides have three names listed on the package: a chemical name, a common name, and a trade name. The user must be able to recognize either the common or chemical name in order to be certain of the active ingredient present, particularly when multiple labels and formulations are available.

Herbicides are classified several ways depending on their characteristics or use at any one time. Classification categories include degree of response among plants, coverage of the plant or soil, length of activity in the soil, coverage of the target area, application time with respect to weed and crop development, methods of soil application, characteristics related to plant response (translocation, mode of action), and chemistry.

Most modern herbicides are organic compounds (contain carbon). The basic structural components of most organic herbicides are carbon chains (aliphatic groups) and rings (aromatic groups). Other elements commonly present include hydrogen, oxygen, nitrogen, chlorine, phosphorus, sulfur, and fluorine.

The chemistry of a compound determines how the herbicide will act in biological and physical systems such as plants, soils, and water.

Many herbicides contain an acid group. The formulated product may be available as the acid; as a salt of sodium, potassium, lithium or ammonium; as an amine salt; or as an ester.

Salts are soluble in water, insoluble in oil, and nonvolatile, whereas esters are insoluble in water, soluble in oil, and are moderately to very volatile. Acids are intermediate between salts and esters with respect to these properties.

Injury to off-target vegetation occurs from volatilization and vapor drift, spray drift, or movement of the herbicide through or on the surface of the soil and frequently is highly visible. Herbicide movement can and should be minimized in all situations.

Volatilization (vapor formation) and vapor movement can be minimized by selecting nonvolatile formulations rather than volatile ones (e.g., salts or amines rather than esters).

Movement of some herbicides (e.g., certain members of the triazine group) into groundwater has resulted in restrictions being placed on their use.

Before a pesticide can be sold in the United States, it must be registered with the EPA.

Some of the data needed for herbicide registration include (1) the amount and kind of toxicity of the herbicide and its stable metabolites to mammals, (2) herbicide and herbicide metabolite levels in feed and food crops, and (3) its fate in the environment.

A tolerance level, which is established for each herbicide, is the maximum amount of herbicide allowed on a crop.

The potential for human health problems is evaluated by tests for acute toxicity, chronic toxicity, carcinogenicity, reproductive problems, and mutagenic properties on test animals.

The value obtained by multipying the tolerance level of the herbicide on the crop by the potential exposure of an individual to that crop is compared to toxicology data when determining the potential for the herbicide to cause human health problems.

Toxicology studies on certain species of wildlife and fish and the potential exposure of wildlife and fish to the herbicide in the environment are used to determine the impact on non-target species.

The effect of a herbicide on the environment is evaluated by studying the rate of degradation under various environmental conditions, the identity of breakdown products, and the likelihood that a herbicide or its metabolites will migrate and/or accumulate in the environment.

Each herbicide label contains sufficient information to allow the product to be properly and safely applied. The label should be consulted each time an application is made.

Each herbicide use must be approved by the EPA and state regulatory agencies. In most cases EPA approval satisfies state requirements. Product uses may be labeled on national, regional, or state bases.

Herbicide uses are designated as either general use or restricted use, depending on the hazards they present to the environment and applicators. Few herbicides are classified as restricted use; high toxicity or highly persistent effects account for most restricted-use designations.

Experimental use permits (EUP), special local need permits [24(c)], and emergency local need permits (section 18) can be provided by the EPA to allow limited use of herbicides when uses are not covered by current federal labels.

The agricultural advisor, applicator, or grower has a moral obligation to use or recommend only those herbicides that have label approval. Failure to do so justifiably invites legal and financial risks and embarrassment.

TERMS INTRODUCED IN THIS CHAPTER

Active ingredient
Acute toxicity
Aliphatic compound
Aromatic (ring) compound
Band treatment
Bordeaux mixture
Broadcast treatment
Carcinogenicity
Chemical name
Chronic toxicity
Common name
Cracking (at emergence)
Directed postemergence
Early postemergence
Early preplant herbicide
Environmental Protection Agency (EPA)
Ester formulation
Estimated Environmental Concentration (EEC)
Experimental Use Permit (EUP)
Exposure
Federal Insecticide, Fungicide and Rodenticide Act (FIFRA)
Foliar application
General-use herbicide
Groundwater
Herbicide registration
Herbicide reregistration
Highly persistent herbicide
Incorporation
IR-4
Late postemergence
Layby application
LC_{50}
LD_{50}
Material Safety Data Sheet (MSDS)
Nonselective herbicide
No-Observable-Adverse-Effect Level (NOAEL)
Organic herbicide

Persistent herbicide
Postemergence herbicide
Preemergence herbicide
Preplant herbicide
Preplow herbicide
Reference dose (RFD)
Residual herbicide
Restricted-use herbicide
Salt formulation
Section 18
Section 24(c)
Selective herbicide
Soil application
Spot treatment
Spray drift
Surface application
Tolerance level
Toxicity
Toxicology
Trade name
2,4-D
Vapor drift
Volatilization

SELECTED REFERENCES ON HERBICIDE HISTORY, CLASSIFICATION, AND ISSUES

Ashton, F. M. and T. J. Monaco. 1991. Weed Science, Principles and Practices, 3rd ed. John Wiley and Sons, New York.

EPA, 1993. *Agricultural Atrazine Use and Water Quality: A CEEPES Analysis of Policy Options.* U. S. Environmental Protection Agency, Office of Program and Policy Evaluation, Water and Agricultural Policy Division, Agricultural Policy Branch.

Gianessi, L. P. 1992. U.S. Pesticide Use Trends: 1966–1989. Quality of the Environment Division, Resources for the Future, 16168 P Street, N.W., Washington, D. C. 20036.

Gianessi, L. P. and C. Puffer. 1991. Herbicide Use in the United States, National Summary Report. Quality of the Environment Division, Resources for the Future, 16168 P Street, N.W., Washington, D. C. 20036.

Herbicide Handbook of the Weed Science Society of America, 7th ed. 1994. Weed Science Society of America, Lawrence, KS.

King, J. 1991. A matter of public confidence—Consumers' concerns about pesticide residues unjustified. Agric. Eng. 72 (14): 16–18.

Radosevich, S., J. Holt, and C. Ghersa. 1997. Chapter 11, Weed Control in a Social Context, In Weed Ecology, 2nd ed. John Wiley and Sons, New York.

Roe, R. M., J. D. Burton, and R. J. Kuhr. 1997. Toxicology, Biochemistry, and Molecular Biology of Herbicide Activity. IOS Press, Amsterdam, The Netherlands.

Information also is available on herbicide labels and in technical literature.

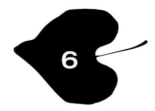

PLANT–HERBICIDE INTERACTIONS

The major prerequisites for effective herbicide use are that the herbicide *come in contact with the plant surface, remain at the plant surface* long enough to penetrate or be absorbed into the plant, and *reach a living, cellular site* where it can disrupt a vital process or structure. Once the herbicidal function is complete, degradation of the compound to simpler nontoxic components occurs either in the plant or in the soil.

In this chapter we will discuss the uptake, movement, mode of action, fate, and selectivity of herbicides in plants. For more detailed information on these subjects, consult Hatzios and Penner (1982), Duke (1985), Böger and Sandman (1989), Hatzios (1991), and Devine et al. (1993).

HERBICIDE UPTAKE

Soil-Applied Herbicides

Most herbicides applied to the soil move readily into the plant with the soil–water solution. Plant structures that take up herbicides generally are those that are actively imbibing or absorbing water. Such structures include germinating seeds, the shoots and roots of seedlings prior to emergence from the soil, and the roots of emerged plants. Soil-applied herbicides do not affect nongerminating seeds.

Some herbicides (e.g., the thiocarbamates and chloracetamides) penetrate the shoots of seedlings prior to emergence from the soil because the shoot cuticle is poorly developed at this stage. In addition, there is no Casparian strip barrier to restrict movement in shoots.

Foliar-Applied Herbicides

It is much more difficult for a herbicide to enter a plant through the shoot than through the root. The major barrier to uptake is the cuticle, which is composed of three distinct materials (Figure 6.1). The first component, the *wax*, occurs on the surface (epicuticular wax) and within the cuticle itself (embedded wax). Waxes consist of long hydrocarbon chains (mostly C_{20} to C_{37}) that include fatty acids; they are nonpolar and oil-like (lipophilic) in nature and form an effective water repellent layer over the shoot surface. The second component, the *cutin,* is composed of intermediate-length (C_{16} to C_{18}) fatty acids and also is lipophilic in nature. However, the cutin also contains free carboxyl (—COOH) and hydroxyl (—OH) groups that can become charged (—COO⁻ and —O⁻, respectively), making the cutin more compatible with water (hydrophilic) than wax. Cutin has the ability to become hydrated in the presence of water. The third component is *pectin*. Pectic strands either are located at the cutin–cell wall interface or are interspersed in the cutin layer. Pectin is composed of polymers of galacturonic acid and thus has charged carboxyl groups. Pectin is the most hydrophilic of the cuticle components and, when hydrated, can provide aqueous pathways for the uptake of water-soluble herbicides. Although the overall chemical nature of the cuticle is

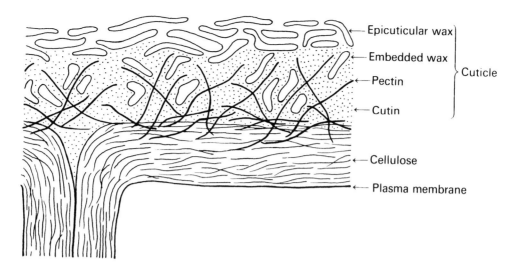

Figure 6.1 Components of the plant cuticle. (F. D. Hess, Novartis Crop Protection, Inc.)

lipophilic, it also carries a negative charge from the charged groups on the cutin and pectin.

The presence of waxes at the exterior of the cuticle results in an extremely lipophilic layer that constitutes a barrier to water-soluble compounds. As a general rule, movement of a herbicide into and through the cuticle is inversely related to its solubility in water. The more water soluble the herbicide, the lower the rate of movement through the cuticle; the more oil soluble the herbicide, the more likely it is to penetrate.

Besides the waxes, a dense covering of hairs (trichomes) on the plant surface also can prevent or lessen the penetration of herbicides (Figure 6.2).

Two important factors affecting the movement of foliar-applied herbicide into the leaf are its degree of retention on the plant surface and its ability to move through the cuticle.

Foliar Retention

If the retention time is too short or if an insufficient amount of herbicide is intercepted by the plant, penetration through the cuticle and eventual control will be unsatisfactory. The type of herbicide carrier and adjuvant used, the amount of spray volume, the amount of shoot growth, and the occurrence of rainfall after application all have an impact on retention and plant coverage.

Herbicide Carrier. Because water is abundant, convenient, and obtainable at minimal cost, most herbicides are applied to foliage as water (aqueous) sprays. The many examples of herbicide formulations sprayed in water include water-soluble powders, water-dispersible granules, water-soluble liquids, wettable powders, flowable liquids, and emulsifiable concentrates (see Chapter 16). Oils are used as herbicide carriers for special purposes such as dormant applications on woody species. Granular formulations frequently are used for foliar applications to turf.

The nature of the carrier is extremely important in determining how well a herbicide will be retained by the plant surface. For instance, oils readily spread out and adhere to plant surfaces. Granules, on the other hand, tend to roll off when applied to foliage. They provide a suitable carrier for foliage uptake only in sites such as turf, in which the weed leaves are close to the soil surface and are dense enough to keep the granules suspended on the leaves. Since granules do roll off leaves, they can be used for some postemergence treatments with greater safety to crops than can liquids sprayed broadcast.

In contrast to oils, water has a high surface tension and tends to bead or "ball up" when it hits the waxy surfaces of leaf and stem cuticles (Figure 6.3A). The subsequent lack of wetting or spread over the plant surface results in lower herbicide penetration. In fact, herbicide often is lost because it can bounce or run off the cuticle.

Adequate retention of aqueous sprays on the plant surface is difficult to obtain and requires that spray solutions be modified to wet the waxy cuticle. Modification can be accomplished by the addition of an adjuvant, such as a wetting agent, to the spray solution. Wetting agents act by reducing the sur-

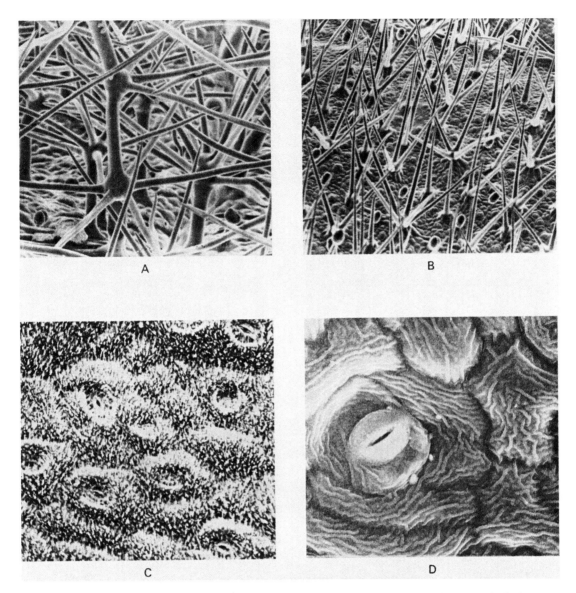

Figure 6.2 Scanning electron micrographs of leaf surfaces: trichomes (hairs) on (A) common mullein and (B) velvetleaf; epicuticular wax on (C) cabbage and (D) field bindweed. The slits in C and D are stomates. (F. D. Hess, Novartis Crop Protection, Inc.)

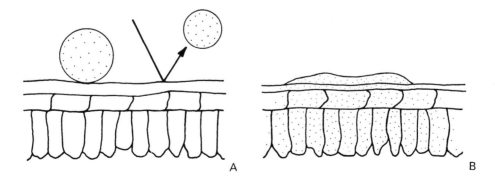

Figure 6.3 Effect of wetting agent on spread of a water-soluble herbicide over the plant surface: (A) without wetting agent, (B) with wetting agent.

face tension of the water droplets, allowing them to spread and make closer contact with the plant surface (Figure 6.3B). Because adhesion is increased, herbicide runoff from the plant also is minimized. Adding wetting agents to the herbicide solution also overcomes the problem of herbicide retention on upright, vertically positioned leaves. Without wetting agents, aqueous spray droplets rapidly roll off the leaves of plants such as grasses, wild onion and wild garlic, and cattails.

Crop oil concentrates and fertilizers are other adjuvants used to improve wetting. In some cases, adjuvants are added to the herbicide during formulation. Sometimes they must be added to the spray solution before use. For more information on wetting agents and other adjuvants, see Chapter 16.

Spray Volume. The application of excessive herbicide solution can result in runoff and loss from the plant surface. Solution volumes should be adjusted to provide good delivery to the target and adequate penetration of the canopy. The proper amount of coverage on the plant surface for optimum performance of the herbicide also is required. This amount varies with the type of herbicide being used. For example, complete coverage of the plant surface is required for herbicides that provide quick kill of plant tissues. In contrast, partial coverage may provide adequate performance of a herbicide that translocates through the symplast and becomes well distributed throughout the plant.

Although symplastically tranlocated herbicides can provide adequate control with somewhat limited coverage, much evidence suggests that their retention and subsequent uptake are improved by using lower carrier volumes and finer spray droplets, leading to improved performance. Adding the same amount of herbicide to a reduced volume of carrier increases the concentration of herbicide per droplet. Since penetration of herbicides into leaf tissue is due to diffusion, presumably the greater herbicide concentration in the droplets increases the concentration gradient across the cuticle, thus leading to the potential for increased diffusion.

Decreasing the spray droplet size at a given volume of carrier can enhance coverage and retention on hard-to-wet plant surfaces, thereby enhancing the performance of translocated herbicides.

Amount of Shoot Growth. There must be enough shoot growth to intercept the amount of herbicide that it takes to kill the plant. Freshly mowed weeds or newly emerging shoots of es-tablished perennial weeds rarely have sufficient leaf area to provide good uptake and movement of symplastically translocated herbicides to underground structures. Several days to a few weeks after emergence may be needed for the development of an adequate amount of top growth for good control. Unfortunately, a perennial weed sometimes can have considerable top growth and yet have such a tremendous volume of underground structures that it is impossible for the shoots to intercept enough herbicide to kill the entire plant.

Rainfall After Application. Rainfall soon after application can wash herbicides from the leaf surface. Retention times of 6 to 24 hr frequently are needed to prevent the loss of water-soluble herbicides by heavy rains. Required retention times are highest for herbicides that are anionic (negatively charged), such as the sodium salt of 2,4-D, because these compounds do not absorb to the cuticle and therefore do not penetrate plant tissue rapidly. On the other hand, herbicides that are cationic (positively charged), such as paraquat, are rapidly absorbed to the cuticle, which carries an overall negative charge. These compounds are less subject to removal from leaves by rain. Retention times for oil-soluble herbicides, which tend to penetrate rapidly into the cuticle, are considerably shorter, possibly as short as 1 hr. Although the required time of retention varies with individual herbicides and environmental conditions, it is wasteful and ultimately disappointing to apply a herbicide to foliage if rainfall is expected before there is adequate time for penetration.

Cuticular Penetration

Even though a herbicide may be evenly distributed over the plant surface, it still must move past the nonliving cuticle into the living tissues of the plant.

Oil-soluble herbicides readily penetrate the cuticle. They can dissolve into and through the cuticle and move rapidly into the plant by the process of simple diffusion. Water-soluble herbicides also move through the cuticle by diffusion; however, they do not penetrate as readily as oil-soluble herbicides and require the addition of a wetting agent for adequate penetration. Wetting agents appear to work on two principles. First, they permit the solution to spread and cover the plant surface (Figure 6.3B). This action increases the chances that the herbicide will come in contact with hydrophilic sites already at the leaf surface that permit the passage of water-soluble materials through the cuticle. Some of these sites include the strands of pectin that may extend through the cuticle, thin areas in the cuticle, and tears or breaks in the cuticle caused by wind, rain, insects, and other agents.

A second and somewhat controversial effect of wetting agents may be to solubilize or dissolve the cuticle. Studies have shown that herbicide penetration is enhanced by increasing the concentration of a wetting agent to a level much higher than that actually necessary just to wet the leaf surface. Although this evidence is indirect, it suggests that wetting agents may allow some of the water-soluble herbicide to move into and through the lipophilic components of the cuticle itself.

Whatever the route, the end result is that wetting agents help water-soluble herbicides get through the cuticle. An excellent demonstration of this result was reported by Freed and Montgomery (1958). When an amitrole-water solution with a wetting agent was sprayed on a plant, 78% of the herbicide was absorbed 24 hr after application. In contrast, only 13% of the herbicide was absorbed by the plant when the wetting agent was deleted from the spray solution.

Penetration of a water-soluble herbicide is more likely to be affected by environmental conditions than is the penetration of an oil-soluble herbicide. Any condition that causes the cuticle to become hydrated (take on water) greatly enhances the ability of a water-soluble material to penetrate. Such conditions include high humidity, warm temperatures, and adequate soil moisture. In one study, for example, leaf absorption of acifluorfen into crotalaria was three to four times greater at 100% relative humidity than at 40% relative humidity (Wills and McWhorter 1981). Glufosinate treatment at 95% relative humidity resulted in complete plant death in contrast to only a 30% inhibition in growth at 40% relative humidity (Anderson et al. 1993). In a study involving johnsongrass, absorption of glyphosate increased when the air temperature was increased from 75 to 95°F and when soil moisture was increased from 12% (near the wilting point) to 20% (field capacity) (McWhorter et al. 1980).

Environmental conditions also affect the amount of cuticle produced by a plant. Plants growing under conditions of low light, high humidity, and adequate soil moisture develop thinner cuticles that can be more easily penetrated than do plants growing under conditions of high light, low humidity, and dry soil.

Conditions that favor stomatal opening (e.g., slight air movement and good soil moisture) may increase herbicide penetration through that route, whereas conditions that cause the stomates to close (e.g., moisture stress) may decrease herbicide penetration. There is some evidence that the organosilicone surfactants can increase stomatal penetration of herbicides (Buick et al. 1992, Field and Bishop 1988), but in general, stomatal opening is considered to be of minor importance in the penetration of water-soluble herbicides when compared to cuticle hydration (Buckovac 1976).

In general, maximum kill of plants treated with a foliar-applied herbicide will be obtained under warm, humid conditions and adequate soil moisture. Minimum kill can be expected when plants experience water stress, cool temperatures, and low humidity. These important factors often are difficult to pinpoint at the time of treatment, yet they account for much of the inconsistency in the performance of foliar-applied herbicides.

Penetration Through the Plasma Membrane

Once a herbicide has penetrated the cuticle (if applied to the foliage) or comes into contact with the cells of the root or shoot (if applied to the soil), it must at some point enter living cells. All target sites for herbicide action are located within the living cell. In addition, symplastically translocated herbicides must enter into the living cell system in order to be moved in the plant. Even soil-applied herbicides that move apoplastically have to move into living cytoplasm, both at the endodermis and at the target site. To enter a cell, the herbicide must cross the plasma membrane of the cell. In some cases, the herbicide must also penetrate the chloroplast envelope if its target site is within the chloroplast.

Most herbicides cross membranes by simple diffusion. The driving force is the concentration gradient across the membrane. The herbicide will move from a region of higher concentration (outside the membrane) to a region of lower concentration (inside the membrane) until the concentrations on both sides are equal. Metabolic energy is not required for this movement. Lipophilic herbicides tend to diffuse more freely across the plasma membrane than hydrophilic herbicides.

Some herbicides diffuse across the plasma membrane but also are concentrated within the cell; in other words, they move against a concentration gradient and accumulate at a higher concentration in the cytoplasm than outside the cell. These herbicides are mostly weak acids, for example, with a carboxylic acid group (—COO⁻), and include bentazon, clopyralid, chlorsulfuron and other sulfonylureas, the imidazolinones, and sethoxydim. Metabolic energy in the form of ATP is expended by the cell to move these compounds through the plasma membrane. The conversion of ATP to ADP by the enzyme ATPase occurs at specific sites in the plasma membrane known as ATPase hydrogen ion pumps. The energy released by the conversion of ATP to ADP is used to pump hydrogen ions (H^+) to the outside of the cell. Cells therefore can maintain a difference in hydrogen ions, and thus pH, across the plasma membrane (Figure 6.4). The pH outside the cells is lower at 5.0 to 5.5 (a higher H^+ concentration) than the pH inside the cells at 7.2 to 8.0 (a lower H^+ concentration). A herbicide that is a weak acid outside the cell becomes protonated (—COOH); in other words, it picks up a hydrogen and loses its negative charge. Thus, the compound becomes uncharged and more

lipophilic, allowing it to diffuse across the plasma membrane. Once inside the cytoplasm, it encounters a higher pH environment in which the hydrogen is stripped from the herbicide (—COO⁻). The herbicide is now charged and more water soluble. Thus, it cannot readily move back through the plasma membrane and becomes trapped in the cytoplasm. Uncharged herbicide molecules continue to diffuse into the cell because the concentration of uncharged herbicide is still higher outside the membrane than inside. This phenomenon is called ion trapping.

Other herbicides also are actively moved across the plasma membrane using metabolic energy (ATP) but pass through the plasma membrane at special sites. Specialized proteins, called carriers, embedded in the plasma membrane are responsible for moving certain natural substances (e.g., sucrose, phosphate) across the membrane. Herbicides that are transported by carriers are paraquat, 2,4-D, and glyphosate. Glyphosate, which has a phosphate molecule as part of its chemistry, crosses plasma membranes via the phosphate carrier.

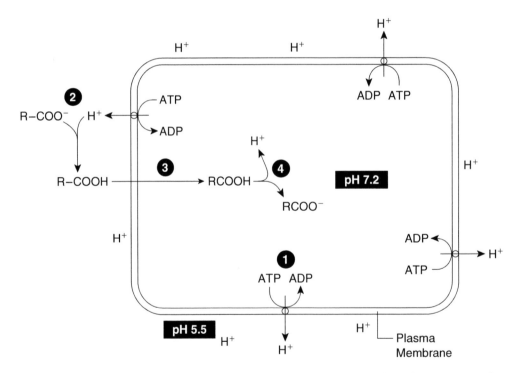

Figure 6.4 An ion-trapping model showing how herbicides that are weak acids can be concentrated within the cell. (1) ATPase hydrogen ion pumps move H⁺ to the outside of the plasma membrane, thus maintaining a low pH (high H⁺ concentration) or acidic environment outside and a high pH (low H⁺ concentration) or basic environment inside the cell. (2) External low pH environments favor the conversion of a charged herbicide (R—COO⁻) to the uncharged, lipophilic form (R—COOH). (3) The R—COOH form readily diffuses across the plasma membrane. (4) When it encounters the more basic internal environment, the R—COOH is converted back to R—COO⁻, which cannot diffuse across the plasma membrane and becomes trapped in the cytoplasm.

HERBICIDE MOVEMENT

Once a herbicide has entered the plant (whether soil or foliar applied), it must reach a specific vulnerable site. In some cases, little herbicide movement is required to kill the plant; in others, the herbicide must move throughout the plant, particularly to the underground portions. The three major groups of herbicides according to kind of movement are (1) symplastically (phloem) translocated herbicides (which also move apoplastically), (2) apoplastically (xylem only) translocated herbicides, and (3) herbicides that show limited or no movement in the plant (called contact herbicides when applied to foliage).

Symplastically Translocated Herbicides

These compounds move in the symplast with the sugars. When applied to the foliage and at the right stage in the life cycle, compounds such as glyphosate, 2,4-D, amitrole, the sulfonylureas, and the imidazolinones become distributed throughout the plant because they move to points of active growth or food storage accumulation (Figure 6.5A). Although immediate effects such as contact burn and twisting can occur, the chronic symptoms at low dosages appear on the new growth (shoot tips, young leaves, buds, etc.). These compounds also can be transported in the apoplast to varying degrees, thus increasing their distribution in the plant.

When applied to the soil, some symplastically translocated herbicides flow through the cells of the root to the stele, where they can move into both the symplast (sieve tube members of the phloem) and apoplast (vessel members of the xylem). Herbicides such as picloram are extremely effective in controlling perennial weeds because they enter the plant through both the foliage and the roots and can "circulate" in the plant through both transport systems. Other compounds such as 2,4-D tend to remain in the symplast upon entry into root tissue. They become incorporated into cortex cells of the root where they disrupt growth or become inactivated. Although some of the compounds may move into the phloem, phloem movement tends to be downward so that distribution of these compounds to the whole plant seldom is optimal when they are applied to the soil.

Symplastically translocated herbicides are used for controlling annual, biennial, and perennial weeds. In fact, these compounds must be used if the major objective of a foliar treatment is to kill all parts of the plant, including the underground portions (e.g., in perennial plants).

Factors that affect the translocation of symplastically translocated herbicides include the stage of plant growth, herbicide distribution on the plant, presence of living tissue, and environmental conditions.

Stage of Plant Growth

The amount of sugar translocation to various parts of the plant changes during the life cycle of the plant and influences the movement of herbicides. For this reason, herbicides for perennial weeds are applied at specific stages of growth. In general, plants are most susceptible to symplastically translocated herbicides when young or when undergoing periods of rapid growth, vegetative reproductive structure development, or food accumulation in storage

Figure 6.5 When applied to a portion of a leaf, symplastically translocated herbicides (A) move back into the stem to points of active growth (meristematic regions) or food storage (underground structures). Apoplastically translocated herbicides (B) move to the edges of the leaf. Limited-movement or contact herbicides (C) do not move from the point of application.

structures. The timing of herbicide applications in relation to life cycles is described in more detail in Chapter 13.

Distribution Pattern

Complete plant coverage with symplastically translocated herbicides is not required; however, it is essential that the herbicide come in contact with and be adequately retained by structures that are actively photosynthesizing and exporting sugars (Figure 6.6). Young leaves (one-fourth or less expanded) and old senescing leaves do not export sugars and therefore do not transport herbicides well. In the case of young expanding leaves, sugars actually must be imported to sustain growth until the leaf achieves full photosynthetic capability.

Placement also determines the direction of herbicide movement (Figure 6.6). The application of a symplastic herbicide to photosynthesizing and sugar-exporting lower leaves results

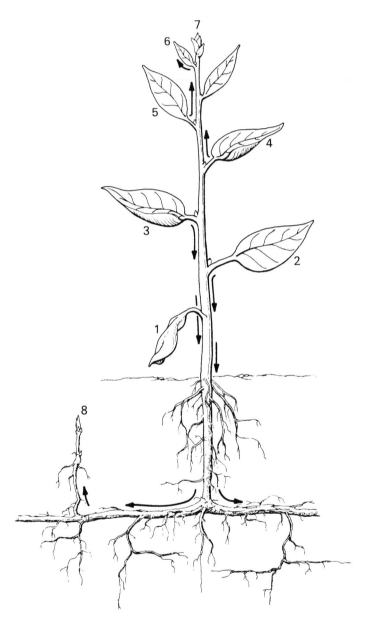

Figure 6.6 The effect of placement on the direction of symplastic herbicide movement. Herbicide placed on sugar-exporting lower leaves (2, 3) moves downward, whereas herbicide placed on sugar-exporting upper leaves (4, 5) moves upward. Senescent leaves (1) do not export herbicide, whereas young expanding leaves (6) and shoot tips (7, 8) act as sinks.

in its movement downward with the sugars to underground structures. Application to fully functioning upper leaves sends the herbicides to the developing shoot tip.

Living Tissue

The conduction of sugars and herbicides through the symplast is dependent on a living, functioning plant. Sucrose is loaded into sieve tube members of the phloem in a manner similar to that described for ion trapping of herbicides that are weak acids (see the section on Penetration through the Plasma Membrane, page 101). Although sucrose is not acidic, it is cotransported with an H^+ across the plasma membrane of the sieve tube members through a specialized carrier protein. The functioning of the ATPase hydrogen ion pump, which pumps H^+ to the outside of the cell so that it can participate in the cotransport of sucrose back into the cytoplasm, can only be accomplished by living cells that are respiring and producing ATP. Once sucrose becomes trapped in the sieve tube members of the phloem, it then is translocated to the sink area(s) where it is stored (as starch) or utilized for growth. Herbicides that move into the symplast move with the flow of sucrose and water to the sinks. *Therefore, active and functional sucrose translocation is required for the optimal movement and distribution of symplastically translocated herbicides.*

Living cells also are required for the uptake of herbicides that are loaded into the symplast by ion trapping (weak acids) or that move through specialized protein carriers (glyphosate, 2,4-D).

Movement of a symplastically translocated herbicide to the underground portions of a plant can take several hours to a few days (depending on the plant and the herbicide). Thus, immediate foliage kill or damage to the symplast, such as the plugging of sieve tubes with callose or death of the cytoplasm of the conducting cells, almost always results in poor herbicide translocation and unsatisfactory control of the underground plant parts. Best results frequently are obtained by applying several low doses of a herbicide that can move underground rather than a single dose that kills the plant too rapidly. Multiple applications also provide time for dormant buds to initiate growth after the initial application and to become susceptible to the second application.

Since almost all herbicides can diffuse through plasma membranes, and since many can accumulate in the cytoplasm of cells by ion trapping, why do not all herbicides enter into and translocate through the phloem? The answer appears to be that herbicides that move primarily in the apoplast are too capable of diffusion across membranes. Atrazine, for example, can diffuse into sieve tube members, but it can diffuse back out again and into adjacent vessel members of the xylem. Since water movement in xylem tissue is moving faster than in the phloem, the net direction of flow of atrazine will be upward with the water stream.

Environmental Conditions

Since movement of symplastically translocated herbicides is associated with sugar production, environmental conditions favoring photosynthesis optimize herbicide movement. These conditions include high light, adequate soil moisture, and moderately warm temperatures.

Apoplastically Translocated Herbicides

These compounds move in the apoplast, the system of the plant that conducts water and minerals. Classic examples of apoplastically translocated compounds are the triazine and phenylurea herbicides.

When soil applied, apoplastically translocated herbicides are taken up by the roots and move into leaves that are actively transpiring. Older leaves are affected first.

When foliar applied, these compounds cannot move downward, since the flow of water and nutrients in the apoplast is into rather than away from the leaves. Apoplastic herbicides applied to the base of a leaf move to the tip (Figure 6.5B). Compounds that are active when foliar applied often work in much the same way as contact herbicides by providing burndown of the shoot tissue. Some internal movement can occur, however, so surface distribution and wetting need not be as thorough as with a contact herbicide.

Contact or Limited-Movement Herbicides

Foliar-applied herbicides in this class often kill the tissue they encounter immediately after penetration (Figure 6.5C). These herbicides cause membrane destruction, burning, and necrosis of leaf and stem tissues, often within hours after application. Since these compounds do not move or redistribute themselves in the plant, complete coverage of the plant surface is required for shoot kill. Application of a contact herbicide usually is sufficient for providing control of annual plants but not of biennials or perennials. Examples of contact herbicides include paraquat, the diphenylether herbicides, glufosinate, and the petroleum oils.

Some soil-applied herbicides also show limited movement in plants. They are most effective when used on seedling plants prior to or at emergence from the soil. There are relatively few barriers to penetration, and the herbicide does not have to move far into the plant to cause damage. Trifluralin, for example, only has to move into a few cell layers of the root tip and meristem to disrupt cell division and cause root growth inhibition. Many of the compounds in this category, such as the dinitroanilines and chloracetamides, appear to be inhibitors or disruptors of cell division of seedling roots and shoots.

HERBICIDE MODE OF ACTION

One or more vital plant processes must be disrupted in order for a herbicide to kill a weed. *Some vital (living) metabolic plant processes that occur at the intracellular level include:*

Photosynthesis. In photosynthesis, the capture of light energy by chlorophyll results in the splitting of water to produce oxygen and electrons. The electrons are used to reduce, or "fix," carbon dioxide (CO_2, which is obtained from the atmosphere) into organic compounds (for example, glucose, $C_6H_{12}O_6$). These simple organic compounds then become the building blocks from which more complex compounds such as proteins, fats, and other carbohydrates are synthesized. Photosynthesis is by far the most common way in which inorganic compounds (CO_2 and H_2O) are converted into organic compounds by living organisms, namely, plants.

The part of photosynthesis that requires light, in other words, the capture of light energy by chlorophyll and the production of electrons (reducing power), is called the light reaction (or light-dependent phase) of photosynthesis. The carbon fixation process is termed the dark reaction (or light-independent phase) of photosynthesis.

Respiration. In respiration, organic compounds (for example, glucose) are oxidized (broken down) into carbon dioxide and water. In the process, the energy present in the bonds of the glucose molecule is transferred to ATP, a chemical form of energy that can be used by organisms to metabolize, grow, and pump ions across membranes (e.g., ion trapping, Figure 6.4). All living organisms respire. If respiration ceases, so does life.

Amino Acid and Protein Synthesis. Proteins are made up of long chains of amino acids. Proteins serve as structural components in the cell (for example, up to 50% of cell membranes are made up of proteins) and enzymes, which catalyze biological reactions. Amino acids are characterized by the presence of an amino group (—NH_2) and a carboxyl group (—$COOH$). Amino acids are synthesized via enzymatically catalyzed reactions that occur either in the cytoplasm or in the chloroplast. Plant growth eventually ceases when amino acid synthesis is inhibited.

Fat (Lipid) Synthesis. Fats are important components of cell membranes (the other 50%) and the outer plant cuticle. Fats are also stored, often in seeds for use during germination. Fats are composed of long-chain fatty acids, and two of the major fatty acids, palmitic acid (C_{16}) and oleic acid (C_{18}), are synthesized in the chloroplast via enzymatically catalyzed reactions. Other fatty acids are then formed in the cytoplasm by the modification of these acids.

Pigment Synthesis. Pigments such as chlorophyll and carotenoids are needed for energy capture from light for photosynthesis and other processes. Both chlorophyll and carotenoids are synthesized in the chloroplast.

Nucleic Acid Synthesis. The nucleic acids include deoxyribonucleic acid (DNA) and ribonucleic acid (RNA). DNA contains the genetic information for the cell; RNA is required to translate that information into the manufacture of proteins, particularly enzymes, which catalyze all important reactions in the cell. Most of the DNA and RNA in a cell is synthesized in the nucleus, but DNA and RNA are also present in the chloroplasts and mitochondria and can encode for chloroplast and mitochondrial proteins, respectively.

Membrane Integrity. All cell membranes (e.g., plasma membrane, tonoplast) must be intact in order to remain alive and functioning. Anything that disrupts membranes will lead to leakage of the cell contents, desiccation, and cell death. Typically, the portion of the membranes destroyed by herbicide activity is the lipid component; this destruction is termed lipid peroxidation and is caused by the interaction of fatty acids with membrane-damaging, unstable molecules such as triplet chlorophyll (^{3}chl), singlet oxygen ($^1O_2$), and hydroxyl radicals (OH·). This interaction (Figure 6.7) results in the production of lipid radicals, which then react with oxygen, eventually resulting in the breakdown of the lipids to short-chain carbon compounds.

Some vital (living) plant processes that occur at the cellular and tissue levels include:

Growth and differentiation

Mitosis (cell division) in plant meristems

Meiosis (cell division resulting in the formation of eggs and sperm)

Uptake and translocation of ions and molecules

The mode of action is the overall manner in which a herbicide disrupts vital processes at the intracellular, cellular, or tissue level. Herbicides with the same mode of action almost always have the same translocation (movement) pattern and similar injury symptoms. Selectivity on crops and weeds, behavior in the soil, and use patterns are less predictable but are often similar for herbicides with the same mode of action.

Two definitions frequently used to describe herbicide symptoms are chlorosis and necrosis.

Chlorosis is the loss of green pigment, and *necrosis* is the death of tissue, usually characterized by browning and desiccation.

Our Mode-of-Action Scheme

We group herbicides into those chemical families that are applied to the foliage and those that almost exclusively are applied to the soil. Foliar-applied chemical families are grouped according to their translocation patterns. Soil-applied chemical families are all cell division inhibitors and are grouped according to whether they cause spindle malfunction in mitosis; cell divisions resulting in shoot malformation; or other cell division abnormalities.

Translocation (long-distance transport through conducting tissues) determines the performance of herbicides applied to foliage. The same is true of herbicides applied to the soil that subsequently move to shoots that have already emerged above the soil line. In contrast, herbicides applied to the soil that act on newly germinating and emerging seedlings only move across a few layers of cells and reach the site of action by simple diffusion, ion trapping, or in some cases by carriers. Thus, long-distance mobility is not integral to their performance.

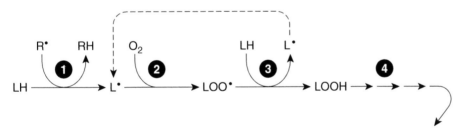

Short-chain carbon compounds
such as ethane (C_2H_6) and malondialdehyde (OHC-CH$_2$-CHO)

Legend:

LH = Lipid, e.g., linoleic acid, $CH_3CH_2CH = CHCH_2CH = CHCH_2CH = CH(CH_2)_7COOH$

R· = Initiating factor (e.g., OH·, $^1O_2$, or ^{3}chl)

L· = Lipid radical

LOO· = Peroxidized lipid radical

LOOH = Lipid peroxide

Figure 6.7 Pathway for lipid peroxidation. Reaction of the lipid with an unstable initiating factor (1) causes oxidation of the lipid to form a lipid radical (L·). The lipid radical reacts with oxygen (2) to form a peroxidized lipid radical (LOO·). This series of reactions is perpetuated by the interaction of the peroxidized lipid radical with an unaltered lipid (3), which now become converted to a lipid radical and can feed back into the series of reactions at step (2). The other product of reaction (3) is a lipid peroxide (LOOH), which is broken down through a series of reactions (4) to alkanes, aldehydes, and other short-chain carbon compounds. This series of reactions effectively destroys the lipid components of membranes.

Herbicide Families with Significant Foliar Use

The translocation pathways for herbicides applied to foliage determine the general pattern of symptom development and the potential to kill weeds. The three categories of movement are (1) translocated herbicides showing initial symptoms on new growth, (2) translocated herbicides showing initial symptoms on old growth, and (3) nontranslocated herbicides showing initial localized injury.

Translocated Herbicides Showing Initial Symptoms on New Growth

These herbicides are frequently described as symplastically translocated, phloem-mobile, downwardly mobile, or basipetally translocated. These herbicides initially move to sites of new growth, for example, the growing points or meristems of both above- and belowground tissues, newly developing leaves, and developing storage organs. Symptoms can include pigment loss (yellow or white), growth stoppage, and distorted (malformed) new growth. Symptoms normally appear only after several days. Plants die slowly from the top down (new to old growth).

These herbicides have the potential to move to and kill underground perennial structures with one or two foliar applications. However, the degree of mobility within each of the groups of herbicides discussed in this category can vary substantially. For example, among the carotenoid pigment inhibitors, amitrole translocates more readily than clomazone.

Auxin-Type Growth Regulators. This group of herbicides causes growth effects similar to those of the naturally produced auxin hormones such as IAA (indoleacetic acid). In fact, it is thought that these herbicides compete with natural auxins for binding to an as yet unknown auxin-binding receptor on plasma membranes of cells. The initial effect of this binding is to cause rapid changes in the elasticity of the plant cell wall, which in turn allows the cell to increase in size and elongate. The unregulated elongation of cells results in unequal and abnormal curvature of the plant tissues. In fact, one of the most immediate symptoms of plants treated with growth regulator herbicides is a downward twisting and curvature (epinasty) of the stems and leaves (Figure 6.8). A secondary, slower effect of the growth regulator herbicides is to cause an increase in RNA and protein synthesis, which causes the plant tissues to become meri-

Figure 6.8 Epinastic effects and twisting caused by 2,4-D on a Canada thistle plant.

stematic. The cells undergo division and growth but in an uncontrolled manner. Shoot and root growth is distorted, callus tissue develops, roots can appear from stem tissue, and the phloem becomes plugged or broken, preventing the movement of sugars from the leaves to the stems. Death occurs in several days or weeks.

Broadleaves are more sensitive to this group of herbicides than monocot species. However, even tissues of monocots that are undergoing active meristematic activity can show symptoms. An example is that of normally tolerant corn, which forms abnormal brace roots when 2,4-D is applied during the period of brace root formation from intercalary meristems.

Selectivity to the growth regulator herbicides is not well understood and is probably the result of several factors, including differences in herbicide retention, translocation, metabolism, and plant anatomy.

Herbicide groups and compounds with auxin-type growth regulator activity are the phenoxy acids, benzoic acids, and picolinic acids.

Aromatic Amino Acid Inhibitors. The herbicide in this group, glyphosate, kills plants by acting on the chloroplast enzyme 5-enolpyruvylshikimate-3-phosphate synthase (EPSPS), thus preventing the conversion of shikimate to the amino acids phenlyalanine, tyrosine, and tryptophan (Figure 6.9). This pathway is often referred to as the shikimic acid pathway. A reduction in amino acids reduces protein synthesis and subsequently reduces growth. Other, less well understood factors related to the inhibition of this pathway may also add to the phytotoxicity observed after glyphosate application.

Because phenylalanine and tyrosine contain a benzene ring in their structure, they are called aromatic amino acids; hence, the name of this group of inhibitors. Tryptophan also contains a ring structure, but it is often grouped with the heterocyclic amino acids. EPSP synthase is, however, also essential for the formation of tryptophan. These three amino acids are synthesized only in the chloroplasts of plants and not by animals; thus, the inhibition of the synthetic pathway by glyphosate poses little problem to nontarget animal species.

Symptoms from glyphosate treatment include inhibition of new growth followed by a gradual yellowing starting with new tissue and progressing to older tissues and leaves. Complete development of symptoms is relatively slow and usually requires one to three weeks depending on the herbicide dose, plant species, and temperature. Very few plant species are tolerant of glyphosate.

Branched-Chain (Aliphatic) Amino Acid Inhibitors. These herbicides interfere with the activity of the chloroplast enzyme acetolactate synthase (ALS), also called acetohydroxyacid synthase (AHAS), by binding very tightly to the substrate–ALS complex (Figure 6.10). This enzyme inhibition prevents the formation of valine, leucine, and isoleucine, amino acids that have a branched carbon chain in their structure. These amino acids are also only synthesized by plants, and again, the reduction in protein synthesis plus less-understood secondary effects reduces growth, resulting in eventual death of the plant.

Four groups of herbicides have this mode of action: the imidazolinones and sulfonylureas that were developed in the 1980s and the newer triazolopyrimidine sulfonanilides and pyrimidinyl oxybenzoates.

Shoot meristems cease growth within a few hours of application. However, symptom development is very slow and occurs over two to

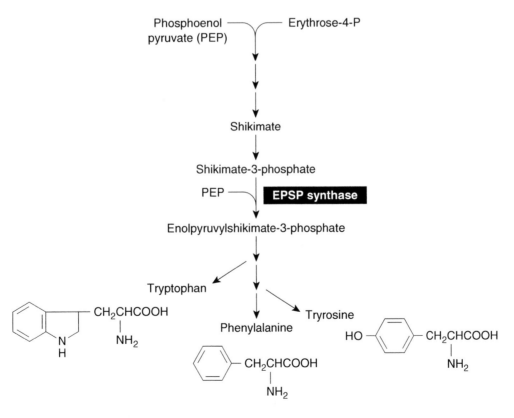

Figure 6.9 Shikimic acid pathway. Glyphosate blocks the activity of the enzyme EPSP synthase (or EPSPS), resulting in a reduction of the biosynthesis of the amino acids tryptophan, phenylalanine, and tyrosine.

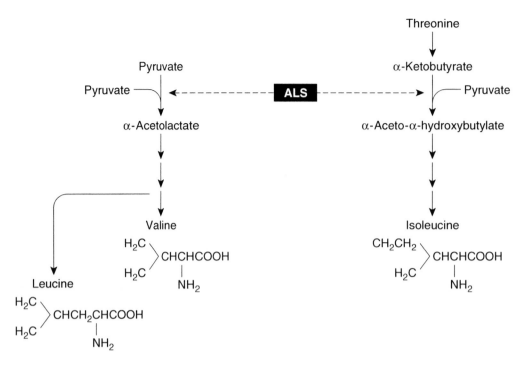

Figure 6.10 Branched-chain amino acid biosynthesis. The imidazolinones (e.g., imazethapyr) and sulfonylureas (e.g., chlorsulfuron) block the activity of the enzyme ALS, resulting in a reduction of the biosynthesis of the amino acids leucine, valine, and isoleucine.

three weeks or more. Meristematic areas become yellow and die, and yellow streaks on leaves and pink and purple veins develop. Root growth is also stunted, and the secondary roots are shortened and all nearly the same length, producing a "bottlebrush" look. Late applications for weed control in corn may result in malformed ears.

Selectivity of the branched-chain amino acid inhibitors is due to metabolism by tolerant species.

Carotenoid Pigment Inhibitors. These herbicides prevent the formation of carotenoids, which is followed by the loss of chlorophyll, resulting in the formation of whitened plants. One of the functions of carotenoids in plants is to prevent photo-oxidation of chlorophylls. During the light reaction of photosynthesis, chlorophyll absorbs light energy and becomes converted to an energized state known as the singlet form of chlorophyll (¹chl). Most of the energy is then transferred from singlet chlorophyll to the other components of the light reaction to form reducing power. However, under bright light, not all of the energy can be used in this way and the excess energy absorbed by chlorophyll causes a conversion from the singlet form to the more reactive triplet form (³chl). Normally, the excess energy of triplet chlorophyll will be transferred

to carotenoids where the structure of the carotenoid molecule allows this excess energy to be dissipated in a harmless way. When carotenoid synthesis is disrupted, the energy in triplet chlorophyll initiates various degrading reactions, among which is the destruction of chlorophyll itself and the peroxidation of cell membranes (Figure 6.7). Thus, when no new carotenoids can be produced, the chlorophyll in the young plant tissue will degrade and the result is a total loss of pigment and a whitening of the tissue. The destruction of cell membranes leads to leakiness and rapid tissue desiccation.

Since carotenoids themselves are not destroyed by these herbicides, only their synthesis, older tissues with carotenoid and chlorophyll pigments formed prior to herbicide application maintain their original color for some time. Eventually, the gradual natural loss of pigments (that can no longer be replaced) results in chlorosis and death.

Depending on the herbicide, one of several enzymatically catalyzed steps in the carotenoid biosynthesis pathway is inhibited. For example, norflurazon and fluridone inhibit the activity of the enzyme phytoene desaturase; amitrole may inhibit the activity of the enzyme zeta carotene desaturase. The exact site of inhibition of another member of this group, clomazone, has not yet been characterized.

Symptoms include vivid white new growth (sometimes tinged with pink or purple) on the youngest portions of the shoots and appear within a few days. This new growth initially appears normal except for the conspicuous lack of green and yellow pigments. Death occurs within one to several weeks. Symptoms can occur on most species of plants; however, whether the plant actually dies from the herbicide probably depends on factors such as the degree of uptake, movement, and metabolism.

Lipid Biosynthesis Inhibitors. The lipid biosynthesis inhibitors block the activity of the enzyme acetyl-coenzyme A carboxylase (ACCase) in grass species. Acetyl-coenzyme A is essential to lipid (fat) synthesis, specifically the synthesis of the fatty acid, palmitic acid. Since growth is dependent on lipid synthesis (e.g., to produce new cell membranes), the plants stop growing and eventually die.

The symptoms of this group of herbicides include discoloration and disintegration of meristematic tissues, resulting in the cessation of plant growth. The leaf sheaths at the nodes become mushy and brown, and the leaves themselves turn yellow and sometimes wilt. Necrotic symptoms gradually spread over the entire plant; a week or more is required for complete plant death. Seedling grasses tend to lodge by breaking over at the soil line. At sublethal doses flowering and seedhead production of perennial grasses may be inhibited.

The selectivity of the lipid biosynthesis inhibitors is clearly at the ACCase site of action and is due to the fact that the structure of the ACCase in grasses differs from that in dicots. The grass ACCase contains a site that allows the herbicide to bind and deactivate it. No such binding site is present in the dicot form of ACCase. Animals also have an ACCase enzyme that is susceptible to this group of inhibitors, but negative effects on nontarget organisms have not been demonstrated. The herbicide groups that are lipid biosynthesis inhibitors are the aryloxyphenoxy propionates and the cyclohexanediones.

Translocated Herbicides Showing Initial Symptoms on Old Growth

These herbicides are also described as apoplastically translocated, xylem mobile, upwardly mobile, or acropetally translocated. Overall symptoms on the shoots develop from the bottom leaves and progress upward. Thus, older leaves show the most injury, and newer leaves show the least injury. On individual leaves chlorosis first appears between leaf veins and along leaf margins. This is followed by necrosis of the tissue. Symptoms usually require several days to develop but are more rapid in appearance than the symptoms from translocated herbicides showing initial symptoms on new growth.

Any potential control of established perennials must come from continued soil uptake and not from movement downward through the plant from the shoots. Foliar application provides only shoot kill.

"Classical" Photosynthetic Inhibitors. The photosynthetic inhibitors act on the light reaction of photosynthesis in the chloroplasts. These compounds traditionally were described as inhibitors of the Hill reaction, the splitting of water that produces oxygen and reducing power in the form of electrons. We now know that the photosynthetic inhibitors do not act at the site where water is split but rather later in the light reaction as electrons flow through a series of acceptors. The light reaction occurs within two reaction centers known as photosystem II (PS II) and photosystem I (PS I). The components of these photosystems are located on proteins that are embedded in membranes within the chloroplast. Photosystem II is specifically embedded in two proteins called D2 and D1 (Figure 6.11). Light energy, which is captured by chlorophylls and carotenoids in special centers called light-harvesting complexes, is transferred to a reaction center within PS II called chlorophyll P680. This concentrated energy is used to transform the electrons split off from water into "excited-state" electrons. These energized electrons are then passed through a series of acceptors starting with pheophytin. The electrons are passed from pheophytin to Q_A and from Q_A to Q_B and eventually to the components of PS I. In order for the acceptor Q_B to function normally it must bind to a portion of the D1 protein. The photosynthetic inhibitor herbicides competitively bind to the D1 protein, thus blocking the ability of Q_B to transport electrons and effectively stopping electron flow through PS II.

The inability to transfer captured light energy in the form of energized electrons to processes that fix carbon dioxide into carbohydrate (the dark reaction) should result in starvation of plant tissues. Since plants treated with the classical photosynthetic inhibitors and left in the light die faster than those placed in the dark, it has been suggested that processes in addition to or rather than starvation result in plant death. Because electrons can no longer be moved

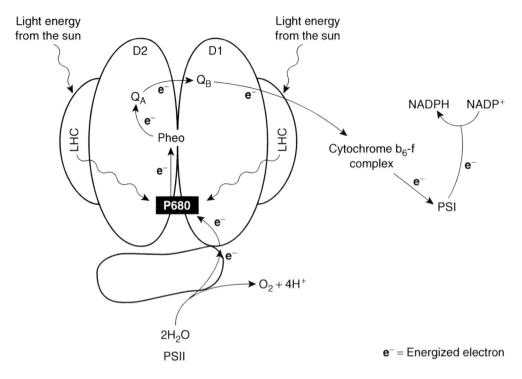

Light energy from the sun

Light energy from the sun

e^- = Energized electron

PSII

Figure 6.11 Electron transport through photosystem II (Photosystem II is embedded in the chloroplast membrane). The splitting of two water molecules results in the release of four electrons (only one is shown). After passing through the P680 reaction center, which has absorbed light energy from the light-harvesting centers (LHC), the electrons are energized and are passed from P680 to pheophytin (Pheo), Q_A, and Q_B. The Q_B acceptor is very loosely bound to the D1 protein. Photosynthetic inhibitors can prevent Q_B from binding correctly to D1, thus stopping the flow of electrons. Normally, the electrons, after passing through PSII and PSI, will be used to generate reducing compounds (NADPH) for the fixation of carbon dioxide.

through PS II, a buildup of energized chlorophyll occurs in the light-harvesting centers. The singlet energy state of the chlorophyll molecule (^{1}chl) is transformed to the more reactive triplet energy state (^{3}chl). Normally this triplet state energy is dissipated harmlessly by carotenoids, but because the energy continues to build up in the chlorophyll, the carotenoid quenching system becomes overloaded. Energy in triplet chlorophyll that is not "quenched" by the carotenoids is transferred to either lipids to produce lipid radicals or oxygen to produce singlet oxygen molecules, which in turn react with lipids to produce lipid radicals. The result is lipid peroxidation and the destruction of membranes (Figure 6.7). This in turn causes leakage of the cell contents and results in the desiccation and death of plant tissues.

The symmetrical triazines have this mode of action and are prone to selection for resistant weed populations when a mutation of the gene that codes for the D1 protein prevents the herbicide from binding with it. The phenlyurea herbicides work at the same site (Q_B binding site at the D1 protein), but they bind in a different way to

the D1 protein. Thus, plants can be resistant to triazine herbicides (e.g., simazine) but still susceptible to phenylurea herbicides (e.g., diuron).

The development of visible symptoms from treatment with photosynthetic inhibitors occurs over a period of several days. Interveinal chlorosis of the leaves and yellowing of the leaf margins is followed by necrosis of these areas. Injury starts with the oldest leaves and progresses to the top so that older leaves show the most injury and newer leaves the least injury. Root growth is not affected. Selectivity is mostly due to differential metabolism of the herbicide.

The "Rapidly Acting" Photosynthetic Inhibitors. These compounds inhibit photosynthesis at the Q_B acceptor site, but kill is much more rapid than with the triazine and phenylurea herbicides. Rapidity of kill is consistent with the destruction of cell membranes due to lipid peroxidation that occurs subsequent to photosynthesis inhibition. Depending on the speed of kill, symptoms range from those associated with "classical" photosynthetic inhibitors to those associated with rapid cell membrane disrupters (see below). Examples of the "rapidly acting" photosynthetic inhibitors include pyridate, desmidipham, pyrazon, bromoxynil, and bentazon.

Nontranslocated Herbicides Showing Initial Localized Injury

These herbicides are frequently described as contact or limited-movement herbicides. The *rapid disruption of cell membranes* by these herbicides provides very rapid kill and prevents translocation. These herbicides penetrate into the cells at the point of contact and inhibit some primary process that results in the rapid loss of cell membrane integrity. Depending on the primary (initial) mode of action, these groups of herbicides can cause production of hydroxyl radicals (OH*), singlet oxygen ($^1O_2$), or triplet state chlorophyll (^{3}chl). All of these unstable molecules can initiate degradation of lipids (lipid peroxidation) in the cell membranes (Figure 6.7). Damage to cell membranes leads to the leakage of cell contents, desiccation, and cell death. The amount of damage to individual plants is proportional to the completeness of coverage during application.

Photosystem I Energized Cell Membrane Destroyers. These herbicides, namely paraquat and diquat, accept electrons from photosystem I. In this reduced state, these molecules become highly unstable free radicals that undergo oxidation back to the original paraquat or diquat ion (Figure 6.12). During this oxidation process, superoxide radicals are produced from molecular oxygen. Superoxide radicals are then enzymatically altered to form hydrogen peroxide. Superoxide radicals and hydrogen peroxide react together to produce hydroxyl radicals (OH*), which are extremely efficient at initiating lipid peroxidation. Cell membrane destruction follows. Severe injury is evident hours after application, and maximum kill is attained in a week or less. Such rapid disruption of cell membranes provides very rapid kill and prevents translocation. Thus, the rapid foliar activity provides only shoot kill.

Partial coverage with these herbicides results in spotting and/or partial shoot kill. Since most plants are susceptible to paraquat and diquat damage, whether or not the plant dies depends on the degree of coverage on the shoot surface. Since no translocation occurs, new growth on surviving plants will be normal in appearance. For example, the treated shoot surfaces of most perennial plants will be injured but new sprouts will be unaffected.

Protoporphyrinogen Oxidase (Protox) Inhibitors. The effect of these compounds is to cause a rapid chlorosis and desiccation of the plant. In some cases, initial effects can be

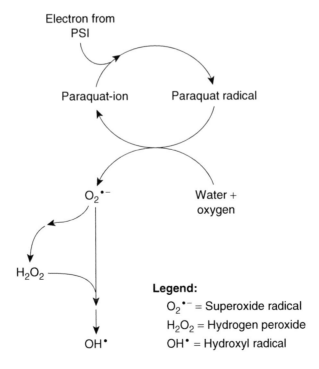

Figure 6.12 Mode of action of paraquat and diquat. The formation of hydroxyl radicals results in lipid peroxidation.

observed within hours of application. The primary target site is an enzyme, protoporphyrinogen oxidase (also called Protox), which is essential for the formation of chlorophyll. Inhibition of this enzyme results in a buildup of protoporphyrin IX, which in turn can react with oxygen and light leading to the formation of singlet oxygen ($^1O_2$). Singlet oxygen then causes rapid lipid peroxidation, membrane destruction, and cell desiccation and death.

Most of the Protox inhibitors are used as contact herbicides. They include the diphenylether herbicides and oxadiazon. Most treated plant surfaces will show symptoms; selectivity is based on differential wetting and penetration of the foliage, metabolism, and the ability to recover from injury.

Glutamine Synthesis Inhibitors. The inhibition of the glutamine synthetase enzyme in plants treated with glufosinate, the only herbicide with this mode of action, results in the buildup of ammonia in the cells. It is uncertain, however, whether the ammonia itself is reponsible for the toxic effects. Inhibition of glutamine synthesis leads to a reduction in amino acid synthesis, and photosynthesis is also inhibited. A result of electrons no longer able to move through the light reaction of photosynthesis appears to be the buildup of triplet chlorophyll molecules (^3chl), which can lead to lipid peroxidation and membrane disruption.

The herbicide is applied only to the foliage, and symptoms include chlorosis within 2 to 5 days after treatment followed by necrosis within 1 to 2 wk. Glufosinate is not selective, but because it is not well translocated, recovery can occur from growing points that do not come into contact with the herbicide.

Direct Solubilization of the Cell Membrane. This direct effect on cell membranes occurs with the aromatic petroleum oils, sulfuric acid and monocarbamide dihydrogen sulfate.

Herbicide Families Applied Almost Exclusively to the Soil

These herbicides are applied mostly to the soil. They do not significantly translocate within the plant but rather affect the plant tissue with which they come into contact. Because of the lack of translocation, they have little or no activity on the foliage of established plants. In most cases, cell division and shoot and root growth are severely inhibited.

Microtubule/Spindle Apparatus Inhibitors. These compounds disrupt the mitotic (nuclear division) sequence (prophase, metaphase, anaphase, telophase) by interfering with the formation or functioning of the spindle apparatus. The spindle apparatus is made up of protein structures called microtubules. These microtubules form the framework on which chromosomes are moved during mitosis. The dinitroaniline herbicides and pronamide prevent spindle formation by slowing or preventing the assembly of the microtubules. The prophase sequence appears normal (Figure 6.13); however, without the presence of the spindle apparatus, the chromosomes are unable to move into the metaphase configuration and daughter chromosomes cannot migrate to their respective poles (anaphase). After a time in the prophase state, the chromosomes coalesce in the middle of the cell and the nuclear envelope reforms causing a polyploid cell. Without the proper distribution of chromosomes to daughter cells, new cell production and eventually growth ceases.

These compounds are most effective on germinating seeds and seedlings. One of the characteristic symptoms of the dinitroanilines and related herbicides is that the roots become swollen and stubby. Microtubules are not only involved in chromosome movement but also regulate the orientation of new cell wall materials (cellulose microfibrils). The loss of microtubules at the cell periphery causes an irregular pattern of microfibril deposition and leads to cells that swell to a spherical rather than a cylindrical shape, hence the swollen root tips. Lateral root production also is inhibited. Selectivity is due to metabolism and to the presence of altered forms of microtubules that are resistant to the activity of the dinitroanilines.

Other herbicides that cause the disruption of mitosis include dithiopyr, thiazopyr, and possibly DCPA; in all cases, the inhibition of microtubule formation or function has been implicated as the cause of toxicity.

Shoot Inhibitors. The cellular mode of action for herbicides placed in this group is unknown, but most inhibit shoot growth of seedling plants and probably affect cell division or cell enlargement processes to some degree. Herbicide groups that inhibit shoot growth are the thiocarbamates and the chloroacetamides. Typical injury symptoms include malformed (twisted), dark green shoots and leaves on injured young plants.

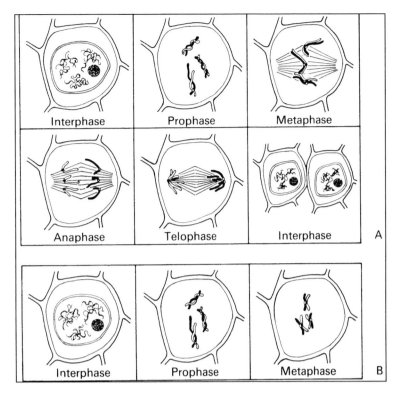

Figure 6.13 (A) Normal mitotic sequence and (B) effect of a dinitroaniline herbicide on mitotic sequence in root cells. (F. D. Hess, Novartis Crop Protection, Inc.)

Miscellaneous Cell Division Inhibitors. Bensulide, napronamide, and siduron inhibit root growth probably by inhibiting cell division or cell elongation processes.

Cell Wall Formation Inhibitors. Several herbicides, dichlobenil, isoxaben, and quinclorac, inhibit cellulose synthesis. The loss of cell wall formation results in abnormal cell formation at the shoot and root tips and eventually death. Quinclorac also has auxin-type growth regulating properties.

HERBICIDE FATE

When herbicides are applied, they make contact with either the soil or foliage. Herbicides intercepted by foliage may (1) be washed from the plant surface and deposited on the soil by rainfall, (2) remain on the plant surface where they are subject to dissipation by processes such as volatilization and photodegradation, or (3) penetrate the cuticle and enter the plant. Herbicides not intercepted by foliage go directly to the soil and, like soil-applied herbicides, are subject to the degradative processes that occur there (see Chapter 7 for a discussion of the fate of soil-applied herbicides).

Once a herbicide has entered the plant, it may move to a site of action where, depending on the degree of sensitivity of the species, it may cause a toxic reaction. Herbicides applied to tolerant species will generally be inactivated so that they are no longer herbicidally active. The terms *herbicide metabolism* and *biotransformation* are used interchangeably to describe the processes within plants that lead to the alteration and inactivation of herbicides. Very few herbicides are broken down all the way to their simplist components (for example, CO_2 and H_2O) by higher plants. In most cases, herbicides are altered chemically to form one or more *metabolites*. These metabolites may be just as toxic as the parent herbicide, but usually they are less toxic or nontoxic to the plant. Often, the metabolites are bound to plant structures or substances (such as glucose) to form *conjugates*. As a conjugate, the herbicide is usually in a nontoxic form and is stored in some part of the plant cell where it cannot affect living processes. The general principles that govern metabolite and conjugate formation follow.

Formation of Herbicide Metabolites

Herbicide molecules typically contain reactive or functional groups or linkages such as hydroxyl (OH), alkyl (C_nH_{2n+1}; for example, CH_3, a methyl group), amino (NH_2), nitro (NO_3), chloro (Cl), carboxyl (COOH), and thio (S) groups. These groups are susceptible to enzymatic or chemical activity and can be altered by processes such as oxidation, reduction, and hydrolysis. Herbicide metabolism primarily involves two types of reactions: oxidations and hydrolyses.

The majority of biotransformation processes are oxidative in nature, and many of those oxidations are catalyzed by a series of enzymes in the plant known as the P_{450} enzymes (the mixed function oxidase enzymes). Interestingly, the P_{450} enzymes are also found in animal systems where they act to detoxify herbicides and other toxic materials.

One type of oxidation reaction catalyzed by the P_{450} enzymes is the hydroxylation of the herbicide ring structure (also called aromatic hydroxylation). Since most herbicides have a ring structure, this reaction, which involves the addition of a hydroxyl group (OH^-) to the ring structure, is very important in plants. For example, in 2,4-D (2,4 dichlorophenoxyacetic acid) metabolism, the #4 chlorine molecule on the phenyl ring is displaced by a hydroxyl group and moved to the #5 or #3 position (Figure 6.14). The resulting compound (2,5 [or 2,3] dichloro-4-hydroxyphenoxyacetic acid) no longer has the auxin-like activity of 2,4-D. In addition, the hydroxyl group readily conjugates with glucose (Figure 6.14), further detoxifying the compound. The 2,4-D molecule also can undergo decarboxylation of the side chain and dechlorination of the phenyl ring. Other examples of ring hydroxylation include the conversion of bentazon to 6-hydroxy bentazon and chlorsulfuron to the 5'-hydroxy derivative (Figure 6.14). However, the latter product is not detoxified until it is conjugated to glucose.

Other types of oxidation processes include *N*-dealkylation (oxidation of alkyl groups), *N*-demethylation (removal of methyl groups), *N*-deamination (removal of amino groups), and sulfoxidation (oxidation of sulfur molecules) (Figure 6.14).

Another way in which plants metabolize herbicides is by hydrolysis, which is the splitting of a molecule through the addition of water. Rice rapidly hydrolyzes propanil to the nontoxic metabolites 3,4-dichloroaniline and propionic acid (Figure 6.15). These reactions are catalyzed by enzymes such as aryl acylamidase or, in the case of atrazine, by a cyclic hydroxamic acid referred to as DIMBOA. The hydrolysis of atrazine in corn roots results in the displacement of the chlorine (Cl) atom by a hydroxyl group (OH), resulting in the production of nontoxic hydroxy atrazine (Figure 6.16).

Formation of Conjugates

Herbicides and herbicide metabolites can be conjugated to sugars, amino acids, proteins, lignin, or other natural plant constituents. When the herbicide has been hydroxylated, the metabolite frequently conjugates with glucose or another sugar in a process called glycosidation. For example, glucose conjugates with the hydroxylated form of 2,4-D (2,5 [or 2,3] dichloro-4-hydroxyphenoxyacetic acid) at the #4 site occupied by the hydroxyl ion (Figure 6.14). Similarly, 6-hydroxy bentazon and the 5'-hydroxy derivative of chlorsulfuron are rapidly conjugated with glucose and detoxified. Members of several important herbicide groups, such as the chloroacetamides, aryloxyphenoxy propionates, triazines, diphenyl ethers, and the sulfoxide metabolites of metribuzin and the thiocarbamates conjugate with the tripeptide molecule, glutathione (glutamate-cysteine-glycine), via an enzymatic reaction involving the enzyme glutathione-S-transferase. Glutathione conjugation is the major way in which corn leaves rapidly detoxify atrazine (Figure 6.16).

Once conjugated, the herbicide is usually rendered immobile, nontoxic, and inactive. It may subsequently be further metabolized or transferred to other conjugants. Because the conjugates tend to be water soluble, they are usually partitioned and stored in aqueous cell components, namely the plant cell wall or vacuole where they can no longer have any impact on the living portions of the cell.

Some herbicides are not metabolized, conjugated, or altered in any way. For example, paraquat, diquat, and glyphosate are extremely stable in plants. However, once the dying plant residues are deposited on the soil, paraquat and diquat are tightly bound to soil particles and are no longer available for plant uptake. They also can undergo photodecomposition on the plant or soil surface. Glyphosate also is absorbed into soil but can be gradually broken down by microbial activity.

It is important to note that the interaction and fate of any herbicide, whether in the plant,

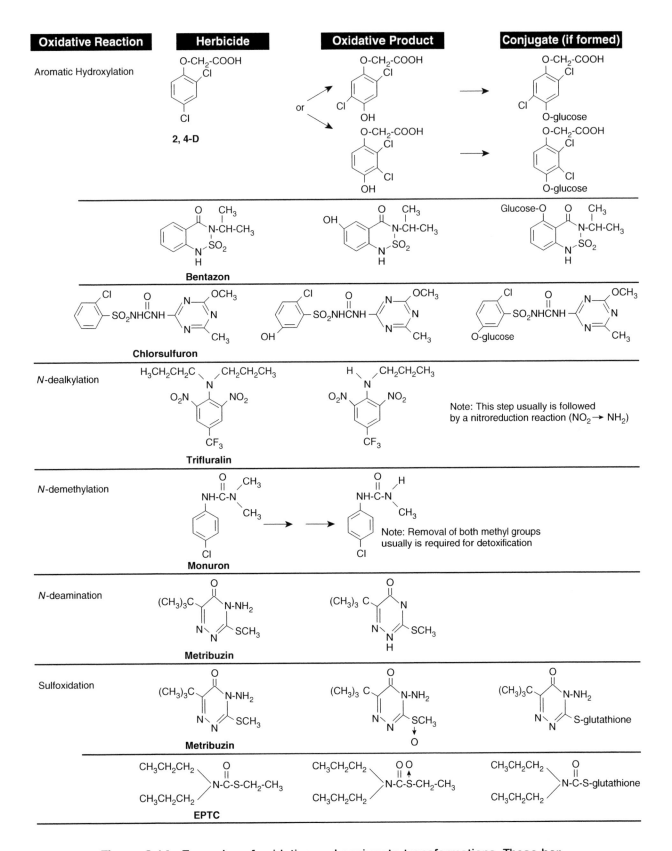

Figure 6.14 Examples of oxidation and conjugate transformations. These herbicides may be further metabolized, or conjugated or metabolized in other ways, depending on the plant species.

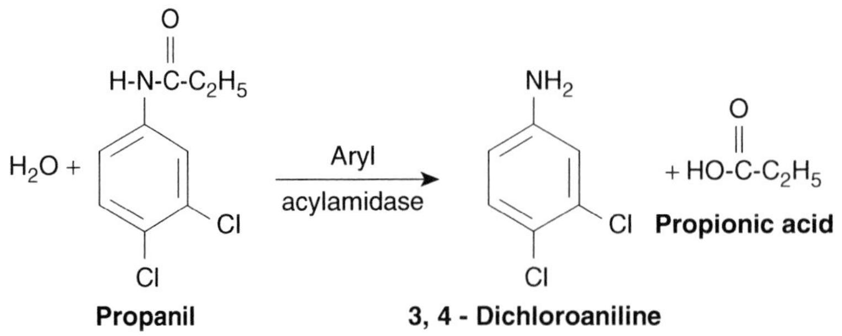

Figure 6.15 Hydrolysis of propanil to the nontoxic metabolites 3,4-dichloroaniline and propionic acid.

soil, atmosphere, water, or other organisms, is determined by the basic principles and laws of physics and chemistry. *Even though herbicides are synthesized by humans, they undergo exactly the same basic chemical reactions and alterations as biologically derived compounds do.* Their fate in both a plant and the environment can be predicted and measured. This is particularly helpful in determining the types and amounts of herbicide residues in crop plants grown for human or animal consumption and which are expected not to have potentially hazardous levels of breakdown products. Compounds that are not degraded to harmless levels are either dropped from consideration as herbicides during development or restricted to noncrop applications.

PLANT-RELATED SELECTIVITY

Some plants can be killed with doses of herbicides that have little or no effect on other plants. This phenomenon is termed *selectivity*. A number of factors determining selectivity are plant related. Age and height differences among plants can be used to successfully remove weeds from some crops. Other plant-related factors include differences in herbicide retention, penetration, translocation, and metabolism. It should always be remembered, however, that selectivity is dose dependent. At high enough concentrations, most herbicides will kill all plants. At sufficiently low dosages, herbicides will not kill any plants. Selectivity occurs at dosages between these two extremes.

Plant Age and Stage of Development

Since plants have definite periods or stages of sensitivity to herbicides, selectivity can be obtained if herbicides are applied at times when the crop plant is most tolerant and the weed is most susceptible. For example, corn is tolerant to many postemergence herbicides if treated before tasseling and after the dough stage of grain development. Many soil-applied herbicides kill germinating and newly emerged seedlings but fail to control plants beyond that stage. Thus, they can be used to control late flushes of emerging weeds in crops that are established.

Plants are also more tolerant when dormant; early-season weeds can be controlled without injuring alfalfa when the herbicide is applied to the crop during its dormant period in the winter or spring. Similarly, herbicides can be applied safely to conifers prior to bud break in the spring and after rapid terminal growth has ceased in the fall.

Plant Height

Height differentials between weeds and crops can be exploited by directing the herbicide spray away from the crop and onto the weeds (Figure 6.17). In this way chemicals with limited selectivity can be used in a number of cropping situations to control weeds not effectively or safely controlled using other means. Examples include the use of 2,4-DB, linuron, or paraquat as directed sprays in soybeans; glyphosate with the rope wick or spongewick applicator in soybeans and cotton; and a number of foliage-applied herbicides used as directed sprays to control herbaceous weeds in established woody species.

Differential Retention

Variations in plant shape, leaf orientation, and leaf surface characteristics can account for differences in foliar retention. Plants with broad, horizontally oriented leaves are more likely to intercept and retain spray than the vertically oriented leaves of plants such as grasses and sedges. Cattails and onions have upright waxy leaves

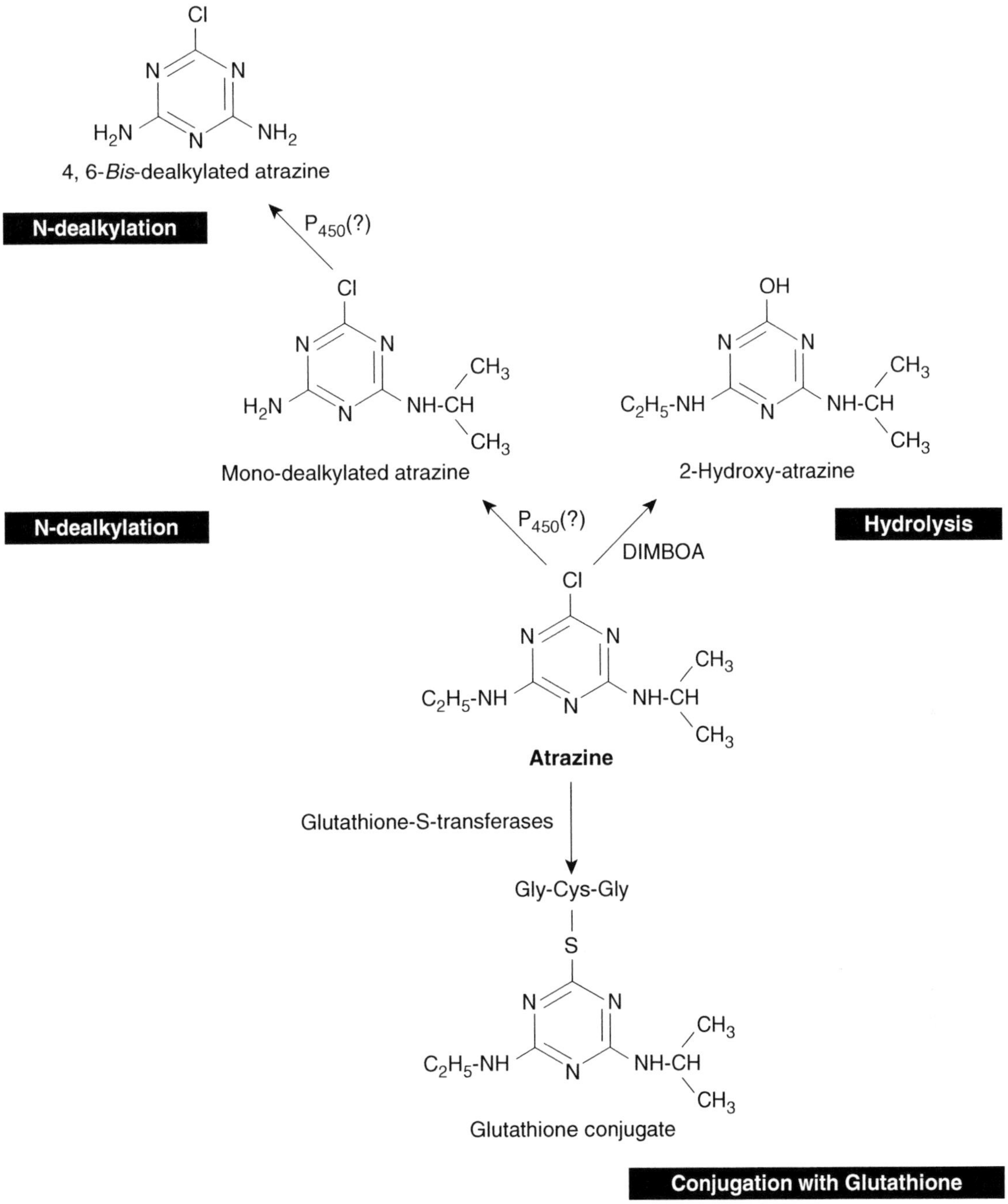

Figure 6.16 Major detoxification pathways for atrazine.

that are difficult to wet. In fact, the former practice of selectively removing broadleaves from onions with solutions of sulfuric acid was largely attributed to differential wetting. The horizontally positioned leaves of coast fiddleneck retained 2.2 times more bromoxynil than the vertically oriented leaves of wheat (Schafer and Chilicote 1970).

Leaves with a moderate degree of hairiness or surface roughness may be better able to trap and retain the runoff of spray than leaves that are smooth and nonhairy. An excessive amount of hairs, on the other hand, may prevent the penetration of the herbicide to the cuticular surface.

Differential Penetration

Differential penetration is probably most important as a selectivity factor among plants of different

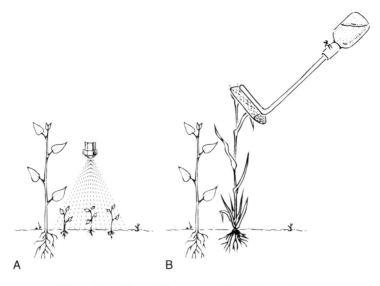

Figure 6.17 Selectivity using height differentials among plants: (A) herbicide directed at the base of a crop plant to kill smaller weed seedlings, (B) herbicide applied through a spongewick applicator to kill tall weeds in a shorter crop.

ages since cuticle thickness tends to increase and cuticle composition may change as the plant matures. This is probably one of the reasons seedling plants are much more susceptible to herbicides than mature plants. There are few documented instances in which differential penetration alone has been the cause of selectivity among herbaceous species at the same stage of growth.

The addition of wetting agents or alterations in herbicide formulation can decrease both retention- and penetration-related selectivity by increasing the amount of herbicide entering the plant. This is often the cause of injury among crops that normally show a high degree of inherent tolerance to a particular herbicide. The entry of massive amounts of the chemical may be more than the plant can effectively and rapidly detoxify. For example, the addition of crop oil to 2,4-D amine can result in injury to corn when 2,4-D amine alone would be nontoxic. Ester formulations of 2,4-D penetrate plant tissue more readily than amine and other salt formulations and are therefore recommended at lower dosages in order to be used safely for broadleaf weed control in corn and small grains.

Differential retention or uptake are seldom major selectivity factors among soil-applied herbicides. As long as the herbicide is present in the root zone and in the soil water solution, it will be taken up by the plant.

Differential Translocation

Upon entry, herbicides may be conjugated, adsorbed, and/or compartmentalized in portions of the plant where no damage can occur. When this happens, the herbicide is no longer available for transport to the site of action. An example in which binding or inactivation at or near the point of entry protects certain plants is linuron in the roots of parsnip and carrots. The inactivation prevents the translocation of the herbicide to the leaves and its site of action in the chloroplast. Another example is norflurazon, which is compartmentalized in the glands of glanded cotton. Also in cotton, fluridone is taken up by the roots but does not translocate into the shoots. However, it is translocated in susceptible weed species. One of the bases for the selectivity of 2,4-D and other phenoxy herbicides is their slower rate of translocation in grasses and other tolerant species. However, these cases are exceptions. Most translocatable materials do move to sensitive cellular sites in quantities large enough to cause plant injury. Unless accompanied by differential metabolism, selectivity is seldom obtained by differential translocation alone.

Differential Metabolism/Conjugation

Alteration of the herbicide within the plant is the major reason for selectivity among most species. Certain herbicides are much more rapidly metabolized and conjugated by tolerant species. This alteration can occur in a number of ways depending on the herbicide and the plant and have been described in the section on herbicide fate.

A classic case in which differential metabolism accounts for the majority of the selectivity is that of the chloro-*s*-triazine herbicides when used for weed control in certain grass crops. The chloro-*s*-triazines can be detoxified by three different pathways: *N*-dealkylation, hydrolysis, and glutathione conjugation (Shimabukuro et al. 1978) (Figure 6.16). *N*-dealkylation of atrazine is a minor metabolic pathway that occurs in the leaves and roots of most plants. Its presence in species such as the broadleaved crops soybean and cotton may account for their partial tolerance to atrazine. DIMBOA-mediated hydrolysis, which converts atrazine to hydroxy atrazine, occurs primarily in plant roots and contributes substantially to detoxification of soil-applied atrazine. Glutathione conjugation occurs in the leaves. DIMBOA-mediated hydrolysis and glutathione conjugation both result in the removal of the chlorine from the atrazine molecule and account for the majority of detoxification processes by tolerant plants.

The ability of corn to utilize all three pathways renders it particularly tolerant to atrazine. Sorghum, which is slightly less tolerant than corn, only detoxifies atrazine via glutathione conjugation. Species susceptible to atrazine lack the ability to metabolize it to hydroxy-atrazine and to produce high enough levels of glutathione-S-transferase to produce glutathione conjugates. For example, glutathione-S-transferase was present in high concentrations in atrazine-tolerant species such as corn, sorghum, Sudan grass, sugarcane, and johnsongrass but was missing in atrazine-susceptible species such as barley, wheat, oat, and pigweed (Frear and Swanson 1970; Lamoureux and Frear 1979). Thompson et al. (1971) found that the order of susceptibility among five grass species was correlated with their ability to metabolize atrazine to hydroxy-atrazine and the glutathione conjugate. The increasing order of susceptibility was corn, fall panicum and large crabgrass, giant foxtail, and oats. Six hours after treatment, these five species had metabolized 96, 44, 50, 17 and 2% of the atrazine to nontoxic forms, respec-

tively. Atrazine in fall panicum and large crabgrass, which are relatively tolerant, appeared to be converted mostly to the glutathione conjugate rather than to hydroxy-atrazine. These selectivity mechanisms work whether the herbicide is applied to the soil for root uptake or to the foliage.

In some cases, the herbicide is in a nontoxic form when applied but is converted to a toxic form in susceptible plants. An example is that of 2,4-DB which is used for weed control in small-seeded legumes. Susceptible species convert nontoxic 2,4-DB to toxic 2,4-D by a process known as beta-oxidation (Figure 6.18), whereas resistant species such as the small-seeded legumes do not. In the case of diclofop, differential detoxification after conversion to a toxic form in the plant accounts for selectivity (Figure 6.19). Diclofop-methyl (the form of the herbicide that is applied) itself is not toxic, but hydrolysis of the herbicide inside the plant results in the formation of diclofop, which is toxic. Tolerant plants such as wheat metabolize diclofop by hydroxylation of the ring structure and conjugation. Although susceptible plants such as wild oats can form a glucose ester with the diclofop, apparently this is a reversible association that results in the release and toxic expression of diclofop.

Role of Varietal Differences in Selectivity

Genetic differences among plants within a species can vary substantially and can account for selectivity. Varieties of sugarbeets, corn, and soybeans all have been shown to vary in their response to individual herbicides (Figure 6.20). Differences also have been detected among weed populations; for example, johnsongrass, Canada thistle, and field bindweed biotypes respond differently to some herbicides. In most cases, the differences are tenfold or less.

The maximum tolerance that plants can develop to herbicides is due to genetic changes that result in herbicide resistance. The development

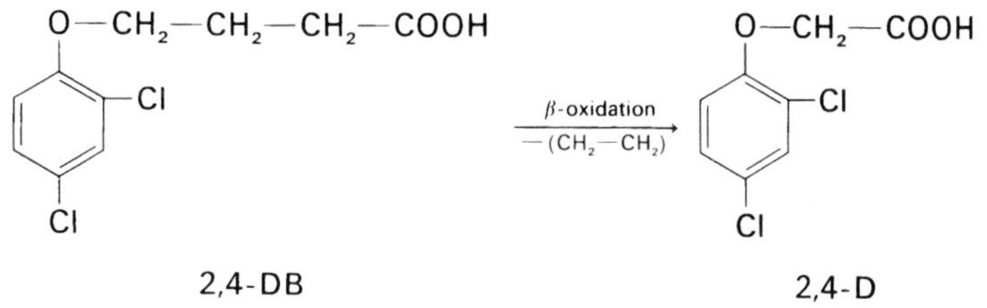

Figure 6.18 Beta-oxidation of 2,4-DB to 2,4-D.

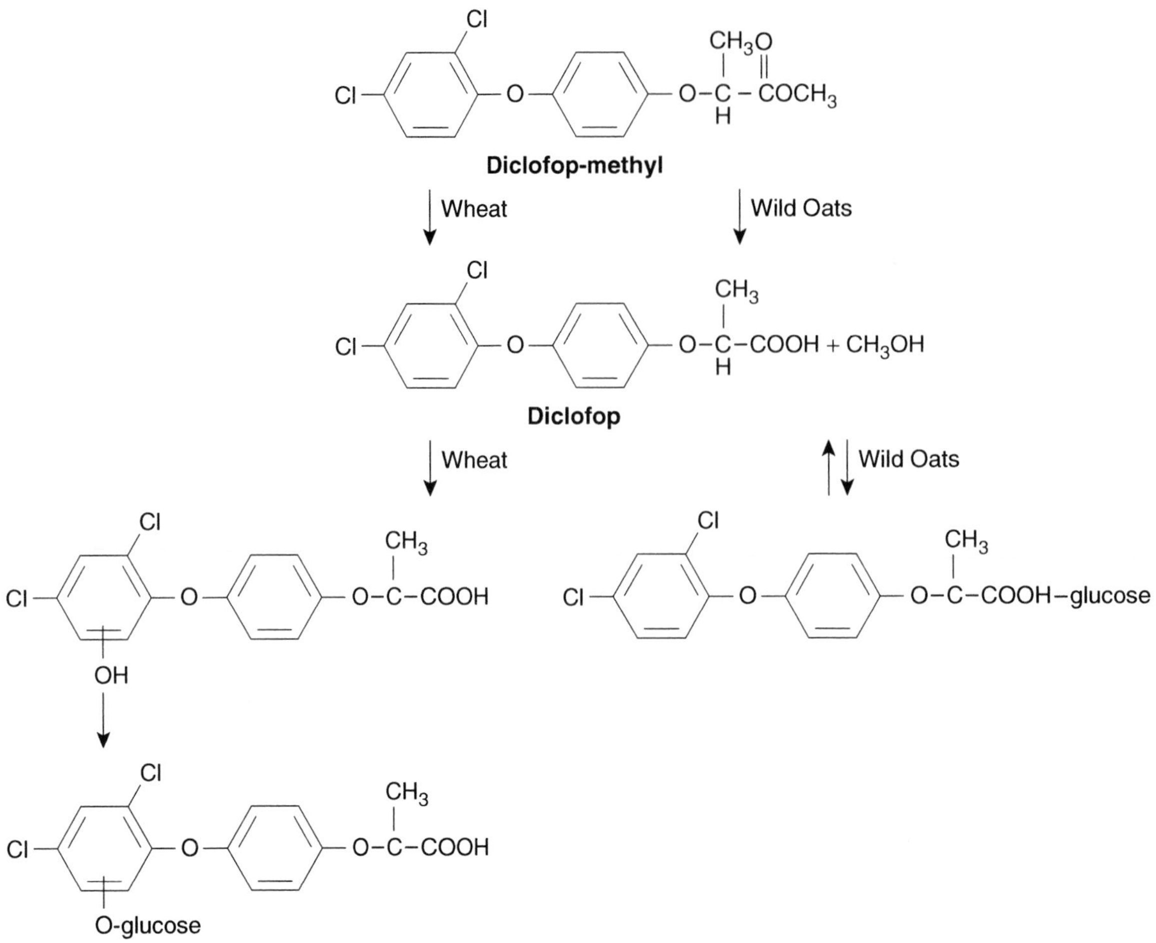

Figure 6.19 Biotransformations of diclofop-methyl when applied to tolerant wheat and susceptible wild oats. Although both plants form glucose esters of diclofop, the reaction is reversible in wild oats, leading to the release and toxic expression of diclofop.

of herbicide resistance within weeds and crops is discussed in Chapter 8.

Role of Other Chemicals in Selectivity

The addition of other chemicals can increase or decrease the activity and selectivity of herbicides. Combining diclofop with phenoxy herbicides results in reduced effectiveness of the diclofop for wild oat control. Some of the other aryloxyphenoxy propionate herbicides also exhibit antagonism when tank mixed with other postemergence herbicides (Table 6.1).

The use of herbicides such as mefluidide and certain thiocarbamate herbicides preconditions some plant species to be more sensitive to later applications of postemergence herbicides. In the latter case, the thiocarbamates reduce the amount of cuticle produced by tolerant plants thus rendering them more susceptible to herbicide uptake and damage when postemergence herbicides are used.

Several chemicals (called safeners or protectants) have been shown to improve the tolerance of partially tolerant grass species (mostly corn and sorghum) to herbicides in the thiocarbamate and chloracetamide groups without increasing the tolerance of the susceptible species. One of the earliest safeners used was naphthalic anhydride, which was applied to the crop seed to confer protection to the thiocarbamate herbicides during germination. Some protectants are packaged with the herbicide. For example, the safener dichlormid is packaged with butylate or acetochlor, and a currently unnamed safener designated R-29148 is packaged with EPTC or butylate. Both safeners increase corn tolerance to the herbicides. Benoxacor (proposed common name for CGA-154281) is prepackaged separately and then added to metolachlor at a 1:30 ratio, also to increase corn tolerance. The Concep products (proposed common names are oxabetrinil and fluxofenim) and flurazole are used as seed treatments to protect sorghum

Figure 6.20 Differences in EPTC tolerance among certain corn inbred lines: (left) less susceptible, (right) more susceptible.

Table 6.1 Antagonism Between Selective Postemergence Herbicides (Haloxyfop, Fluazifop, and Quizalofop) and Bentazon or Acifluorfen When Used for Johnsongrass Control

		% Control of Johnsongrass		
	lb/Acre	Alone	Bentazon	Acifluorfen
Haloxyfop	1/16	89	89	54
Haloxyfop	1/8	97	90	76
Fluazifop	1/8	82	64	48
Fluazifop	1/4	97	90	78
Quizalofop	1/16	74	30	31
Quizalofop	1/8	81	75	42
Average				
Low rates		82	61	44
High rates		92	85	65

from injury by metolachlor and other chloracetamides.

In most cases documented to date, the safener appears to protect the crop by increasing the metabolism of the herbicide. They do this by increasing the levels of either glutathione and glutathione-S-transferases, which increase glutathione conjugation, or the P_{450} enzymes, which enhance the oxidation of herbicides. The use of safeners can result in a marginally safe compound becoming an important and effective herbicide.

MAJOR CONCEPTS IN THIS CHAPTER

For a herbicide to be effective, it must contact the plant surface, penetrate into the plant, and move to reach a living site in quantities sufficient to disrupt a vital process. It also must subsequently degrade to nontoxic components either in plants themselves or in the soil.

Herbicide uptake from the soil is dependent on the availability of the herbicide in the soil solution. Plant structures that are imbibing or absorbing water will take up herbicides. These structures include germinating seeds, shoots and roots of seedlings prior to emergence from the soil, and roots of emerged plants.

Uptake of foliar-applied herbicides is more difficult to achieve than that of soil-applied herbicides and involves the processes of retention, cuticular penetration, and movement.

The more oil soluble a foliar-applied herbicide is, the better it will be retained on the cuticle and the more rapidly it will penetrate. Water-soluble herbicides do not penetrate rapidly; they require the addition of wetting agents for adequate penetration.

Retention times of 6 to 24 hr frequently are needed to prevent loss of activity of water-soluble herbicides from wash off by heavy rain. Retention times needed for oil-soluble herbicides are considerably shorter, possibly as little as 1 hr.

Environmental parameters that affect the cuticle influence penetration. High humidity, adequate soil moisture, and warm temperatures enhance penetration. Minimum kill can be expected when plants are under conditions of water stress, low humidity, and cool temperatures.

Variable weather conditions account for much of the inconsistency encountered with the performance of foliar-applied herbicides.

Herbicides move into cells by diffusion, ion trapping, or with the aid of protein carriers located on the plasma membrane. The latter two mechanisms require metabolic energy (ATP).

Herbicide movement is dependent on the pathway and the amount of herbicide moved. Herbicides are described as symplastically translocated (moving downward), apoplastically translocated (moving upward only), or limited movement or contact.

- Symplastically translocated herbicides move with the sugars to actively growing (young) tissues and to sites of storage. These herbicides must be used at concentrations sufficient to kill susceptible sites but low enough not to kill the phloem tissue that serves as the transport system. They must be used if the major objective of a foliar treatment is to kill all parts of the plant, including the underground portions (e.g., in perennial plants). Injury progresses from new to older growth.
- Apoplastically translocated herbicides move into older leaves and concentrate in the leaf tips and margins. Injury progresses from older to new growth.
- Contact herbicides move little in the plant and cause rapid localized killing. Thus, top kill with a contact herbicide requires good foliage coverage.
- Contact and apoplastically translocated herbicides will not move into underground structures from foliage applications.

The way a herbicide affects a plant is called its mode of action.

- Modes of action of foliar-applied, symplastically translocated herbicides showing initial symptoms on new growth include auxin-type growth regulators, aromatic amino acid inhibitors, branched-chain amino acid inhibitors, carotenoid pigment inhibitors, and lipid biosynthesis inhibitors.
- Modes of action of foliar-applied, apoplastically symplastically translocated herbicides showing initial symptoms on old growth include "classical" photosynthetic inhibitors and low doses of the "rapidly acting" photosynthetic inhibitors.
- Modes of action of foliar-applied, non-translocated (contact) herbicides showing

initial localized injury include photosystem I energized cell membrane destroyers, Protox inhibitors, glutamine synthesis inhibitors, and herbicides that directly solubilize the plasma membrane.

- Modes of action of herbicides applied almost exclusively to the soil include the microtubule/spindle apparatus inhibitors; shoot inhibitors; miscellaneous cell division inhibitors; and cell wall formation inhibitors.

A herbicide that does not penetrate the plant may photodegrade or may move into the soil and be subjected to soil degradative processes. A herbicide that enters the plant may be inactivated (a must for herbicides applied to food and feed crops), activated, or conjugated to an inactive site; remain unchanged; or be exuded from the plant.

Selectivity is the ability of a herbicide to remove one plant without undue injury to another. Herbicide selectivity is related to dose. Young plants from seed usually are most susceptible. Resistance increases with age. Crops tend to be susceptible to herbicide injury when young or when reproductive structures are being formed.

- Differentials in plant height can be exploited to gain selectivity by using directed sprays or wipe-on devices.
- Differential retention and wetting can account for selectivity between plants with markedly different leaf orientation and surface waxiness. Adjuvants added to foliar sprays can alter this type of selectivity.
- Differences in the rates of herbicide penetration, translocation, or deactivation may account for selectivity. In some cases, differential metabolism from an inactive to an active compound accounts for herbicide selectivity.
- Individuals within a species can exhibit a range of heritable differences to herbicides. Sugar beets, corn, soybeans, johnsongrass, and Canada thistle are all species in which individuals can exhibit up to a tenfold difference in herbicide response.
- Some chemicals can increase or decrease activity and selectivity. Herbicides such as haloxyfop, fluazifop, and quizalofop can exhibit reduced activity when tank mixed with phenoxy herbicides, acifluorfen, and bentazon.
- Several chemicals are used to improve the tolerance of large-seeded grass crops to herbicides such as EPTC, butylate, alachlor, and

metolachlor. These compounds are called safeners or protectants. Some are used as seed treatments; others are prepackaged with the herbicide.

TERMS INTRODUCED IN THIS CHAPTER

ACCase inhibitor
ALS inhibitor
Amino acid
Apoplastically translocated herbicide
Aromatic amino acid inhibitor
ATP
ATPase hydrogen ion pump
Auxin-type growth regulator
Beta-oxidation
Biotransformation
Branched-chain amino acid inhibitor
Carotenoid
Carotenoid pigment inhibitor
Carrier
Cell wall formation inhibitor
Chlorophyll
Chlorosis
Classical photosynthetic inhibitor
Conjugates
Contact herbicide
Cutin
D2 and D1 proteins
DNA
Epinasty
EPSPS inhibitor
Fatty acid
Glutamine synthesis inhibitor
Glutathione conjugation
Herbicide metabolism
Hydrolysis
Hydrophilic
Ion trapping
Limited-movement herbicide
Lipid biosynthesis inhibitor
Lipophilic
Metabolites
Microtubule/spindle apparatus inhibitor
N-dealkylation
Necrosis
Oxidation
Pectin
P_{450} enzyme
Photosynthesis
Photosystem I
Photosystem I energized cell membrane destroyer
Photosystem II
Protectant

Protox inhibitor
Rapidly acting photosynthetic inhibitor
Reduction
Respiration
RNA
Safener
Selectivity
Sucrose loading
Symplastically translocated herbicide
Wax

SELECTED REFERENCES ON PLANT–HERBICIDE INTERACTIONS

Ashton, F. and A. S. Crafts. 1981. Mode of Action of Herbicides. John Wiley and Sons, New York.

Böger, P. H. and G. Sandman, Eds. 1989. Target Sites of Herbicide Action. CRC Press, Boca Raton, Florida.

Devine, M. D., S. O. Duke, and C. Fedtke. 1993. Physiology of Herbicide Action. Prentice Hall, Englewood Cliffs, New Jersey.

Dodge, A. D., Ed. 1989. Herbicides and Plant Metabolism. Cambridge University Press, Cambridge.

Duke, S. O., Ed. 1985. Weed Physiology. Vol. II. Herbicide Physiology. CRC Press, Boca Raton, Florida.

Hatzios, K. K. 1991. Biotransformations of herbicides in higher plants. Pages 141–185 in Environmental Chemistry of Herbicides, Volume II, Grover, R. and A. J. Cessna, Eds. CRC Press, Boca Raton, Florida.

Hatzios, K. K. 1991. Enhancement of crop tolerance to herbicides with chemical safeners. Pages 293–303 in Herbicide Resistance in Weeds and Crops, Caseley, J. C., G. W. Cussans and R. K. Atkin, Eds. Butterworth-Heinemann, Oxford, England.

Hatzios, K. K. and R. E. Hoagland. 1989. Crop Safeners for Herbicides: Development, Uses, and Mechanisms of Action. Academic Press, San Diego.

Hatzios, K. K. and D. Penner. 1982. Metabolism of Herbicides in Higher Plants. Burgess Publishing, Minneapolis.

Herbicide Handbook of the Weed Science Society of America, 7th ed. 1994. Lawrence, Kansas.

Kearney, P. C. and D. D. Kaufman, Eds. 1975 and 1976. Herbicides, Chemistry, Degradation, and Mode of Action. Vol. I and II. Marcel Dekker, New York.

Lamoureux, G. L., R. H. Shimabukuro, and D. S. Frear. 1991. Glutathione and glucoside conjugation in herbicide selectivity. Pages 227–261 in Herbicide Resistance in Weeds and Crops, Caseley, J. C., G. W. Cussans and R. K. Atkin, Eds. Butterworth-Heinemann, Oxford, England.

Moreland, D. E., J. P. St. John, and F. D. Hess, Eds. 1982. Biochemical Responses Induced by Herbicides. Amer. Chemical Soc., Washington, D.C.

Sterling, T. M. 1994. Mechanisms of herbicide absorption across plant membranes and accumulation in plant cells. Weed Sci. 42:263–276.

7

SOIL–HERBICIDE INTERACTIONS

Soil is a major factor in the use of herbicides. In this chapter we describe how various soil processes affect the activity and fate of herbicides and then relate this information to weed control. For more information on soil–herbicide interactions, see Kearney and Kaufman (1975, 1976), Hance (1980), Grover (1988), and Grover and Cessna (1991).

FATE OF HERBICIDES IN THE SOIL

The major processes that determine the fate of a herbicide when it reaches the soil are adsorption, movement, and degradation (Figure 7.1). All three of these processes are interrelated and difficult to separate, yet the herbicide user must have a general understanding of how these processes affect herbicides in order to determine how much of the chemical to apply, how far it will move from the site of application, and how long it will persist in the soil.

Adsorption

Adsorption affects all other aspects of herbicidal activity and fate. Molecules tightly bound to soil colloids are not available for weed control, probably will have limited mobility in the soil, and may not be accessible for degradation by soil microorganisms. Loosely bound herbicides, on the other hand, will be available for weed con-

trol if they can be retained in the proper position in the soil for uptake. They may be susceptible to leaching and probably will be available for microbial and other types of degradative processes.

Adsorption is defined as the association of molecules with the surfaces of solids. In soils, the negatively charged particles of clay and organic matter provide most of the surface area for adsorption of ionized particles. Through this kind of adsorption and the reverse process, desorption, herbicide molecules attain an equilibrium between the soil colloid and the soil solution (Figure 7.2). Herbicides with a negative charge tend to be repelled by negatively charged soil colloids, whereas herbicides with a positive charge tend to be attracted to them. Herbicide availability in the soil depends a great deal on the kind and amount of colloids present, the ionization properties of the herbicide, and the amount of soil moisture.

Not all herbicide binding can be explained by ion adsorption, particularly when organic matter is involved. The "active" component of organic matter is *humus,* which coats or partially coats the mineral components of soil. Humus mostly consists of a complex of degraded plant residues and microbially synthesized polymers. A single molecule of humus is very large and is composed of numerous aromatic ring structures to which are attached carboxyl ($-COOH$), hydroxyl ($-OH$), and other ionizable groups (Figure 7.3). At typical soil pHs, the carboxyl and hydroxyl groups are ionized ($-COO^-, -O^-$),

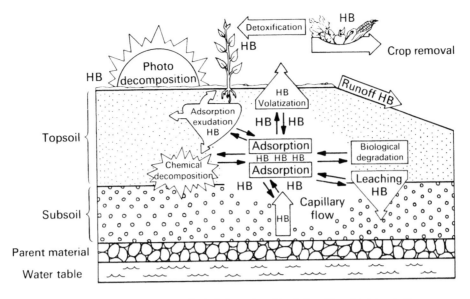

Figure 7.1 Processes that determine the fate of herbicides in the environment (HB=herbicide) (redrawn from Weber et al. 1973).

giving humus the overall negative charge that results in cation adsorption. The presence of the aromatic rings in the molecule, however, also imparts a nonpolar property to humus, which allows it to interact with nonpolar chemicals (such as nonpolar herbicides).

The phenomenon of "like dissolves like," in this case with respect to nonpolar substances, is termed *partitioning*. Partitioning implies movement of one substance into another rather than just adsorption of a substance to a solid surface. *Sorption*, the retention of a substance on or *in* a solid phase, is probably a more appropriate term to describe soil–herbicide interactions than is adsorption.

Factors that determine the amount of sorption include properties of the soil colloids, solubility and ionization properties of the herbicide, and the amount of soil moisture.

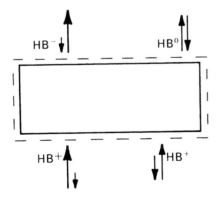

Figure 7.2 Equilibria between representative herbicide molecules and negatively charged soil colloids.

Colloids

Organic matter generally has a greater impact on the sorptive capacity of a soil than does clay. Sorptive capacity, however, can differ greatly where high or low adsorptive types of clay predominate (Table 3.1).

Special considerations may be needed when choosing herbicides and selecting dosages for soils with high colloidal contents (see the section under Soils, Herbicides, and Weed Control entitled, Availability). In fact, the attraction of herbicides to organic matter can be predicted to some extent by analytical measurements of the sorption potential of herbicides to different soil types. The *Herbicide Handbook* of the Weed Science Society of America (1994) lists the sorption coefficients K_d and K_{oc} for most herbicides. Knowing these values can help predict the behavior of herbicides in the soil and specifically to organic matter.

K_d, or the soil sorption coefficient, is measured in a water/soil slurry as the ratio of the amount of herbicide that is sorbed to the soil in relation to the amount of herbicide that is left in the water. Therefore, a low K_d indicates that the herbicide only slightly sorbed to the soil; a high K_d indicates strong sorption to the soil. K_{oc}, the soil organic carbon sorption coefficient, is calculated as K_d divided by the weight fraction of organic carbon present in the soil. Again, a low K_{oc} indicates slight sorption to organic matter; a high K_{oc} indicates strong sorption to organic matter. Therefore, the K_{oc} is a measure of the sorption based on the amount of organic carbon in a particular soil.

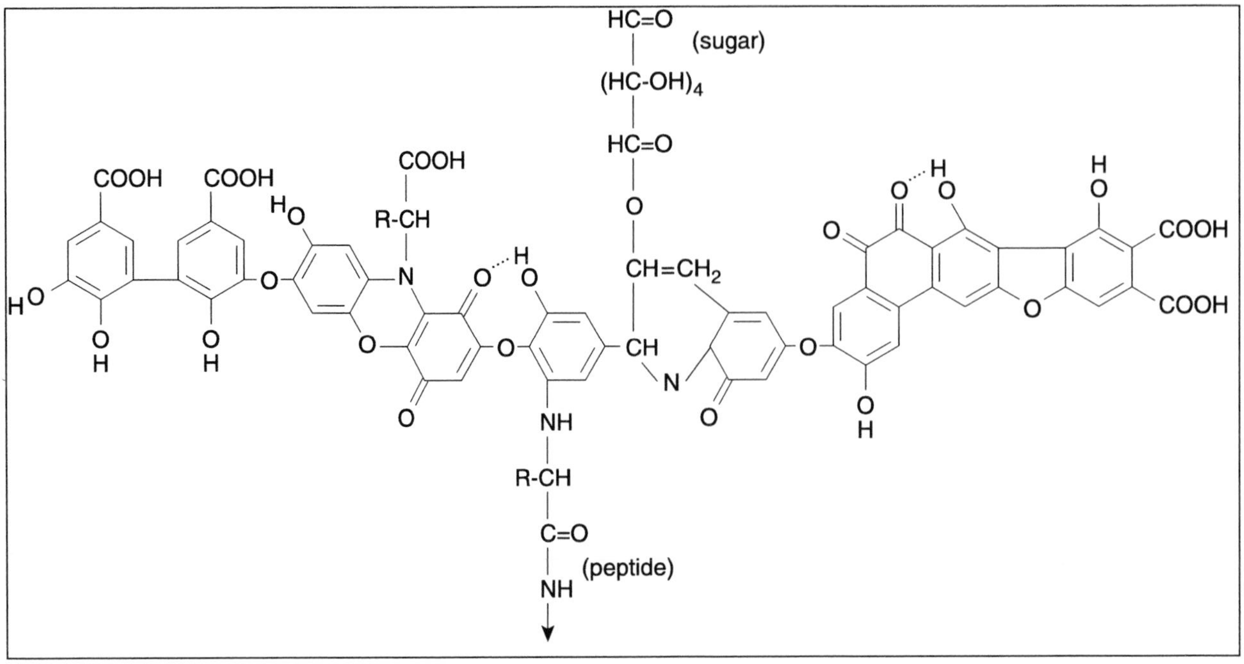

Figure 7.3 Hypothetical structure of humic acid (humus). Note the numerous ionizable (—COOH, —OH) groups and aromatic rings. (From F. J. Stevenson, *Humus Chemistry: Genesis, Composition, Reactions*, Wiley-Interscience, New York, 1982, p. 259)

Examples of how K_d and K_{oc} can be used in a general way to estimate the sorption of herbicides to different soil types are illustrated in Table 7.1. The percentage of clay and the percentage of organic matter in the soils tested for herbicide sorption are roughly within the same range in both the ametryn and pendimethalin examples. Ametryn showed low values for K_d and K_{oc}, a reflection of its moderate sorption to most soils. In contrast, pendimethalin had much higher K_ds and K_{oc}s than ametryn; these values are indicative of the strong sorption of pendimethalin to most soils. Note that it is difficult to use these data for anything other than general trends. The K_ds and K_{oc}s within a data set tend to increase with increasing percentages of clay and of organic matter. However, the relationships are not perfect. For example, the K_{oc} of ametryn was lower in the sandy loam soil (96) than in the sand (205) even though the sandy loam soil had a higher percentage of organic matter than the sand. The lack of direct proportionality in data sets that provide K_ds and K_{oc}s can be due to variations in the soils, such

Table 7.1 K_d (Soil Sorption Coefficients) and K_{oc} (Soil Organic Carbon Sorption Coefficients) Derived for Ametryn and Pendimethalin Mixed with Different Soil Types in a Water/Soil Slurry. The % clay and % organic matter contents of each soil also are provided.

Soil type	% clay	% organic matter	K_d	K_{oc}
Ametryn[a]				
Sand	2.2	0.9	1.1	205
Sandy loam	16.8	1.9	1.1	96
Loam	9.0	0.8	1.2	257
Clay	15.0	4.8	26.2	927
Pendimethalin				
Loamy sand	3.2	0.8	30	15,000
Sandy loam	11.2	1.6	110	13,000
Loam	15.0	3.8	301	13,700
Silt loam	19.2	4.7	380	14,100
Silty clay loam	25.0	5.0	854	29,400

[a]The K_f (Freundlich isotherm) is used in place of K_d for ametryn.
Data taken from the *Herbicide Handbook* of the Weed Science Society of America, Lawrence, Kansas (1994).

as different carbon contents of the organic matter; the presence of different adsorptive types of clays (Table 3.1); other soil components and chemical factors that affect sorption; and analytical errors.

Ionization Properties

Ions will be tightly bound to soil if positively charged (cationic) and repelled if negatively charged (anionic). Cations such as the herbicides diquat and paraquat have no soil activity since they are tightly bound by soil colloids. Most anions such as nitrate are repelled by soil colloids and are readily mobile in the soil.

Some anions, however, can bind to soil. An example is phosphate ($H_2PO_4^-$). Phosphate binds to cationic metals such as calcium, iron, or aluminum (Ca^{+2}, Fe^{+2}, Fe^{+3}, Al^{+3}) that are associated with negatively charged colloids (clays or organic matter). This interaction results in the formation of colloid-metal-phosphate complexes (Figure 7.4), which effectively bind the phosphate to the colloid. Phosphate also binds to cations that are present in the soil solution to form insoluble calcium, iron, or aluminum phosphates. Similarly, glyphosate, which is anionic because of its phosphonic acid group, binds to vacant phosphate sorption sites in the soil. The higher the concentrations of metallic cations such as calcium, iron, and alu-

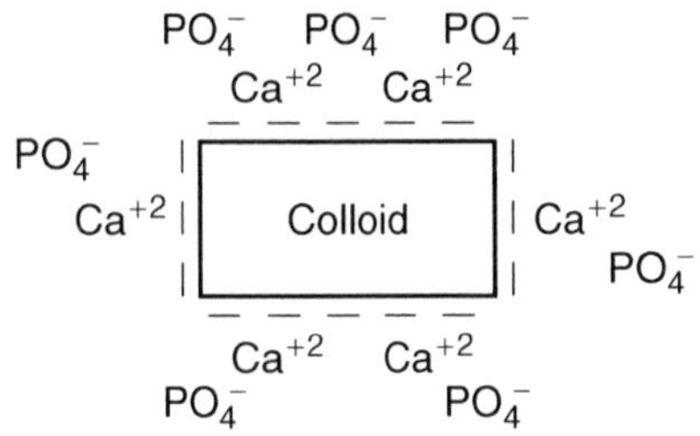

Figure 7.4 A colloid-cationic metal (Ca^{+2})-phosphate complex. Glyphosate, which has a phosphonic acid group, forms a similar complex with colloids.

minum in soils or in water (such as in the spray tank), the greater the glyphosate sorption and inactivation.

Herbicides that are weak acids or bases only partially ionize. The degree of ionization depends on soil conditions such as pH (Figure 7.5). For example, weak acids such as 2,4-D are uncharged molecules at low pHs but are anionic at high pHs. Most soils have pH values sufficient for the dissociation of 2,4-D acid to its anionic state. Thus, 2,4-D acid, once converted to the anion, is moderately leachable in soils. However, 2,4-D formulated as an ester is not affected by pH until the molecule has undergone hydrolysis to the acid form. Then it takes on the anionic charge and becomes susceptible to leaching. Symmetrical triazine, imidazolinone,

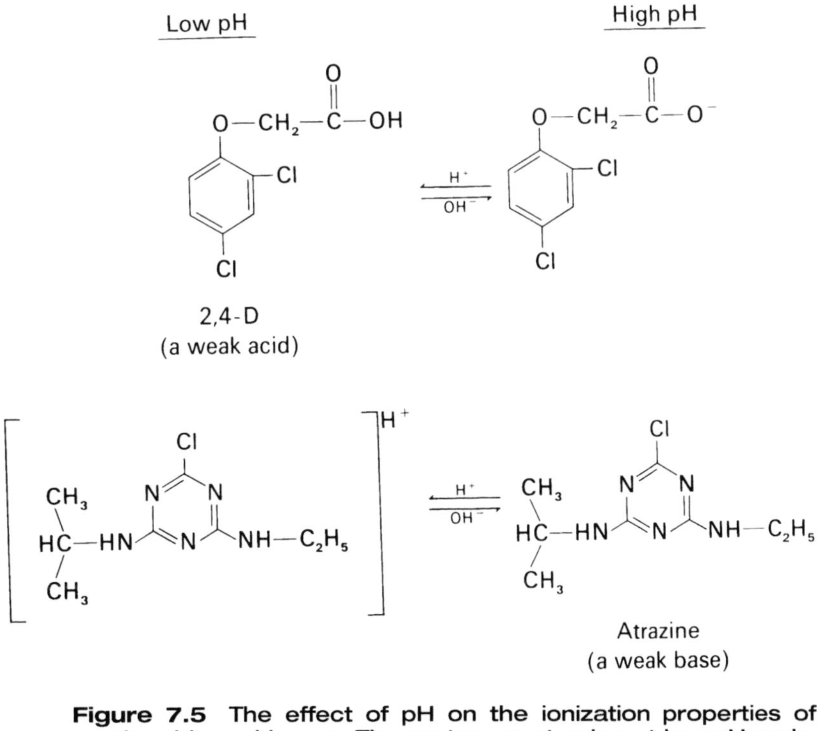

Figure 7.5 The effect of pH on the ionization properties of weak acids and bases. The proton on atrazine at low pH probably resonates among N atoms on the ring and side chains. (Cruz et al. 1968)

and other basic herbicides take on an overall positive charge at low soil pHs, becoming relatively tightly bound under these conditions. These compounds are more readily available to the soil solution when applied to neutral or alkaline soils.

The majority of herbicides have no charge and are unaffected by soil pH or surface charges. These herbicides behave according to other properties such as their volatility, size, and the strength of weak attractive forces that exist between molecules. The lipophilic versus hydrophilic properties of weakly charged or uncharged herbicides also affect sorption. Herbicides that are hydrophilic tend to partition into water and thus have a high water solubility. Herbicides that are lipophilic tend to sorb into the aromatic ring components of organic matter rather than into water. The general trend is for herbicides to sorb to soil in an inverse relationship to their water solubility (i.e., the lower the soil sorption, the greater the water solubility). For example, dicamba is poorly sorbed to soil, with an average K_{oc} of only 2 ml/g. It has, however, a water solubility of 720,000 ppm (Table 7.2). Prodiamine, on the other hand, has a high sorptive capability ($K_{oc} = 13,000$ ml/g) but a very low water solubility at 0.013 ppm.

Soil Moisture

Soil moisture interferes with the adsorption of herbicide molecules because water competes for binding sites on the soil particles. This interference is particularly important in the case of volatile herbicides such as the thiocarbamates, in which adsorption is essential for herbicide retention in the soil. On wet soils, a film of water on the soil particles prevents adsorption and the herbicide is lost to the atmosphere. Higher soil temperatures also increase volatility. The greater the volatility of the herbicide, the more susceptible it is to vapor loss with increasing soil moisture content and temperature.

Soil moisture even affects the sorption/desorption characteristics of herbicides that are not significantly volatile. Herbicides tend to bind tightly to dry soil particles; with increasing moisture, more of the herbicide desorbs from the soil and becomes available in the soil solution. If the soil is too dry, the growth and ability of plants to take up herbicides decreases. Although a herbicide bound to dry soil is not available for uptake, such binding actually may help prevent herbicide loss through leaching or volatilization. The addition of moisture to the soil is often said to

"activate" a herbicide. In other words, more herbicide becomes available in the soil solution and the plants begin to grow and take up substances from the soil solution. Adequate moisture for plant growth ensures a much better chance for good weed control.

Movement

The application of a herbicide accounts for its initial entry onto or into the soil system. The extent of movement from the site of application generally determines the position of the compound in relation to developing plant parts and the amount of herbicide available for plant uptake. The ability of a compound to move is largely determined by its affinity for soil colloids and its water solubility. The major methods by which movement takes place are volatilization and movement with water. Herbicides also can be moved when absorbed into other systems such as vegetation and animals.

Volatilization

Volatilization is the physical change of liquid or solid to a gas. It is responsible for the loss of herbicides on or near the soil surface to the atmosphere and often is responsible for redistribution in the soil. Two problems associated with volatilization are excessive and rapid loss of the herbicide, leading to loss of weed control, and injury to sensitive off-target plants.

The potential of a compound to volatilize is determined by vapor pressure. To the best of our knowledge, problems associated with herbicide loss from the site of application by volatilization has primarily involved herbicides with vapor pressures greater than 1.0×10^{-4} mm Hg at 25°C. Examples include the thiocarbamate herbicides, with vapor pressures that range from 1.1×10^{-4} mm Hg (triallate) to 3.4×10^{-2} mm Hg (EPTC), and trifluralin, with a vapor pressure of 1.1×10^{-4} mm Hg. Losses of almost the total amount of herbicide that was applied have been reported for EPTC (90% loss) and trifluralin (74% loss) when applied to the surfaces of wet soils. Losses over the growing season approaching one-fourth of the original dose have been documented for soil-incorporated applications of trifluralin and triallate.

Injury to nontarget vegetation from vapors has been a continuing and common problem for esters of auxin-type growth regulators and more recently for clomazone (vapor pressure = 1.4×10^{-4} mm Hg).

Most herbicides with vapor pressures above 1×10^{-4} mm Hg are incorporated into the soil

Table 7.2 Water Solubilities and Soil Organic Carbon Sorption Coefficients (K_{oc}) of Selected Soil-Applied Herbicides[a] Arranged According to Decreasing Water Solubilities. As water solubility decreases, sorption to organic matter increases.

Herbicide	Solubility in Water (ppm)[b]	Average K_{oc} (ml/g)
Dicamba (amine salt)	720,000	2
Clopyralid (amine salt)	300,000	6
Picloram (potassium salt)	200,000	16
Hexazinone	33,000	54
Chlorsulfuron	31,800	40
Tebuthiuron	2500	80
Metribuzin	1100	60
Clomazone	1100	300
Molinate	970	190
Triasulfuron	815	65–191
2,4-D (amine salt)	796	20
Prometon	720	150
Terbacil	710	55
Bromacil	700	32
Propachlor	613	112
Metalachlor	488	200
Chlorimuron	450	110
EPTC	370	200
Sulfometuron	300	78
Alachlor	242	124
Ametryn	200	300
Cyanazine	171	190
Fluometuron	110	100
Vernolate	108	260
Cycloate	85	600
Linuron	75	400
Napropamide	73	700
Pebulate	60	430
Imazaquin	60	20
Butylate	45	400
Diuron	42	480
Prometryn	33	400
Atrazine	33	100
Norflurazon	28	700
Bensulide	25	1000
Butachlor	23	700
Dichlobenil	21.2	400
Siduron	18	420
Pronamide	15	800
Fluridone	12	1000
Simazine	6.2	130
Triallate	4	2400
Oryzalin	2.6	600
Isoxaben	1	190–570
Oxadiazon	0.7	3200
DCPA	0.5	5000
Trifluralin	0.3	7000
Ethalfluralin	0.3	4000
Pendimethalin	0.275	17,200
Oxyfluorfen	0.1	100,000
Benefin	0.1	9000
Prodiamine	0.013	13,000

[a]Data from *Herbicide Handbook* of the Weed Science Society of America, Lawrence, Kansas (1994).
[b]Water solubilities at 25°C or within 10°C of 25°C.

or are applied as granules or slow-release formulations to prevent loss from volatilization. It is essential that these compounds be applied to cool dry soils or be incorporated immediately. Even with incorporation, sizable losses of these volatile herbicides into the atmosphere can oc-cur on some coarse-textured soils. Such losses may be due to the porous nature of these soils and the lack of binding sites.

Herbicides vary greatly in volatility, ranging from the most nonvolatile compound listed in Table 7.3, sulfometuron, with a vapor pressure of

Table 7.3 Vapor Pressures of Selected Soil-Applied Herbicides[a] Arranged According to Decreasing Vapor Pressures

Herbicide	Vapor Pressure (mm Hg)[b]
EPTC	3.4×10^{-2}
Butylate	1.3×10^{-2}
Vernolate	1.0×10^{-2}
Pebulate	8.9×10^{-3}
Cycloate	6.2×10^{-3}
Molinate	5.6×10^{-3}
Dichlobenil	1×10^{-3}
Clomazone	1.4×10^{-4}
Triallate	1.1×10^{-4}
Trifluralin	1.1×10^{-4}
Pronamide	8.5×10^{-5}
Ethalfluralin	8.2×10^{-5}
Propachlor	7.9×10^{-5}
Benefin	7.8×10^{-5}
Metalachlor	3.1×10^{-5}
Linuron	1.7×10^{-5} (20°C)
Alachlor	1.6×10^{-5}
Pendimethalin	9.4×10^{-6}
Dicamba (amine salt)	9.2×10^{-6}
Prometon	7.7×10^{-6}
Butachlor	4.5×10^{-6}
Napropamide	4×10^{-6}
Ametryn	2.7×10^{-6}
DCPA	2.5×10^{-6}
Oxyfluorfen	2×10^{-6}
Clopyralid (amine salt)	1.3×10^{-6}
Prometryn	1.2×10^{-6}
Fluometuron	9.4×10^{-2}
Bensulide	8×10^{-7}
Oxadiazon	7.76×10^{-7}
Isoxaben	$<3.9 \times 10^{-7}$
Bromacil	3.1×10^{-7}
Terbacil	3.1×10^{-7}
Atrazine	2.9×10^{-7}
Hexazinone	2×10^{-7}
2,4-D (amine salt)	1.4×10^{-7}
Metribuzin	1.2×10^{-7} (20°C)
Tebuthiuron	1×10^{-7}
Fluridone	$<1 \times 10^{-7}$
Diuron	6.9×10^{-8}
Norflurazon	2.9×10^{-8}
Prodiamine	2.51×10^{-8}
Simazine	2.2×10^{-8}
Imazaquin	$<2 \times 10^{-8}$ (45°C)
Triasulfuron	$<1.5 \times 10^{-8}$
Oryzalin	$<1 \times 10^{-8}$
Siduron	4×10^{-9}
Cyanazine	1.6×10^{-9} (20°C)
Chlorsulfuron	2.3×10^{-11}
Chlorimuron	4×10^{-12}
Sulfometuron	5.5×10^{-16}
Picloram (potassium salt)	Negligible

[a]Data from *Herbicide Handbook* of the Weed Science Society of America, Lawrence, Kansas (1994).
[b]Vapor pressures at 25°C unless otherwise indicated.

5.5×10^{-16} mm Hg at 25°C, to the most volatile compound listed, EPTC, with a vapor pressure of 3.4×10^{-2} mm Hg at 25°C (Table 7.3). More than a trillionfold difference in vapor pressures separates these two compounds. Significant differences also exist among herbicides considered volatile. For example, EPTC is approximately 300 times more volatile than trifluralin.

The vapor pressures of most herbicides are in the range of 10^{-8} to 10^{-6}, and they are much less likely than the thiocarbamates to move off target because of volatilization. Even these compounds, however, can be affected by volatility. Over an entire growing season, appreciable loss (from 10 to 90%) of herbicides from the soil surface may occur because of volatility, photodegradation, or a combination of the two.

There can be some benefits to volatilization. Movement through volatilization can help to even out herbicide distribution patterns in the soil. Thiocarbamates such as EPTC and pebulate can redistribute as much as 4 in. in the soil after application. Trifluralin will redistribute about an inch. Herbicides with appreciable vapor pressures may reach and be taken up by underground plant parts as vapors.

Water Movement

Water is a major factor responsible for the movement of herbicides on or in soils. The major types of water movement are runoff, leaching, and capillary action.

Surface runoff is responsible for the movement of herbicide molecules on the soil surface to areas outside of the target area. Rarely is more than 2 to 3% of a herbicide, however, moved off site by runoff. Surface runoff of herbicides is most likely with heavy rainfall before the herbicides have a chance to move into the soil. Once herbicides penetrate the soil surface, movement with runoff takes place only if the soil is suspended and carried in the water. Movement of herbicides in runoff is most serious when the herbicide can enter bodies of water or is carried to sensitive vegetation. Surface movement and injury to susceptible vegetation frequently is a problem with highly persistent herbicides applied to soil surfaces at relatively high rates.

Leaching is the movement of herbicides with water downward through the soil profile. It is a major method of herbicide delivery to underground plant parts. Leaching of herbicides away from the root zone can account for as much as a 4% loss of herbicide. Problems associated with leaching are movement of the herbicide away from the root zone, resulting in poor weed control, and the potential for contamination of groundwater and other subsurface water supplies such as tile lines that will eventually discharge into surface waters.

The amount of herbicide leached is a function of several factors including water solubility, K_d (sorption capability), susceptibility to degradation, and the timing and amount of water moved. Herbicides with high water solubility and low sorption properties are more likely to leach than herbicides that tightly sorb to soil colloids. On the other hand, highly water-soluble herbicides also are available to microbes, and a herbicide that is susceptible to degradation may not persist long enough to sustain significant leaching. Obviously, timing of the leaching event(s) determines how much herbicide, even a biodegradable one, will leach. Leaching that occurs immediately after herbicide application may move significant quantities of herbicide out of the root zone before it can be degraded, whereas leaching that occurs after the herbicide has degraded will move only small amounts of the herbicide downward. The volume of the water percolating through the soil also influences the amount and rate of leaching. In general, leaching is maximized when highly water soluble, long-lived herbicides are used.

The potential for a herbicide to move through soil can be estimated by determining how much herbicide moves with a wetting front through a soil column or through soil-based thin layer chromatography plates. The relative mobilities of herbicides are expressed as R_F values, which then can be categorized into the five-class system developed by Helling and Turner (1968). Herbicides with a low R_F value are tightly sorbed to soil and show little or no movement, whereas herbicides with a high R_F value are highly mobile (Table 7.4). As a general rule, mobility in the soil is consistent among members of the same chemical group. For example, the bipyridilium (diquat, paraquat), dinitroaniline, and organic arsenical herbicides and glyphosate are not mobile (Class 1) in soil. The diphenylether and imidazolinone herbicides show low to moderate mobilities (Class 2), and the *s*-triazine, sulfonylurea, and thiocarbamate herbicides are in the low to moderate (Class 2) and moderate (Class 3) movement classes. Herbicides that are auxin-type growth regulators (2,4-D, picloram, and dicamba) show moderate to high (Class 4) and high movement (Class 5).

Capillary action is the movement of water upward through small pores in the soil in response to evaporation of water at the soil surface. The herbicide movement that accompanies this process can counteract herbicide movement by leaching. Movement of materials by capillary action is particularly important in areas where subirrigation or furrow irrigation is used in climates with low humidities.

Absorption

Movement of herbicides sometimes occurs when the compound is absorbed into a biological system (crop plants, microorganisms, animals).

Table 7.4 Relative Movement of Selected Herbicides in a Silty Clay Loam Soil Based on Helling's Classification System

Class 1. Little or no movement (0.0–0.9 R_F)
  Bipyridilium herbicides
    Diquat
    Paraquat
  Dinitroaniline herbicides
    Benefin
    Ethalfluralin
    Oryzalin
    Pendimethalin
    Trifluralin
  Organic arsenical herbicides
    DSMA
    MSMA
  Others
    Glyphosate
Class 2. Low to moderate movement (0.1–0.34 R_F)
  Imidazolinone herbicides
    Imazapyr
    Imazaquin
    Imazethapyr
  Diphenylether herbicides
    Fomesafen
    Oxyfluorfen
  Substituted urea herbicides
    Diuron
    Linuron
  Sulfonylurea herbicides
    Sulfometuron
  Symmetrical triazine herbicides
    Prometryn
  Thiocarbamate herbicides
    Pebulate
    Vernolate
  Others
    Bensulide
    Clomazone
    Diclobenil
    Oxadiazon
    Propanil
Class 3. Moderate movement (0.35–0.64 R_F)
  Chloroacetamide herbicides
    Alachlor
    Metolachlor
  Asymmetrical triazine herbicides
    Metribuzin
  Substituted urea herbicides
    Fluometuron
    Tebuthiuron
  Sulfonylurea herbicides
    Chlorsulfuron
    Metsulfuron
  Symmetrical triazine herbicides
    Ametryn
    Atrazine
    Cyanazine
    Prometon
    Simazine
  Thiocarbamate herbicides
    Butylate
    EPTC
  Others
    DCPA
    Endothall
    Ethofumesate
    Fluridone
    Napropamide
    Naptalam
    Norflurazon
    Terbacil

Table 7.4 (Continued)

Class 4. Moderate to high movement (0.65–0.89 R$_F$)
    Asymmetrical triazine herbicides
        Hexazinone
    Chloroacetamide herbicides
        Propachlor
    Phenoxy herbicides
        2,4-D
    Picolinic acid herbicides
        Picloram
    Others
        Amitrole
        Bromacil
Class 5. High movement (0.9–1.00 R$_F$)
    Benzoic acid herbicides
        Dicamba
    Others
        Bentazon

If not degraded, the herbicide may be carried as a contaminant to some other site or remain at the site but be unavailable for other processes.

Examples of herbicides moved in biological systems include crop plants removed from a treated area by humans. Although the herbicides used on most food and feed crops have dissipated by harvesttime, residues sometimes are present. For example, terbacil residues can be carried on spent distilled mint hay. Other examples of movement include grazing animals that feed on newly treated vegetation and then excrete unaltered herbicides when they move to other areas. Grass clippings from a recently sprayed lawn used as a mulch can carry herbicides to sensitive plants. Similar injury can occur when topsoil with herbicide residues is moved to new sites and comes into contact with established or newly planted sensitive species.

Degradation

Herbicides readily degrade in the environment. This characteristic has both desirable and undesirable consequences. It is desirable from the standpoint that herbicide residues will not build up in the environment. It is undesirable from the standpoint that the period of weed control obtained is limited and may be shorter than needed.

Degradation rates for herbicides approximate typical half-life curves (Figure 7.6). This means that a herbicide with a half-life of 50 days is degraded to one-half of its original amount by 50 days after application, to one-fourth of its original amount by 100 days after application, to one-eighth of its original amount by 150 days after application, and so on. For a

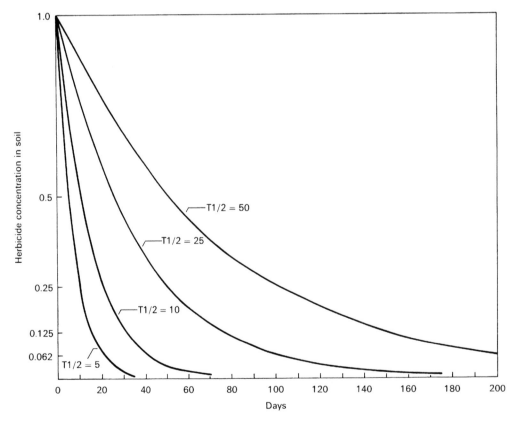

Figure 7.6 Persistence curves for compounds with half-lives of 5, 10, 25, and 50 days. Actual half-life curves for 2,4-D and dicamba in Oklahoma soils approximate the 5- and 25-day curves, respectively. (Altom and Stritzke 1973)

herbicide with a half-life as long as 120 days, and assuming 240 days per year of conditions suitable for herbicide degradation, the amount of herbicide left at the end of 240 days would be approximately one-fourth of the original application amount. The half-life for the majority of herbicides is less than 120 days (Table 7.5), placing them in the residual and persistent herbicide categories, and many of these have half-lives of 3 mo or less.

Herbicides with half-lives of 120 days or more (highly persistent herbicides) have significant potential for carryover; in other words, they can harm susceptible plants during the second season and beyond. Part of this is due not only to the presence of herbicide residues in the soil but also to the fact that certain plant species may be susceptible to extremely low dosages of the herbicide, even though the majority of it has degraded. In fact, herbicides with relatively short half-lives can cause carryover to crops that are extremely sensitive to small amounts of the herbicide. For example, imazethapyr, a herbicide with a half-life of 60 to 90 days, is used for weed control in soybeans and peanuts. However, because of the potential for extremely minute amounts of the herbicide to carry over and cause crop injury, imazethapyr-treated areas should not be planted to cotton for 18 mo after application, and to potatoes for 26 mo after application (see the Pursuit label for other crops affected). In the case of potatoes, which are extremely sensitive to low amounts of the herbicide, imazethapyr is highly persistent.

Since most herbicides are degradable, they do not build up in the soil, even after repeated use for many years. Theoretically, with several consecutive annual applications and degradation to one-fourth, one-third, or one-half the highest level for any given year, the maximum amount of remaining herbicide 1 yr later would approach one-third, one-half, and one time the initial application rate, respectively (Figure 7.7). Among currently available herbicides, only the highly persistent herbicides degrade to just one-half the highest level (Figure 7.7C).

In the eastern corn belt, atrazine has a relatively long half-life (for a selective herbicide), and residues toxic to sensitive plants can be expected to last into a second year following the initial application. Thus, a full year without atrazine usually is required before rotating to a sensitive crop. Atrazine continues to degrade each year that it is used to nearly the same levels as the year before (it approximates curves shown in Figure 7.7A), so only 1 yr without use provides enough time for the herbicide to dissi-

Table 7.5 Half-Lives and Categories for Length of Persistence of Selected Soil-Applied Herbicides. Although these half-lives generally reflect the persistence of herbicides in soil, factors such as soil temperature, moisture, pH, herbicide dosage, and susceptibility of plant species can shift a herbicide from one category into another.

Herbicide	Half-life[a] (in days)
Residual herbicides (harm susceptible species less than one season)	
Dicamba	5
EPTC	6–30
Propachlor	7
2,4-D	10
Butachlor	12
Butylate	13
Cyanazine	14
Pebulate	14
Alachlor	21
Molinate	21
Sulfometuron	20–28
Cycloate	20–30
Persistent herbicides (harm susceptible species during the first and sometimes into the second season after application)	
Clomazone	30
Metolachlor	30–50
Metribuzin	30–60
Benefin	40
Chlorimuron	40
Chlorsulfuron	40
Clopyralid	40
Pendimethalin	44
Trifluralin	45
Ametryn	60
Atrazine	60
Bromacil	60
DCPA	60
Dichlobenil	60
Ethalfluralin	60
Flumetsulam	60
Imazaquin	60
Linuron	60
Oxadiazon	60
Prometryn	60
Pronamide	60
Simazine	60
Thiazopyr	64
Imazethapyr	60–90
Napropamide	70
Triallate	82
Fluometuron	85
Diuron	90
Hexazinone	90
Siduron	90
Highly persistent herbicides (harm susceptible species during the second season and sometimes for longer periods of time)	
Bensulide	120
Isoxaben	(50–)120
Prodiamine	120
Oryzalin	(20–)128
Triasulfuron	(69–)139
Imazapyr	(25–)142
Norflurazon	(45–)180
Terbacil	180
Picloram	(90–)300
Prometon	300
Tebuthiuron	300

[a]Data from the *Herbicide Handbook* of the Weed Science Society of America, Lawrence, Kansas (1994).

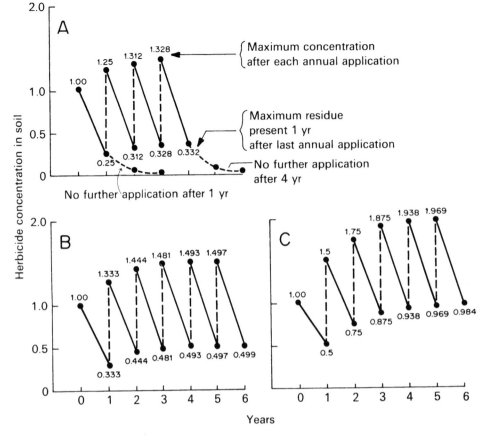

Figure 7.7 The effect of repeated yearly applications on the persistence of a degradable herbicide in which one-fourth (A), one-third (B), and one-half (C) the original concentration remains at the end of the season.

pate, even though the herbicide may have been used over a period of many years.

Degradation processes can be classified according to the type of degradative agent involved. Photodegradation is the breakdown of the compound by light, biodegradation is breakdown caused by biological systems, and chemical degradation occurs without involvement of living organisms.

The soil is an excellent medium for biodegradation. The soil microbial population represents a complex adaptable system capable of producing numerous enzymes that degrade a large number of pesticides. The principal biochemical reactions associated with microbial changes in herbicides include dealkylation (removal of an aliphatic chain), decarboxylation (removal of a carboxyl group to produce carbon dioxide), dehalogenation (removal of fluorine, chlorine, iodine, or bromine), amide or ester hydrolysis (cleavage of a bond by addition of water), oxidation (removal of electrons), reduction (addition of electrons), hydroxylation of an aromatic ring, cleavage of an aromatic ring, and conjugation (combining of two molecules). Several of these same reactions also occur in

photodegradation and chemical degradation processes.

The rate of herbicide degradation is affected by dosage, environmental conditions, affinity for binding to soil colloids, and the microbial population.

Dosage

Herbicides applied at high dosages take longer to dissipate than those applied at low dosages. Thus, when simazine is used at rates for total vegetation control (20 to 40 lb/acre), three or more seasons are required for the compound to dissipate before sensitive plants can be reestablished. When used at selective rates (2 to 4 lb/acre), 2 yr (year of application plus 1 yr) usually are sufficient for dissipation to acceptable levels to occur to allow planting of sensitive crops. Increasing the dosage of a herbicide that is in the persistent category (half-life of less than 120 days) may shift its persistence to that of highly persistent herbicides.

Injury from carryover of selective herbicides on cropland is frequently associated with the application of higher dosages because of double

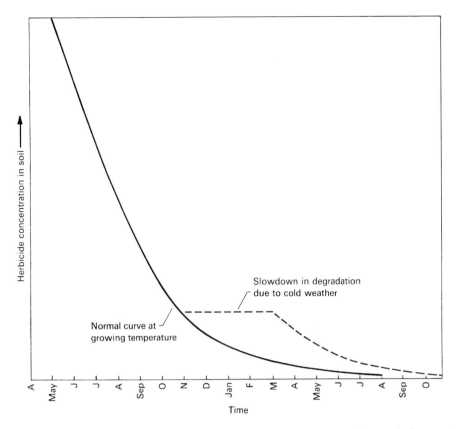

Figure 7.8 Herbicide degradation pattern over a year. Degradation rates during the winter are considerably lower than during the summer.

spraying (overlapping of spray patterns) or mistakes in calculating the dosage or mixing the spray solutions.

Environmental Conditions

The two most important environmental factors are temperature and moisture. Little or no breakdown of herbicides takes place when temperatures approach freezing. Herbicides therefore can persist longer in cold than in warm climates. They also remain stable through the winter; degradation proceeds again in the spring (Figure 7.8). In contrast, periods of extremely warm, moist soil conditions usually accelerate herbicide breakdown.

Affinity for Binding

Herbicides that are strongly adsorbed to soil colloids, such as paraquat and diquat, are not available for rapid microbial breakdown. Herbicides that are highly mobile and leach deep into the soil profile out of the zone of greatest microbial activity also degrade at slower rates.

Microbial Populations

Microbes capable of breaking down the herbicide must be present in the soil in sufficient numbers for microbial degradation to proceed. Microbes respond to herbicides in three ways: by building up their populations in response to the food source, by adapting their enzymes to degrade the compound, or by degrading the herbicide without adaptive changes.

In enzyme adaptation, the chemical acts as a trigger that causes the microbes to produce enzymes capable of degrading that chemical. The period of time required for enzyme induction and the buildup of the induced microbial population is termed the lag phase. After this period, degradation can proceed at its optimal rate. The length of the lag phase varies with the herbicide and with environmental conditions, but it has been reported to be about 2 wk for phenoxy herbicides.

In some cases, the soil microbes may be unable to use the herbicide as a major food source, yet they have the enzymes necessary to degrade the herbicide. Degradation proceeds without the lag phase or population enrichment. This type of degradation apparently occurs when triazine and uracil herbicides are applied to soil.

In recent years, the continued application of certain herbicides has resulted in the buildup of microorganisms capable of degrading those herbicides. This phenomenon, which has caused an increase in the rate of degradation and a subsequent loss of weed control, has been documented for the phenoxy and thiocarbamate herbicides.

SOIL, HERBICIDES, AND WEED CONTROL

Success in controlling weeds with soil-applied herbicides depends on three conditions. The herbicide must be in contact with actively developing plant structures (dormant meristems and seeds are not affected), it must be available to the plant in adequate quantities, and moisture must be present. These conditions are determined largely by the three processes described in the previous section: sorption, movement, and degradation. How these processes interact to affect weed control is described here.

Contact with Developing Plant Structures

Herbicides do not kill inactive seeds or buds, so they must be present in the proper position and at the proper time for optimal exposure to a developing plant part. Positioning usually is determined by the depth of the absorbing or sensitive plant structure. For example, control of most small-seeded weed species that germinate near the soil surface can be accomplished by placing the herbicide in the upper inch or so of soil. Plants developing from deep-germinating seeds or vegetative structures may be more susceptible to herbicides placed deeper in the soil.

Positioning of the herbicide can be achieved by leaching with rainfall or overhead irrigation or by mixing with tillage equipment (incorporation).

Soil-Related Selectivity

Herbicide selectivity in soil sometimes can be achieved by taking advantage of the ability to position a herbicide, the mobility of the herbicide in soil, and the differential root growth between weed and crop. This type of selectivity is called *depth protection*.

The first example of achieving depth protection is to place an immobile, soil-applied herbicide at the soil surface when a deep-rooted crop is present (Figure 7.9A). The herbicide kills shallow-rooted weeds, but the crop roots function effectively below the zone of the herbicide. The control of annual weeds in established woody and herbaceous perennial crops by using simazine or oryzalin is an excellent example of this approach.

This method also is used to control shallowly germinating weeds in large-seeded annual crops and transplants. Large-seeded crops can emerge from depths of 1 in. or more. Therefore, herbicides that normally show little selectivity for a crop but have limited mobility in the soil sometimes can be used safely as a surface application on a deep-planted, large-seeded crop. The tolerance of corn to surface-applied pendimethalin is attributed to this type of depth protection.

When depth protection is the primary method of selectivity, any condition that brings the herbicide into the germinating crop seed zone (or vice versa) will cause injury. Poor seed coverage and shallow planting are two conditions that expose or bring the seed toward the soil surface and increase injury. Similarly, unusually deep placement of the herbicide and crop injury can occur from mechanical incorporation or excessive rainfall or by treating coarse-textured soils.

Selectivity on transplants can be increased by dipping the plants in slurries of activated charcoal to deactivate the herbicide. This precaution is necessary when tolerance is limited or if there is danger of herbicide movement into the root zone. This method has been particularly effective with tomatoes transplanted into trifluralin-treated soil. Selectivity also can be improved by placing a band of activated charcoal in the seed zone of direct-seeded crops.

The second example of achieving differential placement consists of using a mobile herbicide (frequently applied as a spot treatment) and leaching it to kill a deep-rooted species (Figure 7.9B). Injury of shallow-rooted ground cover can be kept to a minimum. This technique commonly is used for removing brush or trees from a grass sod; for example, using tebuthiuron pellets to control brush in a pasture situation.

Availability

Ideally, soil-applied herbicides should be available at a concentration just sufficient to control weeds for the entire season and then disappear before the next crop is planted (Figure 7.10). Herbicides are degradable, however, and begin to disappear as soon as they are applied to the soil, so maintaining a uniform level of herbicide during the season seldom is achieved. In actual practice, high rates of herbicide are applied initially (often with some risk of crop injury) in order to provide the level of control needed later in the season as the herbicide concentration decreases.

Herbicides vary considerably in their rates of degradation. A residual herbicide may be so

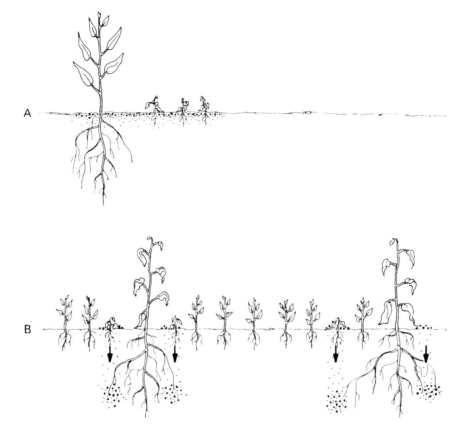

Figure 7.9 Examples of herbicide selectivity by differential soil movement. Placement of an immobile, soil-applied herbicide to the soil surface will control shallow weeds and leave deep-rooted crops unaffected (A). Mobile herbicides can be used as spot treatments to control deep-rooted weeds, leaving most of the shallow-rooted crops unharmed (B).

short lived that its level will drop below that needed for weed control long before the season is over (Figure 7.10). On the other hand, a persistent herbicide may provide desired full-season control but may last into the next year and damage a sensitive crop.

Theoretically, the ideal level of herbicide could be obtained through the use of slow-release herbicides. Unfortunately, this technology has not been accepted widely for use with herbicides. Applications of short-lived herbicides made sequentially at appropriate intervals over the growing season also have the potential to provide season-long weed control while preventing carryover to subsequent crops (Figure 7.10B). Two applications of short-lived compounds such as alachlor can be used to provide season-long control in seed corn.

The availability of soil-applied herbicides for weed control also is dictated by their affinity for clay and organic matter. The greater the affinity for soil colloids, the higher is the dose required for adequate control. It is therefore important that the user know the adsorptive capacity of the soil being treated. Clay content can be determined by mechanical texture analysis, and organic matter by chemical analysis. These tests are conducted routinely for a nominal charge by laboratories that provide soil testing services. Since soil texture does not change over time (except through severe erosion), one determination for a field or portion of a field should be sufficient. Organic matter usually is stable over several years, so a determination at 5- or 10-yr intervals should be adequate. In fields that have one or more soil types, each soil should be sampled and analyzed individually rather than combined into a composite sample.

Charts on the product labels of soil-applied herbicides indicate use rates on the basis of texture and organic matter according to a standard format suggested by the Weed Science Society of America. The rules require that soils be described first by texture. Soil textures are grouped into categories of coarse, medium, and fine (Figure 3.1). Coarse textures include sand, loamy sand, and sandy loam. Medium textures include loam, silt loam, silt, and sandy clay loam. Fine

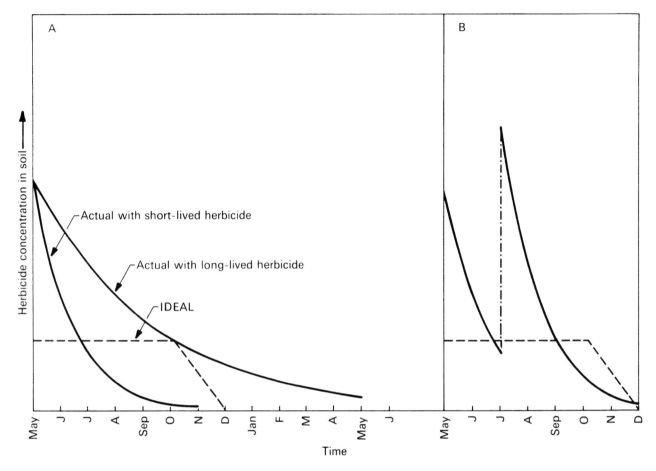

Figure 7.10 Ideal versus actual concentrations of soil-applied herbicides over a season (A). Two applications of short-lived compounds (B) can provide season-long control while preventing carryover to the next year's crop.

textures are silty clay loam, clay loam, sandy clay, silty clay, and clay. Herbicide rates are provided for the appropriate organic matter content for each category. Organic soils (over 10% organic matter) are considered separately but in the same section of the label.

The sample rate chart illustrated in Figure 7.11 is organized with organic matter increasing from left to right and with coarse-textured soils at the top and fine-textured soils at the bottom. Thus, the lowest dosage of herbicide for mineral soils is in the upper left-hand corner of the chart and the highest dosage is in the right-hand corner. The suggestions for organic soils appear below and adjacent to those for mineral soils.

The dosage range on the label provides an indication of the affinity of the herbicide for soil colloids. The range of rates for herbicides with little affinity for soil colloids will be narrow. These compounds also may be suggested for use on peat and muck soils. In contrast, the range of dosages for a tightly bound herbicide will be wider and the compound probably will not be recommended for organic soils. The in-

formation in Figure 7.11 is typical for a herbicide with considerable affinity for soil colloids. The maximum suggested dose is 3.3 times the minimum dose.

Two frequent types of herbicide performance problems are related to soil composition. Coarse-textured soils with low organic matter content lack adequate amounts of colloid to provide a buffering or adsorptive capacity. Crop injury can occur with slight overdoses or with variations in weather such as heavy rainfall that leaches the herbicide into the crop root zone. Many soil-applied herbicides are specifically excluded from use on coarse-textured soils with less than 0.5 or 1.0% organic matter.

Soils high in organic matter, on the other hand, bind most soil-applied herbicides so tightly that inadequate activity is obtained. Effective soil-applied herbicides on these soils are limited to those with minimum affinity for soil colloids. Soil-applied herbicides that do not provide weed control on high organic matter soils include the dinitroanilines, phenylureas, and *s*-triazines. Postemergence herbicides that are

BRAND X HERBICIDE
PREEMERGENCE BROADCAST APPLICATION RATES PER ACRE

Soil Texture Description	Pounds of BRAND X 80W					
	Percent Organic Matter in Soil					
	Less than 1%	1%	2%	3%	4%	Over 4%
Coarse Sand, loamy sand, sandy loam	DO NOT USE	1.5	2.0	2.5	3.0	3.5
Medium Loam, silt loam, silt, sandy clay loam	2.0	2.5	3.0	3.5	4.0	4.5
Fine Silty clay loam, clay loam, sandy clay, silty clay, clay	2.5	3.0	3.5	4.0	4.5	5.0
Peat or muck	NOT RECOMMENDED					

Figure 7.11 Sample use rate chart taken from a product label for a soil-applied herbicide using the Weed Science Society of America format.

not affected by soil can be used effectively on weeds growing in highly organic soils.

Problems associated with variations in soil composition frequently are encountered in areas with uneven topography in which several different soil types can occur in the same field. These fields typically range from coarse-textured high spots with low organic matter to low spots with medium- or fine-textured soils with 3 to 6% organic matter (Figure 7.12). In some cases muck or peat soils may be present in the low areas. Soil-applied herbicides sensitive to soil binding and with limited crop tolerance are difficult to use in such fields because the variation in colloidal content will result in a wide range of herbicide availability. On the fine-textured, high organic matter soils not enough herbicide will be available to kill weeds, whereas on the coarse-textured, low organic matter soils the availability of too much herbicide will cause crop injury or kill.

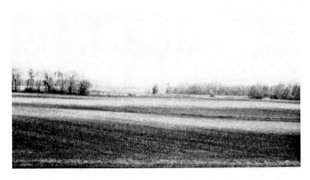

Figure 7.12 Variability in soil types in a single field. High spots with a low organic matter content are readily visible as lighter patches of soil.

Some solutions to a variable soil problem are to use soil-applied herbicides that are not tightly bound to soil colloids, soil-applied herbicides that have excellent crop tolerance at dosages required for weed control on dark soils, or postemergence herbicides that act through the foliage rather than the soil. Another solution is to vary the herbicide rate during application to match soil differences.

Presence of Soil Moisture

Herbicides do not work when left on the dry soil surface. Not only are they unavailable for plant uptake but they are susceptible to dissipation by photodegradation and volatilization. Although less soil moisture is required for adequate performance of mechanically incorporated herbicides, some moisture is required for plant uptake. Even mechanically incorporated herbicides do not work well under extremely dry soil conditions.

Excessive leaching can move the herbicide out of the plant uptake zone, resulting in poor performance, or it can move the herbicide into the crop root zone, causing injury. Compounds that normally leach very little in soils, such as linuron, the diphenylether herbicides, and surface-applied dinitroanilines such as

pendimethalin, most often cause injury to crops following heavy rainfall.

Soil–Herbicide Interactions in Reduced Tillage Systems

The presence of plant residues or cover crops on the soil surface can affect a number of soil characteristics. When compared with conventionally tilled areas, soils in reduced tillage often have higher organic carbon; are more acidic; have higher moisture levels; have more active microbial populations; and have more stable, aggregated structures with large continuous channels for water drainage (reviewed by Locke and Bryson 1997). Since all of these factors can affect herbicide performance and persistence, it is logical to think that soil–herbicide interactions in reduced tillage systems differ from those in conventionally tilled systems. Results of studies, however, are mixed and vary considerably among herbicides; environmental conditions; and the type, condition, and density of crop residues, so that it is difficult to make generalizations. Clearly the major impact of reduced tillage systems has been to reduce the volume of water runoff and consequently the amount of herbicides that move off site.

MAJOR CONCEPTS IN THIS CHAPTER

Adsorption, movement, and degradation are the major interrelated processes that determine the fate of herbicides in soil. A general understanding of these processes allows us to determine how much chemical to apply and predict how far it will move from the site of application and how long it will persist.

Adsorption is the adhesion of molecules to the surfaces of solids; in soils, these solids are the negatively charged clay and organic matter particles. Increasing amounts of clay and organic matter reduce herbicide availability. Organic matter has a considerably greater impact than clay.

Strong cations such as paraquat are bound so tightly to negatively charged soil particles that little or no soil activity is possible. In contrast, most anions are repelled and remain in the soil solution where they are subject to plant uptake and movement with water. An exception is glyphosate, which is inactive in soil. The phosphonic acid group of this herbicide binds to cations, which in turn bind to negatively charged soil particles.

Some herbicide molecules (weak acids and bases) ionize with changing pH. The acidic phenoxy acid herbicides are uncharged at low pHs and become partially ionized at high pHs. The basic triazines take on a positive charge at low pHs and are uncharged at high pHs.

Herbicides with no charge behave according to water and lipid solubility, volatility, size, and strength of weak attractive forces that exist between molecules. They tend to partition into the aromatic ring structures of organic matter. The lower the water solubility, the greater is the tendency for a herbicide to do this.

Entry of a herbicide onto or into the soil is usually a result of application. Except for mechanical mixing, volatilization and movement with water account for most herbicide movement in the soil. Herbicides are sometimes transported away from the site of application after being absorbed into other systems such as vegetation and animals.

Herbicides remaining on or very near the soil surface may be lost into the atmosphere by volatilization or photodegradation. Once in the soil they may redistribute by vapors and reach the plants in that form. The presence of moisture on the soil surface can result in excessive loss of volatile herbicides.

Water movement processes affecting herbicides are surface runoff, leaching, and capillary action. Movement with water largely accounts for the distribution of herbicides in soil.

Most herbicides readily degrade in the soil. Degradation approximates classical (typical) half-life curves, and most reach less than one-third of the original application dose in a growing season. At these levels, 1 yr without use provides enough time for the herbicide to dissipate, whether the herbicide was used for 1 yr or for several consecutive years.

High dosages, lack of moisture, and cold temperatures increase the length of persistence. Lack of soil persistence, however, is a more common problem with soil-applied herbicides. They are frequently degraded to ineffective levels before all of the weed seeds have germinated and been controlled.

Success in controlling weeds with soil-applied herbicides depends on the herbicide being in contact with actively developing structures of the plant, being available in adequate quantities, and being in the presence of adequate moisture.

The herbicides must be in contact with the absorbing or developing plant part at the proper time. Positioning of the herbicide in the proper place may be achieved by leaching with rainfall or by mechanical mixing.

Soil mobility and positioning can contribute to selectivity between plant species and also can be involved in minimizing crop injury.

Selectivity based on depth protection is decreased by any condition that results in the herbicide moving into contact with absorbing structures of the crop plant.

Crop tolerance can sometimes be improved with adsorbants such as activated charcoal. Activated charcoal can be used for dipping roots of transplants or it can be layered over the seed row.

Dosage of soil-applied herbicides should be prescribed on the basis of accurate laboratory analyses of soil texture and organic matter.

Charts on the labels of soil-applied herbicides indicate the proper use rates for individual soil textures and organic matter content. The dosage rate range on the chart usually is an indication of the affinity of that herbicide for soil colloids.

Herbicide performance problems most frequently are associated with low organic matter, coarse textured soils, or with peat and muck soils. Crop injury tends to be a problem with the former and lack of herbicide activity with the latter. Fields with widely differing soil types present difficulties in selecting and using soil-applied herbicides. Where differences are extreme, postemergence herbicides that are not influenced by soil type may be the solution.

Herbicides do not work when left on dry soil surfaces.

TERMS INTRODUCED IN THIS CHAPTER

Absorption
Activated charcoal
Adsorption
Capillary action
Colloid-metal-phosphate complex
Depth protection
Desorption
Enzyme adaptation
Half-life curve
Herbicide positioning
Humus
Incorporation
Ionization
Lag phase
Leaching
Partitioning
Photodegradation
Relative mobility of herbicides (R_F)
Slow-release herbicide
Soil testing
Sorption
Sorption coefficients (K_d, K_{oc})
Surface runoff

SELECTED REFERENCES ON SOIL-HERBICIDE INTERACTIONS

Edwards, C. A. 1989. Impact of herbicides on soil ecosystems. Crit. Rev. Plant Sci. 8:221–257.

Edwards, C. A. 1993. Effects of herbicides on soil and surface-inhabiting invertebrates. Brighton Crop Prot. Conf. Proc. pp. 133–138.

Fawcett, R. S., B. R. Christensen, and D. P. Tierney. 1994. The impact of conservation tillage on pesticide runoff into surface water: a review and analysis. J. Soil and Water Conserv. 49:126–135.

Grover, R., Ed. 1988. Environmental Chemistry of Herbicides, Vol. I. CRC Press, Boca Raton, Florida.

Grover, R. and A. J. Cessna, Eds. 1991. Environmental Chemistry of Herbicides, Vol. II. CRC Press, Boca Raton, Florida.

Hance, R. J., Ed. 1980. Interactions Between Herbicides and the Soil. Academic Press, New York.

Harper, S. S. 1994. Sorption-desorption and herbicide behavior in soil. Rev. Weed Sci. 6:207–225.

Herbicide Handbook of the Weed Science Society of America. 1994. Weed Science Society of America, Lawrence, Kansas.

Kearney, P. C. and D. D. Kaufman, Eds. 1975 and 1976. Herbicides. Chemistry, Degradation, and Mode of Action. Vol. I and II. Marcel Dekker, New York.

Linn, D. M. and T. Carski, Eds. 1993. Sorption and degradation of agricultural chemicals in soils. Special Publication 32. Soil Science Society of America, Madison, Wisconsin.

Locke, M. A. and C. T. Bryson. 1997. Herbicide–soil interactions in reduced tillage and plant residue management systems. Weed Sci. 45:307–320.

Racke, R. D. and J. R. Coats. 1990. Enhanced Biodegradation of Pesticides in the Environment. ACS Symposium Series 426. American Chemical Society, Washington, D.C.

Sawhney, B. L. and K. Brown, eds. 1989. Reactions and Movement of Organic Chemicals in Soils. Special Publication No. 22. Soil Science Society of America, Madison, Wisconsin.

Schnoor, J. L. 1992. Fate of Pesticides and Chemicals in the Environment. John Wiley-Interscience, New York.

HERBICIDE RESISTANCE

The widespread use of selective herbicides was initiated in the late 1940s and expanded rapidly (see Chapter 5). Herbicide resistance was first reported in 1968 in the state of Washington where a triazine-resistant population of common groundsel was identified (Ryan 1970). In the following two-and-a-half decades, more than 110 weed species were reported to be resistant to one or more of some 15 classes of herbicides (LeBaron 1991; Holt et al. 1993). The development of herbicide-resistant weed populations threatens to reduce the tremendous contribution herbicides make to cost-efficient crop production and improved management practices. On the other hand, the introduction of herbicide resistance into crop plants holds the promise for the development of new weed control strategies that will reduce the negative impact of herbicides on the crop and the environment. Both aspects, resistance of weeds and resistance of crops to herbicides, are discussed in this chapter.

WEED RESISTANCE TO HERBICIDES

The response of a plant to herbicides is a heritable trait. Terms used to describe the levels of response are tolerance, susceptibility, and resistance.

Tolerance is the ability of a plant population to remain uninjured by herbicide doses normally used to control other plant species. *Susceptibility* indicates that the plant population dies at herbicide doses that normally do not injure other plant species. Tolerance and susceptibility are what we expect to see when a selective herbicide is used on a mixed population of plant species. Preferably the crop or desired plant will be tolerant and the weeds will be susceptible. If a plant is susceptible, the effects are usually observed quickly, often within days or weeks, and they occur from exposure to a single herbicide application at a one- to sixfold range of doses. The differences in plant response to the herbicide are usually attributed to differential plant age and stage of development, retention, penetration, translocation, or metabolism (see Chapter 6).

Resistance is the ability of a formerly susceptible plant population to survive herbicide doses above those that were once used to control the original plant population. The resistant portion of the plant population is referred to as a biotype; in other words, *resistant biotypes* appear in species of plants that are susceptible to the herbicide. The development of resistance becomes obvious only after several years of repeated applications of the same herbicide and is usually characterized by plants that survive a dose several to many times (20 to 1000 times or more) greater than normally needed for commercially acceptable weed control. These high levels of resistance are most frequently caused by a single gene mutation that modifies the herbicide binding site on an enzyme or protein. If the herbicide can no longer bind to its site of action, it no longer has an inhibitory effect. Occasionally, differential metabolism, uptake, availability, or movement will be responsible for the development of resistance.

Once a plant population develops resistance to a single herbicide it is likely that it will be resistant to other, closely related herbicides as well. The order of probable occurrence of resistance to another herbicide, from greatest to least, is resistance to chemical analogs (e.g., atrazine and simazine are analogs), members of the same chemical family (e.g., atrazine is a chloro-*s*-triazine and prometon is a methoxy-*s*-triazine; both are symmetrical triazines), members of a closely related chemical family (e.g., atrazine is a symmetrical triazine and metribuzin is an asymmetrical triazine), and herbicides from different chemical families but with the same mode of action (e.g., atrazine, which is an *s*-triazine, and terbacil, which is a uracil, have the same mode of action).

Cross resistance is a specific term used to describe weed populations that are resistant to two or more herbicides that have the same mode of action. In some cases the herbicides may be in the same chemical family (for example, atrazine and simazine); in other cases they are in different chemical families but with the same modes of action, such as atrazine and terbacil, or the sulfonylurea herbicides and imidazolinone herbicides. Cross resistance to two such herbicides initially occurs when a plant is exposed to and develops resistance to the first herbicide; because the modes of action are the same, the plant also is resistant to the second herbicide even though it may never have been exposed to that herbicide.

Multiple resistance is used to describe weed populations that are resistant to two or more herbicides from different chemical families and with different modes of action. Multiple resistance occurs when weeds are exposed to and develop resistance to different chemistries; it is much less frequent than cross resistance. The most infamous example of multiple resistance is that of rigid ryegrass (*Lolium rigidum*), a major annual weed in Australia. One biotype of this species is resistant to at least nine chemical families with five different modes of action (Burnet et al. 1994). Multiple resistance is attributed to mechanisms not associated with the sites of action but to differential uptake, retention, translocation, or metabolism.

Weeds have been much slower to exhibit resistance to chemicals than disease organisms (bacteria, fungi) and insects. Weeds have fewer generations per growing season (frequently only one) and maintain a reservoir of the original population due to buried and dormant seed.

In addition, resistant plants that do develop can be less competitive than nonresistant plants. These less competitive plants may produce low numbers of seeds or may never even live long enough to set seed.

Not all herbicide use results in resistance. Even though weeds have been shown to be resistant to 15 different herbicide mode-of-action groups somewhere in the world, certain herbicide classes and mode-of-action groups are much more frequently associated with resistant weed populations than others. Herbicide groups that are frequently associated with resistance and the approximate time periods for the resistance to develop are as follows:

Appearance of First Resistant Populations in as Little as Three to Five Years of Continuous Use:

Branched-chain amino acid inhibitors (ALS/AHAS inhibitors)—Sulfonylureas, imidazolinones, and triazolopyrimidines.

Lipid biosynthesis inhibitors (ACCase inhibitors)—Aryloxyphenoxy propionates and cyclohexanediones.

Appearance of First Resistant Populations After One Decade or More of Continuous Use:

"Classical" photosynthesis inhibitors. Over 60 species are resistant, mostly to the symmetrical triazines atrazine, simazine, and prometon.

Occasional Occurrence of Resistant Populations:

Photosystem I energized cell membrane destroyers—Bipyridiliums (paraquat, diquat). Other cell membrane disruptors rarely are implicated in resistance.

Microtubule/spindle apparatus inhibitors—Dinitroanilines.

Time Period for Development of Resistant Biotypes within a Species

The major source for the development of weed resistance is through the *naturally occurring mutation* of a gene in a weed population. These naturally occurring mutations then can be passed along to the progeny. Gene mutations are *not* induced by the application of the herbicide.

Although mutations occur infrequently, the probability of a herbicide-resistant biotype appearing in a field of many plants of the same species can be high. For example, Jasieniuk et al. (1996) predicted, based on a mutation rate of 1×10^{-6} in a dominant allele of a gene that regulates resistance, that at least one resistant plant is almost certain to occur in a weed population of 5 plants per m² in a 30 ha field (1 plant in a field of 1.5 million plants) of a random-mating species (i.e., not strictly self-pollinated). This plant may never be fit enough to survive to produce seed, but if it does, then the population of resistant plants can increase if the herbicide to which the plant is resistant continues to be used. Continued use of the same herbicide increases *selection pressure* on the plant population; in other words, susceptible plants will be killed while the resistant plants, freed from competition, will survive and increase in number.

The rate of increase of herbicide-resistant weeds is extremely difficult to observe or measure because during the first few years after herbicide application initiates the selection process, the proportion of resistant weeds remains very low (often less than 1% of the population) (Figure 8.1). As long as herbicide applications continue, the resistant portion of the population will continue to increase. At some point, a single additional treatment will result in a major weed control failure. For example, a predicted 8% of the plant population having resistance in year 6 of continuous herbicide applications will increase to 55% of the population in year 7 and 93% of the population in year 8 (Figure 8.1). Thus, it is very common to go from excellent control of a particular weed to very poor control with the next application. Resistance generally is not detected as a problem until approximately 30% of the weed population shows it.

Factors Determining Rapid Development of Herbicide Resistance

The single most important cause of weed resistance is a management system in which repeated use of a single herbicide, or use of several herbicides with the same site of action, provides continuing selection pressure for herbicide resistance. Additional factors, such as herbicide characteristics and weed biology, also play a role in determining how rapidly the resistance will appear in the population.

Herbicide Characteristics

Single Site-Specific Mode of Action. Herbicides that have a single *site-specific mode of action* are more likely to result in resistance than those that have more general herbicidal properties and affect several processes. Most mutations that result in herbicide resistance are single gene mutations; thus, the effect is to produce an extremely small change at the site of action, such as in the enzyme or other protein to which the

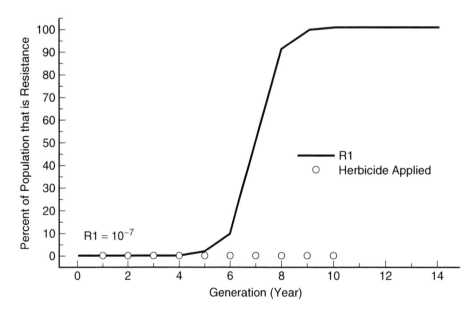

Figure 8.1 Predicted development of an herbicide-resistant weed population (R1) starting from a mutation rate of 1×10^{-7} at generation 0 and exposed to an herbicide with 90% kill of the susceptible population for 10 continuous generations. The herbicide-resistant weed was assumed to be equal in fitness to the susceptible weed. (*Source:* Figure 2 in Maxwell, B. D. and A. M. Mortimer. 1994. Selection for herbicide resistance. In S. B. Powles and J. A. M. Holtum, Eds. Herbicide Resistance in Plants, Biology and Biochemistry. CRC Press, Boca Raton, Florida. pp. 1–25.

herbicide normally binds and exerts its effect. For example, the branched-chain amino acid inhibitor herbicides bind with and inactivate acetolactate synthase (ALS), an enzyme that is critical for the formation of the amino acids valine, leucine, and isoleucine. A mutation in the gene that codes for ALS changes a single amino acid in the enzyme (Figure 8.2) so that the herbicide can no longer bind to it. The appearance of resistant populations to the ALS inhibitor herbicides has been relatively rapid, sometimes within 3 to 5 years of continuous use. This is in contrast to the older, less site-specific auxin-type growth regulators. These herbicides cause general effects on RNA and cell growth and have resulted in relatively little resistance, even after prolonged use.

Unfortunately, the effort to develop herbicides that are site specific in order to avoid harming nontarget species may backfire if it results in the development of herbicides to which plants become resistant within a few years.

Continuous Exposure. Soil-applied herbicides that are persistent and provide control for most of the growing season (for example, the triazine herbicides) further increase the selection pressure for resistance. Susceptible plants of the same and other species do not have an opportunity to grow and provide competition. A similar selection pressure can be obtained with multiple applications over a growing season of a shorter-lived herbicide (for example, paraquat, which is used as a foliar treatment) used for several consecutive years.

Other herbicide-related factors that favor the development of resistant plants include a high level of initial control (96–99%) of susceptible plants and high dosages that eliminate both

```
Susceptible: -Ala-Ile-Thr-Gly-Gln-Val-Pro-Arg-Arg-
Resistant:   -Ala-Ile-Thr-Gly-Gln-Val-Thr-Arg-Arg-
                                      ↑
```

Figure 8.2 The amino acid sequence of a very small portion of the acetolactate synthase (ALS) enzyme in kochia. A change of a single amino acid, proline (Pro), in the susceptible biotype to threonine (Thr) induces resistance to the ALS inhibitor herbicides (Guttieri et al. 1992).

the susceptible portion of the population and other potential plant competitors.

Weed Biology

Life Cycle. Herbicide resistance appears most frequently in plants with an annual life cycle. Some of the most frequently reported plants to show resistance (Table 8.1) include biotypes of pigweed (*Amaranthus* spp.), lambsquarters (*Chenopodium* spp.), kochia (*Kochia scoparia*), barnyardgrass (*Echinochloa crus-galli*), and goosegrass (*Eleucine indica*), all of which are annual plants. Since annuals reproduce from seed only, the potential to transfer resistance to the next generation of plants is relatively high. This is in contrast to perennial plants which can persist for several years; thus, the frequency with which resistance appears in the population may be greatly reduced. Seed production is not high in some perennial species, and seedlings sometimes have a difficult time surviving in an established perennial plant population. The selection pressure from consecutive herbicide applications on annual plants is greater than on perennials since annual plants tend to be relatively susceptible to

Table 8.1 Weeds Reported to Have Herbicide-Resistant Biotypes in the United States, but Note That the Problem of Herbicide Resistance in Weeds is Worldwide. Modified from various sources.

Herbicide group	Weeds
Aryloxyphenoxy propionates and cyclohexanediones	
	Italian ryegrass, wild oat, giant foxtail, green foxtail, large crabgrass, johnsongrass
Dinitroanilines	Goosegrass, johnsongrass
MSMA, DSMA	Common cocklebur
Paraquat, diquat	American black nightshade
Sulfonylureas, imidazolinones and triazolopyrimidines	
	Kochia, prickly lettuce, Russian thistle, Palmer amaranth, common waterhemp, common cocklebur, perennial ryegrass
Triallate	Wild oat
Triazines	Lambsquarters, kochia, water starwort, common groundsel, bur cucumber, black nightshade, pigweeds, common waterhemp, downy brome, velvetleaf, barnyardgrass, witchgrass, annual bluegrass, foxtails

herbicides, in contrast to perennials which, because of their underground structures, often escape total destruction. An exception to resistance only occurring in annual plants is the reported resistance of johnsongrass, a perennial plant, to the lipid biosynthesis (ACCase) inhibitor (Smeda et al. 1993) and dinitroaniline (Wills et al. 1992) herbicides. Johnsongrass is a prolific seed producer, which may explain its ability to develop resistance.

Annual weeds are also characterized by the production of large numbers of seeds per plant and the persistence of those seeds for long periods in the soil. The result is a large plant population and seed bank. These very large numbers increase the genetic material that is present and consequently increase the chance that a mutated gene will appear.

Seed Longevity. If the soil is tilled, long-lived seeds from previous susceptible populations will dilute the population of resistant seeds, and resistance can be delayed. On the other hand, seeds that stay on the soil surface and have a short seed life (little or no dormancy) will either die or germinate within one to several years. Thus, management systems (for example, no-till) that leave weed seeds on the soil surface leave mostly resistant seeds since the susceptible ones have germinated and have been killed by the herbicide application.

Genetic and Fitness Characteristics. In most documented cases, the following genetic principles apply to herbicide resistance:

1. Resistance is controlled by a single, heritable gene.

2. The allele (form of the gene) that confers resistance usually is dominant to the allele that confers susceptibility. *Dominant alleles* are much more likely to become established in a population than recessive alleles. This means that a resistant dominant gene carried by pollen will be expressed in all the progeny that are produced.

3. Resistance is most likely to occur in plants that outcross (random mating), although some are self-fertilizing. *Outcrossing* spreads the resistance to many other plants, in contrast to *selfing* in which resistance will be limited to the progeny of the self-fertilized plant.

4. The higher the mutation rate, the greater the chance that resistance will occur. A high frequency of mutations in a plant coupled with the potential to produce large weed populations probably explains why some weed species are more likely to exhibit resistance than others.

The frequency of mutations also appears to vary with the herbicide group. For example, the frequency of triazine resistance appearing in a typical annual plant population is estimated to be between 10^{-10} and 10^{-20} (Gressel 1991) in comparison to a much higher frequency of 10^{-6} for resistance to the ALS inhibitors (Stannard and Fay 1987). This difference in initial mutated gene frequency may explain why resistance to ALS inhibitors has occurred more rapidly than resistance to triazine herbicides.

5. Relative fitness of the resistant and susceptible plants in the presence of the herbicide is important in determining the number of resistant plants that will survive and reproduce. Since a gene that codes for an enzyme or protein that is essential for plant survival has mutated, there is a good chance that the resistant plant will not be as fit as the susceptible portion of the population. Triazine-resistant weeds tend not to be as fit as susceptible weeds, a fact that also may help explain why triazine resistance has been relatively slow to develop. However, there is evidence that biotypes resistant to other herbicides are as fit as susceptible biotypes in the absence of herbicide treatment. Examples include biotypes of kochia (Thompson et al. 1994) to sulfonylurea herbicides, cocklebur to arsenical herbicides (Haigler et al. 1994), goosegrass to dinitroaniline herbicides (Murphy et al. 1986), and giant foxtail to the ACCase inhibitor herbicides (Wiederholt and Stoltenberg 1996).

6. Most resistance is determined by *nuclear inheritance;* that is, the mutated allele occurs in the DNA of the nucleus. The spread of resistance alleles thus occurs through both pollen (the paternal parent) and ovules (the maternal parent). This means that both pollen and seed can spread the resistance. Pollen is a relatively rapid method of spreading resistance since it is commonly wind or insect borne. Kochia pollen, for example, has been found as far as 50 m (150 ft) from its source (Mulugeta et al. 1992). Some resistance, such as the resistance to triazine herbicides, occurs by a mutation of the DNA present in the chloroplast. Since chloroplasts are inherited strictly through the maternal parent, pollen cannot move the resistance to another population. Triazine resistance is thus spread by seed rather than by pollen. Seeds are dispersed not only by natural methods but by humans as well; thus, the movement of resistance genes to sites far removed from the original source can occur.

Situations

The development of resistant weed populations can be avoided. Resistance that does occur, particularly since so much is now known about its causes, can be considered the result of mismanagement. Some of the most common situations in which resistance has appeared include the following:

Monoculture Cropping. The risk of weed resistance is greatest when the same herbicide is used repeatedly on the same crop, same field, and same weeds over a long period of time. In other words, if all the management practices are the same year after year, the selection pressure for the development of resistance will be high and constant. This cropping pattern represents an ideal environment for the increase of resistant biotypes.

Bare Soil Sites. Resistant weeds can develop in total vegetation control systems (e.g., railroads, roadsides, and industrial sites). Herbicides are usually applied at high dosages and are often highly persistent, effectively eliminating the susceptible portion of the population. The almost total lack of competing vegetation further enhances the potential of the resistant weed to dominate. The lack of tillage in these sites also leads to the proliferation of resistant seed on the soil surface. The development of resistant biotypes of kochia along railroad rights-of-way is the result of this type of management system.

Sites with Little or No Tillage. The loss of the normal benefits of tillage enhances the development of resistant weeds. Tillage kills resistant weeds that escape herbicide applications and provides a mechanism for mixing and diluting resistant seeds with the original weed population. The resistance of common groundsel (*Senecio vulgaris*) to triazine herbicides has occurred in untilled sites such as orchards, nurseries, and roadsides but not in cornfields where tillage is routinely applied (Gressel 1991).

Management Practices to Minimize the Selection of Herbicide-Resistant Weeds

Most of the following management practices will help reduce the selection pressure exerted on a weed population by continued use of a single herbicide:

1. *Alternating nonchemical control measures with herbicides* helps reduce herbicide selection pressure. However, alternative nonchemical weed control practices may be limited in number and/or perform inconsistently.

2. *Mixing herbicides with different modes of action* (either as tank mixes or as sequenced treatments the same year) will delay the development of resistance provided that both herbicides have activity on the target weed. What one herbicide misses, the other will kill. Therefore, very few resistant survivors will remain. Mixtures with one herbicide that controls one weed and a second that controls a second weed but not the first will not delay the buildup of resistance.

3. *Alternating (rotating) herbicides and herbicide mode-of-action groups* from year to year can slow the onset of resistance. This approach is less effective than using tank mixtures and sequences the same year because portions of the population will survive the year in which the herbicide with the other mode of action is used. In weed populations of equal fitness (competitive abililty), however, the appearance of resistance can be substantially delayed. This delay can be estimated by adding the number of years the herbicide causing the problem is not used to the expected time of resistance buildup. For example, if resistance is likely to occur in 5 yr of consecutive use, buildup of resistance would be delayed for 10 yr if the selecting herbicide is used every other year. Resistance still occurs after five applications. Where a less fit resistant weed population is present, buildup would be delayed more (depending on the degree of unfitness) because of competition from the susceptible wild type during off years in the rotation.

4. *Crop rotation* (where herbicide chemistry and other control methods differ for each crop) can delay the buildup of resistance.

5. *Controlling weeds that escape the primary control program* prevents resistant weeds from reproducing. A resistant weed presents the same type of threat as a serious, new, invading weed, and failure to control it will contribute to its spread. Control measures that vary from standard herbicide applications (e.g., hand pulling, limited tillage, spot treatments with other herbicides) should help eliminate escaped plants.

6. *Contain the resistant infestation* once it appears. Weeds spread slowly; thus, initial resistant infestations will be localized and spread can be partially slowed with a good preventive program.

7. *Accept less than excellent weed control levels* where weeds are already well established. One approach to managing pests is to accept economic threshold levels of weed control (e.g., 60

to 80%) rather than 90 to 95% control. Most growers and landowners are not satisfied with this level of control, even though yield losses may be minimal. Additional weeds can increase the difficulty of harvest, and they produce seeds for future infestations. Determining economic threshold levels for weeds is thus complicated by seed production and longevity as well as by new weed introductions which require additional control measures.

The continued use of herbicides for efficient crop production depends on our ability to manage these valuable tools so that buildup of resistance can be minimized or avoided. We currently know enough about managing herbicides to minimize the development of resistance. Unfortunately, instances of weed resistance to herbicides continue to increase. Coordinated efforts among crop producers, the agrichemical industry, and the educational community are needed to convince practitioners that long-term gain is more important than the short-term benefits provided by limited practices.

Period Required for a Field to Recover from Resistant Populations

Once a herbicide-resistant population develops, the impact is long term. Recovery from a resistant weed population to the susceptible wild type can occur by germination of susceptible seed from the seed bank in the soil and from immigration of pollen and seed from adjacent susceptible populations. These processes can take a long time, particularly if the resistant seed in the seed bank is long lived. On the other hand, resistant weeds with a short seed life should be relatively easy to eliminate if new seed production can be prevented. The fitness of the resistant biotype also is a factor in the speed of recovery. If resistant plants are less fit than susceptible plants, discontinued use of the selective herbicide permits natural selection to gradually restore the dominance of susceptible plants in the population (Warwick and Black 1994). If the resistant and susceptible biotypes have equal fitness, the relative proportion of resistant to susceptible plants in a population will remain constant in the absence of the selective herbicide (Jasieniuk and Maxwell 1994). In general, the return to a susceptible population will take a long time. A simulation model for predicting the development of herbicide resistance in weeds suggests that it would take 25 yr for a weed population in which 92% of the plants are resistant to

equilibrate to below 1% of the plants in that field after the suspension of herbicide use (Bruce Maxwell, pers. commun.).

CROP RESISTANCE TO HERBICIDES

Herbicide-resistant crop varieties (cultivars) can be developed using two methods: (1) herbicide-resistant individuals can be selected from within a crop (standard breeding techniques) or (2) resistant genes can be identified in unrelated organisms and transferred to a crop (*genetic engineering*). Both approaches currently are being used.

Herbicide-Resistant Crop Cultivars from Standard Plant Breeding

If herbicide-resistant genes occur in weeds, they should also occur in crops. Resistant crop cultivars have been selected from herbicide-resistant individuals found by screening naturally varying populations of a crop species. Herbicide resistance can also be induced in some plant species by exposure of cells or tissues to mutagenic (mutation-causing) substances. In either case, the resistance must then be transferable from one plant to another and from one cultivar to another by conventional plant breeding methods. This ensures that the resistance can be introduced into the high-yielding crop varieties used by farmers.

Herbicide-resistant cultivars developed using modern plant breeding techniques and available to growers include IR and IT (imidazolinone-resistant and imidazolinone-tolerant) corn, STS (sulfonylurea-tolerant) soybeans, and SR (sethoxydim-resistant) corn. Corn varieties that are designated IR have resistance on both alleles of the gene, whereas in the IT corn varieties, only one of the two alleles confers resistance. IR cultivars provide a higher level of resistance than IT cultivars, but both are currently available on the market.

Herbicide-Resistant Crop Cultivars from Genetic Engineering

With the development of gene transfer technology, our ability to introduce genes from virtually any living organism into plants has greatly increased. The introduction and successful expression of a foreign gene in a plant is called *transformation*, and the genetically altered plant is called *transgenic.* Transfer of a gene that confers glyphosate resistance from the bacterium *Agrobacterium* sp.(strain CP4) to various plants has produced glyphosate-resistant soybeans, corn, wheat, and other crops. Glyphosate-resistant (Roundup Ready) soybeans were first grown commercially by farmers in 1996.

Glufosinate-resistant (Liberty-Link) soybeans, corn, sugar beets, canola, and other crops resulted from the transfer of the *pat* gene from the bacterium *Streptomyces viridochromogenes.* Glufosinate-resistant canola has been available in Canada since 1995. Liberty-Link corn and other crops were available to growers starting in 1997.

Both glyphosate and glufosinate are nonselective postemergence herbicides that control a broad spectrum of annual weeds. Glyphosate also has good activity on a number of perennial weeds. Both herbicides perform consistently, have no soil activity or persistance, and have not been associated with selection of resistant weed populations. Glyphosate in particular has been used since the late 1970s with only two reports (both in annual ryegrass in Australia) of the natural development of glyphosate-resistant weeds or crops. In fact, researchers first tried to generate and select for glyphosate-resistant genes in flowering plants but were unsuccessful. This is the reason that a gene from a bacterium was used to confer glyphosate resistance to crop plants.

Benefits of Herbicide-Resistant Crops

Crops with herbicide resistance provide a number of benefits:

1. An *increased margin of safety* can be obtained with a herbicide-resistant crop, when the herbicide is used in that crop. This greatly lessens the chance of crop loss due to herbicide injury.

2. A herbicide-resistant variety can be used *to overcome injury due to carryover* of that herbicide from a previous crop. For example, IR corn can be used where there is concern that residues of imazaquin, an imidazolinone herbicide that is used on soybeans, may carryover into the next growing season.

3. *Where herbicides are limited in number or effectiveness,* such as on minor crops, the genetic transformation of the crop can increase the options for weed control. An existing herbicide that is already registered can be used on a crop that is resistant to that herbicide at less cost and risk

than developing a totally new herbicide for the individual crop. Much of the herbicide development cost already will have been incurred.

4. *Problem weeds* may be controlled by the use of transgenic crops. For example, planting glyphosate-resistant soybeans allows improved control of creeping perennial weeds. This may allow the shift from conventional tillage to no-till or minimum tillage in sites with perennial weed infestations.

5. Use of resistant crops can *increase the "window of application,"* in other words, the time during which the herbicides can be effectively applied. This is particularly true for glyphosate and glufosinate, compounds that are not as dependent as other herbicides on stage of weed growth to be effective. This is less true for the imidazolinones and sulfonylureas, which mostly are used early postemergence.

6. Crops that are resistant to more environmentally "soft" herbicides (such as glyphosate) or to herbicides that are used at very low dosages (such as the sulfonylureas) can *reduce the potential environmental damage* that might occur from the use of other herbicides.

7. *Lower costs* may be possible with the use of a single herbicide on a herbicide-resistant crop to control a broad spectrum of weeds as compared to using several narrow-spectrum herbicides to accomplish the same objective. Under the best of circumstances, herbicide-resistant crops can lead to a reduction in the total amount of herbicide used.

Concerns Regarding the Use of Herbicide-Resistant Crops

Several concerns have been raised regarding the introduction of herbicide-resistant crops in general and of transgenic herbicide-resistant crops in particular. These concerns include the following:

1. The *selection for herbicide-resistant weed populations* may increase. The use of genetically altered resistant crops is associated with the increased and repeated use of a single herbicide, which could result in the selection of weed species resistant to that herbicide. However, the selection of herbicide-resistant weed populations is typically a consequence of frequent use and is a potential problem with any effective herbicide or group of herbicides. It is not unique to herbicides used with genetically altered crops.

Several of the herbicides to which resistant cultivars were available to growers by 1996 (for example, the imidazolinones, sulfonylureas, and sethoxydim) are herbicide groups with demonstrated weed resistance problems. Herbicides such as the imidazolinones and sulfonylureas with a single mode-of-action site (AHS/ALS inhibitors) need to be managed properly in order to avoid the rapid development of resistant weed populations. Strategies (for example, rotating the herbicide-resistant crop with other crops, using nonchemical control methods, etc.) outlined in the section on avoiding the development of resistant weeds will be necessary.

2. *Transgenic crops could themselves become weed problems.* Concerns include herbicide-resistant crop cultivars becoming more aggressive than standard varieties, volunteer herbicide-resistant crops contaminating subsequent crops, herbicide-resistant crop seeds becoming deposited in noncrop areas, and herbicide-resistant crops invading natural ecosystems.

There is no evidence to suggest that herbicide-resistant crops behave any differently from standard varieties with regard to competitive ability and seed longevity. Thus, increased weediness or invasion into natural ecosystems is unlikely. Problems with volunteer crop plants or deposited seed should be short term and easily controlled with appropriate control measures (for example, mechanical methods, alternative herbicides, etc.).

3. *Outcrossing of genes from resistant crops to closely related weeds* is a problem but mostly in theory at this point. The transfer of genes from crops to weeds within the same species presents the biggest potential problem. Plants within a genus but of different species would be the next most likely to cross. In fact, the transfer of herbicide resistance from canola (*Brassica napus*) to wild mustards (*Brassica* spp.) has been demonstrated. Gene transfer from crops such as sorghum (*Sorghum bicolor*) and rice (*Oryza sativa*) to weeds such as johnsongrass (*Sorghum halepense*) and red rice (*Oryza sativa*), respectively, must be monitored and managed if it does occur. However, the transfer of genes from other major crops such as soybeans (*Glycine*), corn (*Zea*), or cotton (*Gossypium*) is extremely unlikely since there are no weedy species within those genera. Furthermore, gene flow from crops with natural herbicide resistance to closely related weed species has never been identified as a problem associated with herbicide use.

4. *Inferior crop quality or performance* may result from the transformation of a crop plant. Gene mutations and gene transfer do provide some increased chance of a gene imparting an unexpected or undesirable trait, such as loss of disease resistance or lower yield (yield drag).

The potential for *yield drag* on crop cultivars is obviously a major concern to growers. Gene mutations are frequently associated with decreased fitness. In certain cases (for example, triazine resistance), resistant weeds do show a decreased fitness. It is logical then to assume that mutations associated with herbicide resistance could result in decreased fitness in the crop as well. On the other hand, weed resistance to the AHS/ALS inhibitors does not result in loss of fitness. Transgenically derived crop cultivars resistant to these herbicides as well as to glyphosate and glufosinate are reported to be as equally fit as standard commercial varieties. With the possible exception of triazine-resistant cultivars, the appearance of inferior performance or quality is probably not much more of a concern with genetically altered crops than with any newly introduced crop variety.

Seed producers are striving to provide growers with herbicide-resistant premium cultivars, and they claim that newly introduced crops do not sacrifice yields. Should inferior crop varieties be released, they will be weeded out (by economic competition) very soon after introduction.

5. *Public distrust of genetically engineered products* is a potentially negative factor in the development of herbicide-resistant crops. The adverse publicity surrounding transgenic products (including insect-resistant crops and hormonally altered milk products) occurs not only in North America but also in many European countries. To compound the problem, the movement of genetically altered grains mixed with nongenetically altered grains into importing countries can lead to further uncertainty and concern. The solution to these perceived problems is education of the public and regulation, both during the development and field testing of the transgenic crop, and by establishing criteria for importation. Such regulations are currently being developed on a country by country basis. In the United States, field testing of transgenic crops is primarily regulated by the USDA and appropriate state agencies. The final product must be approved for food or animal food use by the FDA and for environmental clearance by the USDA. If the product is pesticidal (for example, the transferred gene produces a toxin that is lethal to an insect or weed), the product also must be approved by the EPA. Crops that are toxic (allelopathic) to weeds because of genetic transformation have not been developed commercially at this point.

Even with negative publicity, genetically transformed crops and animal products will continue to be developed. Their acceptance by the public will depend on clear demonstrations of their safety and benefits to society and the continued education of the public in the scientific principles that govern genetics and biology. The challenge for agricultural professionals is to effectively convey the message that contributions to be gained from herbicide-resistant crop cultivars by far exceed potential problems.

MAJOR CONCEPTS IN THIS CHAPTER

Herbicide-resistant weed populations are a threat to the tremendous contributions herbicides make to efficient crop management, while herbicide-resistant crop plants hold the promise of new weed control strategies that will reduce the negative impact of herbicides on the crop and the environment.

One hundred ten weed species have been reported to be resistant to one or more of 15 classes of herbicides.

The response of a plant to herbicides is a heritable trait.

Tolerance, susceptibility, and *resistance* are terms used to describe response levels of plants to herbicides. Tolerance is the ability of a plant population to remain uninjured, whereas susceptibility indicates that the plant population dies at the herbicide doses normally used. Tolerance and susceptibility are what we expect to see when a selective herbicide is used on a mixed population of plant species and results from differential retention, penetration, translocation, or metabolism.

Resistance is the ability of a formerly susceptible plant population to survive herbicide doses above those that were once used to control the original plant population. Resistance is most frequently caused by a single gene mutation that modifies the herbicide binding site on an enzyme or protein.

Weed populations that are resistant to two of more herbicides with the same modes of action are described as being cross resistant. Multiple resistance describes weed populations that are resistant to two or more herbicides from different chemical groups with different chemistries. Multiple resistance is less frequent than cross resistance and is attributed to differential retention, penetration, translocation, or metabolism rather than to an altered site of action of the herbicides.

Resistance to chemicals has been slower to occur in weeds than in disease organisms (bacteria, fungi) and insects.

Certain herbicide classes and mode-of-action groups are much more frequently associated with the development of resistance than others. These are the branched-chain amino acid (ALS/AHAS) inhibitors, the lipid biosynthesis (ACCase) inhibitors, and the "classical" photosynthesis inhibitors.

Resistance occurs in weed populations by naturally occuring mutations. Mutations are not induced by herbicide applications.

The rate of increase of herbicide-resistant weeds is difficult to observe because the proportion of resistant weeds remains very low initially. At some point, a single additional treatment will result in a major weed control failure.

The single most important cause of weed resistance is the continuous use of a single herbicide or of several herbicides with the same mode of action over several years. Continuous exposure either from soil-applied herbicides that persist all season or from multiple applications of a shorter-lived herbicide over a growing season provides maximal selection pressure and hastens the rate of resistance buildup.

Additional factors that increase the development of resistant weeds include the use of herbicides that target a single site of action; a site of action that is controlled by a single, dominant gene; the control of the susceptible portion of the population which removes potential plant competitors; weeds with an annual life cycle; high mutation rates; and the ability (fitness) of the resistant weed to survive and reproduce.

Monoculture cropping and bare soil sites with little or no tillage are most frequently associated with the development of resistance.

Management strategies that reduce the probability of developing resistance include alternating nonchemical control measures with herbicides, mixing herbicides with different modes of action for the same target species, alternating herbicides and herbicide mode-of-action groups from year to year, rotating crops, and controlling escaping weeds and resistant infestations once they first appear.

Once a herbicide-resistant population develops, the impact is long term (several decades).

Herbicide-resistant crop cultivars developed using modern plant breeding techniques and available to growers include imidazolinone-resistant corn, sulfonylurea-tolerant soybeans, and sethoxydim-resistant corn.

A gene that confers glyphosate resistance was transferred from the bacterium *Agrobacterium* sp. to crops, resulting in glyphosate-resistant soybeans, corn, wheat, canola, and other crops.

Glufosinate-resistant soybeans, corn, sugar beets, canola, rice, and other crops resulted from the transfer of a gene from the bacterium *Streptomyces viridochromogenes*.

The benefits of herbicide-resistant crops include an increased margin of safety to the crop, preventing injury from carryover of a herbicide from a previous crop, increased options for weed control in limited acreage crops, improved control of problem weeds, lower weed control costs, and the ability to use environmentally safe herbicides on crops that were once susceptible to those herbicides.

Several concerns have been raised regarding the introduction of herbicide-resistant crops in general and of transgenic herbicide-resistant crops in particular. The transfer of a herbicide-resistant gene from the crop to closely related weed species (for example, from canola [*Brassica napus*] to the weedy mustards [*Brassica* spp.]) has been reported. Gene transfer is unlikely, however, for several of the most common crops grown in the United States (e.g., soybeans, corn, cotton) since these species do not have closely related species that are weeds.

Other concerns about herbicide-resistant crops include an even greater buildup of herbicide-resistant weed populations (if the herbicide is used continuously in the crop), the potential for the transgenic crop itself to become a weed, and inferior crop performance. To date, none of these concerns have been realized; however, public distrust of genetically engineered products is a potentially negative factor in developing herbicide-resistant crops.

TERMS INTRODUCED IN THIS CHAPTER

Cross resistance
Dominant allele
Gene coding
Genetic engineering
Herbicide-resistant crop
Multiple resistance
Mutation
Nuclear inheritance
Outcrossing
Recessive allele
Resistance
Resistant biotype
Selection pressure
Selfing
Site-specific mode of action
Susceptibility
Tolerance
Transformation
Transgenic plant
Yield drag

SELECTED REFERENCES ON HERBICIDE RESISTANCE

Caseley, J. C., G. W. Cussans, and R. K. Atkin, Eds. 1991. Herbicide Resistance in Weeds and Crops. Butterworth-Heinemann, Oxford, England.

Duke, S. O. 1996. Herbicide-Resistant Crops. Agricultural, Environmental, Economic, Regulatory, and Technical Aspects. CRC Press, Boca Raton, Florida.

Green, M. B., H. M. LeBaron, and W. K. Moberg, Eds. 1990. Managing Resistance to Agrochemicals: From Fundamental Research to Practical Strategies. American Chemical Society, Washington, D.C.

Holt, J. S. and H. M. LeBaron. 1990. Significance and distribution of herbicide resistance. Weed Technol. 4:141–149.

Holt, J. S., S. B. Powles, and J. A. M. Holtum. 1993. Mechanisms and agronomic aspects of herbicide resistance. Annu. Rev. Plant Physiol. Plant Mol. Biol. 44:203–229.

LeBaron, H. M. and J. Gressel, Eds. 1982. Herbicide Resistance in Plants. John Wiley and Sons, New York.

Moss, S. R. and B. Rubin. 1993. Herbicide-resistant weeds: A worldwide perspective. J. Agric. Sci., Cambridge 120:141–148.

Padgette, S. R., K. H. Kolacz, X. Delannay, D. B. Re, B. J. LaVallee, C. N. Tinius, W. K. Rhodes, Y. I. Otero, G. F. Barry, D. A. Eichholtz, V. M. Peschke, D. L. Nida, N. B. Taylor, and G. M. Kishore. 1995. Development, identification, and characterization of a glyphosate-tolerant soybean line. Crop Sci. 35:1451–1461.

Powles, S. B. and J. A. M. Holtum, Eds. 1994. Herbicide Resistance in Plants. Biology and Biochemistry. Lewis Publishers, Boca Raton, Florida.

9

HERBICIDE GROUPS WITH SIGNIFICANT FOLIAR USE: TRANSLOCATED HERBICIDES SHOWING INITIAL SYMPTOMS ON NEW GROWTH

HERBICIDES AND HERBICIDE GROUPS

The properties and uses of a number of herbicides are described in this and the next three chapters. Most of the herbicides can be placed in groups according to their chemistry. Others are of a unique chemistry and are described individually.

In general, herbicides within the same chemical group share many of the same characteristics. Since this often is true for the type of application for which the group is most suitable, herbicide groups and individuals are arranged according to whether they have significant foliar use (Chapters 9, 10, and 11) or whether they are primarily soil-applied herbicides (Chapter 12). The herbicide groups with significant foliar use are further divided into three major categories based on translocation pattern and initial plant symptoms: (1) translocated herbicides showing initial symptoms on new growth (this chapter), (2) translocated herbicides showing initial symptoms on older growth (Chapter 10), and (3) nontranslocated herbicides showing initial localized injury (Chapter 11). Each of these categories is further subdivided according to herbicide mode of action (modes of action are described in Chapter 6). For a listing of the herbicides discussed in this book and their mode-of-action groupings, see Table 9.1. Each herbicide group (or individual) will be described as follows:

1. A general overview of the group (or individual), its characteristics, and the symptoms that it produces in plants is given.

2. A detailed description of one key herbicide from the group (or individual) is given to serve as an example of the group. This compound usually is the first or the most successful or widely used member of the group and the one for which the most information is available. The format for the detailed descriptions of these individual or key herbicides is as follows:

   a. *Common name* of the herbicide.
   b. *Trade name(s) and manufacturer(s)* of the first or most commonly encountered labels. Not all names and manufacturers can be listed because of space limitations. This is particularly true of older herbicides for which the patent has expired. Such products may be available under many trade names.
   c. *Chemistry,* both in written and diagram form. In most cases, the parent compound (for example, the acid) is used rather than salts, esters, or other derivatives.
   d. *Available formulations* (those sprayable in water and those applied in a dry form). The abbreviation *a.i.* indicates active ingredient; *a.e.* indicates acid equivalent. Types of formulations are described in Chapter 16.
   e. *Response in soils,* including specific information on leachability, volatility,

HERBICIDE FAMILIES WITH SIGNIFICANT FOLIAR USE

Translocated Herbicides Showing Initial Symptoms on New Growth (downwardly mobile, symplastically translocated, phloem-mobile herbicides)

Auxin-Type Growth Regulators

Phenoxy acid herbicides

2,4-D	Numerous
2,4-DB	Butoxone, Butyrac
2,4-DP (dichlorprop)	Available only in mixtures
MCPA	Rhonox, Rhomene, Sword, Weedon MCPA
MCPB	Thistrol
MCPP (mecoprop)	MCPP 4K, Mecomec

Benzoic acid herbicides

Dicamba	Banvel, Clarity, Vanquish

Picolinic acid (pyridinecarboxylic) herbicides and relatives

Clopyralid	Lontrel, Reclaim, Stinger, Transline
Picloram	Tordon
Triclopyr	Garlon, Grandstand, Remedy, Turflon

No chemical family recognized

Naptalam	Alanap

Aromatic Amino Acid (EPSPS) Inhibitors

Glyphosate	Accord, Rodeo, Roundup, Roundup Ultra, Touchdown (sulfosate)

Branched-Chain Amino Acid (ALS/AHAS) Inhibitors

Sulfonylurea herbicides

Bensulfuron	Londax
Chlorimuron	Classic
Chlorsulfuron	Glean, Telar
Halosulfuron	Manage, Permit
Metsulfuron	Ally, Escort
Nicosulfuron	Accent
Primisulfuron	Beacon
Prosulfuron	Peak
Rimsulfuron	Matrix
Sulfometuron	Oust
Thifensulfuron	Pinnacle
Triasulfuron	Amber
Tribenuron	Express
Triflusulfuron	UpBeet

Imidazolinone herbicides

Imazamethabenz	Assert
Imazamox	Raptor
Imazapic	Cadre, Plateau
Imazapyr	Arsenal, Chopper, Stalker
Imazaquin	Scepter, Image
Imazethapyr	Pursuit

Triazolopyrimidine sulfonanilide herbicides

Cloransulam	FirstRate
Flumetsulam	Broadstrike, Python

Pyrimidinyl oxybenzoate herbicides

Pyrithiobac	Staple

Carotenoid Pigment Inhibitors

No chemical family recognized

Amitrole	Amitrol-T
Clomazone	Command
Fluridone	Sonar

Isoxazole herbicide

Isoxaflutole	Balance

Pyridazinone herbicide

Norflurazon	Predict, Solicam, Zorial

Lipid Biosynthesis (ACCase) Inhibitors

Aryloxyphenoxy propionate herbicides

Diclofop	Hoelon
Fenoxaprop	Acclaim, Whip 1EC
Fenoxaprop-P	Aclaim Extra, Option II, Whip 360
Fluazifop-P	Fusilade II, Fusilade DX, Ornamec 170
Haloxyfop	Verdict, Gallant
Quizalofop-P	Assure II

Cyclohexanedione herbicides

Clethodim	Envoy, Prism, Select
Sethoxydim	Poast, Poast Plus, Prestige, Torpedo, Ultima, Vantage
Tralkoxydim	Achieve

Table 9.1 (*Continued*)

Organic Arsenicals
    DSMA   Ansar, DSMA Liquid
    MSMA   Ansar, Arsenate Liquid, Bueno, Daconate
Unclassified Herbicides
    Asulam   Asulox
    Difenzoquat   Avenge
    Fosamine   Krenite
    Propanil   Stam, Stampede

Translocated Herbicides Showing Initial Injury on Older Growth (upwardly mobile only, apoplastically translocated, xylem-mobile herbicides)

"Classical" Photosynthesis Inhibitors
    s-Triazine herbicides
        Ametryn   Evik
        Atrazine   Aatrex, Atrazine
        Cyanazine   Bladex
        Hexazinone   Velpar
        Prometon   Pramitol
        Prometryn   Caparol
        Simazine   Princep
    as-Triazine herbicide
        Metribuzin   Lexone, Sencor
    Phenylurea herbicides
        Diuron   Karmex
        Fluometuron   Cotoran
        Linuron   Lorox
        Tebuthiuron   Spike
    Uracil herbicides
        Bromacil   Hyvar
        Terbacil   Sinbar

"Rapidly Acting" Photosynthesis Inhibitors
    Benzothiadiazole herbicide
        Bentazon   Basagran
    Benzonitrile herbicide
        Bromoxynil   Buctril
    Phenylcarbamate herbicides
        Desmedipham   Betanex
        Phenmedipham Spin-Aid
    Pyridazinone herbicide
        Pyrazon   Pyramin
    Phenylpyridazine herbicide
        Pyridate   Tough

Nontranslocated Herbicides Showing Initial Localized Injury (No movement, contact herbicides, rapid cell membrane destruction)

Photosystem I (PS I) Energized Cell Membrane Destroyers
    Bipyridilium herbicides
        Paraquat   Cyclone, Gramoxone Extra, Starfire
        Diquat   Diquat, Reward
Protoporphyrinogen Oxidase [Protox (PPO)] Inhibitors
    Diphenylether herbicides
        Acifluorfen   Blazer, Status
        Fomesafen   Flexstar, Reflex
        Lactofen   Cobra
        Oxyfluorfen   Goal
    Oxadiazole herbicides
        Oxadiazon   Ronstar
        Fluthiacet   Action
    N-phenylphthalimide herbicide
        Flumiclorac   Resource
    Triazolinone herbicides
        Carfentrazone   Affinity, Aim
        Sulfentrazone   Authority, Cover, Spartan
Glutamine Synthesis Inhibitors
    Glufosinate   Finale, Liberty, Rely

HERBICIDE FAMILIES APPLIED ALMOST EXCLUSIVELY TO SOIL

Microtubule/Spindle Apparatus Inhibitors (root inhibitors)
    Dinitroaniline herbicides
        Benefin   Balan
        Ethalfluralin   Sonalan, Curbit
        Oryzalin   Surflan
        Pendimethalin   Prowl, Pendulum, Pentagon, and many other names for the turf market
        Prodiamine   Barricade, Endurance, Factor
        Trifluralin   Treflan, Tri-4, Trifluralin

Table 9.1 (*Continued*)

None generally recognized
   DCPA   Dacthal
Pyridine herbicide
   Dithiopyr   Dimension
Amide herbicide
   Pronamid   Kerb

Shoot Inhibitors
Chloroacetamide herbicides
   Acetochlor   Harness, Surpass, TopNotch
   Alachlor   Lasso, Micro Tech, Partner
   Dimethenamid   Frontier
   Metolachlor   Dual
   Propachlor   Ramrod
Thiocarbamate herbicides
   Butylate   Sutan+
   Cycloate   Ro-Neet
   EPTC   Eptam, Eradicane
   Pebulate   Tillam
   Triallate   Far-go

Miscellaneous Cell Division Inhibitors
None generally recognized
   Bensulide   Bensumec, Betasan, Prefar
Amide herbicide
   Napropamide   Devrinol
Phenylurea herbicide
   Siduron   Tupersan

Cell Wall Formation Inhibitors
Nitrile herbicide
   Dichlobenil   Casoron, Norosac
Benzamide herbicide
   Isoxaben   Gallery
None generally recognized
   Quinclorac   Facet

and persistence. See Chapter 7 for information on the principles of herbicide–soil interactions.

f. *Sensitive species,* which include those weeds we consider to be of major importance. Not all weed species may be listed. It is essential that the label be consulted for an up-to-date listing of sensitive weeds.

g. *Major uses,* including the type of application as well as the major crops and situations.

h. *Special problems* (where applicable), including the potential for injury caused by drift, human toxicity, persistence, and inconsistencies in performance.

3. A summary of other herbicides in the group and the ways in which they differ from the key herbicide and from each other is given.

TRANSLOCATED HERBICIDES SHOWING INITIAL SYMPTOMS ON NEW GROWTH

These herbicides are frequently described as symplastically translocated, phloem-mobile, downwardly mobile, or basipetally translocated. They initially move to sites of new growth, for example, the growing points or meristems of both above- and belowground tissues, newly developing leaves, and developing storage organs. Symptoms can include pigment loss (yellow or white), stoppage of growth, and distorted (malformed) new growth. Symptoms normally appear only after several days. Plants die slowly from the top down (new to old growth).

Figure 9.1 Left: Jimsonweed seedling showing epinasty (downward bending) after being treated with 2,4-D. Right: Untreated plant.

These herbicides have the potential to move to and kill underground perennial structures with one or two foliar applications. The degree of mobility within each of the groups of herbicides discussed in this category, however, can vary substantially. For example, among the carotenoid pigment inhibitors, amitrole translocates more readily than clomazone.

AUXIN-TYPE GROWTH REGULATORS

The auxin-type growth regulators include three types of chemistries: the phenoxy acids, benzoic acids, and picolinic acids (also referred to as the pyridinecarboxylic acids). The single herbicide, naptalam, also will be described here.

Most auxin-type growth regulators are primarily foliar applied and symplastically translocated. They are effective for removal of numerous broadleaves in grass crops and noncrop situations. They can be applied to the soil, in which case they are absorbed by the roots, but symplastic plant movement in roots is limited, so that these compounds generally are less effective as soil-applied materials.

The most obvious effects of auxin-type growth regulators on susceptible plants are the twisting and downward curvature (epinasty) of the stem and leaves immediately after application (Figure 9.1). Although death may not occur for several weeks or even months, the initial curling of the young growth can be seen a few hours after treatment. Tissues that are undergoing active meristematic activity are particularly susceptible to injury. The effects on developing tissue include leaf strapping (Figure 9.2) and fusion of plant parts (Figure 9.3). Cell proliferation increases and is manifested by the development of undifferentiated cell masses (Figure 9.4) and adventitious root formation on stems (Figures 9.4, 9.5).

Slight variations in symptoms can be caused by specific herbicides on certain species. For example, dicamba, a benzoic acid, causes the cupping of leaves on velvetleaf (Figure 9.6) and soybean, and the picolinic acid herbicides (e.g., clopyralid) cause the formation of an extended leaf midrib on composite (aster-type) plants.

Figure 9.2 Leaf strapping on first true leaves of soybean. Upper left: A formative effect caused by 2,4-D. Plant at lower right shows normal trifoliolate leaflet development.

Figure 9.3 Dandelion with fused peduncles (flowering stalks) caused by 2,4-D.

Figure 9.5 Root proliferation in corn caused by soil-applied 2,4-D.

Auxin herbicides are available as amine salts, inorganic salts, low volatile esters, and occasionally as the acid. Amine salts include diethanol-, dimethyl-, isopropyl-, monoethanol-, triisopropanol-, and diglycolamines. Inorganic salts such as sodium, potassium, or ammonium salts also are available. The butoxyethyl ester and the isooctyl (2-ethylhexyl) ester are the most commonly available of the low volatile esters.

Phenoxy Acid Herbicides

The phenoxy acid herbicides are widely used for controlling broadleaved weeds in a large number of grass crops (corn, small grains, sorghum, rice, sugarcane, pasture, rangeland, and turf). They control numerous annual broadleaves and suppress several perennial broadleaves such as Canada thistle and field bindweed. They also are used for controlling woody plants in forest management and noncropland sites and for controlling certain aquatic weeds. The most widely used and best known of the phenoxies is 2,4-D. This compound was one of the first selective organic herbicides synthesized. It was so effective as a broadleaf killer that it virtually revolutionized crop management and gave a tremendous impetus to the pesticide industry (see Chapter 5). The other herbicides in the group have the same general properties as 2,4-D but differ somewhat in species selectivity. 2,4-DB, MCPA, and MCPB can be used safely on certain broadleaved crops. 2,4-DP and MCPP control broadleaf species missed by 2,4-D.

Figure 9.4 Proliferation of cortex tissue in the stem of a soybean plant caused by 2,4-DB. Some of the outgrowths probably are adventitious roots. (J. L. Williams, Jr., Purdue University)

Figure 9.6 Velvetleaf on left shows the upward leaf cupping caused by the benzoic acid herbicide dicamba.

2,4-D

TRADE NAMES AND MANUFACTURERS

Numerous

CHEMISTRY

(2,4-dichlorophenoxy)acetic acid

2,4-D

AVAILABLE FORMULATIONS

A. Sprayable in water
   1. Water-soluble liquids and dry formulations
      a. Amines (most commonly available nonvolatile formulations)
      b. Salts, such as sodium, potassium, and ammonium, occasionally available for special uses
   2. Emulsifiable concentrates
      a. Low volatile esters
      b. High volatile esters (use prohibited in many states and generally not available)
B. Dry application
   1. Granules for specialized uses such as aquatics and turf

RESPONSE IN SOILS

2,4-D is a relatively small organic acid. It dissociates in the soil solution to form a negatively charged ion that is not strongly attracted to soil colloids. Thus, it can be leached, but rapid degradation in the soil and removal from the soil by plants minimizes leaching. Because it is not strongly adsorbed to soil colloids, 2,4-D is used as a preemergence treatment on muck and peat soils.

Persistence in most soils is relatively short. Under good conditions for crop growth, rates of up to 1 lb a.e./acre will disappear in 1 to 4 wk. The average field half-life is 10 days.

SENSITIVE SPECIES

Many annual and perennial broadleaved species are sensitive to 2,4-D.

A. Weed species controlled with one application of 1/2 lb a.e./acre or less include bidens, common burdock, carpetweed, common cocklebur, croton, galinsoga, kochia, common lambsquarters, marsh elder, morningglory (tall and ivyleaf), mustards, pigweeds, plantains, ragweed (common and giant), sunflower, velvetleaf, and witchweed.
B. Weed species controlled at higher rates (1 to 2 lb a.e./acre) or with repeat applications include Jerusalem artichoke, bindweed (hedge and field), chicory, curly dock, dandelion, hemp, tall ironweed, jimsonweed, Venice mallow, stinging nettle, yellow rocket, Canada thistle, waterhemlock, and willow. Smartweeds, chickweeds, legumes, and woody species tend to escape 2,4-D treatments.

Most creeping perennial weeds require applications of 1 1/2 to 3 lb a.e./acre repeated annually or semiannually. Established perennial weeds generally are most susceptible to 2,4-D when in the bud stage or on fall regrowth.

For other broadleaved species consult individual product labels or university recommendations.

MAJOR USES

A. Small grains, corn, and grain sorghum
   1. Postemergence applications for control of annual and seedling biennial and perennial broadleaved weeds when used at 1/4 to 1/2 lb a.e./acre (minimal crop injury)
   2. Postemergence applications for control of large annuals and control or suppression of established perennial broadleaved weeds when used at 1/2 to 1 lb a.e./acre (crop injury should be anticipated)
   3. Applications should be made at stages during which reproductive growth of the crop is minimal.
      a. Avoid the tillering and the boot to dough stages in small grains.
      b. Avoid applying from the tassel to dough stage in corn and sorghum.
      c. Semidirect with drop nozzles after corn reaches 8 in. and on sorghum to minimize injury.
   4. Some specialized uses
      a. Preharvest treatment in small grains for controlling large broadleaved weeds. Apply 1/2 to 1 lb a.e./acre when grains are in the hard dough stage.
      b. Preemergence soil-applied treatment for corn. Apply 1 to 2 lb a.e./acre after

planting and before crop emergence. Do not use on coarse-textured, low-organic-matter soils (used mostly on high-organic-matter soils). Crop injury and inconsistency can be a problem with this treatment.

   c. Early preplant applications prior to soybeans. Use at 1/2 lb a.e./acre 7 days or 1 lb a.e./acre 30 days ahead of no-till planting.

B. Established perennial grasses

   1. Turf grasses: use 1 to 2 lb a.e./acre for control of dandelions, plantains, and other broadleaves. Presence of hard-to-control species will require application of or combination with dicamba, MCPP, triclopyr, or similar herbicide. Avoid use on juvenile grasses. Some desirable species such as St. Augustine, bentgrass, or clover may be injured.

   2. Pastures and rangeland for control of broadleaf weeds: use 1 to 3 lb a.e./acre depending on weed species.

C. Other uses

   1. Rice

   2. Sugarcane (preemergence and postemergence)

3. Conifer release by controlling alder, tan oak, madrone, canyon live oak, and manzanita
4. Weed and brush control in fallow land, rights-of-way, fencerows, and other noncrop areas
5. Control of certain aquatic weeds
6. Control of ditch-bank and shoreline vegetation

Not every 2,4-D formulation is labeled for the uses described. Select the commercial product appropriate to and labeled for the intended use. Apply when soil moisture conditions are good and the weed is at a susceptible stage and is growing rapidly.

SPECIAL PROBLEMS

Unexpected crop injury can occur due to unusual environmental conditions.

Drift to sensitive nontarget species can be a problem. High volatile esters should not be used and generally are not available. If sensitive species are present, special drift control techniques should be used with the other formulations (low volatile esters, amine and inorganic salts).

Other Phenoxy Herbicides
Typical Herbicide: 2,4-D

Herbicide	Specific Characteristics and Important Differences from 2,4-D	Major Uses
2,4-DB (Butoxone, Butyrac) 4-(2,4-dichlorophenoxy) butanoic acid	Can be used safely on seedling and established legume crops. Good control of cocklebur, ragweeds, and morningglories. Foliar applied only.	Early postemergence in seedling small-seeded legumes and peanuts. Postemergence or directed post for soybeans.
2,4-DP or dichlorprop (available only in mixtures) (+)-2-(2,4-dichlorophenoxy) propanoic acid	Improved control of chickweed, smartweeds, and a few other broadleaves. Control of woody species.	Postemergence broadleaved weed control in small grains and turf. Brush control in nonagricultural lands.

Herbicide	Specific Characteristics and Important Differences from 2,4-D	Major Uses
MCPA (Rhonox and others) (4-choro-2-methylphenoxy) acetic acid	More selective than 2,4-D at equal rates on cereals, legumes, and flax. Average persistence of 1 mo in moist conditions and up to 6 mo in dry climates.	Postemergence on small grains, flax, peas, dormant legumes, and grass pastures underseeded with legumes.
MCPB (Thistrol) 4-(4-chloro-2-methylphenoxy) butanoic acid	Improved crop selectivity and control of Canada thistle.	Postemergence control of Canada thistle and annual broadleaved weeds in peas.
MCPP or mecoprop (MCPP 4K, Mecomec) (±)-2-(4-chloro-2-methylphenoxy) propanoic acid	Improved control of many turf weeds such as chickweeds, clover, plantain, knotweed, and ground ivy. Soil persistence no longer than 3 to 4 wk.	Postemergence control of weeds in established turf. Also some use in small grains.

Benzoic Acid Herbicides

The benzoic acid herbicides are chlorinated derivatives of benzoic acid. As early as 1942 these compounds were recognized as producers of epinasty and other growth regulator effects similar to those of the phenoxy acid herbicides. Three benzoic acid herbicides were developed for commercial use: 2,3,6-TBA, chloramben, and dicamba. Dicamba is the only one of the three currently on the market. Dicamba is similar to the phenoxy acid herbicides in that it readily penetrates and translocates in the plant and has both foliar and soil activity; however, dicamba has more activity through the soil than does 2,4-D. It is used for controlling a number of broadleaved weeds in grass crops such as corn, small grains, pastures, rangeland, and turf, some of which 2,4-D misses.

Dicamba is formulated as water-soluble glycolamine, dimethylamine, and potassium salts. In soil solution, these compounds readily dissociate to a negatively charged ion. Dicamba is not attracted to soil colloids and is quite mobile in the soil.

Dicamba

TRADE NAMES AND MANUFACTURER

Banvel, Clarity, Vanquish—BASF.

CHEMISTRY

3,6-dichloro-2-methoxybenzoic acid

AVAILABLE FORMULATIONS

A. Sprayable in water
   1. 4 lb a.e./gal (dimethylamine salt) water-soluble liquid (Banvel)
   2. 2 lb a.e./gal (sodium salt) water-soluble liquid (Banvel SGF), used on small grains, fallow, and cotton
   3. 4 lb a.e./gal (diglycolamine salt) water-soluble liquid (Clarity), used on corn
   4. 4 lb a.e./gal (diglycolamine salt) water-soluble liquid (Vanquish), used on pastures, rangeland, noncropland, and turf
   5. Soluble salts in mixtures with other herbicides
B. Dry application
   1. 0.7% dry formulation (Scotts K-O-G Weed Control)

RESPONSE IN SOILS

Under conditions favorable to rapid metabolism, dicamba has a half-life of less than 14 days. Although it leaches readily in the soil, it degrades rapidly so that leaching is not a major problem.

SENSITIVE SPECIES

Many annual, biennial, and perennial herbaceous broadleaved species and some woody species are controlled with dicamba. It is particularly useful for species not effectively controlled with phenoxy herbicides. The species controlled vary with the herbicide dose and the stage of plant development; check the label for recommendations.

A. Annual weeds controlled include common chickweed, common cocklebur, knotweed, kochia, common lambsquarters, Venice mallow, tall and ivyleaf morningglories, mus-

tards, black nightshade, pigweeds, ragweeds, sicklepod, smartweeds, sowthistle, common sunflower, and Russian thistle.
B. Biennials and simple perennials include wild carrot, dandelion, curly dock, bull thistle, musk thistle, and waterhemlock.
C. Perennials include Jerusalem artichoke, field and hedge bindweeds, hemp dogbane, wild garlic, goldenrod, kudzu, common milkweed, leafy spurge, and Canada thistle.
D. Woody species include ash, eastern red cedar, chinquapin, creosote bush, grape, hawthorn, poison ivy, mesquite, poison oak, rabbitbrush, sassafras, sumac, trumpet-creeper, and many others.

Annual weeds are generally controlled with rates of 1/4 to 1/2 lb a.e./acre and biennials with 1/4 to 1 lb a.e./acre. Perennial weeds can be suppressed with 1/2 to 1 lb a.e./acre. Simple and creeping perennials are controlled or suppressed with 1 to 2 lb a.e./acre.

MAJOR USES

A. Field and silage corn
   1. Preemergence to early postemergence for control of annual and seedling biennial and perennial broadleaved weeds when used at 1/4 to 1/2 lb a.e./acre (depending on soil type). Corn must be treated no later than the five-leaf stage or 8 in. tall, whichever comes first.
   2. Postemergence directed on weeds at 1/4 lb a.e./acre. Use before corn is 36 in. tall or 15 days before tassel emergence, whichever comes first.
B. Small grains, grain sorghum
   1. Postemergence for emerged weeds. Timing of application dependent on crop, e.g., before the jointing stage in fall-seeded wheat, before the five-leaf stage in spring-seeded wheat.
C. Between cropping systems in corn, sorghum, and wheat
   1. Applied to emerged and actively growing weeds after crop harvest and before a killing frost, usually to late summer-early fall regrowth or following a mowing or tillage operation. Particularly useful for perennial weed control, e.g., Canada thistle, wild garlic.
D. Perennial grass seed crops such as bluegrass, lawn-type fescue, and ryegrass

1. 1/4 to 1 lb a.e./acre on established grass crops
2. 1/4 to 1/2 lb a.e./acre on seedling grasses after crop reaches three- to five-leaf stage.

E. Established turf in lawns at 1/4 to 1/2 lb a.e./acre

F. Pasture and rangeland, general farmstead weed and brush control, noncropland areas such as fencerows, roadways, and wastelands. Rates depend on species to be controlled.

G. Woody species after leaf development, or as injection, frill, girdle, or stump treatments

SPECIAL PROBLEMS

Drift to desirable plants, particularly soybeans, cotton, flowers, fruit trees, grapes, ornamentals, tobacco, tomato, and other broadleaved plants, is a major problem. On some species, dicamba is more biologically active than is 2,4-D. For example, epinastic effects on soybeans can be caused with 2,4-D at 1/16 to 1/8 lb a.e./acre, whereas dicamba produces epinasty at 1/50 of those doses.

Relatively high soil mobility and root uptake can cause injury to sensitive woody ornamentals. Do not use where the chemical can be leached into the root zone of sensitive species.

Picolinic Acid (Pyridinecarboxylic) Herbicides and Relatives

The picolinic acid herbicides are useful for controlling many herbaceous broadleaved and woody species. They are applied either to the foliage or the soil, are readily absorbed by all parts of the plant, and are extremely mobile, moving primarily through the symplast but also through the apoplast.

These compounds are highly phytotoxic; picloram is as much as ten times more potent than 2,4-D on some species. The effects of picloram and the other picolinic acid herbicides are similar to those of the phenoxy acid herbicides (e.g., epinasty and tissue proliferation).

Picloram is extremely mobile and persistent in both soil and water. Because of this characteristic, and because of its high phytotoxicity, care must be used in choosing treatment sites that will not result in lateral movement of the chemical away from treated areas to sensitive broadleaf species or in downward movement by leaching into groundwater.

Clopyralid is less persistent and is used selectively in some crops. Triclopyr is also selective (highly selective in grass crops) and less persistent and active in the soil than are either picloram or clopyralid.

Picloram

TRADE NAME AND MANUFACTURER

Tordon—Dow AgroSciences

CHEMISTRY

4-amino-3,5,6-trichloro-2-pyridinecarboxylic acid

AVAILABLE FORMULATIONS

A. Sprayable in water
1. 2 lb a.e./gal water-soluble (potassium salt) liquid (Tordon 22K)
2. Also available as a mixture with 2,4-D (Tordon RTU)

RESPONSE IN SOILS

Picloram is mobile in soil. Average persistence is two to several seasons after use. Average field half-life is 90 days, with a range of 20 to 300 days, depending on the application rate and climatic conditions.

SENSITIVE SPECIES

Many woody plants and annual and perennial herbaceous broadleaf weeds are susceptible. Most grasses are resistant.

A. Herbaceous perennial weeds include field bindweed, docks, Russian knapweed, larkspur, milkweeds, leafy spurge, and Canada thistle.

B. Woody species include brambles, conifers, wild grapes and other vines, kudzu, hawthorn, maple, mesquite, Gambel's oak, wild rose, sumac, chapparal species, and others. Consult the label for specific recommendations.

MAJOR USES

Many uses are labeled for and restricted in certain states. Consult the label prior to use.

A. Noncropland for broadleaved and woody species control
   1. Applied broadcast using aerial or ground equipment
   2. Spot treatments. Liquids are applied to the foliage or as basal-bark or injected treatments for woody species.
B. Rangeland or permanent grass pastures. Treatments are applied as above.
C. Selective postemergence treatment in small grains. Generally applied during the three- to five-leaf stage of the crop. Consult label.

SPECIAL PROBLEMS

Soil persistence enhances the chances of contamination of nontreated areas and limits options for rotations to sensitive crops.

Mobility in soil is high. Do not apply where compound can leach into root zone of susceptible species or into water supplies. Do not remove treated soil or apply where the site is susceptible to runoff.

Drift can cause injury to susceptible species.

Picloram was one of the first herbicides designated as a restricted-use pesticide.

Other Picolinic Acid Herbicides
Typical Herbicide: Picloram

Herbicide	Specific Characteristics	Major Uses
Clopyralid (Lontrel, Reclaim, Stinger, Transline) 3,6-dichloro-2-pyridinecarboxylic acid	Absorbed by foliage and roots and readily translocated. Is relatively mobile in the soil. Half-life ranges from 12 to 70 days. Residues can injure certain crops one year after application. Should not be applied over roots of sensitive woody species.	Sugar beets, field corn, small grains, conifers, fallow, pasture, and rangeland. Used for control of annual and perennial broadleaf weeds including Canada thistle, wild buckwheat, cocklebur, jimsonweed, ragweed, sowthistle, marsh elder, and wild sunflower.
Triclopyr (Garlon, Grandstand, Remedy, Turflon) [(3,5,6-trichloro-2-pyridinyl)oxy]acetic acid	Absorbed by foliage and roots and readily translocated in plants. Mobile in soil but more rapidly degraded than picloram. Average half-life is 30 days, ranging from 10 to 46 days depending on environmental conditions.	Can be used for broadleaf control on rice and turf, but mostly for noncropland vegetation management (rights-of-way, industrial and forestry sites, rangelands, permanent pastures) of tree and brush species.

Naptalam

Naptalam is a selective, pre- or postemergence herbicide for controlling annual broadleaves in cucurbit crops. It usually is classified as a growth regulator herbicide. When applied to emerged broadleaves it causes epinasty. However, it probably acts as an auxin antagonist (blocks the activity of auxin) rather than as an auxin. One of its unique effects is to cause the loss of normal geotrophic responses in seedlings; shoots and roots lose their ability to grow up or down and often corkscrew themselves laterally into the soil. The major herbicidal effect appears to be as a germination and growth inhibitor.

TRADE NAME AND MANUFACTURER

Alanap—Uniroyal, Inc.

CHEMISTRY

2-[(1-naphthalenylamino)carbonyl]benzoic acid

AVAILABLE FORMULATION

A. Sprayable in water
  1. 2 lb a.i./gal water-soluble liquid formulated as the sodium salt (Alanap-L)

RESPONSE IN SOILS

The sodium salt of naptalam has a water solubility of 249,000 mg/L, in contrast to a water solubility of 200 mg/L for the acid. Leaches in soil and can be moved out of the weed seed zone by heavy rains. Average half-life is 14 days. Effective weed control from 3 to 8 wk usually is obtained at 4 lb a.i./acre on typical soils.

SENSITIVE SPECIES

Primarily used for controlling annual broadleaves including carpetweed, common chickweed, common cocklebur, galinsoga, common lambsquarters, pigweeds, common purslane, common ragweed, shepherd's purse, and velvetleaf.

MAJOR USES

A. Preemergence or postemergence on cantaloupe and other cucurbits
B. Soil application for nursery stocks of woody plants

AROMATIC AMINO ACID INHIBITORS

Glyphosate was developed by the Monsanto Company in 1972 and introduced in the late 1970s. Since that time it has gained wide acceptance as a nonselective but extremely effective foliar-applied herbicide for controlling most annual weeds and a large number of herbaceous and woody perennials. Glyphosate is a major postemergence herbicide for the control of johnsongrass and other perennial grasses. It translocates symplastically and causes death at low rates. Translocation to the underground structures of perennial species prevents regrowth from these propagules and results in their destruction. Glyphosate has been particularly effective for use in recirculating sprayers and as a wipe-on treatment for control of tall weeds in shorter crops. Glyphosate has no significant soil activity.

Symptoms on the shoots include yellowing (Figure 9.7) and wilting, which progress from the new to the older tissues. Some malformation of new growth can occur. Visible effects occur on annual plants in 2 to 4 days and on perennial species in 7 to 10 days. Rainfall occurring within 6 hr of application can reduce the effectiveness of the treatment.

The introduction of glyphosate-resistant crops allows more widespread use of the herbicide and may result in improved management of weeds.

Glyphosate

TRADE NAMES AND MANUFACTURERS

Roundup, Roundup ULTRA, Roundup DRYpak, Accord, Rodeo—Monsanto Company Touchdown—Zeneca Ag Products, Inc.

CHEMISTRY

N-(phosphonomethyl)glycine

The products marketed by Monsanto (with the exception of Roundup DRYpak) are the isopropylamine salts of glyphosate.

Touchdown is the trimethylsulfonium (trimesium) salt of glyphosate. This compound was formerly called sulfosate.

Figure 9.7 Initial effects of glyphosate include yellowing of shoot tip and youngest leaves. Top: Corn. Bottom: Canada thistle. Wilting and necrosis then progresses from these tissues to the bottom of the plant.

AVAILABLE FORMULATIONS

A. Sprayable in water
1. 4 lb/gal water-soluble liquid formulation of the isopropylamine salt of glyphosate (3 lb/gal acid equivalent); formulation includes surfactant (Accord, Roundup, Roundup ULTRA). Roundup ULTRA contains an altered adjuvant system.
2. 5.4 lb/gal water-soluble liquid formulation of the isopropylamine salt of glyphosate (4 lb/gal acid equivalent); formulation contains no surfactant (Rodeo).
3. 5 lb/gal water-soluble liquid formulation of the trimesium salt of glyphosate; formulation contains surfactant (Touchdown).

4. 94% water-soluble granule of the ammonium salt of glyphosate (Roundup DRYpak). Packaged as 1-oz packets to be used in 1 gal water, then applied as a "spray to wet."

RESPONSE IN SOILS

Glyphosate rapidly reacts with and is inactivated by most soils. Therefore, crops can be seeded directly into treated areas. Some injury, however, can occur to transplants when they are placed in treated soil immediately after application. Leaching is minimal. Glyphosate is degraded in the soil by microbes and has a typical field half-life of 47 days.

SENSITIVE SPECIES

Glyphosate controls many annual and perennial weeds. It effectively kills foliage and underground structures. Doses vary with weed species and stage of growth. Consult labels on individual products.

A. Annual weeds include downy brome, common cocklebur, crabgrasses, fleabane, foxtails, kochia, common lambsquarters, prickly lettuce, mustards, panicums, pigweeds, ragweeds, sandbur, shattercane, smartweeds, Russian thistle, velvetleaf, volunteer corn, and volunteer wheat.
B. Herbaceous perennials include Jerusalem artichoke, bahiagrass, bermudagrass, field bindweed, annual bluegrass, brackenfern, Reed's canarygrass, cattail, dandelion, curly dock, hemp dogbane, Carolina horsenettle, johnsongrass, wirestem muhly, purple and yellow nutsedges, orchardgrass, quackgrass, Canada thistle, torpedograss, and others.
C. Woody species include alders, berries (*Rubus* sp.), elderberry, honeysuckle, kudzu, maples, oaks, multiflora rose, poison ivy, poison oak, trumpet creeper, and willows.

MAJOR USES

A. Crop uses
1. Selective control of a broad spectrum of weeds in glyphosate-resistant annual crop cultivars including soybeans, corn, canola, and an increasing number of other crops. Only glyphosate-resistant cultivars will tolerate postemergence applications of glyphosate.
2. Control of weeds by timing and selective placement.
   a. Timing includes site preparation prior to planting, control of weeds at plant-

ing and prior to crop emergence, pre-harvest treatment on mature crops, and postharvest treatments.
   b. Selective applications include directed postemergence under tall crops and wipe-on treatments above short crops.
3. Spot treatments of no more than 10% of the area within a crop
   a. These treatments are available for the following crops (check label for specific recommendations and exceptions): alfalfa and clover, small grains, asparagus, corn, cotton, fallow and conservation reserve systems, grain sorghum, grass for seed production, pastures, peanuts, small fruits, soybeans, tree and vine crops, nut crops, and vegetable crops.

B. Noncrop uses
   1. Glyphosate formulations are labeled for use in areas such as airports, ditch banks, dry ditches and canals, fencerows, highways, industrial sites, railroads, and storage areas.
   2. Rodeo is labeled for many of the same annual, perennial, and woody species as other formulations but also is labeled for controlling plants in aquatic sites, including cattails, common reed, spatterdock, and other emergent weeds. A surfactant must be added to the spray solution.

BRANCHED-CHAIN AMINO ACID INHIBITORS

The branched-chain amino acid inhibitors, also called the ALS/AHAS inhibitors, consist of four groups of chemistry. The sulfonylureas and imidazolinones were developed in the 1980s. The newer triazolopyrimidine sulfonanilides and pyrimidinyl oxybenzoates are the other two groups. A major use for these chemistries is as selective postemergent herbicides. Several, however, are applied to the soil. These chemistries control annual broadleaf weeds in both broadleaf and grass crops such as soybeans and corn. Almost all of them suppress nutsedges. As a group they are generally poor on grass weeds, but grass control can vary from excellent to none depending on the grass species and the herbicide.

A major feature of the branched-chain amino acid inhibitors is that they are active biologically at extremely low doses, in most cases in oz/acre. In general, the sulfonylureas are used at lower doses than the imidazolinones. For example, the imidazolinone herbicide, imazethapyr, is used at 1 oz a.i./acre whereas chlorimuron, a sulfonylurea herbicide, is used at 0.2 oz a.i./acre.

These herbicides tend to be persistent (several months) at doses selective in crops. Several herbicides in these groups are used at higher doses to provide total vegetation control and are highly persistent at these rates. Environmental conditions and soil pH influence persistence in soil.

These compounds are taken up by roots and shoots and are readily translocated to the growing points initially; however, subsequent movement may be limited.

Because they are very site specific in their mode of action, interfering with the activity of the acetolactase synthase enzyme (ALS/AHAS), certain weed species have developed resistance to these herbicides. On the other hand, crop cultivars have been selected that are tolerant or resistant to these herbicides, thus broadening their use and increasing their crop safety. Examples include STS (sulfonylurea-tolerant) soybeans and IR and IT (imidazolinone-resistant and imidazolinone-tolerant) corn.

By preventing the formation of the amino acids valine, leucine, and isoleucine, these herbicides reduce protein synthesis in the plant. This plus less-understood secondary effects reduce

Figure 9.8 Dark veins and yellow streaks on the leaves of susceptible plants are symptoms of the ALS inhibitors. (Photo courtesy of T. N. Jordan, Purdue University, West Lafayette, IN)

Figure 9.9 Shortened lateral roots of approximately the same length (bottlebrush appearance) caused by imazaquin. (Photo courtesy of T. N. Jordan, Purdue University, West Lafayette, IN)

Figure 9.10 Left: Cobs (with kernels and without) from untreated corn. Right: Cobs from corn treated late with nicosulfuron. Note the pinching in of the ear and the uneven arrangement of kernels.

growth, resulting in the eventual death of the plant. Shoot meristems cease growth within a few hours of application. Symptom development, however, is very slow and occurs over 2 to 3 wk or more. Meristematic areas become yellow and die, and yellow streaks on leaves and pink and purple veins develop (Figure 9.8). Root growth also is stunted, and the secondary roots are shortened to nearly the same length producing a "bottlebrush" look (Figure 9.9). Late applications for weed control in corn may result in malformed ears (Figure 9.10).

When these herbicides are used as postemergent applications, adjuvants are always added to the spray tank. Adjuvants frequently include crop oil concentrates, nonionic surfactants, and fertilizer solutions.

Sulfonylurea Herbicides

This herbicide group contains a large number of compounds. They typically control a few species very well and miss others, so they almost always are used in conjunction with a complementary herbicide. For example, a combination of thifensulfuron, chlorimuron (both sulfonylureas), and quizalofop (an aryloxyphenoxy) is needed to control the lambsquarters, other annual broadleaves, and grasses, respectively, that commonly are found in soybean fields.

Soil pH is a major consideration when using these compounds because degradation is greatly reduced at neutral pHs and above. Persistence into the next season can cause injury to subsequent crops and limits the choices for crop rotation.

An interaction between sulfonylurea herbicides and some organophosphate insecticides results in unacceptable levels of herbicide injury to corn.

Chlorsulfuron

TRADE NAMES AND MANUFACTURER

Glean, Telar—E. I. du Pont de Nemours & Company

CHEMISTRY

2-chloro-*N*-[[(4-methoxy-6-methyl-1,3,5-triazin-2-yl)amino]carbonyl]benzenesulfonamide

AVAILABLE FORMULATION

A. Sprayable in water
   1. 75% dry flowable (Glean, Telar)

RESPONSE IN SOILS

The water solubility of chlorsulfuron is 587 mg/L at pH 5 and 31,800 mg/L at pH 7. Therefore, the potential to leach varies with soil pH, with greater leaching occurring as pH increases. Chlorsulfuron is subject to hydrolysis and microbial degradation in soil; however, hydrolysis is the more important of the two processes and diminishes with increasing pH. The average field half-life of chlorsulfuron is 40 days, mostly ranging from 4 to 6 wk. Soil pH, amount of rainfall, dose, and sensitivity of the plant to chlorsulfuron all play a role in determining the potential for crop injury in the next season. For example, plant-back restrictions vary from 0 for wheat, rye, and triticale to 46 mo for cotton and grain sorghum, depending on pH and rainfall subsequent to application.

SENSITIVE SPECIES

A. Annual broadleaves sensitive to Glean include mustards in general, henbit, redroot and prostrate pigweeds, pineapple weed, chickweed, purslane, and common groundsel. Biennial and perennial species controlled include wild carrot and curly dock.
B. Spectrum of weeds controlled at the higher doses used with Telar include asters, bull thistle, Canada thistle, teasel, Equisetum, perennial pepperweed, Scotch thistle, white clover, and wild garlic.

MAJOR USES

A. Postemergence on wheat and barley (Glean) at doses of 0.125 to 0.37 oz a.i./acre. Timing is important to improve weed control and reduce crop injury. Broadleaves should be treated before they are 2 in. tall. Crop can be treated anytime after the two- to three-leaf stage but before the boot stage.
B. Noncrop and industrial sites (Telar) at 0.19 to 2.75 oz a.i./acre

SPECIAL PROBLEMS

Chlorsulfuron has a relatively long soil life at high soil pHs.

The first major instance of weed resistance to sulfonylureas occurred with chlorsulfuron.

Other Sulfonylurea Herbicides
Typical Herbicide: Chlorsulfuron

Herbicide	Specific Characteristics	Major Uses
Bensulfuron (Londax) methyl 2-[[[[(4,6-dimethoxy-2-pyrimidinyl) amino]carbonyl]amino] sulfonyl]methyl]benzoate	Controls annual and perennial broadleaves and sedges. Best control is early postemergence. Used at 0.45 to 0.75 oz a.i./acre.	Pre- and postemergence herbicide on rice.

Chlorimuron (Classic) ethyl 2-[[[[(4-chloro-6-methoxy-2-pyrimidinyl) amino]carbonyl]amino] sulfonyl]benzoate	Persistent herbicide that controls a number of annual broadleaves (sunflowers, ragweeds, smartweeds, morningglories, bur cucumber) and yellow nutsedge. Used postemergence at 0.13 to 0.19 oz a.i./acre and soil applied in combination with metribuzin and other herbicides at 0.43 to 0.75 oz. a.i./acre.	Soybeans, peanuts, and noncrop sites such as roadsides, fencerows, and equipment storage areas.

Herbicide	Specific Characteristics	Major Uses
Halosulfuron (Manage, Permit) methyl 5-[[(4,6-dimethoxy-2-pyrimidinyl)amino]carbonylaminosulfonyl]-3-chloro-1-methyl-1-H-pyrazole-4-carboxylate	Controls annual broadleaves and both purple and yellow nutsedge. Used postemergence at 0.5 to 1 oz a.i./acre.	Field corn, grain sorghum, and fallow (Permit) and warm- and cool-season grasses (Manage).

Herbicide	Specific Characteristics	Major Uses
Metsulfuron (Ally, Escort) methyl 2-[[[[(4-methoxy-6-methyl-1,3,5-triazin-2-yl)amino]carbonyl]amino]sulfonyl]benzoate	Ally is used at 0.06 oz a.i./acre in small grains and at 0.18 oz a.i./acre on pastures and rangeland. At the low dose it controls several mustards, chickweed, henbit, common lambsquarters and smartweeds and suppresses Canada thistle. At higher doses it controls Canada thistle, dandelion, musk thistle, wild carrot, common yarrow. Escort is used at high rates (0.2 to 2.4 oz. a.i./acre) to control woody species as well as herbaceous annual and perrennial species.	Ally is used in wheat, barley, pastures, and rangeland, and reduced-tillage fallow in the High Plains. Escort is used in noncrop and industrial sites and for the release of conifers. Both are applied postemergence.

Herbicide	Specific Characteristics	Major Uses
Nicosulfuron (Accent) 2-[[[[(4,6-dimethoxy-2-pyrimidinyl)amino]carbonyl]amino]sulfonyl]-*N,N*-dimethyl-3-pyridine carboxamide	Used at 0.5 to 1 oz a.i./acre for control of most annual grasses (except crabgrasses), johnsongrass, quackgrass, and some annual broadleaves.	Postemergence control in corn.

Herbicide	Specific Characteristics	Major Uses
Primisulfuron (Beacon) methyl 2-[[[[[4,6-bis (difluoromethoxy)-2-pyrimidinyl]amino] carbonyl]amino]sulfonyl] benzoate	Used at 0.29 to 0.57 oz a.i./ acre for control of broadleaves, some grasses, and sedges.	Postemergence control in corn.

Herbicide	Specific Characteristics	Major Uses
Prosulfuron (Peak) 1-(4-methoxy-6-methyl-triazin-2-yl)-3-[2-(3,3,3-trifluoropropyl)-phenyl sulfonyl]-urea	Controls broadleaved weeds at 0.3 to 0.6 oz a.i./acre. Is combined with primisulfuron to provide broadleaf control in corn (sold as Exceed).	Postemergence control in grain sorghum, wheat, barley, rye, oats, triticale, and proso millet.

Herbicide	Specific Characteristics	Major Uses
Rimsulfuron (Matrix) N-((4,6-dimethoxy-pyrimidin-2-yl)amino-carbonyl)-3-(ethylsulfonyl)-2-pyridinesulfonamide	Controls grasses, annual broadleaves, and yellow nutsedge. Used at 0.25 to 0.38 oz a.i./acre.	Pre- and postemergence control in in potatoes. Combined with other compounds and used early postemergence in corn.

Herbicide	Specific Characteristics	Major Uses
Sulfometuron (Oust) methyl 2-[[[[(4,6-dimethyl-2-pyrimidinyl) amino]carbonyl]amino] sulfonyl]benzoate	Controls grasses, broadleaves, and sedges. Used at 0.75 to 6 oz a.i./acre.	Noncropland, unimproved turf, and under asphalt and concrete paving. Can be used selectively in conifers and hardwoods. Used pre- and postemergence.

Herbicide	Specific Characteristics	Major Uses
Thifensulfuron (Pinnacle) methyl-3-[[[[(4-methoxy-6-methyl-1,3,5-triazin-2-yl)amino]carbonyl]amino]sulfonyl]-2-thiophene-carboxylate	Used at 0.06 oz a.i./acre in soybeans for pigweed species, common lambs-quarters, velvetleaf, and mustards. Combined with tribenuron for wild garlic and broadleaved control in small grains (sold as Harmony Extra) at 0.3 oz a.i./acre.	Postemergence control in soybeans only.

Triasulfuron (Amber) 2-(2-chloroethoxy)-N-[[(4-methoxy-6-methyl-1,3,5-triazin-2-yl)amino]carbonyl]benzenesulfonamide	Controls broadleaves and suppresses winter annual bromegrasses at 0.2 to 0.35 oz a.i./acre.	Pre- and postemergence control in wheat, barley, pastures, rangeland, and conservation reserves.

Tribenuron (Express) methyl 2-[[[[(4-methoxy-6-methyl-1,3,5-triazin-2-yl)methylamino]carbonyl]amino]sulfonyl]benzoate	Controls broadleaves at 0.125 to 0.25 oz a.i./acre. Combined with thifensulfuron as Harmony Extra.	Postemergence control in wheat, barley, and oats.

Triflusulfuron (UpBeet) methyl 2-[[[[[4-(dimethylamino)-6-(2,2,2-trifluoroethoxy)-1,3,5-triazin-2-yl]-amino] carbonyl]amino]sulfonyl]-3-methylbenzoate	Controls broadleaved weeds at 0.25 to 0.5 oz a.i./acre.	Postemergence control in sugar beets.

Imidazolinone Herbicides

The imidazolinone herbicides are both foliar and soil applied and are used primarily for broadleaf weed control in soybeans and other legume crops and on forest and industrial sites. As a result of the development of resistant corn cultivars, they are used selectively in corn.

In contrast to the sulfonylurea herbicides, which are more persistent at high soil pHs, the imidazolinones persist under conditions of low soil pH, low moisture, and high organic matter.

Imazethapyr

TRADE NAME AND MANUFACTURER

Pursuit—American Cyanamid Company

CHEMISTRY

2-[4,5-dihydro-4-methyl-4-(1-methylethyl)-5-oxo-1H-imidazol-2-yl]-5-ethyl-3-pyridine-carboxylic acid

AVAILABLE FORMULATIONS

A. Sprayable in water
   1. 2 lb a.e./gal ammonium salt (Pursuit)
   2. 70% a.e. acid in water-soluble packets (Pursuit DG Ecopak)

RESPONSE IN SOILS

Imazethapyr is primarily degraded by microbes. Its typical field half-life is 60 to 90 days. It is weakly bound to the soil except at low pHs; however, mobility is limited. Persistence is greatest at low soil pHs.

SENSITIVE SPECIES

Imazethapyr controls mostly broadleaves and a limited number of grasses.

A. Annual broadleaves controlled include smartweeds, mustards, chickweed, nightshades, pigweeds, and several composites including cocklebur, sunflower, marsh elder, and ragweeds.
B. Grasses controlled include seedling johnsongrass, shattercane, and foxtails.

MAJOR USES

A. Preemergence or early postemergence in alfalfa, soybeans, peanuts, and IMI corn (imidizolinone-resistant corn). A typical dosage is 1 oz a.i./acre.

SPECIAL PROBLEMS

Occasional carryover to sensitive crops.

Other Imidazolinone Herbicides
Typical Herbicide: Imazethapyr

Herbicide	Specific Characteristics	Major Uses
Imazamethabenz (Assert) (±)-2-[4,5-dihydro-4-methyl-4-(1-methylethyl)-5-oxo-1H-imidazol-2-yl]-4 (and 5)-methylbenzoic acid (3:2)	Controls wild oats and some annual broadleaves and grasses.	Postemergence control in wheat and barley.

Herbicide	Specific Characteristics	Major Uses
Imazamox (Raptor) 2-(4-isopropyl-4-methyl-5-oxo-2-imidazolin-2-yl)-5-(methoxymethyl)nicotinic acid	Controls grasses and broadleaved weeds. Shorter persistence time than imazethapyr.	Postemergence control in soybeans.

Herbicide	Specific Characteristics	Major Uses
Imazapic (Cadre, Plateau) ±2-[4,5-dihydro-4-methyl-4-(1-methylethyl)-5-oxo-1H-imidazol-2-yl]-5-methyl-3-pyridinecarboxylic acid	Controls annual grasses, broadleaves, nutsedges, leafy spurge, and rhizome johnsongrass.	Postemergence control in peanuts, unimproved turf, native prairiegrasses, wildflowers and legumes, conservation reserves.

Herbicide	Specific Characteristics	Major Uses
Imazapyr (Arsenal, Chopper, Stalker) (±)-2-[4,5-dihydro-4-methyl-4-(1-methylethyl)-5-oxo-1H-imidazol-2-yl]-3-pyridinecarboxylic acid	Controls annual grasses and broadleaves, perennial broadleaves, and woody species. Species controlled depends on the dose, which ranges from 0.25 to 0.75 lb a.e./acre.	Provides total vegetation control and is used for weed control in unimproved bermudagrass and bahiagrass and under pavement, and for site preparation in forestry and conifer release. Chopper is an emulsifiable concentrate used in oil for basal application and cut stump treatments. Used pre- and postemergence.

Herbicide	Specific Characteristics	Major Uses
Imazaquin (Scepter, Image) 2-[4,5-dihydro-4-methyl-4(1-methylethyl)-5-oxo-1H-imidazol-2-yl]-3-quinoline carboxylic acid	Soil applied with limited foliar activity. Controls annual broadleaved weeds.	Scepter is used in soybeans. Image is used in warm-season turfgrasses and selected landscape ornamentals.

Triazolopyrimidine Sulfonanilides

This group of herbicides has both soil and foliar activity. The two compounds in this group, flu- metsulam and chloransulam, control broadleaves and provide almost no grass control.

Flumetsulam

TRADE NAME AND MANUFACTURER

Python WDG—Dow AgroSciences
Broadstrike (but most generally available as a mixture with other herbicides; e.g., Broadstrike + Dual, Hornet, Scorpion III)—Dow AgroSciences

CHEMISTRY

N-(2,6-difluorophenyl)-5-methyl[1,2,4]tria- zolo[1,5-a]pyrimidine-2-sulfonamide

AVAILABLE FORMULATIONS

A. Sprayable in water
  1. 80% water dispersible granule (Python)
  2. 0.2 + 7.47 lb a.i./gal suspension concen- trate of flumetsulam + metribuzin (Broadstrike + Dual)
  3. 0.25 + 3.4 lb a.i./gal emulsifiable con- centrate of flumetsulam + trifluralin (Broadstrike + Treflan)
  4. 0.20 + 0.54 lb a.i./lb dry flowable of flumetsulam + clopyralid in water- soluble packets (Hornet)

  5. 0.78 + 0.21 + 0.42 lb a.i./lb dry flow- able of flumetsulam + clopyralid + 2,4-D in water-soluble packets (Scorpion III)

RESPONSE IN SOILS

Field half-life is 1 to 3 mo with a shorter persis- tence at high pHs of 7 to 8.

SENSITIVE SPECIES

Flumetsulam controls mostly annual broad- leaved weeds including nightshades, purslane, spurge, sunflower, velvetleaf, pigweeds, lambs- quarters, Venice mallow, and prickly sida. Clopyralid extends the spectrum of broadleaf control, particularly to composites. Trifluralin or metolachlor provide control of annual grasses.

MAJOR USES

A. Python is used on both corn and soybeans as a soil-applied treatment.
B. Broadstrike + Dual is used on both corn and soybeans as a soil-applied treatment.
C. Broadstrike + Treflan is used as a preplant incorporated treatment on soybeans only.
D. Hornet and Scorpion III are used post- emergence on corn only.

Other Triazolopyrimidine Sulfonanilides
Typical Herbicide: Flumetsulam

Herbicide	Specific Characteristics	Major Uses
Cloransulam (FirstRate) N-(2-carbomethoxy-6-chlorophenyl)-5-ethoxy-7-fluoro(1,2,4)triazolo-[1,5-c]pyrimidine-2-sulfonamide	Controls annual broadleaved weeds. Rotation to small grains requires 4 mo interval; corn, sorghum, cotton, and peanuts require 9-mo, and sugar beets, sunflowers, and tobacco require 24-mo intervals.	Soil and foliage applied for soybeans only.

Pyrimidinyl Oxybenzoate Herbicides
Pyrithiobac

TRADE NAME AND MANUFACTURER

Staple—E. I. du Pont de Nemours & Company

CHEMISTRY

2-chloro-6[4,6-dimethoxy pyrimidin-2-yl)thio] benzoate

AVAILABLE FORMULATIONS

A. Sprayable in water

1. 85% water-soluble powder in water-soluble packets

RESPONSE IN SOILS

Plant-back restrictions vary from 4 to 10 mo, indicating that pyrithiobac is a persistent herbicide.

SENSITIVE SPECIES

Pyrithiobac controls annual broadleaves such as composites, morningglories, nightshades, pigweeds, and velvetleaf. It also suppresses yellow nutsedge.

MAJOR USES

A. Pre- or postemergence control in cotton

CAROTENOID PIGMENT INHIBITORS

The members of this group are fairly diverse in their chemistry, the weeds they control, and how they are used. For example, norflurazon is used mostly in cotton and tree fruits. Clomazone is used mostly in soybeans, fluridone is used in aquatic sites, and amitrole is used for general weed control. Their method of application and persistence in soil also vary greatly. Amitrole has almost no soil activity and is foliar applied whereas norflurazon and clomazone are used as preemergence treatments. Norflurazon can be persistent enough to last into a second season whereas clomazone tends to be less persistent.

The feature that unites these herbicides is their mode of action and symptom development. They prevent the formation of carotenoids, which is followed by the loss of chlorophyll, resulting in the formation of whitened plants. New shoot tips and leaves become bleached (sometimes tinged with pink or purple) a few days after treatment and appear normal except for the conspicuous lack of green and yellow pigments (Figure 9.11). Death occurs over one to several weeks.

Norflurazon

TRADE NAMES AND MANUFACTURER

Predict, Solicam DF, Zorial Rapid 80—Sandoz

CHEMISTRY

4-chloro-5-(methylamino)-2-(3-(trifluoro-methyl)phenyl)-3(2H)-pyridazinone

AVAILABLE FORMULATIONS

A. Sprayable in water
1. 80% formulation (Predict)
2. 78.6% dry flowable (Solicam DF)
3. 78.6% wettable powder (Zorial Rapid 80)

RESPONSE IN SOILS

Norflurazon is adsorbed by clay and organic matter; thus, the application rate is determined by clay and organic matter content. Norflurazon does not leach appreciably. It has a moderate to long persistence with a half-life of 45 to 180 days in the southeastern U.S. Losses due to

Figure 9.11 Plants treated with the symplastically translocated herbicide amitrole. New leaves show the chlorotic effect first. Left: Soybean. Right: Wheat.

photodegradation and volatilization can be significant if the herbicide is left on the surface.

SENSITIVE SPECIES

Norflurazon controls a broad spectrum of annual broadleaves, grasses, and sedges.

MAJOR USES

A. Premergence at 1/2 to 1 lb a.i./acre on cotton, peanuts, and soybeans (Zorial Rapid 80)
B. Preemergence (to weeds) at 1 1/4 to 5 lb a.i./acre depending on soil type and crop species on tree fruits, nuts, grapes, asparagus, and farmstead areas (Solicam DF)
C. 2.4 lb a.i./acre as fall or spring applications on field-grown nursery stock and 2 to 4 lb a.i./acre noncropland prior to weed germination (Predict)

SPECIAL PROBLEMS

Persistence of norflurazon can be a problem.

Loss of norflurazon occurs when the herbicide is left for long periods on the soil surface. It must be mixed into the soil either mechanically or with rainfall or irrigation to prevent this from happening.

Other Carotenoid Inhibitor Herbicides
Typical Herbicide: Norflurazon

Herbicide	Specific Characteristics	Major Uses
Amitrole (Amitrole-T) 1*H*-1,2,4-triazol-3-amine	Limited activity in soil. Requires several weeks for complete kill.	Nonselective foliar treatment for annuals and herbaceous and woody perennials in noncrop areas.

Herbicide	Specific Characteristics	Major Uses
Clomazone (Command) 2-[(2-chlorophenyl) methyl]-4,4-dimethyl-3-isoxazolidinone	Residual, soil-applied herbicide with half-life of 24 days. Movement of vapors off target cause bleaching to sensitive, nontarget species. Movement is reduced by mixing into soil or using microencapsuled formulations.	Grass and broadleaf control in soybeans. Particularly good for velvetleaf control. Also can be used in cotton and tobacco.
Fluridone (Sonar) 1-methyl-3-phenyl-5-[3-(trifluoromethyl)phenyl]-4(1*H*)-pyridinone	Average half-life is 20 days in aerobic pond sites, 9 mo in anaerobic sites, and 90 days in pond sediments.	Control of submersed weeds in aquatic sites. Slow-acting but can provide 2 yr of control.
Isoxaflutole (Balance) 5-cyclopropyl-4-(2-methylsulphonyl-4-trifluoromethyl-benzoyl) isoxazole	Controls many broadleaved and grass weeds.	Preemergent control in corn.

LIPID BIOSYNTHESIS INHIBITORS

These herbicides, also called the ACCase inhibitors, consist of two chemical groups, the aryloxyphenoxy propionates and the cyclo-hexanediones. Although the chemistry of these two groups differs, the members are similar in their uses and effects on weeds. They are used for selective postemergence control of grass weeds in nongrass crops. Their ability to translocate to and kill underground structures of perennial grasses makes them a standard treatment for johnsongrass, quackgrass, bermudagrass, and wirestem muhly in broadleaved crops.

Symptoms on susceptible species include discoloration and disintegration of meristematic tissue just at or above the nodes (Figure 9.12), including the nodes of rhizomes. The leaves turn yellow, redden, and sometimes wilt. Seedling grasses tend to lodge by breaking at the soil surface. Maximum control is observed 10 to 21 days after treatment.

Resistance of weeds to both groups has been a problem. Sethoxydim resistance has been incorporated into SR corn.

Mixing these herbicides with postemergence broadleaf herbicides frequently results in reduced grass control (see Table 6.1).

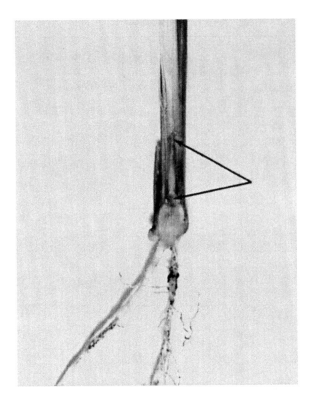

Figure 9.12 Sethoxydim-treated corn cut longitudinally to show discolored meristematic areas (arrows) at base of plant.

Aryloxyphenoxy Propionate Herbicides

An unusual feature of the aryloxyphenoxy propionates is that the inactive ester formulations (e.g., fluazifop butyl ester) are applied to the plant, taken up, and then converted to the herbicidal acid form (e.g., fluazifop acid) within sensitive species. Symptoms and activity, however, are still similar to those of the cyclohexanediones, which do not undergo such transformations.

Frequently, when these herbicides are manufactured, equal amounts of two isomers of the chemical are formed, one of which is active, the other inactive (active and inactive refer to the effects after the chemical has been taken up by the plant). In many cases, the manufacturers have been able to formulate the product so that only the active form, or "para" form, which is designated "P," is present. Sometimes both formulations (with both isomers and with only the para form) are sold. Whip 1EC, for example, contains both isomers and the common name of the active ingredient is designated as fenoxyprop. The inactive isomer has been removed from the chemical that goes into Whip 360, so the common name is designated as fenoxyprop-P. In nearly every case, the first products introduced contained both isomers. In recent years, the tendency has been to remove the inactive isomer and to sell the para forms.

Fluazifop-P

TRADE NAMES AND MANUFACTURERS

Fusilade II, Fusilade DX—Zeneca
Ornamec 170—PBI/Gordon

CHEMISTRY

(R)-2-[4-[[5-(trifluoromethyl)-2-pyridinyl]oxy]phenoxy]propanoic acid

AVAILABLE FORMULATIONS

A. Sprayable in water

1. 2 lb a.i./gal emulsifiable concentrate of the butyl ester of fluazifop-P (Fusilade II, Fusilade DX)
2. 0.125 lb a.i./gal emulsifiable concentrate of the butyl ester of fluazifop-P (Ornamec 170)

RESPONSE IN SOILS

Fluazifop-P is rapidly degraded and has a half-life in soil of 15 days. It has low mobility in the soil.

SENSITIVE SPECIES

Fluazifop-P provides postemergence control of annual and perennial grasses. Chemical that reaches the soil may temporarily suppress or control germinating grass species, but the degree of activity depends on rainfall and soil type.

MAJOR USES

A. Crop uses
1. Cotton, soybeans, asparagus, carrots, onions, garlic, endive, sweet potatoes, yams, tree fruits
2. Woody and herbaceous ornamentals
B. Noncrop uses
1. Fallow land, removal of bermudagrass from zoysia and tall fescue turfs

Other Aryloxyphenoxy Propionates
Typical Herbicide: Fluazifop-P

Herbicide	Specific Characteristics	Major Uses
Diclofop (Hoelon) (±)-2-[4-(2,4-dichloro phenoxy)phenoxy] propanoic acid	Applied as the methyl ester of diclofop. First of the herbicides in this group to be marketed in the U.S.	Control of wild oats and other annual grasses in wheat and barley.

| Fenoxaprop-P(Acclaim Extra, Option II, Whip 360) Fenoxaprop (Acclaim, Whip 1EC) (±)-2-[4-[(6-chloro-2-benzoxazolyl)oxy] phenoxy]propanoic acid | Applied as the ethyl ester of fenoxaprop-P or fenoxaprop. Much less activity on cool-season perennial grasses, such as quackgrass, than most other members of this herbicide group. Therefore, these herbicides can be used for selective grass control in cool-season turfgrasses. | Control of annual grasses and some perennial grasses in soybeans, rice, turf (Kentucky bluegrass, perennial ryegrass, fescues, zoysia, and some creeping bentgrasses), deciduous trees and shrubs, herbaceous ornamentals, and conservation reserves. |

| Haloxyfop (Verdict, Gallant) (±)-2-[4-[[3-chloro-5-(trifluoromethyl)-2-pyridinyl]oxy]phenoxy] propanoic acid | Applied as the methyl ester of haloxyfop. Marketed outside of the U.S. only. | Control of grasses in a large number of broadleaved crops. |

Herbicide	Specific Characteristics	Major Uses
Quizalofop-P (Assure II) (*R*)-2-[4-[(6-chloro-2-quinoxalinyl)oxy] phenoxy]propanoic acid	Applied as the ethyl ester of quizalofop-P.	Annual and perennial weed control in soybeans and cotton.

Cyclohexanedione Herbicides
Sethoxydim

TRADE NAMES AND MANUFACTURERS

Poast, Poast Plus, Vantage—BASF
Prestige, Torpedo, Ultima 160—American Cyanamid

CHEMISTRY

2-[1-(ethoxyimino)butyl]-5-[2-(ethylthio)-propyl]-3-hydroxy-2-cyclohexen-1-one

AVAILABLE FORMULATIONS

A. Sprayable in water
   1. 1.5 lb a.i./gal emulsifiable concentrate (Poast)
   2. 1 lb a.i./gal emulsifiable concentrate (Poast Plus, Prestige, Torpedo, Vantage)
   3. 1.3 lb a.i./gal emulsifiable concentrate (Ultima 160)

RESPONSE IN SOILS

Sethoxydim has little to no activity in soil. It is rapidly degraded by microbes, and its average field life is 5 days.

SENSITIVE SPECIES

Sethoxydim provides postemergence control of annual and perennial grasses.

MAJOR USES

Sethoxydim is used on broadleaved and nongrass monocot crops (check labels for specific crops).

A. Crop uses
   1. Field crops such as canola, cranbe, rapeseed, cotton, flax, mint, peanuts, soybeans, sugar beets, sunflowers, tobacco
   2. Small-seeded legumes such as clover, alfalfa, and sainfoin
   3. Broadleaved cover crops such as alfalfa, birdsfoot trefoil, clover, lespedeza, vetches
   4. Vegetable crops such as artichoke, asparagus, beans, cole crops, bulb vegetables (garlic, leek, onion, shallot), carrots, celery, cucurbits, endive, solanaceous vegetables (tomato and others), lettuce, peas, rhubarb, spinach, and others
   5. Fruit crops such as strawberries and tree nuts
   6. Herbaceous and woody ornamentals, forest nurseries, Christmas trees
B. Noncrop uses
   1. Tall fescue growth suppression in non-food crop areas
   2. Fallow land, fences and hedgerows, around public buildings, recreation areas, etc.
   3. Some turfgrasses (centipedegrass and fine fescue turf)

ORGANIC ARSENICAL HERBICIDES

The organic arsenicals are used as foliar-applied herbicides. They translocate symplastically in some species and have been used successfully for controlling johnsongrass and nutsedge. They have limited selectivity, however. Except for postemergence treatments on cotton and on turf

Herbicide	Specific Characteristics	Major Uses
Clethodim (Select, Prism, Envoy) (E,E)-(±)-2-[1-[[(3-chloro-2-propenyl)oxy] imino]propyl]-5-[2-(ethylthio)propyl]-3-hydroxy-2-cyclohexen-1-one	Similar to sethoxydim.	Postemergence for control of grasses in many broadleaved and nongrass crops (similar listing but less extensive than sethoxydim). Field crops include cotton, soybeans, and sugar beets.

Herbicide	Specific Characteristics	Major Uses
Tralkoxydim (Achieve) 2-[1-(ethoxyimino)propyl]-3-hydroxy-5-(2,4,6-trimethylphenyl)-2-cyclohexen-1-one	Controls grasses such as northern jointvetch, wild oats, green foxtail, reed canary grass.	Postemergence use on wheat and barley.

for crabgrass control, they are used either as directed sprays in certain crops or in noncropland situations.

The effects on plants are the wilting and yellowing of the leaf tissue, cessation of growth, desiccation, and eventual defoliation.

The chemical properties of arsenic are similar to those of phosphorus. Like phosphorus, arsenic is tightly bound to soil and resists leaching by water. Elemental arsenic does not break down, so there is some potential for the accumulation of harmful residues in soil if it is used repeatedly. There is little evidence, however, that significant amounts of arsenic are released from soil and

available for plant uptake, even with large dosages and after repeated use. Standard use rates of the organic arsenicals contribute relatively insignificant amounts of elemental arsenic to the soil.

In contrast to the organic arsenicals, the inorganic arsenicals are no longer used in the United States and most other countries. They are extremely toxic and were discontinued because of the large number of accidental fatalities to both humans and livestock. The oral LD_{50} of arsenic acid (the parent compound of the inorganic herbicide, sodium arsenite) to young rats is 48 mg/kg, whereas the LD_{50} of MSMA, an organic arsenical, is 1800 mg/kg.

DSMA, MSMA

TRADE NAMES AND MANUFACTURER

Ansar 8100, DSMA Liquid (DSMA)—Zeneca
Ansar 6.6, Arsenate Liquid, Bueno 6, Daconate 6, Daconate Super (MSMA)—Zeneca

CHEMISTRY

DSMA: disodium methanearsonate

MSMA: monosodium methanearsonate

AVAILABLE FORMULATIONS

A. Sprayable in water
   1. Available as water-soluble liquids and water-soluble powders.

RESPONSE IN SOILS

Arsenate behaves like phosphate in the soil. Therefore, it is tightly bound and inactivated by soil adsorption. Average field life of DSMA and MSMA is 180 days.

SENSITIVE SPECIES

Organic arsenicals control annual and perennial grasses and some annual broadleaves.

A. Weeds controlled include barnyardgrass, chickweed, cocklebur, crabgrasses, dallisgrass, goosegrass, dandelion, johnsongrass, morningglories, nutsedge, pigweeds, puncturevine, sandbur, and spurges.

MAJOR USES

Selective postemergence treatments in cotton, turf (primarily for crabgrass and yellow nutsedge control), and citrus. Nonselective removal of weeds in noncropland.

SPECIAL PROBLEMS

Slight potential exists for the persistence of elemental arsenic in soils.

UNCLASSIFIED HERBICIDES

These herbicides (asulam, difenzoquat, fosamine, and propanil) are not classified with a specific herbicide group, and their modes of action vary. They all, however, show initial symplastic movement to the growing points and thus are described here briefly.

Herbicide	Specific Characteristics	Major Uses
Asulam (Asulox) methyl [(4-aminophenyl) sulfonyl]carbamate	Very short lived in soil with field half-life of 2.5 to 7 days. Appears to inhibit cell division and expansion in meristematic areas.	Postemergence control of annual and perennial grasses in sugarcane, turf, ornamentals, Christmas trees, and noncropland.
Difenzoquat (Avenge) 1,2-dimethyl-3,5-diphenyl-1H-pyrazolium	Strongly adsorbed to clay; average half-life is less than 4 wk. Mode of action is not well understood.	Early postemergence control of wild oat in barley and wheat. Not all wheat varieties have good tolerance to difenzoquat.
Fosamine (Krenite) ethyl hydrogen (aminocarbonyl) phosphonate	Spray is directed to foliage but foliage does not show symptoms immediately after treatment. However, treated susceptible plants fail to grow the following season. May inhibit cell division.	Postemergence treatment of weedy brush and tree species in noncropland including highway, railroad, and utility and pipeline rights-of-way, industrial sites, and storage areas. Also controls herbaceous perennials such as leafy spurge and field bindweed.

Herbicide	Specific Characteristics	Major Uses
Propanil (Stam 4E, Stam 80 EDF, Stam M-4, Stampede 80 EDF) N-(3,4-dichlorophenyl) propanamide	Very short lived in soil with a field half-life ranging from 1 to 3 days under warm, moist conditions. May inhibit photosynthesis.	Applied postemergence on rice for annual grass and broadleaf control. Also controls curly dock and spikerush.

HERBICIDE GROUPS WITH SIGNIFICANT FOLIAR USE: TRANSLOCATED HERBICIDES SHOWING INITIAL SYMPTOMS ON OLD GROWTH

These herbicides are frequently described as apoplastically translocated, xylem mobile, upwardly mobile, or acropetally translocated. The only direction of movement is up, in contrast to symplastically translocated herbicides, which can move in the symplast in both upward and downward directions according to the location of the sink and in the apoplast when taken up from the soil. Herbicides that translocate only in the apoplast move with water to fully expanded, transpiring leaves. As a result, the older leaves show the symptoms first.

Two mode-of-action groups are included in this chapter: the "classical" photosynthetic inhibitors and the "rapidly acting" photosynthetic inhibitors. These two groups of herbicides are described together primarily because they both inhibit photosynthesis. Only the "classical" photosynthetic inhibitors, however, show the typical symptomology associated with herbicides that move apoplastically. In other words, symptoms first appear on old growth and then move upward to new growth. When applied at normal use rates, the "rapidly acting" photosynthetic inhibitors kill so quickly that their symptoms are more typical of the nontranslocated herbicides, showing localized injury and rapid membrane destruction. What movement occurs in this group, however, does appear to be apoplastic. For example, when the "rapidly acting" photosynthetic inhibitors are applied at marginal doses, symptoms develop first on old growth.

"CLASSICAL" PHOTOSYNTHETIC INHIBITORS

These herbicides move upward in the apoplast to the fully expanded leaves. Chlorosis first appears between leaf veins (Figure 10.1) and along the margins (Figure 10.2), later followed by necrosis of the leaf tissue. The older portions of the plants die first and death progresses upward (from old to new growth). If any tissue appears uninjured, it will be the growing points and younger leaves.

Most of the "classical" photosynthetic inhibitors are applied to the soil as preplant incorporated or preemergence treatments where they control germinating seedlings. Some also are used as postemergence treatments.

Roots of perennial plants that take up soil-applied herbicides from this mode-of-action group are not killed directly, but the destruction of shoot growth can lead to the eventual starvation and death of underground structures. This can occur from a soil application that continually kills and prevents shoot regrowth or from repeat foliar applications that kill each new flush of shoot growth.

This group of apoplastically translocated herbicides consists of three groups of chemistry: triazines, phenylureas, and uracils.

Figure 10.1 Left: Soybeans treated with a range of concentrations of linuron, a substituted urea herbicide. Injury progresses from the older to the younger leaves. Right: Injury symptoms most evident along the veins.

Figure 10.2 Left: Soybeans treated with the triazine herbicide metribuzin. Injury progresses from the older, lower leaves to the younger leaves. Right: Injury symptoms most evident along the leaf margins.

Triazine Herbicides

The triazine herbicides, particularly atrazine, have been the major herbicides used for weed control in corn over the past three decades. Simazine has had widespread use in woody species in addition to corn, and a number of the s-triazines have been used for total vegetation control. Thus, this group has a broad spectrum of uses and has played an important role in reducing crop losses due to weeds over the years.

The triazines are widely used as soil-applied herbicides. They selectively control annual broad-leaved and annual grass weeds in crops. Some triazines also can be used as foliar treatments with surfactants or oils (an important exception is simazine, which has little foliage activity on terrestrial plants).

The basic structure of the triazine molecule is a ring composed of nitrogen and carbon atoms (a heterocyclic nitrogen structure). Most of the triazines are symmetrical; that is, the carbons and nitrogens alternate in the ring. They are designated as s-triazines. The ring structure of metribuzin, however, is asymmetrical, and this herbicide is called an as-triazine.

When soil applied, triazines are readily taken up by the roots of seedlings and move into the emerging foliage. Seedlings typically emerge from the soil and grow until they exhaust the food stored in the cotyledons. They then become chlorotic and die back.

The triazines are used on a wide variety of crops. Selectivity is determined by placement, depth protection, or by internal detoxification mechanisms. Simazine, for example, is applied to the soil surface to control shallow-rooted weeds in deep-rooted established crops such as strawberries and tree crops. When depth protection is impossible, internal mechanisms are required for crop selectivity. The resistance of corn to atrazine and simazine is a classic case of differential metabolism as a means of selectivity (see Chapter 6).

The triazines are relatively persistent in soils and can cause carryover problems in susceptible crops. Dosage and the amount of rainfall, soil type, pH, and other factors all affect persistence. The triazines tend to be more persistent under arid conditions and in high pH soils. The triazines are adsorbed by soil colloids and are not recommended for use in soils with high organic matter.

Atrazine

TRADE NAMES AND MANUFACTURERS

Aatrex—Novartis
Atrazine—several manufacturers

CHEMISTRY

6-chloro-N-ethyl-N'-(1-methylethyl)-1,3,5-triazine-2,4-diamine

AVAILABLE FORMULATIONS

A. Sprayable in water
   1. 90% a.i. water-dispersible granule (Aatrex Nine-O, Atrazine 90WDG, Atrazine 90WDF)

   2. 4 lb a.i./gal dispersible liquid (Aatrex 4L, Atrazine 4L)

RESPONSE IN SOIL

The affinity for atrazine adsorption to soil colloids is moderate to strong. Therefore, rates must be adjusted for soil texture and organic matter. In soils with a high peat, or muck content, atrazine should be used only as a postemergence treatment.

Because of its low water solubility (33 mg/L) and affinity for adsorption, leaching of atrazine in most soils is limited, but extremely low amounts have appeared in groundwater. Therefore, restrictions have been placed on the total amounts of atrazine that can be applied and the sites that can be treated (see Chapter 5).

Atrazine provides weed control for most of a full growing season. The average field half-life is 60 days. Most rotational crops can be planted 1 yr after application of atrazine at selective rates. Crops that are especially sensitive to atrazine (sugar beets, vegetable crops, small-seeded legumes, oats) should not be planted sooner than 2 yr after atrazine has been applied. Arid climates and high soil pHs extend the persistence.

SENSITIVE SPECIES

Mostly controls annual broadleaf species.

A. Soil applied at 1 to 2 lb a.i./acre.
   1. Annual broadleaves include common lambsquarters, kochia, jimsonweed, morningglories, mustards, nightshades, pigweeds, common purslane, and ragweeds.
   2. Wild oats is the only grass species listed.
   3. Some suppression of giant foxtail, cocklebur, sicklepod, and velvetleaf can be obtained.
B. Postemergence application at 1 to 2 lb a.i./acre with emulsifiable oil or oil concentrate in water on broadleaves less than 4 to 6 in. in height controls common cocklebur, common lambsquarters, jimsonweed, annual morningglory, mustards, pigweeds, ragweeds, smartweeds, and wild buckwheat.

MAJOR USES

A. Corn when atrazine is applied preplant incorporated or preemergence; early postemergence with an emulsifiable oil or oil concentrate. Tolerance in corn is excellent under most conditions.
B. Other uses
   1. Sorghum
   2. Sugarcane
   3. Warm-season turfgrasses for sod production (St. Augustine, centipede, and zoysia)
   4. Macadamia nuts and guava
   5. Conifers
   6. Chemical fallow
   7. Roadsides and conservation reserves

SPECIAL PROBLEMS

Carryover of atrazine to susceptible crops is possible.

Unexpected injury can occur when atrazine is used with oil and the crop is under stress from prolonged cold, wet weather or other adverse conditions.

The development of atrazine-resistant weed biotypes is possible after prolonged periods of use.

The presence of minute amounts of atrazine in groundwater and surface water is a concern.

Other Triazine Herbicides
Typical Herbicide: Atrazine

Herbicide	Specific Characteristics	Major Uses
s-triazines:		
Ametryn (Evik DF) N-ethyl-N'-(1-methylethyl)-6-(methylthio)-1,3,5-triazine-2,4-diamine	Foliar and root uptake. Water solubility 200 mg/L. Limited leaching. Field half-life is 60 days.	Selective for annual weeds in bananas and corn when used directed postemergence. Can be applied broadcast to pineapples and sugarcane.
Cyanazine (Bladex) 2-[[4-chloro-6-(ethylamino)-1,3,5-triazin-2-yl]amino]-2-methylpropanenitrile	Foliar and root uptake. Water solubility 171 mg/L. Degree of leaching approximately same as atrazine. Corn tolerance is limited on coarse-textured low organic matter soils. Provides better activity than atrazine on problem annual grasses and poorer activity on pigweed and velvetleaf. Short field half-life of 14 days. Mammalian LD_{50} 334 mg/kg; about ten times more toxic than most triazines.	Selective for annual weeds in corn. Particularly useful where the use of simazine or atrazine is likely to result in carryover injury to sensitive crops. Used preplant incorporated, preemergence, or sometimes early postemergence.

Herbicide	Specific Characteristics	Major Uses
Hexazinone (Velpar) 3-cyclohexyl-6-(dimethylamino)-1-methyl-1,3,5-triazine-2,4(1H,3H)-dione	Foliar and root uptake. Water solubility 33,000 mg/L; very mobile in soil with potential for leaching into groundwater at high doses. Half-life in soil of 1 to 6 mo.	Soil or foliage application for general vegetation control including shrubs and trees. Selective on established alfalfa, pineapple, sugarcane, and many Christmas tree species. Useful for conifer site preparation and release.
Prometon (Pramitol) 6-methoxy-N,N'-bis(1-methylethyl)-1,3,5-triazine-2,4-diamine	Foliar and root uptake. Water solubility 720 mg/L. Relatively mobile in soil. Persists one or more seasons depending on rate and climatic conditions.	Nonselective preemergence and postemergence herbicide for general vegetation control.
Prometryn (Caparol) N,N'-bis(1-methylethyl)-6-(methylthio)-1,3,5-triazine-2,4-diamine	Foliar and root uptake. Water solubility 33 mg/L. Limited leaching. Average field life is 60 days.	Selective for annual weeds in cotton and celery. Used preplant incorporated, preemergence, and directed postemergence in cotton.
Simazine (Princep) 6-chloro-N,N'-diethyl-1,3,5-triazine-2,4-diamine	Root uptake only. Water solubility 6.2 mg/L. Very limited leaching. Average field life is 60 days.	Selective for fruit and nut crops, corn, Christmas trees, shelter belts, nursery crops.

Herbicide	Specific Characteristics	Major Uses
as-triazine: Metribuzin (Sencor, Lexone) 4-amino-6-(1,1-dimethylethyl)-3-(methylthio)-1,2,4-triazin-5(4H)-one	Foliar and root uptake. Water solubility 1100 mg/L. Relatively mobile in soil. Half-life is 30 to 60 days.	Selective for annual weeds in soybeans, potatoes, tomatoes, established asparagus and alfalfa, and sugarcane. Applied preplant incorporated, preemergence; and (on corn, potatoes, tomatoes, small grains) early postemergence. Burn-down treatment in no-till for corn and soybeans.

Phenylurea Herbicides

Most of the phenylurea herbicides are soil applied for controlling germinating weeds selectively. Some (linuron, diuron, fluometuron) also have foliage activity and can be used as directed postemergence treatments for controlling newly emerged weeds. A surfactant is usually added to enhance foliage activity. At high rates, most phenylurea herbicides are nonselective and several are used for general vegetation control.

The selectivity of the ureas is due to differential translocation and metabolism or to depth protection in cases in which leaching is limited.

This group of herbicides tends to be relatively persistent in soils, although at rates selective to crops they seldom cause carryover problems. Persistence is 3 to 6 mo at normal crop use rates. At nonselective rates for general vegetation control, phytotoxic residues can persist for as long as 24 mo. The ureas are broken down in the soil by microbial activity.

Linuron

TRADE NAME AND MANUFACTURER

Lorox DF—E. I. du Pont de Nemours & Co.

CHEMISTRY

N'-(3,4-dichlorophenyl)-N-methoxy-N-methylurea

AVAILABLE FORMULATIONS

A. Sprayable in water
   1. 50% a.i. water-dispersible granule (Lorox DF)

RESPONSE IN SOILS

Linuron has a strong affinity for soil organic matter and to a lesser extent clay. Use rates must be closely matched with soil texture and organic matter when using linuron as a soil treatment. Fields with variable soil types are difficult to treat without observing either some crop injury (on low organic matter soils) or poor weed control (on high organic matter soils). Mechanical incorporation results in a reduction of herbicidal activity. Water solubility is 75 mg/L. Relatively large amounts of rainfall (at least 1/2 in. depending on soil type and moisture content) are needed to move and initiate herbicide activity in the soil. Losses by photodecomposition or volatility are insignificant except when linuron is exposed on the soil surface for several days or weeks under hot, dry conditions. Linuron persists for most of a growing season. When used at normal rates, phytotoxic concentrations disappear within approximately 4 mo after application.

SENSITIVE SPECIES

Linuron controls germinating and newly established annual broadleaved weeds and many grasses. Grasses tend to be more tolerant than broadleaves.

A. When used preemergence, linuron controls barnyardgrass, carpetweed, common chickweed, crabgrass, foxtails, goosegrass, Florida pusly, galinsoga, common lambsquarters, mustards, fall panicum, pigweeds, common purslane, wild radish, common ragweed, and smartweeds. Partial control is obtained on common cocklebur, annual morningglory, prickly sida, sicklepod, and velvetleaf.
B. When used postemergence, linuron controls many of the previously listed species including those only partially controlled preemergence.
C. At rates used in combination with other herbicides (l/2 that when linuron is used alone), grasses and partially susceptible broadleaves tend not to be controlled.

MAJOR USES

A. Preemergence in carrots, celery, corn, parsnips, potatoes, sorghum, and soybeans.

Usually applied in combination with other complementary herbicides. Suggested only in combination with another herbicide for corn and sorghum.
B. Postemergence to the weeds prior to crop seeding or emergence in no-till soybeans and on conventionally prepared seedbeds after weed emergence. In no-till, linuron is combined with paraquat or glyphosate. A surfactant is added to increase foliar penetration.
C. Early postemergence in carrots, celery, asparagus. No surfactant is added for this use.
D. Directed postemergence in field and sweet corn, cotton, sorghum, and soybeans. Crop tolerance is obtained by directed placement. Therefore, a good height differential between the crop and weed is essential for success. A surfactant is added to enhance foliar activity.

SPECIAL PROBLEMS

Variable performance and the potential for crop injury exist when linuron is used on soils of varying soil type and colloid content.

The potential for crop injury exists when linuron is used directed postemergence with a surfactant if placement is not accurate.

Other Phenylurea Herbicides
Typical Herbicide: Linuron

Herbicide	Specific Characteristics	Major Uses
Diuron (Karmex) N'-(3,4-dichlorophenyl)-N,N-dimethylurea	Water solubility 42 mg/L. Resists leaching and can be used safely on many deep-rooted crops. Used selectively at low rates and for general vegetation control at higher rates. Requires surfactant when used postemergence. At low rates, phytotoxic concentrations disappear within one season. At high rates, activity usually lasts longer than one season.	Used selectively as preplant, preemergence, or directed postemergence treatment in cotton and postemergence in pineapple and sugarcane. Used preemergence to weeds in established crops (small-seeded legumes, asparagus, grass seed crops), directed postemergence in fruit and nut crops. General vegetation control at higher rates.
Fluometuron (Cotoran) N,N-dimethyl-N'-[3-(trifluoromethyl) phenyl] urea	Same general properties as linuron. Water solubility 110 mg/L. Average field half-life is 85 days.	Preemergence and postemergence treatments in cotton.

Herbicide	Specific Characteristics	Major Uses
Tebuthiuron (Spike) *N*-[5-(1,1-dimethylethyl)-1,3,4-thiadiazol-2-yl]-*N*,*N*'-dimethylurea	Water solubility 2500 mg/L. Moves well through soil and is persistent. Therefore, controls deep-rooted perennial shrubs, trees, and annual weeds. Estimated half-life of 12 to 15 mo in areas receiving 40 to 60 in. of annual rainfall. Half-life greater in low rainfall areas and high organic matter soils.	General vegetation control in noncrop areas and for brush control in rangeland and pastureland.

Uracil Herbicides

The uracils are closely related to the phenylurea herbicides and have many similar properties. They tend, however, to be more readily leached, less selective, and more persistent than most ureas. Therefore, they are not used for as broad a spectrum of crops as are the phenylureas. The uracils are used primarily on perennial crops such as citrus, certain other tree crops, and mint. High nonselective rates can be used for the control of perennial weeds in noncrop situations.

Terbacil

TRADE NAME AND MANUFACTURER

Sinbar—E. I. du Pont de Nemours

CHEMISTRY

5-chloro-3-(1,1-dimethylethyl)-6-methyl-2,4-(1*H*,3*H*)-pyrimidinedione

AVAILABLE FORMULATIONS

A. Sprayable in water
   1. 80% wettable powder (Sinbar)

RESPONSE IN SOILS

Water solubility is 710 mg/L. Terbacil is moderately mobile in the soil. Field half-life is 120 days, which makes it relatively persistent.

SENSITIVE SPECIES

Terbacil controls a wide variety of annual and perennial weeds. High rates are used to control established perennial weeds.

A. Rates of 1/2 to 1 1/2 lb a.i./acre control chickweed, crabgrass, foxtails, henbit, common lambsquarters, mustards, shepherd's purse, downy brome, wild barley, and others.
B. Rates of 1 to 4 lb a.i./acre control annual bluegrass, dandelion, groundsel, horsenettle, knotweed, nightshades, pigweeds, ragweeds, jungle rice, sandbur, smartweeds, yellow nutsedge, and others.
C. Rates of 4 to 10 lb a.i./acre control perennial ryegrass, quackgrass, and red sorrel.

MAJOR USES

A. Rates of 1/2 to 1 1/2 lb a.i./acre control many winter annual weeds in established alfalfa. Use before alfalfa breaks dormancy in spring.
B. Rates of 2 lb a.i./acre or more are used pre-emergence to weeds or during the early seedling stage of weed growth in certain fruit, citrus, nut crops, and caneberries.
C. Rates of 1 to 2 lb a.i./acre are used pre-emergence and postemergence in mint (peppermint and spearmint).
D. Rates of 1 to 2 1/2 lb a.i./acre are used in sugarcane.

SPECIAL PROBLEMS

Terbacil has limited crop selectivity (with the exception of mint). Selectivity on deep-rooted crops is due to differential placement.

Carryover to susceptible crops can occur.

Some varieties of sugarcane are sensitive to terbacil.

Herbicide	Specific Characteristics	Major Uses
Bromacil (Hyvar) 5-bromo-6-methyl-3-(1-methylpropyl)-2,4 (1H,3H)pyrimidinedione	Residual to highly persistent depending on rate. Water solubility is 815 mg/L. Moderately mobile. Field half-life of 5 to 6 mo when 4 lb/acre is applied to silt loam.	General vegetation control in noncrop areas. Preemergence to weeds or during early seedling stage of weed growth in citrus crops and pineapple.

"RAPIDLY ACTING" PHOTOSYNTHETIC INHIBITORS

This mode-of-action group consists of diverse chemistries. Bentazon is a benzothiadiazole, bromoxynil is a benzonitrile, desmedipham and phenmedipham are phenylcarbamates, pyrazon is a pyridazinone, and pyridate is a phenylpyridazine.

These herbicides are united by the fact that they inhibit photosynthesis, but plant death is much more rapid than that caused by the "classical" photosynthetic inhibitors, usually occurring within 1 to 5 days. Symptoms include water-soaking and/or chlorosis followed by foliar desiccation and necrosis. The rapid burndown is similar to that produced by nontranslocated herbicides that cause initial localized injury (Figure 10.3).

All of these herbicides are used early postemergence to control seedling or young broadleaved weeds; pyrazon also is used preemergence. When translocation occurs in these young plants, it is apoplastic.

All are very short lived in the soil.

Figure 10.3 The plant treated with the rapidly acting photosynthetic inhibitor, bromoxynil (BRM), dies back as quickly as a plant treated at the same time with paraquat (PQT), a nontranslocated herbicide that causes localized injury.

Bentazon

Bentazon is a postemergence herbicide used for the control of broadleaved weeds and sedges in many important grass and large-seeded legume crops. It is used for difficult-to-control broadleaves such as cocklebur, ragweeds, and sunflower in soybeans and other crops. Chlorosis begins 3 to 5 days after application followed by desiccation and necrosis. Translocation of the herbicide appears to be limited because thorough coverage is needed for plant kill and new buds develop normally if shoot kill is not total.

Optimum control is obtained when weeds are small and actively growing.

Crop tolerance is generally excellent. Although bronzing and leaf burn may occur on soybeans, beans, and peanuts, new growth is normal and crop vigor is not reduced.

In order to broaden the spectrum of broadleaved species controlled, bentazon has been used in tank mix combinations with other postemergence herbicides.

Bentazon

TRADE NAMES AND MANUFACTURER

Basagran, Basagran T/O—BASF Corporation

CHEMISTRY

3-(1-methylethyl)-(1*H*)-2,1,3-benzothiadiazin-4(3*H*)-one 2,2-dioxide

AVAILABLE FORMULATIONS

A. Sprayable in water
   1. 4 lb/gal water-soluble liquid (Basagran, Basagran T/O)

RESPONSE IN SOILS

Bentazon has no activity in soil at postemergence rates. It is rapidly metabolized by soil microbes and reaches undetectable levels within 6 wk.

SENSITIVE SPECIES

Broadleaved and sedge species

A. Annual broadleaves include spurred anoda, beggarticks, common cocklebur, jimson-weed, Venice mallow, mustards, purslane, giant ragweed, sesbania, prickly sida, ladys-thumb smartweed, Pennsylvania smartweed, wild sunflower, and velvetleaf.
B. Perennial weeds include yellow nutsedge and shoot removal of Canada thistle.

MAJOR USES

A. Postemergence on soybeans, corn, rice, peanuts, dry or succulent beans and peas, and established peppermint and spearmint.
B. Postemergence treatment for control of broadleaved weeds and yellow nutsedge in established turf (bluegrass, fescue, bentgrass, bermudagrass, bahiagrass, carpetweed, centipede-grass, zoysiagrass, ryegrass, St. Augustine grass).
C. Postemergence treatments on ornamentals, nursery crops, and in noncrop sites.
D. Application of single rate plus a second single rate 7 to 10 days later for control of Canada thistle and yellow nutsedge in all crops listed except rice.
E. Late rescue treatments of soybeans, corn, and peanuts from cocklebur not treated early postemergence. Only partial control may be obtained on plants up to 24 in. tall.

SPECIAL PROBLEMS

Some varieties of dry and succulent beans may have limited tolerance to bentazon.

Other "Rapidly Acting" Photosynthetic Inhibitors
Typical Herbicide: Bentazon

Herbicide	Specific Characteristics	Major Uses
Bromoxynil (Buctril) 3,5-dibromo-4-hydroxybenzonitrile	Postemergence for annual broadleaved weed control. Particularly effective on nightshades, mustards, and smartweeds. Average field half-life in soil is 7 days.	Grass crops such as corn, sorghum, small grains, grass crops grown for seed, nonresidential turfgrass. Other crops include garlic and onions, mint, seedling alfalfa, and flax.

Herbicide	Specific Characteristics	Major Uses
Desmedipham (Betanex) ethyl[3-[[(phenylamino) carbonyl]oxy]phenyl] carbamate	Postemergence for annual broadleaved weed control. Average half-life in soil is less than 1 mo.	Sugar beets.

$$CH_3-CH_2-O-\overset{\overset{\textstyle O}{\|}}{C}-HN\text{—(ring)—}O-\overset{\overset{\textstyle O}{\|}}{C}-HN\text{—(ring)}$$

Herbicide	Specific Characteristics	Major Uses
Phenmedipham (Spin-Aid) 3-[(methyoxycarbonyl) amino]phenyl(3-methyl-phenyl)carbamate	Postemergence for annual broadleaved weed control. Average half-life in soil is about 25 to 30 days.	Spinach and red beets.

$$CH_3-O-\overset{\overset{\textstyle O}{\|}}{C}-HN\text{—(ring)—}O-\overset{\overset{\textstyle O}{\|}}{C}-HN\text{—(ring)—}CH_3$$

Herbicide	Specific Characteristics	Major Uses
Pyrazon (Pyramin) 5-amino-4-chloro-2-phenyl-3(2H)-pyridazinone	Preemergence or early postemergence for annual broadleaved weed control. Average half-life in soil is 21 days.	Sugar beets and red table beets in the U.S.

(structure: phenyl attached to N-N pyridazinone ring bearing NH₂, Cl, and =O)

Herbicide	Specific Characteristics	Major Uses
Pyridate (Tough) O-(6-chloro-3-phenyl-4-pyridazinyl)S-octyl carbonothioate	Postemergence for annual broadleaf weed control. Average half-life in soil is 7 to 21 days.	Peanuts and corn in the U.S. Additional crop uses in Europe.

$$O-\overset{\overset{\textstyle O}{\|}}{C}-S-(CH_2)_7-CH_3$$

(structure: Cl-substituted pyridazine ring fused/attached to phenyl with the carbonothioate side chain above)

11

HERBICIDE GROUPS WITH SIGNIFICANT FOLIAR USE: NONTRANSLOCATED HERBICIDES SHOWING INITIAL LOCALIZED INJURY

All of the compounds discussed in this chapter cause rapid disruption of cell membranes and very rapid kill of the plant. Rapid destruction of cell membranes prevents translocation of the herbicide to other regions of the plant. Severe injury is evident within hours to a few days after application. Symptoms first appear as water-soaking and/or chlorosis, followed by desiccation and necrosis (Figure 11.1). Maximum kill usually is attained in a week or less. An exception is glufosinate, which is somewhat slower in its symptom development.

These herbicides often are referred to as contact herbicides. Symptom development is hastened by bright sunlight, high humidity, and moist soil.

Good coverage of the target plant is essential for control. Partial coverage results in spotting (Figure 11.2) or only partial shoot kill (Figure 11.1). New growth on surviving plants is normal in appearance.

The three mode-of-action groups that cause rapid disruption of cell membranes are the photosystem I energized cell membrane disrupters, the protoporphyrinogen oxidase (Protox) inhibitors, and the glutamine synthesis inhibitors. Herbicides in these groups cause the production of unstable molecules (OH^-, $^1O_2$, or ^3chl) that degrade the lipid components of cell membranes (see Chapter 6).

PHOTOSYSTEM I ENERGIZED CELL MEMBRANE DISRUPTERS

The chemical group with this mode of action, the bipyridilium group, consists of two herbicides, diquat and paraquat. These two herbicides share several important characteristics. They are strong cations and adsorb quickly and tightly to soil colloids. Therefore, they lack soil activity and are used only as foliar applications. Both are contact herbicides, have limited translocation, require the addition of a surfactant to increase wetting and penetration of leaf surfaces, and cause visible water-soaking and wilting of the plant within hours of treatment.

Neither compound is very selective. Paraquat is used for weed control prior to crop emergence and is particularly useful in minimum tillage programs for the burndown of weeds prior to or at planting. It can be used as a directed spray and in noncrop situations. Diquat is registered in the United States for aquatic weed control. Each is labeled as a harvest aid desiccant.

Even though these compounds are highly water soluble, they are rainfast in an hour or less after application because they are rapidly absorbed into the plant.

The mammalian toxicity of paraquat is high and accounts for the restricted-use status of this product. Diquat is slightly less toxic to mammals; when diluted in water for aquatic weed control it has very low toxicity to mammals and fish.

Figure 11.1 Wheat treated with paraquat. Plant on left treated with sublethal dose shows almost complete recovery.

Paraquat

TRADE NAMES AND MANUFACTURER

Cyclone, Gramoxone Extra, Starfire—Zeneca

CHEMISTRY

1,1'-dimethyl-4,4'-bipyridinium ion (dichloride salt)

$$H_3C - \overset{+}{N} \quad \quad \overset{+}{N} - CH_3 \quad 2Cl^-$$

AVAILABLE FORMULATIONS

A. Sprayable in water
   1. 2.5 lb a.i./gal water-soluble liquid (Gramoxone Extra)
   2. 1.5 lb a.i./gal water-soluble liquid (Starfire Herbicide)
   3. 3 lb a.i./gal water-soluble liquid (Starfire Concentrate, Cyclone Concentrate)
   4. 2 lb a.i./gal water-soluble liquid (Cyclone Herbicide)

RESPONSE IN SOILS

Paraquat is rapidly inactivated by soil due to adsorption of the cation to negatively charged clay minerals. When bound to the soil it is persistent but biologically unavailable.

SENSITIVE SPECIES

Paraquat is nonselective, although it may have slightly more activity on grasses than on broadleaves.

MAJOR USES

A. Preplant or preemergence (prior to crop emergence) for control of emerged weeds

Figure 11.2 Localized lesions on corn caused by contact with the bipyridylium herbicide paraquat. Crop yield will not be affected significantly. (J. L. Williams, Jr., Purdue University)

in asparagus; dry beans; chemical fallow; all types of corn, cotton, onions, and garlic; peanuts; pineapple; potatoes; rice; safflower; small grains; soybeans; sorghum; sugar beets, and vegetable crops.

B. Used for kill of emerged weeds where corn, sorghum, soybeans, and vegetable crops will be planted directly into a cover crop, established sod, or in previous crop residues (minimum tillage systems).

C. Suppression of existing sod and weeds to permit pasture and range reseeding.

D. Dormant treatment on alfalfa, clovers, and mint.

E. Directed spray in corn, cotton, soybeans, small fruits, orchards, vineyards, windbreaks, and shade and ornamental trees.

F. Used as a harvest aid to defoliate cotton and sunflowers and to desiccate weeds in soybeans.

SPECIAL PROBLEMS

Paraquat is a restricted-use pesticide with high mammalian toxicity (oral LD_{50} to rats is 150 mg/kg). This has resulted in exaggerated fear and reduced use of the product.

Other Photosystem I Energized Cell Membrane Disrupters
Typical Herbicide: Paraquat

Herbicide	Specific Characteristics	Major Uses
Diquat (Diquat, Reward) 6,7-dihydrodipyrido [1,2-a:2',1'-c] pyrazinediium ion	Formulated as bromine salt. Same general properties as paraquat: strongly cationic, inactivated by clay fraction of soils, nonselective, and a contact herbicide. Oral LD_{50} to rats is 230 mg diquat ion/kg.	Limited crop uses. Primarily used on noncrop or nonplanted areas, as an aquatic herbicide, and as a preharvest aid to defoliate potatoes.

PROTOPORPHYRINOGEN OXIDASE INHIBITORS

This mode-of-action group consists of four chemical families: the diphenylethers, oxadiazoles, N-phenylphthalimides, and triazolinones. These herbicides mostly control broadleaves, and several are used for broadleaf control in soybeans. Although bronzing or burning of soybeans can occur after application, yield is rarely affected. Depending on the compound, they can be applied to the foliage or to soil.

Diphenylether Herbicides

These compounds are used on a wide variety of crops, including soybeans, rice, cotton, mint, several vegetables, and tree crops. They exhibit both soil and foliar activity and are used either pre- or postemergence for controlling annual weeds. They are effective primarily on broadleaved weeds but will control some grass species.

Light is required for these compounds to work. Thus, when the diphenylethers are applied as soil treatments, the seedlings must emerge from the soil before symptoms develop. Preemergence applications cause severe burn to the seedling tissue, particularly at the soil line.

These compounds are adsorbed to soil colloids, but not nearly as much as paraquat or diquat. They resist leaching and must be moved into the upper soil surface with rainfall or irrigation in order to maximize contact with seedlings and to prevent loss from photodegradation. Mechanical soil incorporation results in excessive dilution and loss of activity. Soil peristence varies, but these compounds seldom cause carryover problems. Only fomesafen has restrictions for crops planted the next season.

Acifluorfen

TRADE NAMES AND MANUFACTURERS

Blazer—BASF Corporation
Status—American Cyanamid

CHEMISTRY

5-[2-chloro-4-(trifluoromethyl)phenoxy]-2-nitrobenzoic acid

AVAILABLE FORMULATIONS

A. Sprayable in water
   1. 2 lb a.i./gal water-soluble liquid (Blazer, Status)

RESPONSE IN SOILS

Acifluorfen resists leaching and is susceptible to photodegradation. Its field half-life is 14 to 60 days. It should not persist into the subsequent cropping season at normal use rates.

SENSITIVE SPECIES

Primarily broadleaved weeds are sensitive to acifluorfen, and some top kill of perennial vines may be obtained.

A. Broadleaved weeds controlled include carpetweed, common cocklebur, crotalaria, tropic croton, galinsoga, jimsonweed, nine species of annual morningglories, wild mustard, black nightshade, pigweeds, common purslane, Florida pusley, ragweeds, hemp sesbania, smartweeds, and prostrate spurge.

MAJOR USES

A. Early postemergence control of seedling weeds in soybeans
B. Postemergence control of hemp sesbania in rice
C. Preemergence, cracking, or postemergence treatments in peanuts

Diphenylether Herbicides
Typical Herbicide: Acifluorfen

Herbicide	Specific Characteristics	Major Uses
Fomesafen (Reflex, Flexstar) 5-[2-chloro-4-(trifluoromethyl)phenoxy]-N-(methyl-sulfonyl)-2-nitrobenzamide	Moderately mobile. Average field half-life is 100 days. Can persist long enough to injure susceptible crops such as sugar beets, sunflowers, and sorghum 1 yr after application.	Postemergence control of broadleaved weeds in soybeans.

Herbicide	Specific Characteristics	Major Uses
Lactofen (Cobra) (±)-2-ethoxy-1-methyl-2-oxoethyl 5-[2-chloro-4-(trifluoromethyl)phenoxy]-2-nitrobenzoate	Water solubility 0.1 mg/L. Resists leaching. Very short lived in soil; average field half-life is only 3 days. Tends to cause high levels of initial burn to soybeans.	Postemergence control of broadleaved weeds in soybeans, cotton, and southern pine seedlings.

Oxyfluorfen (Goal) 2-chloro-1-(3-ethoxy-4-nitrophenoxy)-4-(trifluoromethyl)benzene	Water solubility 0.1 mg/L. Resists leaching. Average field half-life is 35 days.	Preemergence control of broadleaved weeds in cole crops; pre- or postemergence on conifer seedbeds, transplants, container stock, tree fruits, nuts, vines; directed postemergence on cotton; over-the-top postemergence on onions; dormant applications to mint.

Oxadiazole Herbicides

The chemical structure of this group of herbicides is much different than that of the diphenylethers, but its mode of action and symptom development are similar. Oxadiazon must be applied before the weeds germinate and must be moved into the soil by rainfall or irrigation.

Oxadiazon

TRADE NAME AND MANUFACTURER

Ronstar—Rhone-Poulenc Inc.

CHEMISTRY

3-[2,4-dichloro-5-(1-methylethoxy)phenyl]-5-(1,1-dimethylethyl)-1,3,4-oxadiazol-2-(3H)-one

FORMULATIONS AVAILABLE

A. Sprayable in water
    1. 50% wettable powder (Chipco Ronstar 50 WP)
    2. 50% water-soluble packet (Chipco Ronstar 50 WSP)
B. Dry application
    1. 2% granule (Chipco Ronstar G)

RESPONSE IN SOIL

The water solubility of oxadiazon is about 0.7 mg/L. It is strongly adsorbed to soil colloids and resists leaching. It is moderately persistent in soil, giving season-long control. Field half-life is 60 days.

SENSITIVE SPECIES

Annual broadleaf and grass species.

A. Broadleaves include carpetweed, groundsel, common lambsquarters, several mustards, common purslane, redroot pigweed, Pennsylvania smartweed, sowthistle, speedwell, yellow wood sorrel, and others.
B. Grasses include crabgrass, goosegrass, Poa annua, fall panicum, and green foxtail.

MAJOR USES

A. Preemergence control of weeds in established turf (bluegrass, bentgrass, bermuda- grass, perennial ryegrass, St. Augustine grass, tall fescue, zoysia).
B. Preemergence control of weeds in a wide variety of newly transplanted and established ornamental shrubs, vines, and trees. These can include container- grown, field-grown, and forest nur- sery plants.

SPECIAL PROBLEMS

Temporary discoloration of bentgrass, bermuda- grass, and St. Augustine grass can occur. Re- seeding of turfgrasses should be delayed until 4 mo after application.

Oxadiazole Herbicides
Typical Herbicide: Oxadiazon

Herbicide	Specific Characteristics	Major Uses
Fluthiacet (Action) methyl [[2-chloro-4-fluoro-5-[(tetrahydro-3-oxo-1*H*-[1,3,4]thiadiazolo[3,4-a]pyridazin-1-ylidene)amino]-phenyl]thio]acetate	Half-life of 1 to 2 days.	Primarily for velvetleaf control in soybeans. Also has activity on lambsquarters, pigweeds, morningglories, and several other broadleaves. Also being developed for corn.

N-Phenylphthalimide Herbicides

Herbicide	Specific Characteristics	Major Uses
Flumiclorac (Resource) [2-chloro-4-fluoro-5-(1,3,4,5,6,7-hexahydro-1,3-dioxo-2*H*-isoindol-2-yl)phenoxy]acetic acid	Strongly adsorbed to clay and organic matter; does not leach. Very short lived with half-lives ranging from less than 1 to 6 days.	Postemergence treatment for annual broadleaved weeds in soybeans and corn.

Herbicide	Specific Characteristics	Major Uses
Carfentrazone (Affinity, Aim) ethyl 2-chloro-3-[2-chloro-4-fluoro-5-[4-(difluoromethyl)-4,5-dihydro-3-methyl-5-oxo-1H-1,2,4-triazol-1-yl]phenyl]-propanoate	Control of broadleaved weeds, including many sulfonylurea-resistant weeds. Half-life less than 1.5 days.	Postemergence control in corn, small grains, and rice.

Sulfentrazone (Authority Cover, Spartan) N-[2,4-dichloro-5-[4-(difluoromethyl)-4,5-dihydro-3-methyl-5-oxo-1H-1,2,4-triazol-1-yl]phenyl] methanesulfonamide	Control of annual broadleaves, some grasses, and sedges. Replant restrictions vary from 4 to 30 mo depending on the crop. Subject to rapid degradation if left on soil surface. Low affinity for organic matter.	Preemergence control in soybeans, sugarcane, peas, and tobacco.

GLUTAMINE SYNTHESIS INHIBITORS

This mode-of-action group consists of only one herbicide, glufosinate. Glufosinate is used postemergence only, is nonselective, and controls annual broadleaved and grass weeds. It can be used on perennial weeds, but because translocation is limited, it mostly kills the aboveground portions of the plant. Regrowth of the plant from underground structures can be expected.

Inhibition of the glutamine synthesis enzyme in the affected plant results in the decrease of several amino acids. This and the inhibition of photosynthesis eventually leads to cell membrane disruption and cell death. Symptoms in the plant include a bronzing of the foliage, followed by chlorosis and necrosis 3 to 5 days after herbicide application. Symptom development is somewhat slower than that produced by other nontranslocated contact herbicides.

Glufosinate resistance has been introduced into a number of crops including canola, corn, cotton, rice, soybeans, and sugar beets.

Glufosinate

TRADE NAMES AND MANUFACTURER

Finale, Liberty, Rely—AgrEvo

CHEMISTRY

2-amino-4-(hydroxymethylphosphinyl)-butanoic acid

OH–C(=O)–CH(–NH₂)–CH₂–CH₂–P(=O)(CH₃)–OH

$$OH-\overset{O}{\underset{}{C}}-\underset{NH_2}{CH}-CH_2-CH_2-\overset{O}{\underset{CH_3}{P}}-OH$$

FORMULATIONS

A. Sprayable in water
   1. 1 lb/gal ammonium salts of glufosinate

RESPONSE IN SOILS

Glufosinate has no activity through the soil. It is rapidly degraded by soil microbes. The typical field half-life is 7 days.

SENSITIVE SPECIES

Glufosinate controls a broad spectrum of annual broadleaved and grass weeds.

A. Annual broadleaved weeds include chickweed, cocklebur, jimsonweed, kochia, marestail, mustards, purslane, smartweeds.
B. Annual grasses include barnyardgrass, foxtails, fall panicum, goosegrass, shattercane, and stinkgrass.
C. Perennials that are controlled or suppressed with one application include clovers, dandelion, poison ivy, and johnsongrass.

MAJOR USES

A. Nonselective control of emerged weeds in noncrop areas (Finale).
B. Selective control of weeds in glufosinate-resistant crops (Liberty).
C. Used as a directed spray for the control of emerged weeds in apples, grapes, and tree nuts (Rely).

12

HERBICIDE GROUPS APPLIED ALMOST EXCLUSIVELY TO THE SOIL

With very few exceptions, the herbicides in this group lack activity on emerged foliage. They are used to inhibit germinating seeds. They are taken up by either shoots or roots or by both shoots and roots of the germinating plants before the plants establish leaves above ground.

These herbicides are divided into four groups: microtubule/spindle apparatus inhibitors (root inhibitors), shoot inhibitors, miscellaneous cell division inhibitors, and cell wall formation inhibitors. The effect of all four groups is to prevent normal cell division and development, thereby inhibiting growth.

MICROTUBULE/SPINDLE APPARATUS INHIBITORS

The major chemical family of microtubule/spindle apparatus inhibitors is the dinitroaniline group. Others with the same mode of action are DCPA, a herbicide for which no chemical family is generally recognized; dithiopyr, a pyridine herbicide; and pronamide, an amide herbicide. These herbicides have little or no foliar activity and are applied mostly preplant incorporated or preemergence for the control of seedling grasses and some broadleaves.

These compounds are readily taken up by germinating seedlings but do not translocate significantly in the plant. Their major effects are on root growth, which they inhibit by interfering with mitosis. The inability of the cells to form a normal spindle apparatus results in swollen cells with multiple sets of chromosomes. Cell division is effectively inhibited.

Highly susceptible grasses and broadleaves fail to emerge from the soil. Symptoms on emerged plants and on injured crops include a characteristic swelling of the root tips and the inhibition of lateral or secondary root development (Figure 12.1). The shoot portion of the plant may emerge and appear normal for some time, but the inhibition of root growth eventually leads to the death of the entire plant. Shoot symptoms include swollen stems and brittleness.

Dinitroaniline Herbicides

The dinitroaniline herbicides are used as soil-applied herbicides for controlling seedling weeds, especially annual grasses. They also control a limited number of broadleaved weeds.

A wide range of major crops are tolerant to the dinitroanilines. These include soybeans, dry and snap beans, cotton, tobacco, peanuts, and many fruits, vegetables, ornamentals, and established turfgrasses. Many of the dinitroanilines also are labeled for noncrop sites.

Some dinitroanilines (e.g., trifluralin) are so volatile and susceptible to photodegrada-

206

Figure 12.1 Effect of the dinitroaniline herbicide trifluralin on corn roots.

tion that they require soil incorporation. Others (e.g., oryzalin) can be applied to the soil surface without appreciable loss. All tend to adsorb to soil colloids, and their water solubilities are less than 1 mg/L (except for oryzalin). Therefore, they show limited leaching. Either mechanical incorporation or percolating water is required to move these compounds into the soil.

The dinitroanilines provide season-long control and usually are decomposed in soil in less than 12 mo. Higher than normal rates can cause carryover problems in succeeding susceptible crops. Normal use may delay the reseeding of turfgrasses.

Selectivity can be achieved by placement; for example, through shallow incorporation or surface application of the herbicide above a deep-planted crop. In other cases, however, the natural tolerance of the crop to the herbicide must be relied on. Some tolerant broadleaved plants that germinate in the herbicide zone show initial root inhibition, but the rapid growth of the taproot out of the treated zone eventually permits normal root development.

Trifluralin

TRADE NAMES AND MANUFACTURERS

Treflan—Dow AgroSciences and other manufacturers
Tri-4—American Cyanamid
Trifluralin—Riverside/Terra

CHEMISTRY

2,6-dinitro-N,N-dipropyl-4-(trifluoromethyl)benzenamine

$$CH_3-CH_2-H_2C \quad CH_2-CH_2-CH_3$$

(structure of trifluralin: benzene ring with N(dipropyl) at top, O_2N and NO_2 at the 2,6 positions, and CF_3 at the bottom)

AVAILABLE FORMULATIONS

A. Sprayable in water
   1. 4 lb a.i./gal emulsifiable concentrate (Treflan EC, Treflan HFP, Treflan M.T.F., Tri-4 HF, Trifluralin 4EC)

B. Dry application
   1. 5% granule (Treflan 5G)
   2. 10% granule (Treflan TR-10)

RESPONSE IN SOILS

Soil incorporation of trifluralin is required because the compound is moderately volatile (vapor pressure is 1.1×10^{-4} mm Hg at 25°C) and also subject to breakdown by ultraviolet light. Trifluralin has a moderate to strong affinity for soil colloids, and leaching is negligible. Solubility in water is less than 1 mg/L. Average field half-life is 45 days on most soils but can be as high as 120 days in cool, dry areas. This compound can be expected to give full-season weed control. At higher than normal use rates, it may cause some damage to subsequent susceptible crops such as corn, sorghum, and sugar beets.

SENSITIVE SPECIES

Most seedling grasses and some annual broadleaved weeds are sensitive to trifluralin. It kills germinating seedlings, *not* established plants.

A. Grasses controlled include barnyardgrass, annual bluegrass, brachiaria, bromegrasses

(winter annuals), crabgrasses, foxtails, goosegrass, johnsongrass, fall panicum, Texas panicum, sandbur, shattercane (wild cane), sprangletop, and stinkgrass.
B. Annual broadleaved weeds include carpetweed, common chickweed, goosefoot, henbit, knotweed, kochia, common lambsquarters, pigweeds, common purslane, Florida pusley, Russian thistle, and stinging nettle. Many commonly encountered annual broadleaved weeds are not controlled adequately by trifluralin.

MAJOR USES

A. Preplant incorporated for direct-seeded crops such as soybeans, canola, cotton, beans (castor, dry, guar, mung, lima, snap), carrots, cole crops (broccoli, brussels sprouts, cabbage, cauliflower), greens, okra, peas (English, dry), safflower, and sunflowers.
B. Postplant preemergence for direct-seeded crops or seed pieces (shallow incorporation) in cotton, potatoes, and sugarcane.
C. Preplant incorporated on transplants of cole crops (broccoli, brussels sprouts, cabbage, cauliflower), peppers, tomatoes, grapes, citrus, pecans, and ornamentals.
D. Directed postemergence to crops and preemergence incorporated to weeds on corn, cotton, cucurbits (cantaloupes, cucumbers, watermelons), sugarcane, direct-seeded tomatoes, citrus, and pecans.
E. Surface application for transplanted and established ornamental woody shrubs,

trees, roses, and other established flowers. Should be incorporated with rain or irrigation for optimum weed control.
F. Special uses
1. Double rate program for rhizome johnsongrass control that includes
a. Double rate 2 yr in soybeans or cotton
b. Single rate 3rd yr to permit normal crop rotation
c. Conventional tillage system with as much preplant tillage as can be utilized
2. Fall application. Trifluralin may be applied between October 15 and December 31, and the site can be planted in cotton, soybeans, safflower, or dry beans and peas the following spring.
3. Shallow preplant incorporated or postplant preemergence for deep-planted winter wheat in Idaho, Oregon, and Washington and preplant incorporated for barley and spring-seeded wheat in Minnesota, North Dakota, and South Dakota.

SPECIAL PROBLEMS

Trifluralin is limited to systems in which it can be incorporated or placed into soil, or to technology that will prevent volatilization (e.g., slow-release formulations).

Carryover to sensitive crops can be a problem when trifluralin is used at higher than normal rates.

Trifluralin is toxic to fish when placed directly in water. Its strong adsorption to soil particles and soil incorporation preclude the possibility of hazard to fish when used as recommended.

Other Dinitroaniline Herbicides
Typical Herbicide: Trifluralin

Herbicide	Special Characteristics	Major Uses
Benefin (Balan) N-butyl-N-ethyl-2,6-dinitro-4-(trifluoro-methyl)benzeneamine	Vapor pressure 7.8×10^{-5} mm Hg at 25°C. Requires incorporation except when used as granules on turf. Half-life is 40 days.	Preplant incorporated in lettuce and small-seeded legumes. Granules used for annual grass control in established turf.

$$CH_3—H_2C \quad \quad CH_2—CH_2—CH_2—CH_3$$

O_2N \quad NO_2

CF_3

Herbicide	Special Characteristics	Major Uses
Ethalfluralin (Sonalan, Curbit) *N*-ethyl-*N*-(2-methyl-2-propenyl)-2,6-dintro-4-(trifluoromethyl)benzeneamine	Vapor pressure 8.2×10^{-5} mm Hg at 25°C. Requires incorporation. Half-life is 60 days.	Preplant or postplant incorporated in cotton, soybeans, peanuts, sunflowers, dry beans, and peas (Sonalan). Surface applied after planting and prior to weed emergence for cucurbits (Curbit).
Oryzalin (Surflan) 4-(dipropylamino)-3,5-dinitrobenzene sulfonamide	Vapor pressure $<1 \times 10^{-8}$ mm Hg at 25°C. Can be applied to the soil surface as a preemergence treatment. Requires 1/2 in. or more of rain or irrigation for optimum performance. Somewhat more leachable than trifluralin. Water solubility is 2.6 mg/L. Half-life is 20 to 128 days depending on dose.	Directed preemergence to weeds in fruit, nut crops, and vineyards, and directed or broadcast preemergence to weeds in field- and container-grown woody ornamentals, field-grown roses, and Christmas trees. Also used on a limited number of herbaceous ornamentals, noncropland, and warm-season turfgrasses.
Pendimethalin (Prowl, Pendulum, Pentagon, and numerous others in the turf market) *N*-(1-ethylpropyl)-3,4-dimethyl-2,6-dinitrobenzeneamine	Vapor pressure 9.4×10^{-6} mm Hg at 25°C. Mostly incorporated but can be used pre- or postemergence on corn and in minimum till. Most effective preemergence when followed by rain or irrigation. Half-life is 44 days.	Preplant incorporated or preemergence in beans, cotton, soybeans, sunflowers. Preplant incorporated in peanuts. Preemergence or early postemergence in corn, potatoes, and rice. Preplant incorporated or layby in transplanted tobacco. Preemergence in established turf, field- or container-grown ornamentals, ornamental nurseries and gardens, ground covers, nonbearing fruit and nut trees, and conifer and hardwood plantations.
Prodiamine (Barricade, Endurance, Factor) 2,4-dinitro-N^3,N^3-dipropyl-6-(trifluoromethyl)-1,3-benzenediamine	Vapor pressure 2.5×10^{-8} mm Hg at 25°C. Is longer lived (half-life of 120 days) than other dinitroanilines.	Preemergence on established turfgrasses, landscape ornamentals, wildflower plantings.

TRADE NAME AND MANUFACTURER

Dacthal—Zeneca

CHEMISTRY

dimethyl 2,3,5,6-tetrachloro-1,4-benzenedicarboxylate

AVAILABLE FORMULATIONS

A. Sprayable in water
  1. 6 lb a.i./gal flowable (Dacthal Flowable Herbicide Turf Care)
  2. 75% wettable powder (Dacthal W-75, Dacthal W-75 Turf Care)

RESPONSE IN SOILS

A very low solubility in water of 0.5 mg/L and a strong affinity for soil colloids result in no soil movement by leaching. DCPA can be incorporated into the soil to a depth of 1 in. without appreciable loss of activity. Effective performance is limited to mineral soils. Half-life in most soils is 60 to 100 days.

SENSITIVE SPECIES

Primarily seedling grasses and some broadleaves. Weeds that have emerged at the time of application will not be controlled.

A. Seedling grasses include crabgrasses, foxtails, lovegrass, and witchgrass. Sandbur, barnyardgrass, goosegrass, browntop panicum, annual bluegrass, and johnsongrass from seed are moderately susceptible.
B. Broadleaves include carpetweed, common chickweed, common lambsquarters, Florida pusley, and common purslane. Pigweed, dodder, and nodding spurge require higher rates.

MAJOR USES

A. Preemergence control of crabgrass, other annual grasses, and some broadleaf weeds on established ornamental turf. Not suggested for golf greens or bentgrass mowed at golf green height. Apply in spring before grass seeds germinate. Do not reseed turf sooner than 2 mo after treatment.
B. Preemergence control in vegetable and field crops
  1. Direct-seeded crops such as cole crops (broccoli, brussels sprouts, cauliflower, cabbage), large-seeded legumes (field beans, snap beans, mung beans, black-eyed peas), collards, kale, mustard greens, turnips, radishes, garlic, and onions
  2. At planting of seed pieces or transplanting of cole crops, garlic, horseradish, onions, potatoes, sweet potatoes, yams, and strawberries
  3. Established annual plants after seeding or transplanting
    a. Seeded melons (cantaloupe, honeydew, watermelons), cucumbers, summer and winter squash when plants have four to five true leaves
    b. Tomatoes, eggplant and peppers 4–6 wk after transplanting or on direct-seeded plants at 4–6 in. in height
  4. Layby applications: onions, potatoes, sweet potatoes, and yams
C. Control of seedling grasses and some annual broadleaf weeds in established perennial crops with clean soil underneath
  1. Established strawberries (fall and early spring)
  2. Nursery stock and ornamental plantings (numerous species)
D. Control of seedling grasses and some annual broadleaf weeds in established annual ornamental plantings (numerous species). DCPA should not be used on bugleweed, carnation, geum, germander, mesembryanthemum, pansy, phlox, sweet william, and telanthera.

SPECIAL PROBLEMS

A high volume of water and good agitation of sprayable formulations in solution are essential for adequate dispersal and suspension. High concentrations cause the material to settle out rapidly.

A breakdown product of DCPA has been detected in groundwater.

Herbicide	Special Characteristics	Major Uses
Dithiopyr (Dimension) S,S-dimethyl 2-(difluoro-methyl)-4-(2-methylpropyl)-6-(trifluoromethyl)-3,5-pyridinedicarbothioate	Vapor pressure 4×10^{-6} mm Hg at 25°C. Water solubility 1.4 mg/L. Resists leaching and requires moisture for activity. Controls crabgrasses, other seedling grasses, and some annual broadleaves. Controls more species when used preemergence than postemergence. Average persistence is short with a half-life of 17 days.	Preemergence and very early postemergence on established lawns and ornamental turfs. Can be used on putting greens.
Pronamide (Kerb) 3,5-dichloro (N-1,1-dimethyl-2-propynyl) benzamide	Vapor pressure 8.5×10^{-5} mm Hg at 25°C. Water solubility 15 mg/L. Resists leaching and requires moisture or mechanical incorporation to reach weed seed zone. Controls annual and perennial grasses. Very good on chickweeds and other winter annuals. Average persistence 2 to 9 mo depending on soil type and climate. May carry over to small grain and grass crops.	Preplant incorporated or preemergence in lettuce, endive, escarole. Used preemergence or early postemergence to weeds in legumes grown for forage or seed; for annual bluegrass control in bermudagrass turf, artichokes, cane fruits, fallow land, dodder control rhubarb, tree fruits and vines; and in established woody ornamentals, nursery stock, and Christmas trees. Used for quackgrass control in forage legumes.

SHOOT INHIBITORS

The shoot inhibitors consist of two major groups of chemistry: the chloroacetamides and the thiocarbamates. These herbicides control germinating grass seeds and some seedling broadleaves. They have activity on nutsedges germinating from tubers. They have little foliar activity and are used mostly as preplant incorporated or preemergence treatments. The chloroacetamides also are labeled for very early postemergence treatments which extend the application period and help control newly emerged weeds.

Susceptible weeds usually do not emerge from the soil, but grasses that do emerge show tightly twisted leaves that appear to be unable to unfold properly (Figures 12.2, 12.3). Susceptible broadleaves also show distorted leaves and do not grow beyond the seedling stage (Figure 12.2).

Chloroacetamide Herbicides

Most chloroacetamides provide annual grass control and control of some broadleaved weeds. All are soil applied (preemergence) for germinating seedling control. Preplant incorporated applications of the chloroacetamides also are valuable for the suppression of yellow nutsedge developing from tubers.

The chloroacetamides are used on a wide range of crops, including corn, sorghum, soybeans, tobacco, peanuts, cotton, rice, vegetables, ornamentals, fruit and nut trees, and turf. These compounds have been the major preemergence herbicides for the control of seedling grasses and small-seeded broadleaves in corn and soybeans.

Chloroacetamides readily enter germinating seedlings through the shoots and/or roots but show limited translocation within the plant. They primarily affect shoot tips where they retard cell division and meristematic growth. Shoot activity in grasses is observed as the inhibition of leaf emergence from the coleoptile and malformed leaves.

The chloroacetamides readily move into the soil with rainfall and overhead irrigation. All can be used in no-till systems. The chloroacetamides provide 8 to 16 wk control depending on weather conditions and the compound.

Figure 12.2 Effects of the thiocarbamate herbicide EPTC at a range of concentrations on soybeans (left) and corn (right). Soybeans show leaf malformation and corn shows deformed leaf growth typical of grass responses to thiocarbamates.

Alachlor

TRADE NAMES AND MANUFACTURER

Lasso, Micro-Tech, Partner—Monsanto Company

CHEMISTRY

2-chloro-N-(2,6-diethylphenyl)-N-(methoxymethyl)acetamide

AVAILABLE FORMULATIONS

A. Sprayable in water
 1. 4 lb a.i./gal emulsifiable concentrate (Lasso)

2. 4 lb/gal liquid suspension of microencapsulated alachlor (Micro-Tech)
3. 65% a.i. alachlor in a water-dispersible, nonliquid microencapsulated formulation (Partner)

B. Dry application
 1. 15% granule (Lasso II)

RESPONSE IN SOILS

Alachlor has a vapor pressure of 1.6×10^{-5} mm Hg at 25°C. It is slightly soluble in water (242 mg/L) and has a low to moderate affinity for soil colloids. Alachlor should be incorporated if inadequate moisture is available to move the compound into the germinating weed seed zone. Alachlor provides 6 to 10 wk of weed control, and its half-life is 21 days. It does not persist long enough to injure crops the following season.

Figure 12.3 Thiocarbamate-injured corn showing typical leaf looping.

SENSITIVE SPECIES

Primarily annual grasses and a limited number of annual broadleaves. Reasonably good control of the perennial weed yellow nutsedge is obtained with preplant incorporated treatments.

A. Annual grasses include barnyardgrass, crabgrass, foxtails, goosegrass, fall panicum, red rice, signalgrass, and witchgrass.

B. Annual broadleaved weeds include carpetweed, black nightshade, galinsoga, pigweeds, and common purslane.

MAJOR USES

A. Preplant incorporated and preemergence applications
 1. Preplant incorporated soil applications in corn, soybeans, and peanuts. Can be used in dry beans and lima beans west of the Mississippi River.
 2. Preemergence soil applications in corn, soybeans, and peanuts
 3. In combination with paraquat or glyphosate for no-till planting of corn and soybeans
 4. With safener-treated seed for grain sorghum
B. Directed postemergence to crop (preemergence to weeds)
 1. Corn (all)
 2. Woody ornamentals

SPECIAL PROBLEMS

The potential for movement into groundwater is a concern.

Other Chloroacetamide Herbicides
Typical Herbicide: Alachlor

Herbicide	Special Characteristics	Major Uses
Acetochlor (Harness, Surpass, TopNotch) 2-chloro-*N*-(ethoxymethyl) -*N*-(2-ethyl-6-methylphenyl) acetamide	Vapor pressure 3.4×10^{-8} mm Hg. TopNotch is microencapsulated; all are packaged with a safener. Water solubility 223 mg/L. Limited leaching. Provides 8 to 12 wk control. Can be used on high organic matter soils.	Preplant incorporated and preemergence on corn. Has potential for use on soybeans and other large-seeded legume crops
Dimethenamid (Frontier) 2-chloro-*N*-[(1-methyl-2-methoxy)ethyl]-*N*-(2,4-dimethyl-thien-3-yl)-acetamide	Vapor pressure is relatively high (2.8×10^{-4} mm Hg) but is not susceptible to photodegradation and does not have to be incorporated. Limited leaching. Half-life up to 5 to 6 wk in northern U.S.	Preplant incorporated and preemergence on corn and soybeans.

Herbicide	Special Characteristics	Major Uses
Metolachlor (Dual) 2-chloro-*N*-(2-ethyl-6-methylphenyl)-*N*-(2-methoxy-1-methylethyl) acetamide.	Vapor pressure 3.1×10^{-5} mm Hg at 25°C. Water solubility 488 mg/L. Slightly more leachable in soil than alachlor. Half-life is 3 to 5 mo.	Preplant incorporated or preemergence in field corn, soybeans, peanuts, potatoes, sorghum, and dry beans. Can be used with safener-treated seed in grain sorghum.

Herbicide	Special Characteristics	Major Uses
Propachlor (Ramrod) 2-chloro-*N*-(1-methylethyl)-*N*-phenylacetamide	Vapor pressure 7.9×10^{-5} mm Hg at 25°C. Water solubility 613 mg/L. Requires less moisture for movement into the weed seed zone than alachlor. Effective on high organic matter soils and under dry conditions. Half-life approximately 7 days.	Preemergence in corn and grain sorghum.

Thiocarbamate Herbicides

The thiocarbamates are soil-applied herbicides used for seedling weed control. They have high vapor pressures and must be incorporated into the soil immediately following application to prevent vapor loss.

In some cases, the margin of crop safety has been quite narrow, particularly when these compounds are used at higher rates for perennial grass control. For this reason, EPTC and butylate are available with safeners added to the formulation (sold as Eradicane and Sutan +, respectively) that substantially reduce the frequency of crop injury to corn.

The thiocarbamates are shoot growth inhibitors. The exact cellular mode of action of these compounds is unknown, but they do produce well-defined symptoms in susceptible grass seedlings. The first leaves emerging from the coleoptile appear twisted, distorted, dark green, and appear to be unable to unfold properly (Figure 12.2). The affected seedlings emerge from the soil surface before dying back. Susceptible broadleaves also show distorted leaves and do not grow beyond the seedling stage.

The thiocarbamates are readily degraded in soil. They are susceptible to microbial activity and generally persist less than a growing season (1 to 3 mo) in most soils. They do not cause carryover problems to crops grown in the next season. In fact, this group of herbicides is so susceptible to soil microbial activity that in some areas, particularly in the plains states, a buildup of EPTC-degrading microbes has caused the accelerated breakdown of the compound, greatly decreasing the length of effective weed control.

EPTC

TRADE NAMES AND MANUFACTURER

Eptam, Eradicane—Zeneca

CHEMISTRY

S-ethyl dipropyl carbamothioate

AVAILABLE FORMULATIONS

A. Sprayable in water

1. 7 lb a.i./gal emulsifiable concentrate (Eptam 7E)
2. 6.7 lb a.i./gal emulsifiable concentrate with safener (Eradicane 6.7E)

B. Dry application
   1. 20% granule (Eptam 20G) and 25% granule (Eradicane 25G)

RESPONSE IN SOILS

EPTC is volatile (vapor pressure = 3.4×10^{-2} mm Hg at 25°C) and must be incorporated to prevent loss. A dry soil surface is required for effective mechanical incorporation. Vapor movement is probably responsible for most short-distance redistribution of EPTC in soil. Water solubility is 370 mg/L so that overhead irrigation can be used to move the chemical into the soil, but excessive moisture can leach the compound out of the weed seed zone. The half-life is 6 days, and weed control can be expected for 2 to 6 wk.

SENSITIVE SPECIES

Most seedling grasses and some annual broadleaf weeds are sensitive to EPTC. It kills germinating seedlings, *not* emerged plants. It also inhibits shoot growth from underground vegetative structures of some perennial grasses and nutsedges.

A. Grasses from seed include annual bluegrass, barnyardgrass, crabgrass, foxtails, goosegrass, johnsongrass, fall panicum, sandbur, shattercane, and wild oats.
B. Annual broadleaves from seed include black nightshade, common chickweed, henbit, lambsquarters, purslane, and pigweeds.

C. Suppression and/or control of weeds from vegetative reproductive structures on perennials: bermudagrass, johnsongrass, quackgrass, purple nutsedge, and yellow nutsedge.

MAJOR USES

A. Eptam
   1. Preplant incorporated for direct-seeded crops: small-seeded legumes, dry beans, castor beans, corn, flax, potatoes, safflower, sugar beets, and sunflowers.
   2. Preplant or postplant incorporated on conifer nursery stock
   3. Postemergence on established crops after clean cultivation: beans, potatoes, sugar beets, citrus
B. Eradicane
   1. Preplant incorporated for corn
   2. The presence of the safener permits higher application rates without corn injury from EPTC. Thus, improved control of perennial weeds can be obtained. Chopping up vegetative reproductive structures to stimulate bud sprouting prior to treatment enhances performance.

SPECIAL PROBLEMS

EPTC must be incorporated immediately.

Possible leaching of EPTC can occur with heavy rainfall.

EPTC has a relatively short life in soil and is susceptible to accelerated breakdown from microbial buildup.

Other Thiocarbamate Herbicides
Typical Herbicide: EPTC

Herbicide	Special Characteristics	Major Uses
Butylate plus safener (Sutan +) S-ethyl bis (2-methylpropyl) carbamothioate	Safener added for greater corn tolerance. Vapor pressure 1.3×10^{-2} mm Hg at 25°C. Water solubility 44 mg/L. Leaches less than EPTC. Average field half-life is 13 days and provides 4- to 7-wk weed control.	Preplant incorporated in corn. Also can be used for suppressing nutsedges.

$$CH_3-CH_2-S-\overset{\overset{O}{\|}}{C}-N\overset{CH_2-\overset{\overset{CH_3}{|}}{CH}-CH_3}{\underset{CH_2-\underset{CH_3}{\overset{|}{CH}}-CH_3}{}}$$

Herbicide	Special Characteristics	Major Uses
Cycloate (Ro-Neet) *S*-ethyl cyclohexylethyl-carbamothioate	Vapor pressure 6.2×10^{-3} mm Hg at 25°C. Water solubility 85 mg/L. Leached less than EPTC or pebulate. Average half-life is 30 days.	Preplant incorporated in sugar beets, table beets, and spinach.

$$CH_3-CH_2-S-\overset{\overset{O}{\|}}{C}-N\overset{}{\underset{CH_2-CH_3}{}}$$

| Pebulate (Tillam) *S*-propyl butylethylcarbamothioate | Vapor pressure 8.9×10^{-3} mm Hg at 25°C. Water solubility 60 mg/L. Leached less than EPTC but more than cycloate. Average half-life is 2 wk. | Pretransplant incorporated in tobacco; preplant, pretransplant, and posttransplant directed incorporated in tomatoes; preplant incorporated in sugar beets. |

$$CH_3-CH_2-CH_2-S-\overset{\overset{O}{\|}}{C}-N\overset{CH_2-CH_3}{\underset{CH_2-CH_2-CH_2-CH_3}{}}$$

| Triallate (Far-go) *S*-(2,3,3-trichloro-2-propenyl) bis(1-methylethyl) carbamothioate | Vapor pressure 1.1×10^{-4} mm Hg at 25°C. Water solubility 4 mg/L. Limited leaching. Average field half-life is 82 days. | Preplant incorporated for wild oat control in winter wheat, spring and durum wheat, barley, green peas, field-dried peas, lentils. May be applied in fall or spring depending on crop. |

$$\underset{\overset{\|}{Cl}}{\overset{Cl\ \ Cl}{C}}=C-CH_2-S-\overset{\overset{O}{\|}}{C}-N\overset{\overset{CH_3}{\overset{\|}{CH}-CH_3}}{\underset{\underset{\underset{CH_3}{\|}}{CH}-CH_3}{}}$$

MISCELLANEOUS CELL DIVISION INHIBITORS

This group of soil-applied herbicides consists of three compounds, each from a different chemical family. Bensulide has no generally recognized chemical group, napropamide is an amide, and siduron is a phenylurea.

Bensulide

Bensulide is a soil-applied herbicide for the control of seedling crabgrass and other weeds in turfgrass and dichondra lawns and for weed control in several vegetable crops. The compound is marketed as Betasan for turf and Prefar for vegetables. Like DCPA, water solubility, soil mobility, vapor pressure, and movement into plant tissue are low. To compensate for lack of movement into soil, the herbicide is used at high rates preemergence (10 to 25 lb a.i./acre) on turf and incorporated mechanically or with water on cropland. This compound inhibits root growth, presumably by disrupting either cell division or cell enlargement.

TRADE NAMES AND MANUFACTURERS

Bensumec 4 LF, Pre-San Granular 7 G—PBI/Gordon
Betasan 4 E—United Horticulture Supply Company
Prefar 6-E—Gowan
Sold under other trade names.

CHEMISTRY

O,O-bis(1-methylethyl) S-[2-[(phenylsulfonyl)-amino]ethyl]phosphorodithioate

AVAILABLE FORMULATIONS

A. Sprayable in water
  1. 4 lb a.i./gal flowable (Bensumec 4 LF)
  2. 4 lb a.i./gal emulsifiable concentrate (Betasan 4 E)
  3. 6 lb a.i./gal emulsifiable concentrate (Prefar 6-E)
B. Dry application
  1. 7% granule (Pre-San Granular 7 G)

RESPONSE IN SOILS

Water solubility is 25 mg/L. The compound has a strong affinity for soil colloids and resists leaching in the soil. Therefore, it should be mechanically incorporated or watered into the soil to obtain activity. Bensulide is nonvolatile. Average field half-life is 120 days. Carryover to sensitive crops can occur.

SENSITIVE SPECIES

Bensulide primarily controls seedling annual grasses and some broadleaves. Weeds that have emerged at the time of application are not susceptible.

A. Seedling grasses include crabgrass, foxtail, goosegrass, jungle rice, fall panicum, sprangletop, and watergrass.
B. Broadleaved weeds include common lambsquarters, redroot pigweed, and common purslane.

MAJOR USES

A. Preemergence on established turf and dichondra lawns prior to crabgrass and annual bluegrass emergence. Temporary yellowing can occur to bermudagrass turf. Do not reseed turf sooner than 4 mo after treatment.
B. Preplant incorporated or preemergence on cucurbits, cole crops, eggplant, peppers, lettuce, garlic, bulb onions, shallots, and grass seed.

SPECIAL PROBLEMS

Carryover from relatively long residual activity can be a problem with bensulide.

Other Miscellaneous Cell Division Inhibitors

Herbicide	Special Characteristics	Major Uses
Napropamide (Devrinol) N,N-diethyl-2-(1-naphthalenyloxy)propanamide	Vapor pressure 4×10^{-6} mm Hg. Requires moisture or mechanical incorporation to reach weed seed zone. Water solubility is 73 mg/L. Typical half-life is 70 days. Rotational crops not listed on label should not be planted for 12 mo after treatment.	Incorporated or surface applied prior to weed emergence in newly planted or established crops: citrus, nuts, pome fruits, small fruits, stone fruits, tobacco, and vegetables (cole crops, eggplant, tomatoes, sweet potatoes)
Siduron (Tupersan) N-(2-methylcyclohexyl)-N'-phenylurea	Is resistant to leaching. Water solubility is 18 mg/L. Average field half-life is 90 days. Provides weed control during season of application but does not carry over into the next season.	Preemergence control of annual grasses in newly seeded and established turfgrasses (bluegrass, fescue, perennial ryegrass certain strains of bentgrass). Effective for crabgrass, foxtail, and barnyardgrass control.

CELL WALL FORMATION INHIBITORS

These herbicides inhibit cellulose formation, thus preventing the formation of new cell walls following cell division. Because growth ceases, susceptible seedlings do not emerge from the soil. Three compounds are in this group: diclobenil is a nitrile herbicide, isoxaben is a benzamide, and quinclorac has no generally recognized grouping.

Quinclorac is unusual in that it shows some auxin-type symptoms on broadleaves in addition to growth inhibition, but it is included here because it is mostly soil applied and may inhibit cell wall formation in grasses.

Dichlobenil

TRADE NAMES AND MANUFACTURERS

Casoron—Uniroyal Chemical, Inc.
Norosac—PBI/Gordon

Chemistry

2,6-dichlorobenzonitrile

Available formulations

A. Dry application
   1. 4% granule (Casoron 4G, Norosac 4G Dichlobenil)

Response in Soils

Dichlobenil has a relatively low water solubility of 21 mg/L and is moderately adsorbed by soil organic matter. Movement through the soil is limited; the compound tends to stay in the upper soil layers. The vapor pressure of dichlobenil is high (1.0×10^{-3} mm Hg at 25°C), so soil incorporation sometimes is recommended to prevent soil loss. The average field half-life is 60 days, and weed control can be obtained for 2 to 6 mo.

Sensitive Species

Many annual and certain herbaceous perennial species are sensitive to dichlobenil.

A. Annual weeds controlled include wild barley, annual bluegrass, carpetweed, common chickweed, crabgrass, foxtails, henbit, common lambsquarters, wild mustards, Texas panicum, peppergrass, purslane, ragweeds, redroot pigweed, and smartweeds.
B. Perennial and biennial species controlled include dandelion, evening primrose, groundsel, horsetail, rosary pea, and bull thistle.

Major Uses

A. Preplant soil incorporated treatments in orchards and nurseries. Applied preemergence to weeds in established almond, apple, blueberry, cherry, grape, pear, blackberry, and raspberry. Also used on cranberries.
B. Woody ornamentals, shelterbelts, and forest plantings, both established and in nurseries. Plants include arbor vitae, beauty bush, camellia, dogwood, elm, euonymus, forsythia, honeysuckle, juniper, magnolia, maple, oak, privet, rose, and yew.
C. Under asphalt for general weed control with 100 to 120 lb a.i./acre applied immediately prior to asphalting.

Other Cell Wall Formation Inhibitors
Typical Herbicide: Diclobenil

Herbicide	Special Characteristics	Major Uses
Isoxaben (Gallery) N-[3-(1-ethyl-1-methylpropyl)-5-isisoxazolyl]-2,6-dimethoxy benzamide	Water solubility is 1 mg/L. Adsorbed to soil; leaching is minimal. Average field half-life is 50 to 120 days, and weed control is obtained for 5 to 6 mo.	Preemergence control of annual broadleaves in established warm- and cool-season turfgrasses, herbaceous and woody ornamentals, nonbearing fruit and nut trees, Christmas tree plantations, and noncropland.

Herbicide	Special Characteristics	Major Uses
Quinclorac (Facet) 3,7-dichloro-8-quinolinecarboxylic acid	Residues may injure susceptible crops 1 yr after application.	Preemergence control of annual grasses and certain annual and perennial broadleaved weeds in rice.

WEED LIFE CYCLES AND MANAGEMENT

The life cycle of a weed determines its adaptability to various management systems and its susceptibility to control measures. Herbaceous weeds are grouped as summer annuals, winter annuals, biennials, simple perennials, or creeping perennials. Certain control principles are unique to each life cycle group. Once the life cycle of a weed has been determined, a general body of information regarding the weed group and methods for its control becomes available. Coupling this information with specific details regarding the individual weed should provide a sound basis for applying weed control practices.

This chapter begins with a brief review of the life cycle groups (see Chapter 2 for definitions). This is followed by a discussion of controlling weeds that germinate from seed. The chapter concludes with a detailed description of creeping perennial weeds and their control.

WEED LIFE CYCLES

Annuals

Annual weeds survive on disturbed sites. They flourish in any area where adequate sunlight, moisture, warmth, and time (usually considerably less than a full growing season) are available for germination, growth, and seed maturation. These weeds are well adapted to annual crops in areas where disturbed open sites are created by annual tillage. They also invade perennial crops and noncrop areas. In perennial crops they grow in open spaces, during periods of crop dormancy, and in areas where the crop canopy has been opened by disturbance or crop death.

Annual weeds with a growth habit similar to that of the crop complete their life cycle during the crop growing season. For example, summer annuals grow simultaneously with spring-planted crops such as corn, sorghum, and tomatoes. Winter annuals grow at the same time as winter annual crops such as the fall-seeded small grains.

Weeds with appreciably different habits than the crop may do best during periods of crop dormancy. Winter annuals, for example, become established in perennial hay and turf crops when these crops are dormant in late fall, winter, and early spring. By midspring they have sufficient growth to suppress the crop when it starts growing again.

Special considerations in controlling annual weeds are covered in the section on Controlling Weeds That Germinate from Seeds.

Biennials

Biennial weeds produce a rosette the first year, overwinter, respond to cold, then bolt and seed the second year. They are not well adapted to sites that are severely disturbed annually in the fall, winter, or early spring since one effective tillage per year both destroys rosettes and prevents these plants from setting seed. Thus, biennials seldom are a problem in conventionally tilled annual crops. Biennials persist best in undisturbed noncrop and roadside areas, in perennial crops, and in areas that are mowed infrequently. For example, biennials are more likely to survive along rights-of-way that are mowed two or three times a year than in turf that is mowed weekly.

Biennials are most susceptible to herbicides as young rosettes (during late summer, fall, and spring). After bolting, sensitivity to herbicides diminishes drastically. Close mowing when bolted stalks are in the bud stage can significantly reduce seed production of some biennials. Mowing after full flowering merely disperses the seed crop.

Simple Perennials

Simple perennials usually require most of a season or more for establishment and reproduction. They survive best, and therefore cause problems, in undisturbed sites such as in perennial crops, no-till sites, noncrop areas, and rights-of-way. Simple perennials with leaves close to the ground (e.g., dandelions, plantains) can persist in frequently mowed turf. Herbicide applications are least effective during and following flowering and during cold weather. Repeated tillage will destroy established plants.

The success of methods that remove only the rosettes or tops of simple perennials is variable. Most herbaceous simple perennials generate shoots from buds on crowns, some of which can extend several inches below the ground (e.g., dandelion, curly dock). Such plants tolerate very shallow tillage, hoeing, contact herbicides, and mowing.

Creeping Perennials

Creeping perennials are the best equipped of the various weed types for survival under a wide range of conditions. The production of new independent plants from vegetative reproductive structures coupled with seed formation allows these plants to survive, multiply, and compete in both annual and perennial cropping situations. In addition, the relatively robust growth habit of plants that develop from vegetative propagules provides them with a competitive advantage over plants that develop from seed.

Creeping perennials compete in most management systems including annual crops, mowed herbaceous crops, orchards, vineyards, cane fruits, nurseries, and ornamental plantings. Most management systems provide niches for invasion by creeping perennial weeds.

Although the shoot portion of a creeping perennial may be killed by single tillage operations or applications of a herbicide with limited movement, new shoots produced from vegetative reproductive structures render these methods ineffective unless they can be repeated often enough to starve the underground structures.

CONTROLLING WEEDS THAT GERMINATE FROM SEEDS

All life cycle groups, including perennial weeds, produce seed, persist because of seed-related characteristics, and reestablish from seed. Dormancy and long-term survival of seeds buried in soil result in the potential for seed germination through most of the growing season as well as over a period of years. Many annual plants can emerge, flower, and set seed in a short period of time, resulting in the continual replenishment of weed populations. In fact, weeds often complete total life cycles between control operations within a single season, most frequently between the last weeding operation and harvest. Although biennials and simple perennials require longer periods to grow and set seed, many of these plants can reappear from seeds stored in the soil when effective weed control practices are stopped.

These seed-related characteristics make weed eradication nearly impossible. A more realistic goal is to control or delay the growth of weeds to give the crop a competitive advantage. Control methods for weeds that develop from seed, regardless of life cycle, are similar and are based on the following principles:

1. Dormant seed is not destroyed by most weed control practices.

2. Germinating seeds and very young plants are the most susceptible to control methods. Because of this susceptibility, the emphasis in controlling weeds that develop from seeds usually is directed at the germinating seed and young plant. Weeds tend to germinate in largest numbers at the beginning of the season as soon as climatic conditions are favorable for growth. Thus, weed populations and subsequent pressure on the crop can be decreased by preparing the seedbed for planting, allowing the first crop of weeds to germinate, and then killing the weeds either with shallow tillage or herbicides. More choices of control measures are available if the weeds can be stimulated to germinate before the crop is planted. When weeds are allowed to emerge with the crop, dependence must be placed on using selective weed control measures within the crop itself.

3. The most resistant stage of plants developing from seed occurs after flowering. In fact,

if plants are allowed to reach full flowering, not only do they achieve maximum resistance but the main objective of preventing seed production and seedbank replenishment has been lost. In the case of creeping perennials, new vegetative reproductive structures also have been produced by the time of flower maturity.

4. Most plants that develop from seed can be killed by removing the shoot portion of the plant. In contrast to established creeping perennials that develop from underground reproductive structures, plants that arise from seed can be killed by a single, effective tillage operation or herbicide application (Table 13.1). Annual plants that reroot at the nodes (e.g., smartweed, purslane) are somewhat more difficult to kill with tillage because cut pieces can reroot when left in moist soil. Seedling annual grasses tend to be more resistant to contact herbicides, flaming, and mowing than broadleaves because of their lower, more protected growing points.

5. Soil fumigants or steam applications can be used to kill weed seeds, but these methods are limited primarily to high value crops, potting soils, plant beds, and greenhouses. Steam is more effective than fumigation for killing weed seeds.

6. Most weed seedlings establish poorly in shade. Thus, shading by a competing crop frequently constitutes the major control pressure once the crop is taller than the weed. The portion of the season the crop does not shade the soil is the portion during which the control measures listed in Table 13.1 should be used.

CREEPING PERENNIAL WEEDS

Creeping perennial weeds cause particularly serious weed control problems because of their impact on crops and their resistance to control methods.

Factors That Contribute to the Success of Creeping Perennial Weeds

Creeping perennial weeds such as Canada thistle, field bindweed, johnsongrass, quackgrass, bermudagrass, yellow nutsedge, and leafy spurge have been recognized as serious weed problems for years. Other commonly encountered creeping perennial weeds include horsenettle, wild garlic, common milkweed, hedge bindweed, Jerusalem artichoke, hemp dogbane, purple nutsedge, bigroot morningglory, and honeyvine milkweed. Interestingly, the majority of these plants are introduced species and have been able to establish themselves successfully in North America. Some may lack the natural enemies that would otherwise keep them in check.

Creeping perennial weeds frequently are the dominant weeds in perennial crops because of their similarities in growth patterns. Understanding why these weeds are so successful is essential to combating them effectively. One reason for their success is their reproductive and survival mechanisms. Another is the fact that they have been favored by the substantial changes in cropping and vegetation management practices that have taken place in recent decades.

Reproductive and Survival Mechanisms

Vegetative reproductive structures commonly found on perennial weeds are rhizomes, tubers, bulbs, stolons, and creeping roots (see Chapter 2). These structures serve as major food storage organs and possess numerous buds capable of generating new plants. These characteristics make it extremely unlikely that a single weed control treatment will effectively kill an established creeping perennial weed (Table 13.1).

Table 13.1 Likelihood of Substantial Kill or Control from One Treatment on Established Individual Plants with Different Life Cycles[a]

| Control Measure | Annual | | Biennial | Simple Perennial | Creeping Perennial | |
	Grasses	Broadleaves			Seedling	Established
Hoeing, scraping, or shallow tillage	Good	Good	Good	Poor–Good	Good	Poor
Subsurface tillage	Good	Good	Good	Good	Good	Poor
Contact herbicide	Fair–Good	Good	Poor–Good	Poor–Good	Fair–Good	Poor
Residual soil-applied herbicide	Good	Good	Poor–Good	Poor–Good	Good	Poor–Fair
Limited translocated herbicide	Good	Good	Good	Good	Good	Poor–Fair
Well-translocated herbicide	Good	Good	Good	Good	Good	Fair–Good
Long residual soil-applied herbicide	Good	Good	Good	Good	Good	Good

[a]Reestablishment from seed was not a consideration in assigning these ratings.

Advantages that vegetative reproductive structures give to perennial weeds include the following:

1. Plants that develop from vegetative structures are larger, grow more quickly, and compete sooner than plants that develop from seed. This characteristic partially explains the competitive advantage of established creeping perennial weeds over crops and other weeds. It also explains the popularly held notion that only plants from vegetative structures rather than plants from seeds contribute significantly to the success of creeping perennial weeds.

2. Apical dominance regulates the activity and growth of buds on vegetative structures. The apex (growing tip) of a developing structure inhibits the sprouting of nearby buds, so most of the buds on intact vegetative structures are inactive or dormant. Removing the growing point or tearing or cutting the vegetative structures apart removes the domination by the apex and stimulates sprouting of previously inactive buds. Thus, fragmentation of an underground system can result in the emergence of more shoots than were present when the plants were intact. This process of bud stimulation can serve as an effective mechanism for the rapid reestablishment of a perennial weed after it has been severely disturbed or even partially destroyed. Small pieces of vegetative structures also provide a mechanism for transplant and spread, particularly in localized areas.

3. Many vegetative buds are deeply dormant and in this state escape injury from cold, drought, herbicides, or tillage. Herbicides that move readily into growing meristems bypass dormant buds.

4. Vegetative reproductive structures, even when not dormant, are capable of resisting conditions of adversity such as short drying periods. For the most part, vegetative structures can be expected to survive 1 to 5 yr if left undisturbed in the soil. Even the use of clean fallow (tillage every 3 to 4 wk through the growing season) frequently requires at least two seasons to provide more than 90% control. Once an acceptable level of control has been achieved, failure to maintain adequate subsequent control can result in the total reestablishment of the weed from seed or from surviving buds in one to three seasons.

5. Substantial portions of underground vegetative structures lie below the plow layer and are protected from the effects of tillage. This is particularly true of many of the perennial broadleaved weeds with creeping roots because roots tend to penetrate deeper than rhizomes. These underground structures also provide an enormous amount of tissue in which downwardly translocating foliar-applied herbicides can be diluted. Such extensive systems are not easily exhausted of buds and food reserves even when continuous removal of tops is attempted. The depth of underground systems is influenced by a number of factors. Coarse-textured soils, deep soils, dry climatic conditions, and occasional tillage encourage deep penetration of underground structures. In contrast, frequent close mowing, tight soils, impervious soil layers, high water tables, and waterlogged soils tend to restrict downward development.

Finally, many creeping perennial weeds also produce seeds that can greatly add to their already considerable reproductive and survival advantages. In fact, some perennial weeds such as johnsongrass are competitive just on the basis of seed production alone.

Changes in Cropping and Management Practices

A number of cultural practices have contributed to the increase in perennial weed problems:

1. Numbers of crops in rotation have been reduced or eliminated. A continuous annual crop or two annual crops in rotation is more often the rule than the exception. This regimen reduces the number of control alternatives and is ideal for the buildup of adapted annual and creeping perennial weed species.

2. There has been a trend toward less tillage. Elimination of a tillage operation results in the loss of its associated increment of control. Of nine species of perennial weeds found in field corn, sorghum, and soybeans in 12 midwestern states, eight species occurred more frequently in no-till systems than in conventional till systems (1992 survey available from David R. Pike, Department of Crop Science, University of Illinois). For example, hemp dogbane was reported to occur in almost 80% of the no-till systems but only in 40% of the conventional till systems.

3. The recent trend toward earlier planting dates favors maximum crop yields. Unfortunately, it also favors weed development by limiting the number of weed control operations that can be employed prior to planting.

4. Many current programs for controlling weeds in annual crops target annual weeds but are relatively ineffective for controlling many creeping perennials.

5. Modern combine harvesters are extremely efficient at spreading weed seed over large areas. This is frequently coupled with larger acreages per operator sometimes scattered over several sites.

6. Economic and political pressures have resulted in fewer spraying, mowing, and weed control operations in noncrop areas. Many of these areas are major sources of seed and vegetative propagule production for perennial weeds.

7. Less attention has been paid to new infestations. In recent years, scattered individual weeds or small patches of weeds have been ignored rather than hand pulled, spot sprayed, or mowed. Control measures are not considered until the weed becomes a serious problem.

8. Dwarf crop varieties (e.g., turf) and crop varieties selected for the maximum penetration of light to the lower leaves of the crop favor weed growth in general.

Control of Creeping Perennial Weeds

Successful control of creeping perennial weeds seldom is achieved using a one-shot control effort. The person confronted with these weeds must realistically approach the control of perennial weeds by being aware of the many different aspects that go into developing a successful control program. These aspects include (1) adopting a realistic approach, (2) being aware of susceptible growth stages, (3) recognizing levels of control, (4) evaluating levels of control, (5) understanding the effects of individual control and management practices, and (6) developing a multicomponent program.

Adopting a Realistic Approach

Disappointment and failure in dealing with perennial weeds frequently are the results of overoptimistic expectations, too limited an understanding of the interacting components, and inadequate control efforts. To avoid these problems, we propose adopting the following five management concepts for dealing with established creeping perennial weeds:

1. Recognize the survival capabilities of the weed. Creeping perennial weeds are hardy, difficult-to-kill species. Once established, they usually persist.
2. Know when and how to evaluate the level of control.
3. Establish the necessary management flexibility to adopt practices that are effective for the control of perennial weeds. The buildup of extensive stands of noxious perennial weeds frequently indicates a lack of willingness to adopt new management practices to meet the problem. Control practices for a creeping perennial may not fit standard practices or be convenient to use, but failing to adopt them usually results in serious consequences in subsequent years.

4. Be on the alert for new infestations spreading toward you. Consider the approach of a serious weed your problem rather than someone else's. Identify and destroy new infestations as soon as possible. Serious perennial weed problems usually start as unnoticed or unattended small infestations that gradually spread.

5. After a substantial level of control has been achieved, establish a program for escapes and for new plants that reestablish from seeds and dormant vegetative buds. The results of an initially effective control program can be negated completely by inadequate follow-up treatments and poor control in subsequent years.

Being Aware of Susceptible Growth Stages

The time to initiate control practices for established perennial weeds (those that develop from vegetative reproductive structures) is related to the development of the plant over the season. Plants are most subject to control from herbicides when they are both actively growing and moving carbohydrate (sugars) downward to the underground structures.

Growth is initiated when shoots begin to develop from vegetative buds and start to use carbohydrate reserves from the underground structures. The depletion of carbohydrate reserves continues until sufficient aboveground photosynthetic surface is formed to permit the net downward movement of sugars. Although the plants are actively growing at this early stage of development, they are not moving sugars downward; thus, efforts to control young shoots as they are emerging are not productive.

In many perennial weeds, the period in which underground reserves have been maximally depleted and sugars are beginning to move back down to form new underground structures occurs when the plant has attained approximately one-fourth of its height or is approaching the early flower bud stage. This time is ideal for initiating treatments such as clean cultivation or close mowing, which result in top removal and the exhaustion of storage materials. It also is a good time for using foliar-applied symplastically translocated herbicides. Continued herbicide movement downward can be expected until the flowers open. Once fertilization occurs, translocation downward decreases (sugars begin moving to the developing seeds) and resistance to herbicides increases rapidly. Viable weed seed can be produced in a week or less after fertilization, so waiting until flowering to treat also allows seed production to occur.

Late summer or fall, when plants again begin to move sugars downward, can provide a good time to treat with symplastically translocated herbicides. Fall treatments are effective on plants that maintain healthy leaves and are not senescing, on plants that produce new shoots with the onset of cool weather, and on plants that are regrowing following shallow tillage or mowing. In the last two cases, the development of several fully expanded leaves on the new growth is required to ensure that sugar movement is again downward. Properly executed fall treatments can be more effective than those at the flower bud stage.

Herbicide applications to the foliage of plants at first shoot emergence, or during seed development, or to plants that are under stress due to cold, lack of moisture, damaged leaves (e.g., from hail, insects), or senescence are likely to provide disappointing results. An exception is the morningglories. These plants flower continuously and should be treated during flowering.

Recognizing Levels of Control

Different practices or combinations of practices provide varying degrees of effectiveness ranging from total eradication to no control at all. It is important to recognize the various levels of control and the practices by which they are obtained. Some frequently encountered levels of control, proceeding from the most effective to the least effective, are as follows:

Eradication. Eradication is difficult to achieve because it requires prolonged efforts or extreme measures such as fumigation. It is most likely to succeed on new or limited infestations where seeds have not been stored in the soil. Eradication can be expensive to accomplish on a per unit area basis, but on limited infestations the high initial expense can be offset by the probable losses in the future.

Top Kill Coupled with Substantial Kill of Underground Structures. This level of control can be achieved with (1) two or more applications of a well-translocated (symplastically) foliar herbicide, (2) one or more applications of a well-translocated foliar herbicide in conjunction with a crop providing canopy closure over the perennial for most of the growing season,

(3) a highly persistent soil-appplied herbicide, or (4) a continuing combination of sequential treatments that result in shoot removal prior to sucrose transport from the leaves to the belowground structures. In the last case, the goal is to starve the plant and prevent it from producing underground structures.

Top Kill Accompanied by Partial Kill of Vegetative Structures. Frequently, only partial kill of vegetative structures is obtained with an application of a symplastically translocated herbicide, even though substantial kill is anticipated. Although the degree of control may be less than expected, an acceptable result can be obtained if the practice is coupled with additional control measures or the treatment is repeated at appropriate intervals.

Top Kill Only. Top kill can be achieved with a single tillage operation, a contact herbicide, hand hoeing, or close mowing of a tall-growing species. Although plant development may be delayed, creeping perennial weeds have the capacity to overcome top removal if nothing else is done. When such treatments are repeated at appropriate intervals (usually every 2 to 4 wk), they can lead to depletion of underground reserves of stored food and vegetative buds and provide substantial overall kill. They also permit a competitive crop to overtake and then outcompete the weed during the growing season.

Reduction of Vigor or Competitive Ability. Reduction in weed vigor can be provided by a treatment that has limited effectiveness or by the introduction of a competitive crop. Properly utilized, this degree of control can minimize the impact of a creeping perennial weed in cropping situations.

Prevention of Formation of Reproductive Structures. This outcome is the major benefit of some types of control efforts. The prevention of seed production is the *minimum* effective effort that should be used to prevent the spread of a noxious species. It can be accomplished with three to five mowings for a tall-growing species in a sod crop, by a well-timed herbicide treatment that provides top removal and partial destruction of underground structures, or by treatments that are selective in competing crops. If the vigor of established stands is to be reduced, vegetative structures also must be kept from developing.

No Control. Lack of control may result when adverse conditions cause a normally effective control measure to fail. It is much more likely to be the result of the use of incorrect, untimely, or inadequate control measures in a given situation. Obviously, no effort at all guarantees no control.

Evaluating Levels of Control

Knowing when and how to evaluate the effectiveness of a control practice or program for creeping perennial weeds is essential in planning further control measures. Evaluation is complicated by the regenerative capabilities of the vegetative structures. Thus, immediate top kill is not a good indicator of long-term control. Having new plants reestablish from seed further complicates the evaluation process.

Evaluation usually is made by checking for top (shoot) kill after treatment. We place the following interpretations on the time required for shoot recovery:

1 mo or less	Shoot kill only
2 to 4 mo	Shoot kill and partial injury to underground structures
6 mo to 1 yr	Substantial kill of underground structures
2 to 3 yr	Total control of underground vegetative structures

Understanding the Effects of Individual Control and Management Practices

Descriptions of the contributions of individual practices to the control of creeping perennial weeds follow.

Tillage. Tillage can contribute significantly to the control of creeping perennial weeds. Prior to the development of modern herbicides, clean fallow was the major method of control on heavily infested cropland. About 1 yr of fallow was required for every 3 yr of crop production. The loss of profit from not growing a crop, the cost of tillage, and the increased susceptibility of cropland to erosion limit the practicality of clean fallow in many situations today. However, in those instances in which clean fallow with tillage is desired, the tillage must be conducted at 3- to 4-wk intervals over one or two full growing seasons to achieve 90% or better control of perennial weeds. A combination of a herbicide treatment at the flower bud stage followed by tillage is more effective than either used alone.

Conventional tillage (plowing and secondary tillage during seedbed preparation) suppresses or delays weed development by cutting off the tops. Rhizomes and roots near the soil surface are broken up, and many are subjected to desiccation. After the crop has been planted, one to three cultivations between the rows can remove one-half to two-thirds of those weeds that escape earlier tillage or herbicide treatments or that emerge from deeper in the soil. Tillage is so important that without it (e.g., in no-till systems), perennial weed pressure often becomes severe enough to prevent a crop from being grown unless additional compensatory control measures are employed.

Running tillage equipment through isolated plants or patches may actually spread the weed rather than contain or eliminate it. Breaking the vegetative structures can stimulate buds to sprout into new shoots, and pieces of vegetative structures can be dragged to new locations.

Subsequent treatments for control should be planned if spread does occur. However, if the infestation is widely distributed through the field, the benefit from repeated tillage far outweighs the disadvantages of any additional spreading that might take place.

Hand Removal. Hand removal by pulling or hoeing usually is limited to a few isolated plants, small areas, or high value situations. If used immediately after a perennial weed has been first observed, it can effectively prevent an infestation. In established infestations, this method must be used every 3 to 4 wk for a year or so to be effective. Hand weeding can be used to reach places inaccessible to equipment (e.g., weeds in ornamental plantings, next to structures, in solid-planted crops, and in the row of a row crop). The careful application (spot treatment) of a well-translocated herbicide frequently provides far better control and may be a more appropriate choice for these areas.

Herbicide Treatments. Herbicides have proven effective and economical for controlling problem creeping perennial weeds. Without modern herbicides for control of these plants, the potential for the adoption of reduced tillage practices and the large acreages managed per farm operator would be greatly reduced.

Highly Persistent Herbicides. These herbicides move into the soil and are taken up by the roots to achieve top kill. When foliar applied, some often translocate downward. When soil applied, they are long-lived, moving into the plant every time new growth is initiated from seeds or dormant buds. They thus inhibit growth and provide season-long pressure on the weed. Many carry over one to several seasons, preventing vegetation reestablishment, and few are labeled for use in or ahead of food crops. Seeds and dormant buds are unaffected by these herbicides; once the herbicide has dissipated to a subtoxic level, the weed can reestablish itself.

Single Season Soil Residual and Persistent Herbicides. These herbicides primarily provide suppression of perennials that develop from vegetative structures. They persist for a portion of one season or for one full season with some residue into the next season. They generally provide good control of seedlings. Some will suppress shoots developing from vegetative structures (usually from tubers and rhizomes).

Foliar-Applied Contact Herbicides. These chemicals provide top kill of vegetation that is equivalent to hand hoeing or very shallow cultivation. Each treatment delays the development of the perennial plant 2 to 4 wk. When selective, these herbicides can allow the crop to reach a stage at which it can shade the perennial weed and provide additional control. Contact herbicides also can be effective when applied to prevent seed production.

Foliar-Applied Symplastically Translocated Herbicides. The degree of movement of these compounds affects the amount of control obtained. Movement depends on the perennial weed species, the extensiveness of the underground structures, the amount of foliage in relation to the underground structures, the degree of bud dormancy, the stage of weed development, and growing conditions. When excellent movement and distribution are obtained, these herbicides can provide substantial kill of underground structures as well as top kill. One or two treatments alone or one treatment with tillage and a competitive crop (or mowing and a sod crop) can provide good control of perennial weeds. When downward movement of these herbicides is limited, they still can provide top kill plus a partial kill of underground structures. Under conditions of limited movement they can provide reasonably good control when applied two to three times by themselves or one to two times with a competitive crop and tillage or mowing.

Several of the symplastically translocated herbicides are selective and, when used in tolerant crops, can provide spectacular control results. In those cases in which the herbicide is not selective or is only marginally selective, selectivity can be attained by directed placement of the herbicide. If the weeds are shorter than or under the crop (e.g., perennial vines in a tall crop), the spray is directed under the crop. Tall weeds in shorter crops can be treated effectively with these herbicides by using wipe-on devices or recirculating sprayers.

Another method of using nonselective herbicides or herbicide rates in a cropping situation is to apply them in the fall. Fall treatments are easily accomplished with a crop harvested by midsummer, such as a small grain, since there is plenty of time for most perennial weeds to regrow and develop enough foliage for treatment. Fall applications sometimes can be used when crops are harvested late in the season. Harvesting, however, may destroy the integrity of the perennial weed shoots so that a period of regrowth usually is required before treatment. In addition, the herbicide spray must reach the weed, which may be difficult if excessive crop residues are matted down over the weeds.

Preharvest treatment presents an attractive alternative to postharvest ones. Herbicides can provide effective control when applied after the crop has matured but before harvest. Several herbicides are labeled for this use on grain crops. These treatments may allow the weed to be in a susceptible stage and permit the herbicide to penetrate the drying crop canopy. Such preharvest treatments require high-clearance ground equipment or aircraft application.

Foliar-applied herbicides can be used as spot treatments with handheld nozzles in areas where individual weeds and clumps are present. Crop plants intermingled with the weeds in the sprayed area may be sacrificed. These treatments are more effective than hand hoeing or pulling because they are likely to destroy much more of the underground system. In addition, herbicide treatments require less time.

Fumigants. Most vegetative structures (nutsedge tubers can be an exception) are susceptible to fumigation. Fumigation also can be effective on weed seeds and other types of pests such as soil-borne disease organisms, soil insects, and nematodes. Disadvantages of this method include the high cost per unit area, the escape of some types of weed seeds, and special handling procedures.

Grazing Animals. The impact of grazing animals on perennial weeds primarily depends on the palatability of the weed in comparison to other plant species. If the animal prefers the weed, then proper grazing management should result in effective reduction. Johnsongrass can be controlled in a pasture situation since it is effectively grazed. On the other hand, unpalatable species such as Canada thistle and tall ironweed will increase with grazing. This effect will be accentuated when overgrazing of desirable species occurs. Selective grazing of grass weeds in crops has been achieved with geese.

Other Biological Controls. Several biological control agents have provided effective, low-cost control for individual weeds in low value lands. Examples include the insect controls of St. Johnswort and prickly pear cactus in rangelands. For other biocontrol examples, see Chapter 4.

Developing a Multicomponent Program

By using specific performance data and the principles described previously, a sound control program for perennial weeds can be developed. The major considerations in planning a program include the effectiveness of a treatment in certain crops, the desired or acceptable level of control, the vulnerable stages of the weed, and cost effectiveness. The program should be planned for both the immediate season and for future years as well. Remember that an individual treatment rarely constitutes a control program. Successful programs consist of multiple treatments and are accomplished either by using individual treatments repeatedly over a period of several years or by sequencing different treatments over the same period. The number of individual treatments required to provide substantial reduction of underground vegetative structures in one to three growing seasons is given in Table 13.2.

The effects of sequencing treatments is illustrated in the following section on the biology and control of seven major perennial weeds. All but multiflora rose are creeping perennials. Johnsongrass and quackgrass are perennial grasses. Two other monocot groups, yellow and purple nutsedge and wild garlic, also are described. We conclude the chapter with a discussion of the biology and control of three broadleaved perennial plants, Canada thistle, field bindweed, and multiflora rose.

CONTROLLING CREEPING PERENNIAL GRASSES

The examples we use to illustrate creeping perennial grasses are johnsongrass, which is a major problem in warm climates, and quackgrass, which is a major problem in cool climates. Control measures for other perennial grasses are similar to those used for one or both of these species. Although tillage, mowing, and crop shading can contribute significantly to the control of these weeds, programs that incorporate the use of effective herbicides have proven to be the most successful.

Herbicides

Herbicides used for the control of established perennial grasses are applied to the emerged foliage and include the cyclohexanedione (clethodim, sethoxydim), aryloxyphenoxy propionate (fluazifop-P, quizalofop-P, fenoxaprop-P), sulfonylurea (nicosulfuron, primisulfuron), and imidazolinone (imazapyr, imazethapyr) herbicides, and glyphosate, glufosinate, and asulam. The performance of these herbicides (several are listed in Table 13.3) varies with the herbicide and the perennial weed species. For example, glyphosate tends to give poor control of bermudagrass but good to excellent control of johnsongrass. Several of these herbicides are used on broadleaved crops, but two of them, nicosulfuron and primisulfuron, are used selectively to remove perennial grass weeds in corn, which is a grass crop.

With the exception of johnsongrass, most perennial grass weeds do not compete significantly from seed. Nevertheless, control programs should be designed to prevent the reestablishment of the perennial population from germinating seeds. Seedling control is absolutely essential for effective johnsongrass control.

JOHNSONGRASS
Biology

Johnsongrass (*Sorghum halepense* [L.] Pers.) is a tall perennial grass (Poaceae family) that spreads rapidly by seeds and rhizomes. It is likely that it was introduced into the United States in the 1780s (McWhorter 1989) and reportedly was planted specifically as a forage crop near Selma, Alabama, in 1830 (or 1840; see Mitich 1987) by Col. William Johnson. Although johnsongrass is a warm-season perennial, it is found today in almost every state of the union. It is a major problem in California and throughout the Southwest and the South, extending as far north as St. Louis; Indianapolis; Columbus, Ohio; and Washington, D.C.

Johnsongrass grows to a height of 4 to 8 ft and has a large open panicle type of inflorescence (Figure 13.1). The leaves are broad with a prominent white midrib and a membranous ligule that has a narrow fringe of hairs on the upper edge. The rhizomes are about 1/2 in. in diameter and have tan to black scale-like leaves encircling the nodes (Figure 13.2).

Table 13.2 Number of Treatments Expected to Provide Substantial Reduction of Underground Vegetative Reproductive Structures in One to Three Growing Seasons

Control Practice	Treatments per Year	General Comments
Tillage (broadcast 3–8 inches deep)	4–8	Initiated when levels of carbohydrate are low. Allow shoots to develop several days after emergence. Follow with tillage. Continue practice until end of growing season. May not be best choice.
Hoeing or scraping (0 to 2 inches)	5–9	Time consuming and expensive except for limited numbers of plants. May be the only choice in some situations.
Mowing in conjunction with a competitive cover crop	3–6	For hay crops, sodded roadsides, and waste areas. Most tall-growing species more easily controlled than low-growing or sod-forming ones. Weeds should not be permitted to flower. Minimum program for control of creeping perennials. The 15–30 mowings used on well-managed fine-leaved grasses in lawns, parks, and fairways greatly exceed this number. Some decumbent or sod-forming species can survive.
Foliar-applied herbicides; contact or apoplastically translocated	4–8	Assume no additional activity through soil. Provides top removal to or slightly below soil level. Action similar to scraping or hoeing. (Many apoplastically translocated herbicides do have soil residual activity.)
Foliar-applied herbicides; limited symplastic translocation	2–4	Frequently the degree of control achieved with so-called translocated herbicides. Typical of 2,4-D and related compounds. More effective per treatment than tillage.
Foliar-applied, well-translocated herbicides	1–3	Degree of translocation will depend on weed species, stage of growth, growing conditions. Glyphosate on johnsongrass and quackgrass appears to be in this category. Extremely effective where available. Has allowed effective directed applications. Results proportional to percentage of plants accessible.
Highly persistant herbicides	1	Persist for one or more full growing seasons. Herbicide level is maintained at a level to kill perennial plants and any new plant emerging from seed or vegetative buds. Use only where bare soil can be tolerated. Long residual can cause an environmental hazard.
Soil-applied fumigants (dazomet, metam)	1	Effective on most vegetative perennial weed structures. High cost limits use to small areas or high revenue crops (gardens, plant beds, initial infestations). Vegetative propagules below the plow layer will probably escape.

One of the characteristics of johnsongrass that separates it from most other perennial grasses is its prolific seed production and highly competitive seedlings. Heavy production of seeds provides the plant with mechanisms for both long-term survival and spread. Johnsongrass has been reported to produce up to 10 bu of seed/acre (McWhorter 1981). Viability may be 90% or better, and seeds can survive burial in the soil for a decade or more. Even when heavy infestations of established plants are present, emerging seedlings can outnumber plants emerging from rhizomes ten to one. Although not as competitive initially, the seedlings compete effectively with most annual crops and acquire perennial characteristics by midseason.

Consequently, an effective control program requires that plants from seeds as well as those from rhizomes be held in check.

Seeds also are a serious problem in flood-susceptible bottomlands, where each overflow provides a new supply of seeds. In these areas, seedlings provide a continuous source of new plants, even when established plants have been controlled effectively.

The ability of johnsongrass to compete as vigorously as it does with crop plants and to resist control methods can be attributed largely to the emergence of plants from its extensive rhizome system. Plants developing from rhizomes establish more quickly in the spring than seedlings do. One reason is the more vigorous

Table 13.3 Expected Performance of Some Herbicides Labeled for Selective Postemergence Control of Creeping Perennial Grasses[a,b]

Herbicide	Corn	Cotton	Soybeans	Peanuts	Flax	Forage crops	Canola	Fruit and nut trees[d]	Christmas trees	Woody ornamentals	Herbaceous ornamentals	Lawn & turf	Vegetable crops	Sugar beets	Mint	Dry bulb onions & garlic	Strawberries	Sunflowers	Bermudagrass	Johnsongrass	Quackgrass	Wirestem muhly
	Crops																		Weeds			
Clethodim		X	X					X	X	X	X		X	X	X	X	X	X	G-E	G-E	G	F-G
Sethoxydim	R	X	X	X	X	X	X	X	X	X	X			X		X	X		G	G	F-G	G
Fenoxaprop-P												X							P-F	F	N[c]	F
Fluazifop-P			X					X		X	X		X			X			G-E	G-E	G	G
Quizalofop-P		X	X							X	X								G-E	G-E	G	G-E
Nicosulfuron	X																		—[c]	G	G	F-G[c]
Primisulfuron	X																		—[c]	F-G	F-G	F[c]
Glyphosate (0.5–1 lb a.i./acre)	R	R	R				R												P	G-E	G	G

[a]Crops: X = labeled for crop; R = for resistant cultivars only. Weeds: N = none; P = poor; F = fair; G = good; E = excellent.

[b]Dosages for perennial creeping grasses generally are at the high end of the labeled dosage range. A second application frequently is required for acceptable control.

[c]Not labeled at the current time.

[d]Consult label for registration on fruit- or nut-bearing crops.

Figure 13.1 Johnsongrass plants in a soybean field.

growth typical of plants that arise from vegetative buds as compared to those that arise from seed (Figure 1.13). In addition, rhizome buds can sprout at lower temperatures than seed. A study conducted in Kentucky (Hanson et al. 1976) showed that plants from rhizomes emerge over a period from early May to late June. Rhizome plants started to grow at soil temperatures of 60°F, in contrast to seeds, which germinated at temperatures of 70°F. This study suggested that the first plants from seed emerge about a month after those from rhizomes.

Johnsongrass thrives under high temperature and moisture, so initial shoot growth is slow until warm temperatures prevail. Rhizome initiation occurs when the plant reaches the seven-leaf stage, but rhizome development is slow until flowering occurs in the summer. Approximately 90% of annual rhizome production occurs after flowering takes place. More than 200 ft of rhizomes can be produced by a single plant in a growing season (McWhorter 1961). In Louisiana, more than 7 tons of johnsongrass rhizomes per acre were produced in a sugarcane field (McWhorter 1981).

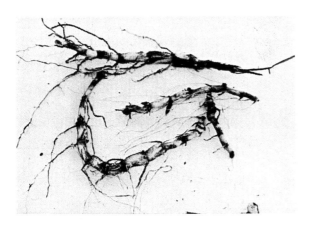

Figure 13.2 Johnsongrass rhizomes. Note clear demarcation of nodes and internodes.

Most rhizomes produced during a growing season can overwinter, but few live more than 1 yr. Rhizomes exhibit little true dormancy. Late-season resprouting of buds is suppressed by apical dominance and cool temperatures. Cutting the rhizome into pieces removes apical dominance and permits sprouting of lateral buds; rhizome pieces with a single node can sprout into new plants.

Response to Cultural and Mechanical Methods

Competitive Crops

Crops that initiate growth prior to the emergence of johnsongrass and provide thorough shading of the soil can be combined with appropriate management practices to weaken established stands. Alfalfa hay, alfalfa–grass hay mixtures, grass pastures, winter small grains, and frequently mowed cool-season turfgrasses all compete well with johnsongrass. Summer annual crops such as corn and soybeans also can compete actively with and shade out johnsongrass when properly managed (providing johnsongrass is controlled until crop canopy closure). Crop rotation can reduce the losses caused by moderate infestations of johnsongrass. Rotations that provide crops in which johnsongrass can be controlled and rhizome formation prevented will aid substantially in control for the subsequently planted crop.

Tillage

Fallow (clean cultivation) was the major method for reducing established stands of johnsongrass in the past. Control of established stands of rhizome johnsongrass can be accomplished with one season of clean cultivation. Land should be plowed in late spring when johnsongrass reaches 18 to 24 in. in height and then tilled with a disk- or sweep-type field cultivator at 3- to 4-wk intervals to keep all growth under 12 in. high (six leaves) and break up the rhizome system. Cultivation should be continued until frost in the fall.

An alternate procedure to clean fallow utilizes a small grain (preferably wheat) that is seeded in the fall. After the crop is harvested the following summer, the land is plowed and then cultivated at regular intervals until fall. Wheat is planted and the postharvest fallow is practiced a second season.

Tillage is an effective component of production systems for annual crops grown in johnsongrass-infested fields. The moldboard plow or chisel plow followed by a disk to level the soil and then two additional diskings to incorporate pre-plant herbicides for seedling control is the amount of preplant tillage recommended for johnsongrass-infested fields. Repeated tillage breaks up the rhizomes and improves the overall effectiveness of the herbicides. Reducing the amount of preplant tillage results in less control of johnsongrass. Cultivations after the crop has emerged also help with control and should be used in as timely and effective a manner as possible.

Plowing in the fall exposes rhizomes to freezing conditions and can result in improved control of established stands.

Symplastically translocated postemergent herbicides introduced since 1980 have allowed growers to reduce the amount of tillage needed in johnsongrass-infested fields. Even no-till is a viable option at this time.

Mowing

When properly executed, mowing can prevent seed formation and in many instances can reduce the vigor of established johnsongrass stands. To prevent seed formation, johnsongrass should be mowed at 3- to 4-week intervals starting in early June and continuing through September (at a latitude corresponding to the Indiana–Kentucky border). In most years, this schedule requires four to five mowings, each done prior to seedhead emergence. Mowing to reduce vigor is most effective when used in conjunction with a competitive forage crop or turfgrass.

Systems for Control

Substantial advances in the control of johnsongrass in crops have been made over the past 25 yr, primarily because of the development and use of herbicides that control the weed effectively at different stages of its life cycle. These advances include preplant soil-incorporated herbicides for controlling germinating seeds, the use of higher rates of these herbicides to give increased seedling control and suppression of plants from rhizomes, translocated herbicides that can be applied to the foliage of established plants and that move into the rhizomes, the development of the recirculating sprayer and wipe-on application devices for selective placement of herbicides, the introduction of selective herbicides that control johnsongrass when applied over the top of established broadleaved crops including cotton and soybeans, the introduction of selective postemergence herbicides in corn, and more recently, the introduction of herbicide-resistant crop cultivars that allow the use of formerly nonselective herbicides such as glyphosate, sethoxydim (in corn), and glufosinate.

The following programs are available for various cropping and noncropping situations. They incorporate postemergence herbicides used for established plants (Table 13.3) and soil-applied herbicides that are used to control or suppress johnsongrass seedlings (Table 13.4). When more than one treatment is listed, note that each treatment represents an alternative.

I. Cropping situations
   A. Soybeans in conventional and conservation tillage systems
      A burn-down treatment (e.g., glyphosate, 2,4-D, paraquat) is frequently needed on conservation tillage systems to control weeds that emerge prior to planting. However, johnsongrass frequently does not become established until after burn-down treatments have been applied.
      1. Postemergence applications of fluazifop-P, quizalofop-P, fenoxaprop-P, clethodim, or sethoxydim when rhizome johnsongrass is 12 to 18 in. tall. This treatment can be preceeded with a single dose of soil-applied herbicide. A preplant incorporated dinitroaniline herbicide is preferred in conventional tillage systems. The preplant herbicide provides initial and residual control of pigweed, lambsquarters, annual grasses, and seedling johnsongrass. On no-till, a preemergence treatment of pendimethalin, clomazone, or a chloracetamide herbicide provides similar control of annual grasses and small-seeded broadleaves but somewhat less control of johnsongrass seedlings.
      2. Postemergence applications of fluazifop-P, quizalofop-P, fenoxaprop-P, clethodim, or sethoxydim when rhizome johnsongrass is 12 to 18 in. tall. If seedlings or regrowth from rhizomes reappear, make a second application with these herbicides.
      3. Glyphosate in glyphosate-resistant soybean cultivars. Treat when rhizome johnsongrass reaches 12 to 18 in. If seedlings or regrowth from rhizomes reappear, make a second application of glyphosate or other selective postemergence herbicide.

Table 13.4 Partial List of Soil-Applied Herbicides Labeled for Control or Suppression of Johnsongrass Seedlings and Plants Emerging from Rhizomes[a]

	Crops																									Weeds	
	Corn	Cotton	Soybeans	Cole Crops	Cucurbits	Peanuts	Small grains	Grain sorghum	Dry beans	Collards	Small-seeded legumes	Tree fruits	Small fruits	Woody ornamentals	Herbaceous ornamentals	Lawn & turf	Vegetable crops	Melons	Sugar beets	Onions & bulbs	Tomatoes	Tobacco	Strawberries	Sunflowers	Potatoes	Seedling Johnsongrass	Rhizome Johnsongrass
Acetochlor	X																									S	
Alachlor	X	X	X					X	X																	S	
Benefin	X																X									C	
Butylate	X						X																			S	
Clomazone		X	X	X										X						X		X	X			S	
DCPA		X	X	X					X											X		X	X			S	
Dimethenamid	X	X	X														X							X	X	C	S
EPTC	X	X	X						X			X					X		X		X			X	X	C	
Ethalfluralin			X		X	X																				C	
Imazaquin	R		X																							C	
Imazethapyr[b]			X			X					X															S	S
Metolachlor	X	X	X			X X		X													X	X	X	X	X	C	
Napropamide												X	X	X							X	X	X			C	S
Norflorazon		X										X	X													C	
Oryzalin		X										X		X												C	
Pendimethalin	X	X	X	X	X	X	X	X	X	X	X						X					X		X	X	C	S
Trifluralin		X	X			X	X			X	X						X							X	X	C	C

[a]Crops: X = labeled for crop; R = for resistant cultivars only. Weeds: C = control; S = suppression.
[b]Early postemergence application provides maximum benefit for johnsongrass control or suppression.

4. Single dosage of a soil-applied herbicide followed by glyphosate applied with a recirculating sprayer or wipe-on device. This program is best suited to light infestations.

B. Cotton

Programs of johnsongrass control similar to those described for soybeans also can be used in cotton. The dinitroaniline herbicides and glyphosate are used with cotton the same as they are used with soybeans. In addition, DSMA and MSMA can be applied directed postemergence prior to the first bloom of cotton.

C. Corn in conventional and conservation tillage systems

1. Postemergence applications of nicosulfuron or primisulfuron when rhizome johnsongrass is 12 to 18 in. tall. This treatment should be preceeded with a single dose (used for annuals) of a soil-applied herbicide that gives good seedling grass and small-seeded broadleaved weed control. Preplant and preemergence herbicides provide initial and residual control of pigweed, lambsquarters, and annual grasses, and suppress seedling johnsongrass. Crabgrasses are not readily controlled by either nicosulfuron or primisulfuron.

2. Postemergence application of sethoxydim on sethoxydim-resistant corn cultivars when johnsongrass is 12 to 15 in. tall. If seedlings or regrowth from rhizomes reappear, make a second application with sethoxydim or other effective selective postemergence herbicide.

3. Glyphosate in glyphosate-resistant corn cultivars. Treat when rhizome johnsongrass reaches 12 to 15 in. and repeat if johnsongrass recovers, or treat with another effective selective postemergence herbicide.

D. Winter wheat

1. Glyphosate applied to johnsongrass regrowth following wheat harvest, or wheat harvest followed by mowing and then treatment. Johnsongrass should be about 24 in. tall but not in seedhead. If johnson-grass reappears, retreat or till. Johnsongrass shoots should be removed anytime johnsongrass reaches 12 in. or the six-leaf stage. Maintain the control program until frost or until seedbed is prepared

for the next crop. This program allows planting of wheat that fall or corn or soybeans the next spring without fear of chemical carryover.

E. Hay (alfalfa)

1. Maintain a good stand of hay and mow often enough to prevent seedhead formation on johnsongrass. Requires three to four mowings a season. Crop competition and mowing combine to reduce the vigor of established johnsongrass stands. Postemergence applications of sethoxydim can be used if johnsongrass persists in alfalfa.

F. Pastures (tall fescue, Kentucky bluegrass, smooth bromegrass, and orchardgrass)

1. Graze intensively enough to keep stress on the johnsongrass. Additional occasional mowing may be required. Do not allow seedhead formation.

G. Most crops (new infestions of isolated plants or small patches)

1. Avoid seed burial in the soil.

2. Spot spray with selective postemergence herbicides.

3. Spot spray with glyphosate after plants are 18 in. tall and before seedheads emerge. Chemical application is less time consuming and provides more effective control of rhizomes than hand pulling or hoeing. Repeat as needed.

4. Do not permit seedheads to form.

H. Established turfgrasses

1. Mow frequently to prevent seedhead formation. Four to five mowings usually are required. More frequent close mowing will provide even better control. Weekly mowing usually will eliminate johnsongrass. When johnsongrass persists, fenoxaprop-P can be used to discourage its survival.

II. Noncropping situations

A. New infestations of isolated plants or patches

1. Avoid seed burial in the soil.

2. Spot spray with selective postemergence herbicides.

3. Spot spray with glyphosate after plants are 18 in. tall and before seedheads emerge. Chemical application is less time consuming and provides more effective control of rhizomes than hand pulling or hoeing. Repeat as needed.

4. Do not permit seedheads to form.

B. Where bare soil is wanted or can be tolerated
   1. Apply highly persistent herbicides such as imazapyr, prometon, and bromacil. Consult labels for dosages and weeds controlled.
   2. Repeat applications of short residual foliage-applied herbicides such as glyphosate, glufosinate, the aryloxyphenoxy propionates, the cyclohexanediones, or asulam. Spray before seedhead emergence to prevent new seed formation.
C. Where ground cover must be maintained to prevent soil erosion
   1. Spot treat with short-lived foliage-applied herbicides such as glyphosate after johnsongrass is 18 in. tall and prior to seedhead formation. Then establish a ground cover.
   2. Maintain a perennial grass sod and treat 15- to 24-in. tall johnsongrass with fenoxaprop-P or fenoxaprop-P plus fluazifop-P.
   3. For selective removal of perennial grasses from broadleaved ground covers, use appropriately labeled aryloxyphenoxy propionate or cyclohexanedione herbicides.

QUACKGRASS

Biology

Quackgrass (*Elytrigia repens* [L.] Beauv.) is a rhizomatous grass (Poaceae family) that is widely distributed in cool, moist climates. In North America this weed thrives in the northern half of the United States (Figure 13.3) and in most agricultural areas of Canada. Severe infestations occur in Minnesota, Wisconsin, Michigan, New York, Pennsylvania, and Ohio. Quackgrass is not found in tropical climates because it apparently requires an exposure to low temperatures during the dormant period for initiation of vigorous growth.

Quackgrass was originally called *Triticum repens* by Linnaeus because the inflorescence is similar to that of wheat (*Triticum aestivum*). Two to eight florets per spikelet are arranged alternately on the rachis (Figure 13.4). Quackgrass leaves are about 1/4 in. wide, are somewhat hairy on the upper surface, and have clasping auricles. The plants are 1 to 3 ft in height.

The plant spreads by rhizomes and seeds (Figure 13.5). Flowering begins in mid-June and can continue into the summer. The plant is wind

Figure 13.3 Quackgrass infestation (background) of beans and corn in Wisconsin. (J. D. Doll, University of Wisconsin)

pollinated and is reported to be self-sterile (Palmer and Sagar 1963). Since a patch (clone) of quackgrass that arises through vegetative propagation from a single plant would tend to be self-sterile, seed set may be concentrated at the edges of the patch, where plants are fertilized with pollen from adjacent patches. The amount of seed production and degree of viability vary from one site to another, depending on growing conditions.

In contrast to many other perennials, seeds do not appear to provide an important mechanism for long-term survival; they have little dormancy at harvest. Seeds have been reported to

Figure 13.4 Flowering heads of quackgrass. (J. D. Doll, University of Wisconsin)

Figure 13.5 Quackgrass plants with rhizomes. (J. D. Doll, University of Wisconsin)

survive dry storage for 1 to 6 yr (Holm et al. 1977) but in the field probably survive no longer than 1 or 2 yr. Their survival in water and waterlogged soils is low (3% survival after 3 mo in water and none after 27 mo).

Rhizome production is the primary means of spread and survival for this plant. Shoots developing from overwintering rhizomes begin active growth in the early spring (Stobbe 1976). Seedling plants develop 3 to 5 wk later than rhizome plants. At the three- to four-leaf stage, the plants begin to tiller and new rhizomes form, with the first flush being produced by mid-May. Each primary shoot typically bears three tillers and three to four rhizomes when growing in a dense patch. Each tiller can produce two to three rhizomes, and each rhizome can produce lateral rhizomes so that as many as 50 rhizomes per plant can be produced in a single season. The plant continues to spread throughout the growing season as the spreading rhizome tips turn up to produce shoots that in turn form more tillers and rhizomes.

The growth of the rhizomes can be so prolific that a dense mat of these structures usually is found just beneath the soil surface. The majority of rhizomes are found in the upper 6 in. of soil. The following statistics taken from a single plant at the end of the growing season (September) give some indication of the vigor of vegetative growth (Raleigh et al. 1962): the diameter of rhizome spread was 10 ft, 14 rhizomes from the original plant had grown a total length of 400 ft, 206 aerial shoots had been produced, and 232 additional growing points (for next season's growth) were found on the rhizomes in the soil.

In the fall, the rhizome tips may develop into rudimentary aerial shoots while the parent shoots die. The rudimentary aerials and other rhizome pieces overwinter and provide new growth the following spring.

Rhizomes can survive for several years. The terminal bud exerts a strong apical dominance over the lateral buds. Unless the rhizome is disturbed (e.g., cut up), the majority (approximately 95%) of the lateral buds remain dormant during the entire life of the rhizome.

The rhizomes of quackgrass are considered to have an allelopathic effect on other plants. Establishment of crops in quackgrass-infested ground is difficult. Although the mechanism of inhibition is not understood, decaying rhizomes and roots of the plant may have a direct inhibitory effect, whereas living quackgrass plants may have more of a competitive effect for available water and nutrients.

Response to Cultural and Mechanical Methods

Competitive Crops

Shading by a taller crop can help to minimize the development of quackgrass through the season.

Tillage

Tillage can contribute significantly to quackgrass control. Prior to the development of modern herbicides, extensive tillage was a major method for control. In fact, the spring-tooth harrow, which tends to pull rhizomes to the soil surface and shake the soil from them, was a preferred tillage tool in some quackgrass-infested regions. Two seasons of clean fallow are needed to ensure good control. The rhizomes resist drying, so best results are obtained when low humidity and dry soils follow the tillage operations. Tillage followed by freezing also is effective.

In conventional cropping systems, primary, secondary, and interrow tillage contribute significantly to control but by themselves are inadequate for complete control. Planting row crops has been reported to be more successful at preventing quackgrass reestablishment after glyphosate treatment than a cover crop such as alfalfa, possibly because of the additional tillage pressure (Harvey 1976).

Mowing

Mowing at 14-day intervals has been reported to deplete rhizome carbohydrates. Quackgrass, however, survives quite well in areas mowed only three to five times during the season (e.g., pastures, mowed roadsides, and hay crops). It frequently is found as a weed in lawns.

Systems for Control

Several postemergence herbicides (Table 13.3) have been introduced that, if properly used and sequenced in crop management systems, can provide excellent control of established quackgrass. If this control can be maintained over a period of two to three seasons, the relatively short life and lack of dormancy in the seeds should make it difficult for the weed to reestablish itself.

Some effective herbicide programs and treatments are listed below. Two applications of a symplastically translocated postemergence herbicide per season may be needed to allow cropping in a no-till situation on quackgrass-infested land.

I. Cropping situations
   A. Spring-planted crops
      1. Glyphosate provides excellent control of quackgrass when applied before plowing and seedbed preparation either in the spring or fall during active growth. Spring treatments can be used ahead of late May- or June-planted crops; fall treatments can be applied ahead of most spring-planted crops. Tillage and cultivation subsequent to herbicide treatment help control escapes.
      2. Glyphosate should be applied for burndown at planting for no-till crops. A herbicide to selectively control quackgrass regrowth should be used in the crop itself.
   B. Corn
      1. Postemergence applications of nicosulfuron or primisulfuron when quackgrass is 4 to 8 in. tall.
      2. Postemergence application of sethoxydim on sethoxydim-resistant corn cultivars when quackgrass is 4 to 8 in. tall.
      3. Glyphosate in glyphosate-resistant corn cultivars when quackgrass is 6 to 8 in. tall.
   C. Broadleaved annual crops
      1. Fluazifop-P, quizalofop-P, clethodim, and sethoxydim are labeled for quackgrass; however, high rates and multiple applications may be required for consistent control.
      2. Glyphosate in glyphosate-resistant soybean cultivars.
   D. Alfalfa and small-seeded legumes
      1. Pronamide can be effective for controlling quackgrass. Season-long control can be obtained with fall applications.
      2. Sethoxydim is labeled for selective control in pure alfalfa stands.
   E. Woody species
      1. Fluazifop-P, clethodim, and sethoxydim can be used; however, high rates and multiple applications may be required for consistent control. Can be used on new and established plantings.
      2. Dichlobenil, simazine, and terbacil can be used as soil applications in some established woody species.
      3. Glyphosate, glufosinate, and amitrole can be used as directed sprays in a number of woody species.
   F. Total vegetation control can be obtained with persistent and highly persistent herbicides such as imazapyr, bromacil, prometon, simazine, and hexazinone.

YELLOW AND PURPLE NUTSEDGE

Biology

Yellow nutsedge (*Cyperus esculentus* [L.]) and purple nutsedge (*Cyperus rotundus* [L.]) are members of the sedge (Cyperaceae) family. Both are creeping perennials that produce rhizomes, tubers, and sometimes seed. Yellow nutsedge is a problem from southern Canada southward to Florida and Texas, but it is most serious in the Upper Midwest and the Northeast. Purple nutsedge is confined to the southern states from North Carolina to southern California. Purple nutsedge was designated by Holm et al. (1977) as the world's worst weed because it is a serious competitor in more crops and more countries worldwide than any other weed. Yellow nutsedge is found on low, moist areas, whereas purple nutsedge is adapted to well-drained soils. When growing in mixed populations, purple nutsedge is the more competitive of the two species.

Although yellow and purple nutsedges appear similar to grasses, they differ from grasses by having three-ranked leaves, triangular stems (in cross section), closed leaf sheaths, and no ligules. Yellow nutsedge, which ranges in height from 1 ft to 2.5 ft., tends to be taller than purple nutsedge (4 in. to 2 ft).

Yellow nutsedge has pale green leaves and yellowish inflorescences. The basal leaves are as long or longer than the stem, and the bracts subtending the inflorescence are longer than the stalks of the seed heads (Figure 13.6). Single, light-colored tubers, which are produced at the terminals of rather fragile rhizomes, are the overwintering structures.

When the tubers sprout, they produce a vertical rhizome (Figure 13.7A), which terminates in a slender shoot. As the rhizome and shoot grow upward through the soil, the shoot forms a small, solid, bulblike structure called a basal bulb (Figure 13.7B). The basal bulb is a very active structure, producing roots and additional rhizomes. These rhizomes grow laterally and can produce more shoots, which spread the plant in a horizontal direction (Figure 13.7C). Other rhizomes eventually produce the single tubers (Figures 13.6, 13.7D).

Purple nutsedge has dark green leaves and reddish brown or purplish inflorescences. The basal leaves are shorter than the stems, and the subtending bracts are about the same length as the stalks of the seed heads (Figure 13.8B). The life cycle of purple nutsedge is similar to that of yellow nutsedge except that the rhizomes produce chains of dark, shaggy-looking tubers on

tough, wiry rhizomes. When these tubers sprout, they also form vertical rhizomes, basal bulbs, and shoots (Figure 13.8A). The basal bulbs are larger and more conspicuous than those of yellow nutsedge. Later, the basal bulbs produce new rhizomes and new plants (Figure 13.8B).

Yellow nutsedge tubers survive freezing in undisturbed soils, whereas purple nutsedge tubers are killed under these conditions. In the laboratory, purple nutsedge tubers were reported to germinate at a minimum temperature of 20°C, whereas yellow nutsedge tubers germinated at 12°C (Stoller and Sweet 1987). Thus, seasonal temperatures probably account in part for the distribution of both nutsedge species.

Flowering is erratic; many populations fail to flower or flower and fail to produce seed. Formation of tubers can be expected if these weeds grow for a full cropping season. Tubers are the major reproductive structures, and densities of 1 to 10 million per acre have been reported for moderate to heavy infestations of yellow nutsedge (Tumbleson and Kommedahl 1961). In general, seed production is not considered to be a major means of propagation in infested areas but probably is important in the spread of these weeds to noninfested sites.

Yellow nutsedge, which is more widespread in the United States than purple nutsedge, will be the focus of the remainder of this section. More than 10.5 million acres of corn and soybeans were reported to be infested with yellow nutsedge in the north central United States alone (Armstrong 1975). A more recent survey taken in 1992 (available from David R. Pike, Department of Crop Science, University of Illinois) indicated that yellow nutsedge was the weed most often reported by growers after common milkweed and Canada thistle. The plant is most abundant on low, poorly drained areas with high organic matter soils, but it also spreads into adjacent drier areas.

Tubers of yellow nutsedge germinate in the spring as the soil warms. In the Corn Belt, the emergence of the shoot from the basal bulb occurs around May 1. Tuber sprouting continues maximally until mid-June and at a slower rate into July.

Several weeks after shoot emergence, rhizomes begin to grow from the basal bulb. Shoot, root, rhizome, and new shoot growth continues until approximately late July, when tuber formation on the rhizomes occurs in response to decreasing day lengths. In the northeast, a good sized nutsedge plant will have a few tubers in early July, 20 to 30 in early August, and 200 to

Figure 13.6 Seed heads and tubers (arrow) on a yellow nutsedge plant. This stage corresponds to Figure 13.7D.

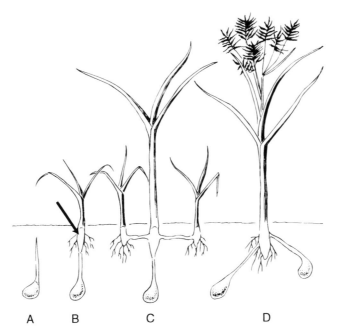

Figure 13.7 Yellow nutsedge. A tuber sprouts to produce a vertical rhizome and shoot (A). The shoot forms a basal bulb (arrow) and emerges from the soil (B). The basal bulb produces rhizomes from which new plants originate (C). Rhizomes from the basal bulb also can produce new tubers (D). In purple nutsedge, the tubers are produced in chains rather than singly.

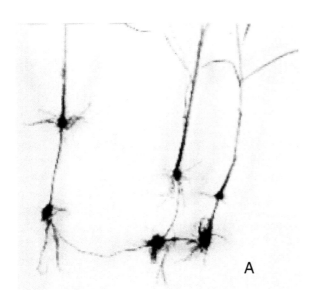

Figure 13.8 Purple nutsedge. (A) Three tubers in a chain. Each tuber has produced a basal bulb and shoot. This stage corresponds to Figure 13.7B. (B) Seed heads. The main flowering stem of this plant has produced rhizomes from its basal bulb, resulting in the production of three new shoots, each with its own basal bulb (arrows). This stage corresponds to Figure 13.7C. This plant is no longer attached to the tuber that produced it.

300 by early September (Sweet 1975). Tuber production ceases with killing frosts.

Under extremely good growing conditions, a single plant can produce several thousand tubers in a season. In one study, a single tuber produced 1900 plants and almost 7000 tubers and covered an area about 6 ft in diameter within a year (Tumbleson and Kommedahl 1961).

Maximum tuber dormancy occurs at the end of the growing season and can persist for several months. Dormancy is broken by an extended cold period and by leaching of the tubers by precipitation. Several buds on a single tuber can sprout at one time, but others can remain dormant for sprouting later. Tubers can survive for several years buried in the soil. Stoller et al. (1979) reported 15% survival of tubers after three seasons of full season control of yellow nutsedge in corn. Tubers near the soil surface, however, can winterkill. They also are susceptible to desiccation when brought to the soil surface by tillage.

Response to Cultural and Mechanical Methods

The following practices are used for the suppression and control of yellow nutsedge.

Competitive Crops

Partial control of yellow nutsedge can be obtained with a good competitive crop. The weed has little tolerance to shade and therefore is susceptible to fast closing crop canopies. Low growing crops usually fail to provide adequate competition. Fast growing crops managed to provide rapid canopy closure for maximum shading inhibit nutsedge growth. Good stands, solid seeding, narrow rows, and increased plant populations help provide quick canopy closure. Rotating crops and herbicides also help in improving yellow nutsedge control.

Yellow nutsedge often produces its densest stands in low-lying wet areas. Its dominance in these sites may be due to a preference for moist sites, an inherent resistance to flooding, or to the lack of crop competition because the crop has been flooded out or is subjected to poor growing conditions. Yellow nutsedge flourishes where crop stands are thin or lacking. Excessive moisture favors maximum development of yellow nutsedge in crops such as corn and soybeans, which need drier soils for maximum growth. Increasing the drainage of these sites can help improve crop stands, which in turn can better compete with yellow nutsedge.

Tillage

A major objective of tillage is to reduce the number of tubers by inducing sprouting or by exposing tubers on the soil surface, where they can be killed by desiccation or low temperatures during the winter. Repeated tillage operations along with delayed planting can be useful in cropping situations. Normal tillage procedures in the preparation of the seedbed, cultivation after planting, and herbicides can prevent the weed from competing until the crop is large enough to produce a shading canopy. Between-row cultivation helps eliminate plants escaping other treatments and removes plants that otherwise would produce new tubers. Tillage also facilitates the use of effective preplant incorporated soil-applied herbicides for nutsedge control. Incorporation usually requires two or more secondary tillage operations. When tillage alone is used as part of a fallow system, tuber numbers can be reduced substantially. Two or more seasons of clean fallow are needed to reduce tuber levels significantly.

Mowing

Nutsedge is low growing and generally survives mowing. Close mowing at weekly intervals probably will reduce tuber formation.

Systems for Control

Control programs for yellow nutsedge depend on multiple control practices spread over the entire growing season. A typical program consists primarily of early season suppression by tillage and herbicides followed by shading from competitive crops until harvest (Glaze 1987). Tuber populations in the soil can be reduced gradually by preventing tuber formation in the late summer and early fall. The herbicides used early in the season for suppression seldom last past midseason, so tuber formation must be prevented by late-season shading by the crop, by postharvest tillage, or by late-season herbicide applications. Integrated programs resulting in season-long control can provide gradual but substantial reduction of the infestation over a period of several years. Eradicating yellow nutsedge in one or two seasons virtually is impossible, however, because of the tuber dormancy and because most herbicides kill the shoots but not the tubers.

As with most creeping perennial weeds, herbicides are an integral component of most consistently effective and economical control programs for nutsedge (Table 13.5).

Table 13.5 Partial List of Herbicides Labeled for Nutsedge Control or Suppression[a]

	Crops																								Weeds		
	Corn	Cotton	Soybeans	Alfalfa	Peanuts	Rice	Sorghum	Sugarcane	Dry beans	Spinach	Legumes	Woody species	Ornamentals	Lawn & turf	Grass crops	Sugar beets	Table beets	Tobacco	Citrus	Mint	Potatoes	Noncrop	Directed	Site preparation—all crops	Yellow nutsedge	Purple nutsedge	Nutsedge species
Soil Applied																											
Acetochlor	X																								C		
Alachlor	X		X									X													C		
Cycloate	X									X						X	X								S	S	S
Dimethenamid	X		X		X		X		X																C		
EPTC				X					X		X		X	X	X	X	X		X		X				S	S	
Imazaquin	R		X		X						X														S	S	S
Imazethapyr	X		X	X	X						X														C	S	
Metolachlor	X	X	X		X		X					X				X		X			X	X			S	S	S
Norflurazon		X										X							X			X			S		
Pebulate																		X		X					S	S	
Terbacil				X				X			X	X								X					C	C	C
Foliar Applied																											
2,4-D	X							X							X[b]							X			C	C	C
Ametryn	X																								C	C	C
Bensulfuron						X																					C
Bentazon	X		X		X	X	X		X		X		X	X						X					C	C	C
Bromacil																			X								C
Chlorimuron			X		X						X														C	C	
Dichlobenil												X							X			X					C
DSMA		X												X									X		S	S	S
Ethofumesate																X											C
Fomesafen	R		X								X			X											S	S	S
Glufosinate	R		R																			X	X		S	S	C
Glyphosate 1 lb a.i./acre	R		R									X		X					X				X	X	C	C	C
Glyphosate 3 lb a.i./acre																									C	C	
Halosulfuron	X						X								X							X			S	S	S
Imazethapyr	R	X	X	X	X							X		X		X			X			X	X	X	C	C	S
MSMA																							X		S	S	
Paraquat	X				X						X														S		S
Primisulfuron	X																					X			S		
Pyridate	X																					X		X	S	S	
Rimsulfuron																X					X				S		
Terbacil				X			X					X		X					X	X					S		
Fumigant																											
Vapam																								X			C

[a] Crops: X = labeled for crop; R = for resistant cultivars only. Weeds: C = control; S = suppression.
[b] Not on all 2,4-D labels.

I. Soil-applied herbicides for control or suppression of yellow nutsedge in crops
   A. Thiocarbamates are used on several crops to provide suppression of nutsedge and grass weeds. EPTC is used on corn and legume crops; cycloate on sugar beets, table beets, and spinach; and pebulate on tobacco and tomatoes. These herbicides require soil incorporation or injection to prevent loss by volatilization. They provide better control when temperatures are adequate for good tuber germination and where prior tillage has been used to stimulate tuber germination.
   B. The chloracetamides, acetochlor, alachlor, metolachlor, and dimethenamid, suppress or control nutsedge in corn, soybeans, and peanuts when used preplant incorporated or preemergence. Preemergence application can be effective if followed by adequate rainfall or overhead irrigation to leach the herbicides into the root zone of the developing nutsedge plants.
   C. Other soil-applied herbicides providing selective suppression of yellow nutsedge include imazaquin, imazethapyr, and norflurazon.
II. Postemergence herbicides for control or suppression of yellow nutsedge in crops are most effective when treated prior to tuber formation. A single postemergence herbicide application, however, will probably not provide adequate control of nutsedge all season.
   A. Selective applications of glyphosate in glyphosate-resistant crop cultivars should provide suppression of nutsedge. Repeat applications of glyphosate may be needed.
   B. Postemergence applications of bentazon to 6- to 8-in.-tall nutsedge provide shoot kill in a number of crops including soybeans, dry beans, green beans, rice, peanuts, corn, spearmint, peppermint, and established turf. Split applications 7 to 14 days apart are more consistently effective than single applications.
   C. Triazine herbicides are somewhat inconsistent for control of yellow nutsedge; however, early postemergence applications of atrazine with oil-based adjuvants help provide nutsedge suppression in corn. Control of yellow nutsedge is enhanced when row cultivation is used following the herbicide treatments.
   D. Several branched-chain amino acid inhibitor herbicides suppress or control yellow nutsedge. Herbicides in the sulfonylurea group include halosulfuron, which is used for corn, sorghum, and turf. Halosulfuron may be the most effective of the sulfonylurea herbicides listed. Others include bensulfuron, chlorimuron, primisulfuron, rimsulfuron, and sulfometuron. Imazethapyr in the imidazolinone group and pyrithiobac in the pyrimidinyl oxybenzoate group also provide suppression.
   E. Postemergence applications of the organic arsenicals (MSMA and DSMA) are used for control of yellow nutsedge in turfgrasses. 2,4-D is sometimes combined with these herbicides on turf. Two or more applications will suppress the nutsedge long enough to allow the turf to compete and gain the advantage in the infested area. The organic arsenicals are used as directed postemergence treatments in cotton as well. Although 2,4-D alone historically has been used for nutsedge control, it tends to be erratic and nearly always requires two or more sequential applications.
III. Suppression in noncropping situations
   A. The organic arsenicals, glyphosate, glufosinate, norflurazon, sulfometuron, bromacil, and terbacil, have potential for nutsedge control in noncrop areas.

WILD GARLIC

Biology

Wild garlic (*Allium vineale* [L.]) is a bulb-forming weed in the lily family (Liliaceae). Its weedy effects differ from those of the other perennials described in that it has more of an impact on crop quality than on crop yield. Its objectional pungent odor contaminates small grains and appears in the milk of grazing livestock. Wild garlic is a cool-season plant, so it is especially troublesome in fall-sown small grains and cool-season pastures. Wild garlic is distributed widely throughout the United States but is most often a problem in areas of the Midwest and East where soft red winter wheat is grown.

The biology of this weed has been described in detail by Peters et al. (l965). The plants grow in clumps that originate from underground bulbs (Figure 13.9). The leaves are hollow and round in cross section, in contrast to those of wild onion, which are solid and flat in cross section. Wild onion (*Allium canadense* [L.]) is easier to control than wild garlic because it does not form dormant hardshell bulbs. Stands of wild onion can be eliminated if the plants are killed before they reproduce.

Seed production in wild garlic occurs only as far north as Virginia, Delaware, and Tennessee. The flowers are borne in umbels at the apex of the plant, and the seeds usually are viable. In the northern states, reproduction is entirely vegetative.

A wild garlic population consists of two types of plants (Figure 13.10). One type is the scapigerous plant, so called because the stem elongates (a leafless stem is called a scape) and terminates in a head that may contain as many as 300 aerial bulblets (or flowers when produced) (Figure 13.11). Aerial bulblets are the major form of spread of the weed. They are harvested along with the wheat and are difficult to separate from the grain, which is similar in size. The underground system of a scapigerous plant consists of one large soft offset bulb and one to six smaller hardshell bulbs (Figure 13.12). Each of these bulbs can form a new plant.

The nonscapigerous plants do not form an elongated stem or scape but consist of leaves that emerge from the base of a large central bulb. These plants also can form one or two underground hardshell bulbs in the axils of the lower leaf bases (Figures 13.10, 13.12).

The factor that determines whether the plant will be scapigerous or nonscapigerous is the amount of food reserves in the bulb. Large bulbs have a greater potential to produce the scapigerous plant and aerial bulblets. The order of propagule size from smallest to largest (and the proportion that forms scapigerous plants) is as follows: seeds (0: all form nonscapigerous plants), aerial bulblets (one-third), hardshell bulbs (one-third), central bulbs (two-thirds), and soft offset bulbs (100%).

Approximately 20 to 30% of the hardshell bulbs produced by either scapigerous or nonscapigerous plants sprout the year after formation. The remainder break dormancy gradually and can remain viable in the soil to provide a continuing source of reinfestation for as long as 5 yr.

Bulbs and seeds initiate sprouting to produce plants in late August, and emergence continues through September and October. The plant continues growth during the winter and spring as long as temperatures are sufficiently warm. Underground bulb formation occurs in February and March and is mostly completed by April, whereas aerial bulblets and seeds are not formed until May and June. The shoots die back in June. The cycle is completed when sprouts reappear in the fall from seeds and bulbs.

Response to Cultural and Mechanical Methods

Competitive Crops

The management of a competitive crop to provide maximum shading can enhance other control efforts. A good stand of winter wheat is required for spring treatments of herbicides to be effective in controlling aerial bulblets. Thin stands of nitrogen-deficient wheat plus a herbicide treatment probably will *not* give adequate control.

Tillage

Wild garlic flourishes when no tillage is practiced from October through March (the period of active growth and bulb formation). Late fall, winter, or early spring tillage will retard plant establishment and underground bulb formation. Tillage in April and May is too late for disrupting underground bulb formation but can prevent the formation of aerial bulblets.

Figure 13.9 Clump of wild garlic. (E. J. Peters, USDA/ARS, University of Missouri)

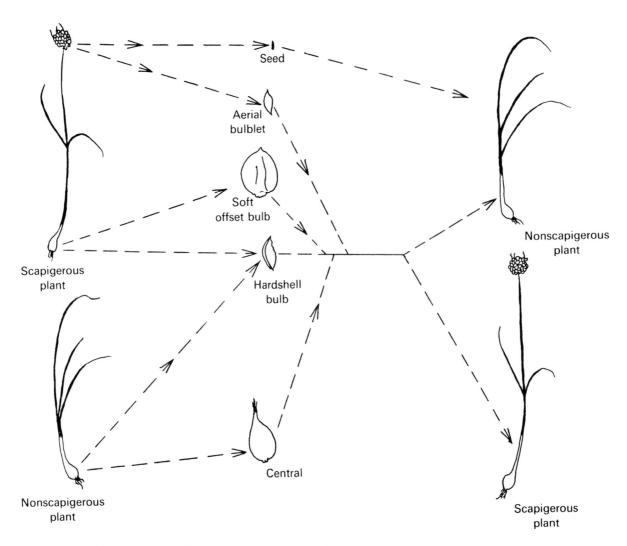

Figure 13.10 Types of bulbs formed by scapigerous and nonscapigerous plants of wild garlic. (E. J. Peters, USDA/ARS, University of Missouri)

Dormancy periods of up to 5 yr for hardshell bulbs and the resistance of bulbs to desiccation make control by tillage difficult.

Mowing

Several close mowings in the spring and early summer have little impact on wild garlic other than the possible reduction of aerial bulblets.

Systems for Control

I. Branched-chain amino acid inhibitors
   A. Herbicides claiming control: Prosulfuron and the mixture of thifensulfuron and tribenuron offer the most reliable control of wild garlic in small grains. Prosulfuron is labeled for grain sorghum, wheat, barley, rye, oats, triticale, and proso millet. Thifensulfuron and tribenuron are labeled for wheat, barley, and oats.

   B. Herbicides claiming suppression: Triasulfuron is labeled for wheat, barley, pastures, rangeland, and conservation reserves; metsulfuron for wheat, barley, pastures, and rangeland; chlorsulfuron (Lesco's TFC Despersible Granule) for Kentucky bluegrass and bentgrass; and imazaquin for warm-season turfgrasses.

II. Auxin-type growth regulator herbicides
   A. High doses for substantial reduction of established plants with one or more applications.
      1. Apply 2,4-D (2 to 4 lb a.e./acre) or dicamba (1 to 2 lb a.e./acre) after and between annual grass crops (corn or sorghum) and on established perennial pasture and turfgrasses. Applications in November or March will substantially reduce underground bulb formation and kill plants emerged from the soil. Care

Figure 13.11 Aerial bulblets formed on scapigerous plants of wild garlic. (E. J. Peters, USDA/ARS, University of Missouri)

should be taken with these high doses to avoid crop injury. Two seasons of this treatment should provide substantial reduction of the stand.

B. Suppression of aerial bulblet formation in wheat (control often is inadequate for

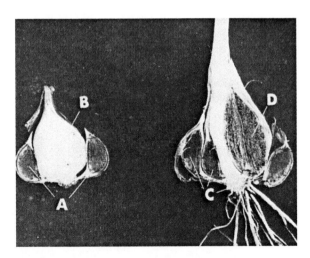

Figure 13.12 Left: Longitudinal section of the base of a mature nonscapigerous plant showing (A) hard-shell bulbs and (B) the central bulb. Right: Longitudinal section of the base of a mature scapigerous plant showing the positions of (C) hardshell bulbs and (D) a soft offset bulb. (E. J. Peters, USDA/ARS, University of Missouri)

inhibiting aerial bulblets enough to prevent dockage at the market place)

1. 2,4-D at 1/2 to 1 lb a.e./acre applied at the tillering to boot stages in fall-planted wheat.

2. 2,4-D at 1/2 to 1 lb a.e./acre and dicamba at 1/8 lb a.e./acre applied at the tillering to joint stage in fall-planted wheat. Vigorously growing wheat is needed to ensure success. Some crop injury can occur with both (1) and (2) treatments. Substantial control prior to the establishment of the wheat may be needed to provide acceptable reduction of aerial bulblets.

C. Control of wild garlic in established turf

1. 2,4-D at 1 to 2 lb a.e./acre or 2,4-D at 1 1/2 lb a.e. plus dicamba at 1/2 lb a.e./acre applied when garlic is actively growing and 4 to 8 in. tall.

D. Control of wild garlic in perennial grass pastures and other perennial grass sods

1. 2,4-D at 2 lb a.e./acre or dicamba at 2 lb a.e./acre; or 2,4-D at 1 to 2 lb a.e. and dicamba at 1/2 lb a.e./acre applied on actively growing garlic that is 4 to 8 in. tall.

CONTROLLING CREEPING PERENNIAL BROADLEAVED WEEDS

The majority of creeping perennial broadleaved weeds have creeping roots (e.g., Canada thistle, hedge bindweed, field bindweed, horsenettle, honeyvine milkweed, common milkweed, hemp dogbane, leafy spurge, smooth groundcherry, trumpetcreeper, kudzu, and poison ivy). Deep penetration into the soil and the presence of dormant buds make control of established plants with creeping roots particularly difficult.

Not all creeping perennial broadleaved weeds have creeping roots. Jerusalem artichoke, swamp smartweed, and goldenrod have rhizomes. Jerusalem artichoke also has tubers. Ground ivy has stolons.

The lateral growth of creeping roots and rhizomes from the mother plant results in the dense circular patches that characterize infestations of creeping perennial broadleaves.

As with the creeping perennial monocots, a realistic approach for dealing with creeping broadleaves must recognize the survival capabilities of these weeds. Control programs must be planned with multifaceted, multiyear components that include appropriate follow-up to prevent reinfestation. Successful programs for established stands must result in depletion of stored foods and destruction of underground buds. Breaking apart the plants followed by drying, repeated top removal, shading by crop plants, and inhibition of growth processes with herbicides are the most common approaches for control of these weeds. These objectives are achieved through tillage, improved crop management, mowing, and herbicide applications. A program that incorporates a sequence of herbicide treatments, timely tillage, and crop shading is more apt to provide control than a more limited approach.

Herbicides

Currently more herbicide options exist for controlling or suppressing creeping perennial broadleaves in grass crops than in broadleaved crops. Table 13.6 lists the herbicides labeled for control or suppression of perennial broadleaved weeds. This list is divided into herbicides used for weeds that are herbaceous creeping perennials, herbaceous simple perennials, and woody species. Most of the selective herbicides are used postemergence.

Foliar-applied herbicides that do not translocate can be used, but only top kill can be expected with these compounds. When applied repeatedly or used in conjunction with a fast closing crop canopy, however, the target perennial can be held in check effectively for most or all of a growing season. A single application by itself will result in a two- or three-week delay in development of the target plant.

Ideally, the best option is to find a symplastically translocated herbicide that moves extensively in the weed and is selective for the crop. Foliar-applied herbicide groups that injure rapidly growing and active meristems have the potential to provide this kind of control. Aromatic amino acid inhibitors (glyphosate), auxin-type growth regulators (e.g., 2,4-D, dicamba), branched-chain amino acid inhibitors (e.g., metsulfuron, primisulfuron), and carotenoid pigment inhibitors (amitrole) all contain one or more herbicides that translocate symplastically in some creeping perennial broadleaves.

Unfortunately, good control with symplastically translocated herbicides is less frequently achieved on perennial broadleaved weeds than on perennial grasses. The herbicide doses often used selectively in crops are not high enough to provide good kill of the underground system; in some cases, translocation may be reduced even though the herbicide penetrates into the tissue. Shoots may be killed, but damage to the underground system is limited. Repeated herbicide applications help solve this problem.

Highly persistent soil-applied herbicides provide continued control pressure, but they are mostly limited for total vegetation control, rights-of-way, permanent pastures, and rangeland. These herbicides come from any mode-of-action group that has compounds providing shoot kill, good soil activity, and adequate mobility in the soil and plant to be taken up and moved to the site of action.

CANADA THISTLE

Biology

Canada thistle (*Cirsium arvense* [L.] Scop.) is a member of the composite (Asteraceae) family. It reproduces by seed and by an extensive system of horizontal and vertical creeping roots. Aerial shoots develop from the horizontal roots; fragments of either root type can give rise to new plants. Canada thistle is a major weed problem in Canada and in the northern half of the United States.

Mature plants are 2 to 4 ft tall (Figure 13.13). The leaves are 4 to 8 in. long with irregularly lobed crinkled edges and spiny-toothed margins. The base of the leaves surrounds the stem, giving the impression that the stem also is

Table 13.6 Partial List of Herbicides Labeled for Perennial Broadleaved Weeds (Herbaceous Creeping Perennials, Herbaceous Simple Perennials, and Woody Species)[a][b]

Herbicide	Herbaceous Creeping Perennials											Herbaceous Simple Perennials					Woody Species							
	Canada thistle	Common milkweed	Field bindweed	Ground ivy	Hemp dogbane	Honeyvine milkweed	Horsenettle	Jerusalem artichoke	Leafy spurge	Russian knapweed	Kudzu	Buckhorn plantain	Chicory	Curly dock	Dandelion	Pokeweed	Brambles	Elm	Juniper	Multiflora rose	Poison ivy	Trumpet creeper	Willow	Honeysuckle
No Translocation (Shoot Removal Only)																								
Acifluorfen		X	X			X																		
Bentazon	X		X																					
Fomesafen			X			X																X		
Lactofen	X	X				X																X		
Paraquat	Individual species are not listed; shoot kill only on most species.																							
Limited Symplastic Translocation (Shoot Removal and Occasional Damage to Underground System)																								
Fosamine		X						X									X	X		X			X	
Glufosinate	X		X	X				X				X		X	X		X		X	X	X	X		X
Imazapyr	X	X	X					X	X	X				X	X	X	X		X	X	X	X		
Symplastic Translocation (Shoot Removal to Substantial Damage to Underground System)																								
Amitrole	X	X					X	X		X		X							X					X
Clopyralid	X							X		X			X	X										
Chlorimuron	X							X																
2,4-D	X		X	X	X	X		X				X	X	X	X							X		
Dicamba	X	X	X		X	X	X	X	X	X	X	X	X	X	X	X	X	X	X	X	X	X	X	X
Glyphosate	X	X	X		X		X	X	X	X	X			X	X		X	X		X	X	X	X	X
MCPA	X											X		X	X									
MCPP (mecoprop)			X	X								X	X	X	X									
Metsulfuron	X	X								X		X	X	X	X		X	X		X				X
Primisulfuron	X					X	X																	
Prosulfuron	X	X																						
Triasulfuron	X													X										
Tribenuron	X																							
Triclopyr	X											X	X	X	X		X	X	X	X	X	X		
Persistent to Highly Persistent (Substantial Damage to Shoots and to Underground Systems)																								
Bromacil					X							X		X			X						X	
Hexazinone	X	X	X									X	X	X			X	X	X		X		X	X
Prometon			X									X												
Tebuthiuron											X						X	X		X	X	X		
Chlorsulfuron	X								X					X	X									
Imazapyr	X	X	X			X			X	X	X	X	X	X	X		X			X	X	X	X	X
Picloram	X	X	X				X		X	X		X					X		X	X	X	X		
Sulfomenturon	X									X			X	X			X							
Barrier or Surface																								
Dichlobenil	X											X	X	X	X									

[a]X = labeled for weed.
[b]Dosages for perennial broadleaved weeds generally are at the high end of the labelled dose range. These dosages frequently are higher than those selective in annual crops. Dosages used selectively in annual crops frequently provide only shoot kill.

Figure 13.13 Canada thistle plants in a wheat field.

spiny. The flowers are borne in heads and range in color from purple to rose to white. Distinctly different biotypes of Canada thistle have been collected from different parts of the United States. Substantial differences in leaf shape, degree of spininess, flower color, plant height, and response to herbicides have been recorded. Several biotypes may be present within the same field (Figure 13.14).

Individual patches tend to have either male or female flowers. Casual inspection of a patch with male flowers may lead to the incorrect conclusion that seed is not produced. The number of seeds produced per plant varies, but as many as 5000 have been recorded with viabilities as high as 95% (Hay 1937). The seeds are plumed and are dispersed readily by wind. They can survive burial in the soil for at least 20 yr, so continuing reestablishment from seed should be expected. Canada thistle seedlings start slowly and compete poorly, particularly if they are shaded, so seedling establishment occurs primarily in open areas. Seedling plants can reproduce vegetatively when they are 7 to 9 wk old.

The spread of horizontal creeping roots (Figure 13.15) from the parent plant results in the dense round patches of shoots typically encountered in the field. Roots from an established plant can spread over a circular area 10 to 20 ft in diameter in 1 yr. The amount of root growth from a single plant was demonstrated by growing Canada thistle in a box 2 ft × 4 ft × 8 ft filled with field soil. One thistle plant with 1 ft of root was planted in West Lafayette, Indiana, on May 11, 1994 and allowed to grow until June 17, 1995. At this time, the box was taken apart and the soil was washed from the underground portion of the plant (Figure 13.16). Roots had reached all portions of the box and totalled 450 ft in length. There also were 150 ft of underground stems and 71 shoots aboveground. Simultaneous demonstrations in Colorado and North Dakota resulted in even more growth than this.

The vertical roots can extend to depths of 10 or more ft if not restricted by water tables or impervious soil layers. Although roots are more numerous near the soil surface, over one-half are located below the plow layer. In a study cited by Hoefer (1981), 40% of the root weight was located below the 2-ft depth. These deep roots escape disturbance from tillage operations and serve as important sources of reinfestation.

Shoots regenerating from creeping roots emerge from the soil from March through May, depending on climatic conditions. Flower buds

Figure 13.14 Canada thistle biotypes from the same field.

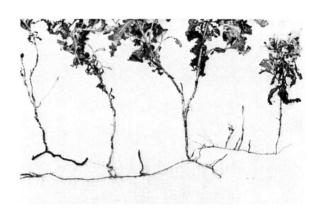

Figure 13.15 Canada thistle shoots originating from horizontal creeping roots.

Figure 13.16 Creeping root system of Canada thistle being washed free of soil.

appear 4 to 6 wk after emergence. In the southern part of its range, bud stage occurs in late May and early June, with full bloom occurring in late June. Farther north, the bud and flower stage can occur in July and August.

Following seed maturation (early July in the southern part of the range), the flowering stalks become inactive. A flush of new shoots emerges in September and growth continues until the tops are killed by freezing temperatures. Canada thistle takes advantage of most of the growing season, since it can tolerate frost and continues to grow later into the fall than most broadleaved species. In the southern part of its present range, Canada thistle can grow into November, some 4 to 8 wk after maturing crops have ceased to provide shading. This growth period can provide major replenishment of depleted underground food reserves and probably contributes significantly to its survival.

Response to Cultural and Mechanical Methods

Competitive Crops

A competitive crop will help keep Canada thistle in check as long as the management system permits the crop to get ahead and stay ahead of the thistle. Three or four years of vigorously growing alfalfa can control Canada thistle. Warm-season annual crops such as corn or solid-seeded soybeans, when coupled with herbicide treatments, can provide suppression until harvest.

Tillage

Allowing 1 to 2 yr of clean fallow with tillage provides substantial reduction of established stands. Control is enhanced by tillage operations that cut off the tops of the plants and tear apart the roots.

Mowing

Mowing associated with the harvest of the hay crop in well-managed alfalfa can reduce Canada thistle stands by 90% in 3 yr. Repeated mowing can limit seed production if properly timed (bud stage). Substantial control can be achieved when mowing is combined with sod or hay crops.

Systems for Control

A partial list of herbicides labeled for control of Canada thistle is given in Table 13.6.

I. Annual crops
   A. Selective control in cropping situations that provides maximum damage to thistle
      1. Glyphosate in glyphosate-resistant crop cultivars applied to thistle approaching the flower bud stage. Repeat treatments if thistle recovers. This program should substantially reduce established stands in one or two seasons.
      2. Clopyralid in grass crops applied at high doses to thistle approaching the flower bud stage. Repeat treatments are used each time the thistle reaches the late rosette stage.
      3. Chlorsulfuron in wheat when chemical residual is not a problem to a subsequent crop.
   B. Selective control in cropping situations that provides shoot control during the cropping season *plus* some damage to the underground root system
      1. This objective can be accomplished with a number of herbicides including auxin-type growth regulators (2,4-D, MCPA, dicamba, triclopyr, and low doses of clopyralid) and sulfonylureas (chlorimuron, metsulfuron, primisulfuron, prosulfuron, triasulfuron, and tribenuron).
   C. Selective control in cropping situations that provides only shoot control during the cropping season with *no* damage to the underground root system
      1. Shoot damage plus a rapidly closing crop canopy can keep Canada thistle in check until the summer annual crop matures. Two herbicide applications may be needed to provide good control until harvest. Herbicides that have the potential to do this include the Protox inhibitor

herbicides (acifluorfen, lactofen, and fomesafen), a rapidly acting photosynthetic inhibitor (bentazon), and glufosinate in conjunction with glufosinate-resistant crop cultivars. This level of control sometimes is all that is gained from treatments listed in B.
   D. Between crop applications. In any of the following practices, an appropriate follow-up treatment should be used to prevent recovery of the thistle if thistle emerges in the crop during the growing season.
      1. Glyphosate (for most crops) at doses suggested on the label. Apply to vigorous fall regrowth and follow with a crop that provides good shade.
      2. Glyphosate (for most crops) at doses suggested on the label. Apply to vigorous spring growth and follow with a late-planted crop that provides good shade.
      3. Clopyralid or dicamba at high doses on vigorous fall regrowth followed by grass crops that provide good shade the next season.
II. Perennial crops
   A. Site preparation (or renovation) to remove Canada thistle prior to planting a perennial crop
      1. Two or more glyphosate or clopyralid applications at 3- to 6-mo intervals to kill established Canada thistle a season or more ahead of crop establishment. Glyphosate provides little or no selectivity in an established crop but gives optimal flexibility regarding the interval between herbicide application and planting of the crop to be established. In contrast, clopyralid is selective in grass crops but restricts choice and planting interval of the next crop to be established.
   B. Directed postemergence applications of glyphosate over Canada thistle in the early bud stage and under leaves of woody species with mature bark (orchards, nurseries, etc.). Consider spring and fall applications.
   C. Spring or fall applications of clopyralid on established grass sods. Two or more applications spread over 2 or more yr will likely be needed.
III. Noncropping situations
   A. Substantial reduction without herbicide persistence

      1. Glyphosate in two applications, one in the spring and one in the fall when the plants have reached bud stage or 12 in. in height. This chemical does not leave a residual in the soil.
   B. Highly persistent herbicides
      1. Picloram (check label for approved regions). Chemical residue will persist into subsequent seasons; however, perennial grass cover crops frequently can be maintained.
      2. Herbicides for general vegetation control (e.g., chlorsulfuron, imazapyr, and bromacil). Chemical residue will persist into subsequent seasons.
IV. Prevention of seed production
   A. Any herbicide capable of shoot removal and approved for the situation, e.g., one application of an auxin-type growth regulator (2,4-D, dicamba) or nearly any other herbicide mentioned previously. These steps are the least that can be done to prevent spread by seed.

FIELD BINDWEED

Biology

Field bindweed (*Convolvulus arvensis* [L.]) is a member of the morningglory family (Convolvulaceae) and spreads by seeds and creeping roots. Although widely distributed, it is a particularly serious problem in the western half of the United States. Its deep, extensive root system makes it an excellent competitor in dry climates.

The plant produces twining or climbing stems that may be as long as 5 to 9 ft (Figure 13.17). The alternately arranged leaves vary considerably in shape and size but generally are arrowhead-shaped with rounded or blunt tips and spreading, downward-pointed basal lobes.

Figure 13.17 Field bindweed plants.

The flowers are about 1 to 1 1/2 in. long and 1 in. in diameter and vary in color from white to pink.

The flowers last only 1 day, but flower production is continuous from early summer through the growing season. Seeds are an important mechanism for both spread and long-term survival. Heavy seed production has been recorded in the Great Plains states, where the warm, dry weather favors pollination. Field bindweed was reported to produce more than a half million seeds per acre in a wheat field and 4.5 million seeds per acre in road ditches in western Texas (Wiese and Phillips 1976). Buried seeds remain viable for decades.

The plant grows vegetatively from spring through light frosts in the fall. Abundant and continuous shoot development occurs from buds on the creeping roots, and a single plant can spread to form a patch 20 ft in diameter in a season (Figure 13.18). The roots contain sufficient stored materials to support plants for long periods even when top growth is removed or destroyed repeatedly. Part of the tremendous survival capability of this plant is due to the penetration of its roots deep into the ground where they can obtain sufficient moisture and are protected from tillage operations. Penetration to 8 ft is common, but penetration to depths of 20 and 30 ft has been reported (Phillips 1978).

Response to Cultural and Mechanical Methods

Competitive Crops

Competitive crops combined or sequenced with tillage or herbicides help to suppress field bindweed. In dryland fields of the Great Plains, bindweed can be controlled with a fallow-wheat-fallow rotation using sweep tillage and 2,4-D applications. Established perennial grasses

Figure 13.18 Field bindweed clumps in an unplowed field in California.

treated repeatedly with 2,4-D can provide effective control after several seasons.

Tillage

Tillage resulting in repeated top removal gradually will deplete the plant of underground food and bud reserves. A single tillage operation delays field bindweed shoot development about 3 wk. Cultivation at about 3-wk intervals over two consecutive seasons has reduced established perennial stands substantially. Conventional tillage for annual crop production helps delay development and reduces the competitive effects of field bindweed.

Mowing

Mowing alone has limited utility because field bindweed tends to lie on the soil surface. However, mowing associated with the harvest of the hay crop in well-managed alfalfa can reduce field bindweed stands by 90% in 4 yr.

Systems for Control

The extensive root system and long-lived seeds make control of well-established stands of field bindweed difficult. Control methods available for cropping situations basically provide top removal or top removal and partial damage to underground structures but do not often kill the major portions of these systems. Field bindweed is a tough, persistent, and extremely difficult-to-kill perennial weed.

Field bindweed is more difficult to control as the climate becomes more arid. Two factors that contribute to this phenomenon are the more extensive root systems produced in dry climates and the fact that plants under moisture stress and low humidity are less responsive to herbicide applications. Another factor to consider in controlling field bindweed is that biotypes can vary considerably in selectivity to herbicides (DeGennaro and Weller 1984), so repeated applications of a chemical remove susceptible plants and leave tolerant ones. Tillage or altering herbicides in the control sequence can help to minimize the selection for more tolerant clones.

A partial list of herbicides labeled for field bindweed control is given in Table 13.6.

I. Herbicides for cropping situations
    Programs for controlling field bindweed in crops should include a competitive crop, one or more herbicide applications per year, and where appropriate, tillage. Such programs can permit nearly optimum crop return and reduce field bindweed simultaneously when used over a period of several

years. Herbicides that result in substantial kill of well-established field bindweed in a single treatment are not available for cropping situations.

    A. Herbicides with potential for movement into the root system and available for selective use in crops include 2,4-D, dicamba, glyphosate (in glyphosate-resistant crops), and prosulfuron.

    B. In grass crops, 2,4-D and dicamba should be used as labeled for the crop being treated. One herbicide treatment should be more effective than a single tillage operation if applied when the field bindweed shoots are at least 6 to 10 in. long, are initiating flower buds, and have good growing conditions. More than one application per season will be needed for maximum control.

    C. Glyphosate is less effective on field bindweed than it is on perennial grasses or Canada thistle.

        1. For most crops, glyphosate can be used only as a directed postemergence (tree crops) treatment or between crops. Glyphosate applications usually need to be repeated.

        2. For glyphosate-resistant crops, one or more applications of glyphosate per season may be applied selectively in the crop itself.

    D. Selective herbicides that provide only top kill include acifluorfen, bentazon, fomesafen, and glufosinate (in glufosinate-resistant crops).

    E. Subsurface layering of dinitroaniline herbicides currently is cleared for use in certain crops such as cotton and grapes on a regional and state basis.

  II. Herbicides for noncrop situations

    A. Highly persistent herbicides labeled for the control of field bindweed include picloram, imazapyr, and hexazinone. These herbicides provide season-long pressure on field bindweed and should result in effective stand reduction.

    B. Short-lived herbicides such as glyphosate, 2,4-D, and dicamba can be used where highly persistent herbicides are undesirable.

MULTIFLORA ROSE

Biology

Multiflora rose (*Rosa multiflora* Thumb.) is a brushy species that reproduces primarily by seeds rather than from vegetative reproductive structures. It is a major weed problem in pastures, rangelands, and fencerows. A native of Japan, the plant was once promoted and widely planted throughout the United States as a "living fence," as a wildlife cover, and as an erosion control measure. It rapidly spread into pastures and other uncultivated areas, drastically reducing livestock productivity in these areas. Not only does it compete with desirable forages, but in continuous hedges it can prevent the movement of livestock.

The plant is a thorny bush 6 to 10 ft tall with long arching stems (Figure 13.19). The leaves have prominent, fringed stipules and are pinnately compound with five to eleven (usually nine) leaflets. The leaflets are 1 to 1 1/2 in. long. The flowers are white or pinkish, numerous, and fragrant. They measure 1/2 to 3/4 in. across. Flowering usually occurs in May or June. The red fruits are globular to ovoid and are about 1/4 in. in diameter. The seeds provide an excellent means of spread because they are readily eaten by birds and pass intact through the digestive tract.

A virus disease, rose rosette disease or RRD, is lethal to multiflora rose in 2 to 5 yr after symptoms are first observed (Epstein et al. 1997). The symptoms include yellowing of the leaves, reddening of the veins, and reduced petiole lengths, which results in a brushy or "witches broom" effect. Although considered a potential biological control, the natural spread of the disease has caused major damage to multiflora rose stands in the midwestern and southeastern United States.

Response to Cultural and Mechanical Methods

Competitive Crops

Invasion of most woody species, including multiflora rose, into pasture and rangeland situ-

Figure 13.19 Multiflora rose in a pasture.

ations is aided by overgrazing or other mismanagement of the forage species present.

Tillage

Multiflora rose cannot survive tillage and is
not a problem in crops that are tilled annually. A
bulldozer may be the most practical means of removing large well-established bushes. Some resprouting can occur from crowns, and seeds will
germinate in the disturbed soil.

Mowing

Mowing keeps the plants shortened but does
not kill them. Mowing several times a year discourages seedlings from becoming established.
Mowing may be impossible in the rough or
woody terrain where the plant often is found.

Systems for Control

A partial list of herbicides labeled for multiflora
rose control is given in Table 13.6.

I. Control of seedlings
   A. Broadcast applications of 2,4-D in pas-
      tures and rangeland will kill small
      seedlings and prevent establishment.
II. Control of established plants with spot
    treatments
   A. Permanent pasture and other perennial
      grass sods
      1. Picloram as a foliar spray is effective
         for selective control in pastures. Estab-
         lished grasses are tolerant, but grass
         seedlings may be suppressed or killed
         for up to 2 yr after application. The
         herbicide should be limited to perma-
         nent pastures. It should not be used
         on fields where crops may be grown
         in the future. Care must be taken not
         to treat areas where the herbicide will
         move into streams or ponds or into
         the vicinity of desirable tree species
         that might be killed or injured.
      2. Dicamba, 2,4-D, and metsulfuron
         also can be used selectively in pasture
         situations. These compounds kill or
         injure most broadleaved plants pre-
         sent but should not kill perennial
         grasses. Application should be made
         between full leaf expansion to fall
         color. Two or more applications may
         be needed for complete control.
      3. Glyphosate can be used nonselec-
         tively in grass pastures. It should be
         applied to fully expanded leaves.

4. Tebuthiuron can be used nonselec-
   tively as a soil application. Care
   should be taken to minimize damage
   to pasture grasses.
B. Noncropland
   1. Fosamine is used as a late summer or
      fall spray treatment prior to the turn
      of leaf colors. The compound is ab-
      sorbed by the foliage, stem, and buds.
      The effects do not become apparent
      until the following spring, when the
      plant fails to refoliate. Good coverage
      of the plant plus the addition of a
      surfactant is essential for optimum
      performance.
   2. Glyphosate can be applied when the
      plant is actively growing; that is, in
      the spring, from early to full bloom
      stage, and in the fall before killing
      frosts. Good coverage is required and
      repeat treatments may be necessary.
   3. Picloram and dicamba also can be
      used in noncrop situations where sod
      cover of perennial grass is to be
      maintained.
   4. Hexazinone, imazapyr, and tebuthi-
      uron can be used where bare soil can
      be tolerated.

MAJOR CONCEPTS IN THIS CHAPTER

Herbaceous weeds can be grouped by life cycle as
summer annuals, winter annuals, biennials,
simple perennials, and creeping perennials.

Life cycle determines the adaptability of weeds
to management systems and susceptibility of
weeds to individual control practices.

Annual weeds complete the life cycle in less
than one growing season. Any practice that
completely removes the shoot or dislodges
the entire plant should result in complete kill.
Later flushes of germinating plants typically
result in reinfestation; thus, control practices
must be provided over most of the season.

Annual grasses are more difficult to control with
flaming or foliar treatments of contact or
apoplastically translocated herbicides than
broadleaved plants are because of their protected growing points.

Summer annuals compete most in spring-
planted crops before canopy closure or in
perennial crops when loss of stand occurs.
Winter annuals compete in fall- or early-
spring-seeded crops or in perennial crops that
go dormant in winter.

Biennial weeds form a rosette the first year and send up a flowering stalk the second year. They do not persist in management systems that are severely disturbed seasonally. Repeated close mowings discourages most biennials. Biennials are most susceptible to herbicides when in the rosette stage.

Simple perennials survive best in perennial crops or undisturbed areas. Those with leaves near the ground can survive mowing, whereas taller growing ones cannot. Simple perennials that generate shoots from crowns under the soil surface tolerate single treatments of shallow tillage, hoeing, and contact herbicides as well as mowing.

Creeping perennials reproduce by seeds and vegetative structures. They are capable of surviving in nearly all management systems. Plants from vegetative buds provide the major competition; the seeds serve in spread and reestablishment (seedlings of johnsongrass are an exception in that they are competitive the first year).

Vegetative structures provide numerous survival and competitive advantages to a creeping perennial.

Plants that develop from seed (including creeping perennials) are most susceptible to herbicides during the early stages of development (germination and seedling). Resistance to control increases with age. Plants from seed become most resistant during and after flowering.

Inactive seeds and dormant buds generally are not killed by herbicides. Germinating seeds and actively developing buds are vulnerable.

Creeping perennials with established underground vegetative structures must be permitted to expend energy in shoot growth and develop adequate top growth for carbohydrate export underground before they become vulnerable to herbicide treatments. This susceptible period commences at early bud stage (when the perennial achieves about one fourth of its potential top growth) and continues until flowering.

Most creeping perennials become more resistant to control after flowering. Morningglories, which flower more or less continuously, are exceptions and should be treated during flowering.

Creeping perennials can be reduced substantially by repeatedly removing the tops and depleting underground food reserves and vegetative buds.

For many creeping perennials, fall regrowth presents an opportunity for effective chemical application.

The increase in perennial weeds has been favored by fewer crops in rotation, trends toward less tillage, programs of control aimed at annual weeds, earlier planting dates, reduced mowing and spraying of noncrop areas, less attention to new infestations, and failure to exploit preventive control methods.

A realistic approach for dealing with creeping perennial weeds should include recognition of the survival capabilities of the weed, the ability to predict the probable degree of control for each practice, the required management flexibility to utilize available control practices properly, and the ability to anticipate and prevent the introduction of new weeds. Control programs should be planned as multifaceted multiyear programs and include appropriate follow-up to prevent reinfestation.

Levels of control of creeping perennial underground systems need to be estimated by observing shoot development at intervals subsequent to treatments.

Tillage, crop shading, mowing, and various herbicides can contribute to the control of perennial weeds. An effective program may include several of these individual practices.

A program that utilizes a sequence of herbicide and tillage typically is more effective for controlling creeping perennial weeds than either used alone.

TERMS INTRODUCED IN THIS CHAPTER

Aerial bulblet
Basal bulb
Central bulb
Hardshell bulb
Nonscapigerous plant
Scapigerous plant
Soft offset bulb

SELECTED REFERENCES ON WEED LIFE CYCLES

Bhowmik, P. C. 1994. Biology and control of common milkweed (*Asclepias syriaca*). Rev. Weed Sci. 6:227–250.

DeGennaro, F. P. and S. C. Weller. 1984a. Differential susceptibility of field bindweed (*Convolvulus arvensis*) biotypes to glyphosate. Weed Sci. 32:472–476.

DeGennaro, F. P. and S. C. Weller. 1984b. Growth and reproductive characteristics of field bindweed (*Convolvulus arvensis*) biotypes. Weed Sci. 32:525–528.

Donald, W. W. 1990. Management and control of Canada thistle (*Cirsium arvense*). Rev. Weed Sci. 5:193–250.

Donald, W. W. 1994. The biology of Canada thistle (*Cirsium arvense*). Rev. Weed Sci. 6:77–101.

Elmore, C. D., H. R. Hurst, and D. F. Austin. 1990. Biology and control of morningglories (*Ipomoea* spp.). Rev. Weed Sci. 5:83–114.

Field Bindweed. 1978. Special Session on Field Bindweed. Proceedings of the North Central Weed Control Conf. 33:140–158.

Forcella, F. and J. M. Randall. 1994. Biology of bull thistle, *Cirsium vulgare* (Savi) Tenore. Rev. Weed Sci. 6:29–50.

Harvey, R. F. 1973. Quackgrass: friend or foe? Weeds Today 4(4):8–9.

Holm, L., J. Doll, E. Holm, J. Pancho, and J. Herberger. 1997. World Weeds, Natural Histories and Distribution. John Wiley & Sons, New York.

Holm, L. G., D. L. Plucknett, J. V. Pancho, and J. P. Herberger. 1977. The World's Worst Weeds, Distribution and Biology. Univ. Press of Hawaii, Honolulu.

Keeley, P. E. 1987. Interference and interaction of purple and yellow nutsedges (*Cyperus rotundus* and *C. esculentus*) with crops. Weed Technol. 1:74–81.

McWhorter, C. G. 1981. Johnsongrass . . . as a Weed. USDA/SEA, Farmer's Bulletin No. 1537.

McWhorter, C. G. 1989. History, biology, and control of Johnsongrass. Rev. Weed Sci. 4:85–121.

Mitich, L. W. 1987. Colonel Johnson's grass: Johnsongrass. Weed Technol. 1:112–113.

Peters, E. J. 1975. Wild garlic, a tough pest. Weeds Today 6(4):13–15.

Quackgrass. 1976. Special Session on Quackgrass, Proceedings of the North Central Weed Control Conf. 31:151–160.

Radosevich, S., J. Holt, and C. Ghersa. 1997. Weed Ecology, 2nd ed. John Wiley & Sons, New York.

Stoller, E. W. 1981. Yellow Nutsedge—A Menace in the Corn Belt. USDA/ARS, Tech. Bull. No. 1642.

Stoller, E. W. and R. D. Sweet. 1987. Biology and life cycle of purple and yellow nutsedges (*Cyperus rotundus* and *C. esculentus*). Weed Technol. 1:66–73.

Watson, A. K., Ed. 1985. Leafy spurge. Monograph Series of the Weed Science Society of America.

Wiese, A. F. and W. M. Phillips. 1976. Field bindweed. Weeds Today 7(1):22–23.

Wills, G. D. 1987. Description of purple and yellow nutsedge (*Cyperus rotundus* and *C. esculentus*). Weed Technol. 1: 2–9. See also articles published from a symposium on the nutsedges in this issue (Vol. 1, No. 1) of *Weed Technology*.

Also see the series on the "Intriguing World of Weeds," descriptions of the biology and management of many different weeds by Larry W. Mitich, originally published in *Weeds Today* and continued in *Weed Technology*.

WEED MANAGEMENT SITUATIONS

The purpose of Chapter 13 (Weed Life Cycles and Management), this chapter (Weed Management Situations), and Chapter 15 (Aquatic Plant Management) is to illustrate the application of previously discussed principles to practical situations. These chapters also illustrate the application of current (1998–1999) state-of-the art technology and provide a basis for solving weed control problems that will arise in the future.

Weed control situations in this chapter are arranged according to management system and crop life cycle. This arrangement minimizes repetition because the same weed control principles generally apply to crops under similar management systems. This arrangement also allows for comparisons of practices among various management systems. The major crop designations discussed are annual crops, perennial crops (herbaceous and woody), and bare soil sites. The term *crop* is used in the broadest sense to include any desired or managed vegetation.

GENERAL CONSIDERATIONS

The number of weed control practices available for annual crops, perennial crops, and bare soil sites varies (Table 14.1). Annual crops grown in rows with primary and secondary (conventional) tillage provide more weed control options than do cropping systems with limited or no tillage and systems with established perennial crops. The more restricted the number of operations in a man-agement system, the more fully the practices that remain (e.g., mowing, herbicides, hand-removal) must be utilized.

In all management systems, the contribution of crop competition always should be fully exploited, for it usually can be obtained at minimal cost (see also Chapter 4). Crop competition, however, cannot solve all weed control problems. Anytime the crop permits full sunlight to reach the soil, weeds have an opportunity to develop. Some weeds that develop simultaneously with the crop can be difficult to control. *The key to a successful weed control program is the judicious insertion into the crop management program of those control techniques that will minimize the impact of weeds not controlled by the competing crop.*

ANNUAL CROPS

Annual crops are designated as row (20 to 60 in. between rows) or solid-seeded (rows 14 in. or less) crops depending on whether the distance between rows is wide enough to allow between-row operations. Row width for an individual crop is determined by available equipment, custom, crop growth habit, weeds present, and harvest needs. Corn, cotton, and potatoes (a perennial plant that is grown as an annual crop) nearly always are grown in rows. Small grains almost always are solid seeded. Some crops such as soybeans are grown in both wide rows and solid-seeded stands.

256

Table 14.1 Weed Control Practices Potentially Available for Various Crop Management Systems

	Annual Crops					Perennial Crops								Bare Soil
	Conventional Till Rows	Conventional Till Solid	Stale Seedbed	No-Till Rows	No-Till Solid	Perennial Crops at Establishment Conventional Till	Perennial Crops at Establishment No-Till	Herbaceous Mowed Weekly	Herbaceous Mowed Two to Five Times	Herbaceous Not Mowed	Herbaceous Limited Tillage	Woody Plants in Rows	Woody Plants Random	Bare Soil
Control Practices for Weeds Established Prior to or at Planting														
Primary tillage	XX	XX	XX			XX								
Secondary tillage	XX	XX	X			XX						X	X	
Early preplant herbicides	XX	XX	XX	XX	XX	X	XX							
Burn-down herbicides		XX	XX	XX			XX							
Selective Control of Germinating Weeds Emerging with the Crop														
Preplant incorporated herbicides	XX	XX	XX			XX								
Preplant herbicides	XX	XX	XX	XX	XX	X	XX							
Preemergence herbicides	XX	XX	XX	XX	XX	XX	XX	X			X	XX	X	XX
Allelopathic crop residues		XX	XX	XX			XX							
Sheet mulches	XX						X					XX	XX	
Rotary hoes or light harrows	X	X									X			
Selective Control of Emerged or Late-Emerging Weeds in an Emerged Crop														
Crop shading	X	XX	XX	X	XX	X	X	XX	XX	XX	X	X	X	
Postemergence herbicides														
Early postemergence herbicides	XX	XX	XX	XX	XX	X	X							
Postemergence herbicides	XX	XX	XX	XX	XX	XX	XX	XX	XX	XX		X	X	
Directed foliar	XX		XX	XX								XX	XX	
Directed soil	XX		XX	XX								XX	XX	
Wipe-on or RCS	X	X	XX	X	X	X	X		X	X	X	X		X
Spot	XX	XX	XX	XX	XX	XX	XX	XX	XX	XX		XX	XX	XX
Preharvest mature crop herbicides	X	X		X	X	X			X					
Cultivation between rows	XX		X			X					X	X		
Hand removal														
Limited	X	X	X	X	X	X	X	X	X	X	X	X	X	X
Unlimited	X	X	X						X					X
Mowing														
Two to six times per year						X	X		XX		X	X		
Weekly								XX						
Long residual soil-applied herbicides											X			XX

XX = routinely available.
X = sometimes available.

The major advantage of growing crops in rows is the accessibility of the crop to passage of equipment during the growing season. Rows 30 in. or more apart offer easy access; those less than about 20 in. apart make it difficult or im- possible to get equipment through without trampling the crop. Weed control operations facilitated by rows include cultivation, flame cultivation, directed postemergence herbicides, semidirected postemergence herbicides, wipe-on herbicide applications, and high clearance sprayer applications for late-season postemergence treatments. These practices generally are not available for crops planted in narrow rows.

Agronomic and Horticultural Crops

Crops frequently are designated as either agronomic (grain or forage) crops or horticultural (vegetable, fruit, flower) crops, depending on their culture and eventual use. Horticultural crops tend to require more care than agronomic crops. They frequently are less competitive with weeds, are less easily harvested mechanically, and do not store well for extended periods of time without being processed (canned, dried, or frozen). They are grown on fewer acres than agronomic crops but return higher revenues per acre (Table 14.2). For example, the average return from the four crops that are grown on the most acreage in the United States (corn, wheat, soybeans, and hay, all of which are agronomic crops) ranges from $156 to $343 per acre. The return from four horticultural crops (broccoli, cucumbers, carrots, and strawberries), each of which is grown on fewer than 120,000 acres, ranges from $1349 to almost $16,000 per acre.

In general, fewer herbicides are available for horticultural crops than for agronomic crops. The potential liability associated with the high value (per acre) of horticultural crops, coupled with the limited acreage, results in high risks for companies interested in developing and marketing herbicides for these crops. Herbicides used on horticultural crops often were developed first for agronomic crops to ensure a reasonable market return. Factors such as the weeds controlled, application methods, and interactions with soil and plants, however, remain the same for a herbicide whether it is used in an agronomic or a horticultural crop.

Horticultural crops may require weed control inputs over much or all of the growing season because they tend to be shorter in height and grow more slowly than agronomic crops. Weeds not controlled provide more severe yield reductions in these types of crops than in more competitive ones. The requirement for hand labor has been reduced by machine harvest and herbicides, but it has not been eliminated completely. Thus, weeds that escape chemicals and tillage in horticultural crops frequently are removed by hand. The high value of horticultural crops often makes hand weeding a viable weed control option. In contrast, weeds that escape

Table 14.2 Acreage and Value of Some Agronomic and Horticultural Crops Grown in the United States in 1996[a]

Crop	Total Acres	Average Production/Acre	Average Price	Average Return/Acre
Corn (for grain)	79,487,000	127.1 bu	$ 2.70	$ 343
Wheat (all)	75,639,000	36.3 bu	4.30	156
Soybeans (for beans)	64,205,000	37.6 bu	6.85	258
Hay (all)	61,029,000	2.45 ton	93.00	228
Cotton	14,666,000	709 lb	0.70	496
Sorghum (for grain)	13,188,000	67.5 bu	2.35	159
Rice	2,819,000	61.2 cwt	9.50	581
Dry beans	1,813,000	15.9 cwt	24.20	385
Potatoes	1,455,000	349 cwt	5.11	1783
Sugar beets (1995)	1,444,600	19.8 ton	38.10	754
Peanuts	1,413,000	2619 lb	0.26	681
Sugarcane (1993)	948,300	32.8 ton	28.50	935
Tobacco	733,900	2133 lb	1.88	4010
Sweet corn (for processing)	492,000	6.95 ton	78.50	546
Tomatoes (for processing)	345,370	33.64 ton	63.50	2136
Tomatoes (fresh)	122,830	260 cwt	28.50	7410
Lettuce	280,250	304 cwt	16.70	5077
Green peas (for processing)	259,000	1.67 ton	284.00	474
Onions	169,630	383 cwt	9.58	3669
Broccoli	116,000	123 cwt	27.70	3407
Cucumbers (for processing)	111,340	5.44 tons	248.00	1349
Carrots (fresh)	96,320	291 cwt	12.80	3725
Strawberries	48,470	336 cwt	47.30	15,892

[a] Data from Statistical Highlights of U.S. Agriculture Crops, U.S. Department of Agriculture (http://www.usda.gov/nass/pubs/stathigh/1997/sthi96-c.htm).

tillage and herbicides in agronomic crops are likely to be left in the field.

Timing of Weed Control Practices to Fit Weed Emergence and Annual Crop Development

Weeds often appear during the etablishment and production of annual crops at three critical times: (1) prior to crop planting, (2) as the crop emerges, and (3) as the crop is growing.

The timing of weed control practices is focused on these three important periods of weed presence. Available practices, which depend on the crop, management system, and weeds present, are listed for each of these three periods (also see Table 14.1).

Control Practices for Weeds Established Prior to or at Planting

These weeds are eliminated prior to planting or crop emergence. The objective of these practices is to provide a weed-free environment for the emergence of the crop.

Primary and secondary tillage are used in management systems in which the whole area is tilled. For example, a moldboard plow or chisel plow provides primary tillage; this step is followed with a tandem disk or field cultivator and harrows to provide secondary tillage and seedbed preparation. The final secondary tillage operation usually is conducted a day or two ahead of planting or on the same day as planting.

Early preplant herbicides are applied to the soil several weeks ahead of planting and prior to weed emergence. They are used in management systems without secondary tillage at planting (e.g., no-till). The herbicides must have a long enough residual in the soil to provide control of germinating weed seedlings through crop emergence. This approach, therefore, involves the use of some of the persistent and longer-lived residual selective herbicides.

Burn-down herbicides are used in stale seedbed and in no-till and other conservation tillage systems to kill all emerged vegetation prior to the emergence of the crop. These herbicides are applied ahead of, at, or just after planting but before the crop emerges.

Selective Control of Germinating Weeds Emerging with the Crop

The objective of these practices is to remove the weeds that emerge at the same time as the crop. These practices must be selective in order to prevent crop injury.

Preplant incorporated herbicides are used to control germinating weeds in management systems that use secondary preplant tillage. These herbicides can be applied either before or with the secondary tillage operation. Nonvolatile herbicides can be applied a week or more before the tillage operation because they can remain on the soil surface without appreciable loss, whereas volatile herbicides must be incorporated immediately.

Preplant or preemergence herbicides applied to the soil surface control germinating weeds in any management system in which weeds have not germinated.

Allelopathic residues can result from cover crops that are killed before or at the time of planting. Cover crops and allelopathic chemicals can provide continued suppression of shallow-germinating weeds in stale seedbeds and in management systems without preplant tillage (e.g. no-till). The weed control obtained from cover crops complements herbicides but unfortunately is not reliable enough to be used alone.

Sheet mulches are limited to high-value crops where they can provide excellent season-long control of germinating weed seedlings.

Shallow selective tillage (harrows, drags, rotary hoes) is used to control shallow-germinating weeds after planting in deep-rooted crops such as corn, large-seeded legumes, potatoes, and sugar beets. This method is used in management systems in which the soil has been loosened by a secondary tillage operation prior to planting.

Selective Control of Emerged or Late-Emerging Weeds in an Emerged Crop

The objective of these practices is to remove the weeds that are present as the crop grows. Several of these techniques also are effective in controlling weeds that emerge late in the growing season. These practices also must be selective in order to prevent crop injury.

Shading from the crop canopy can provide more than half of each season's weed control. Crop shading is inexpensive, and its benefits should be maximized by maintaining a good crop stand; planting the crop at the right time; and using adapted crop varieties, appropriate row spacings, and optimal plant populations.

Several types of postemergence herbicide treatments effectively control weeds that are growing with the crop:

Early postemergence herbicides are applied over both the crop and weeds to control small weeds in crops soon after planting and before crop losses due to competition become significant. Early postemergence treatments are appropriate for all management systems as long as selective postemergence herbicides are available for the crop.

Postemergence herbicides are applied over both the crop and weeds to selectively remove weeds later than early postemergence but prior to canopy closure. Weeds surviving previous control efforts (e.g., perennials) or late-emerging weeds frequently are the targets. Postemergence applications are appropriate for all management systems as long as selective postemergence herbicides are available for the crop.

Directed foliar postemergence herbicides are applied over the weeds and under most of the crop leaves. Annual crops must be grown in rows, and the weeds must be shorter than the crop at the time of application. Maximum crop safety is obtained with this technique. Herbicides used as *directed soil* applications have no foliar activity. They are applied under the crop and directly onto the weed-free soil surface. Weeds are killed as they emerge. Directed postemergence treatments of either kind can be used as alternatives to cultivation between rows.

Wipe-on, spot treatments, and preharvest mature crop herbicides provide additional flexibility in controlling late-season weeds.

Cultivation between rows can remove up to 75% of the weeds emerging in the crop. The crop must be in rows, and this technique is most easily done in management systems with secondary tillage.

Hand weeding is appropriate for high-value crops or scattered weeds.

Management Systems for Annual Crops

Management systems that can be used to grow annual crops vary in the amount of tillage operations conducted. The number of operations range from no fewer than three, but more typically five or six, tillage operations for moldboard plow (conventional) systems to no tillage operations at all in no-till systems. Fewer tillage operations are required when tillage and herbicides are combined than when only tillage is used.

Several tillage systems for annual crops are described here. They include crops grown under conventional tillage systems and crops grown under four conservation tillage systems (no-till, chisel/disk, chisel/sweep, and ridge-till). A comparison of the weed control options available in conventionally tilled systems versus the four conservation tillage systems is provided in Table 14.3.

Whether agronomic or horticultural, several important similarities that relate to weed control exist among the annual crops within these management systems. These similarities include the general types of herbicides used, their application methods, and the timing of application. Weed species may differ somewhat, but their life cycles tend to be identical and they usually pose similar problems within a system. Differences among crops within a management system include the specific herbicides used, some of the weed species present, the period of time during which weed control is required, and the time of planting. The management system will determine the weed species present and limit the control techniques that can be used.

Crops Grown Under Conventional Tillage

Row Crops. Tillage practices and equipment for annual agronomic and horticultural row crops are similar. In fact, climate, local custom, and soil conditions probably account for more differences in specific tillage methods than does individual crop choice.

The tillage sequence (Figure 14.1) is initiated using the moldboard plow (primary tillage). It is followed by one or more passes with the tandem disk, field cultivator, or other secondary tillage implements to prepare a seedbed that is suitable for planting or transplanting and is free from growing vegetation and surface residues. The crop thus has an opportunity at an even start with the weeds. On a few large-seeded crops such as corn and soybeans, rotary hoeing or light harrowing is used after planting and until shortly after crop emergence (a similar operation called dragoff is used on potatoes prior to emergence).

A cultivator is used to remove weeds between the rows after the crop emerges and before it gets too large for equipment to pass through without causing damage. Herbicide options include preplant incorporated, preplant surface, preemergence, early postemergence, postemergence, directed postemergence, and postemergence preharvest treatments. Spot treatments with herbicide and hand-removal are also available. Maximum operational flexibility and choice of weed control methods are provided with the moldboard plow (conventional) tillage system.

As effective as the conventional tillage system is for weed control, annual and creeping perennial weeds can survive depending on the species present in the individual field and the time of year the crop is established. Winter annuals and creeping perennials that grow during

Table 14.3 Effect of Tillage System on Available Weed Control Practices for Annual Crops

Crop Stage	Conventional Tillage	Conservation Tillage Systems			
		Chisel/Disk	Chisel/Sweep	Ridge-Till	No-Till
	MOLDBOARD PLOW[a] DISK/FIELD CULTIVATOR DISK/FIELD CULTIVATOR Preplant incorporated	CHISEL PLOW DISK DISK/FIELD CULTIVATOR Preplant incorporated	CHISEL PLOW SWEEPS		Early preplant Burndown
Planting	**Planting** Preemergence ROTARY HOE	**Planting** Preemergence ROTARY HOE	Burndown (optional) **Planting** Preemergence	Burndown (optional) **Planting** Preemergence	**Planting** Preemergence
Crop emergence	Early post-emergence CULTIVATOR CULTIVATOR[b] Directed postemergence Selective postemergence Spot spray	Early post-emergence CULTIVATOR CULTIVATOR[b] Directed postemergence Selective postemergence Spot spray	Early post-emergence CULTIVATOR[b] CULTIVATOR[b] Directed postemergence Selective postemergence Spot spray	Early post-emergence CULTIVATOR CULTIVATOR[b] Directed postemergence Selective postemergence Spot spray	Early post-emergence Directed postemergence Selective postemergence Spot spray
Layby or canopy closure	Wipe-on LIMITED HAND WEEDING CROP SHADING	Wipe-on LIMITED HAND WEEDING CROP SHADING	Wipe-on LIMITED HAND WEEDING CROP SHADING	Wipe-on LIMITED HAND WEEDING CROP SHADING	Wipe-on LIMITED HAND WEEDING CROP SHADING
Harvest	Between-crop herbicides CROP ROTATION	Between-crop herbicides CROP ROTATION	Between-crop herbicides CROP ROTATION COVER CROPS	Between-crop herbicides CROP ROTATION	Between-crop herbicides CROP ROTATION COVER CROPS

[a]Nonherbicide practices are shown in capital letters.
[b]Operation available if needed.

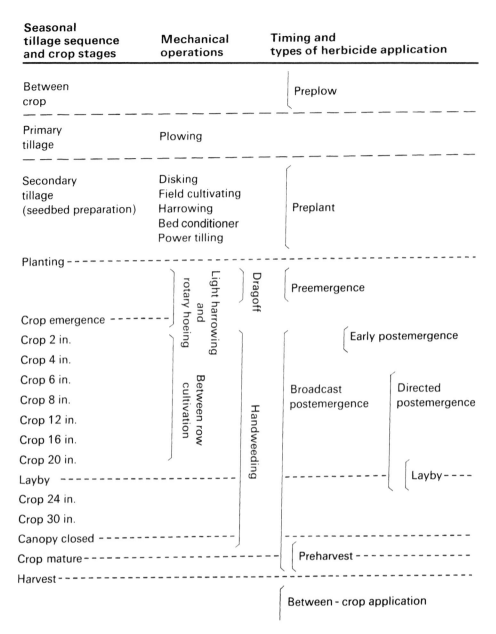

Seasonal tillage sequence and crop stages	Mechanical operations	Timing and types of herbicide application

Between crop — Preplow

Primary tillage — Plowing

Secondary tillage (seedbed preparation) — Disking, Field cultivating, Harrowing, Bed conditioner, Power tilling — Preplant

Planting

Light harrowing and rotary hoeing / Dragoff — Preemergence

Crop emergence

Crop 2 in. — Early postemergence
Crop 4 in.
Crop 6 in.
Crop 8 in. — Broadcast postemergence / Directed postemergence
Crop 12 in.
Crop 16 in.
Between row cultivation / Handweeding
Crop 20 in.

Layby — Layby

Crop 24 in.
Crop 30 in.

Canopy closed

Crop mature — Preharvest

Harvest

Between - crop application

Figure 14.1 Management operations for annual row crops grown under conventional tillage and chisel/disk management systems.

cool weather are major problems in fall-seeded crops. Winter annuals, summer annuals, and creeping perennials are problems in early-spring-planted crops, whereas summer annuals and creeping perennials are problems in later-planted crops.

Differences in size of crop seed also can have an impact on weed populations. Small-seeded crops may require smoother and finer seedbeds than large-seeded crops. Such fine seedbeds also favor germination of small weed seeds.

A standard practice for some small-seeded crops such as sugar beets and direct-seeded tomatoes is to plant a thicker population than needed and then to thin the plants mechanically or by

hand to the desired population. Crop thinning by hand is combined with the first hand weeding.

In situations in which direct seeding will not provide dependable stands, or length of the growing season is limited, seeds are germinated in greenhouses (e.g., transplant tomatoes) or plant beds (e.g., tobacco), and the established plants are subsequently transplanted to the field. A period of recovery from transplant shock frequently is required before applications of postemergence herbicides can be made.

A specialized type of conventional tillage is called the *stale seedbed* method. Stale seedbed also utilizes primary and secondary tillage; however, the last secondary tillage operation is completed weeks or months ahead of planting. In other words, the soil is not disturbed at planting. Thus, a fresh crop of weed seeds is not brought to the soil surface to germinate and produce a new flush of weeds. Two options are available for weed control at planting. A cover crop can be planted at the time of seedbed preparation. A burn-down herbicide is used at planting to kill the cover crop and emerged weeds to eliminate interference with the emerging crop seedlings. Stale seedbed with a cover crop is excellent for areas subject to erosion. It is a frequent practice in vegetable crops, such as transplant tomatoes. The other option is to leave the soil surface bare and to burn down the emerged weeds at planting with a herbicide. Lack of a cover crop reduces protection against soil erosion and removes the weed control benefits provided by residues from cover crops.

Solid-Seeded Crops. Solid-seeded crops (drilled in rows closer than 6 to 14 in. apart) depend on rapid shading from closure of the crop canopy to provide mid- to late-season weed control. Accessibility within the crop provided by rows is lacking so that cultivation between rows; flame cultivation; directed postemergence herbicides; and mid-season or later postemergence herbicides by ground equipment, wipe-on herbicide applications, and high-clearance ground sprayers generally are not available for solid-seeded crops. Mid- to late-season broadcast postemergence sprays and wipe-on applications sometimes are used on solid-seeded crops by sacrificing the crop where the wheels of the equipment run.

With proper crop management and weed control practices, a number of crops now grown in rows could be solid seeded to maximize yield potential if canopy closure can be obtained by midseason. When solid seeding is used continuously, however, the potential exists for selection of shade-tolerant weed species or shade-tolerant individuals within a species.

Conservation Tillage Systems

Conservation tillage (no-till, chisel/disk, chisel/sweep, and ridge-till) systems, which retain crop residues on the soil surface, were initially developed to prevent soil erosion in annual crops. Annual crops historically have been grown in conventional tillage systems, which are accompanied by considerable soil erosion. Although solid-stand perennial hay and pasture crops can be planted to reduce erosion, rarely are they as profitable or marketable as are annual crops such as corn and soybeans. Thus, the financial incentive along with the desire to reduce erosion have made conservation tillage systems for annual crops attractive to the U.S. farmer. Another benefit of conservation tillage has been the ability to devote more time to planting rather than to seedbed preparation ahead of planting. Timely planting is particularly important when inclement weather shortens the period available for seedbed preparation and crop establishment.

Without the introduction of herbicides, no-till could never have become the viable crop management system that it has over the last 20 years. In some Corn Belt states no-till acreage is approaching 50% of that planted to annual crops. The chisel/disk system is more popular than the moldboard plow system on the remaining 50% of the acreage. Limited acreage is farmed with the ridge-till system.

Although no-till is a relatively new development, versions of the chisel/disk, chisel/sweep, and ridge-till systems have been used in the Great Plains to conserve soil since the 1930s.

No-Till. In contrast to the conventional tillage system, the no-till system approaches the lower limit of operational options (Table 14.3). The tillage provided by the no-till coulter is sufficient to place the seed in the soil, but there is no full width cutting or stirring of the soil. No weed control benefits are derived from this limited amount of tillage. At planting time, the soil is covered with residues from the previous season along with weed seeds and any crop seeds missed during harvesting. The soil is firm rather than loose, and weeds usually are growing. The introduction and use of paraquat to burn down this vegetation made no-till a viable crop management option. However, erratic crop stands and poor weed control hindered the initial adoption of no-till. Improved planting equipment and effective selective postemergence herbicides have largely overcome planting and weed control limitations.

Crops most frequently no-till planted are corn and soybeans, although other crops have been grown successfully with this method. The crop may be planted into a perennial grass sod (sod plant), a winter cover crop, the residues of the previous season's crop, or newly harvested small-grain stubble. In the last case, a quickly maturing second crop such as soybeans often is planted into the stubble for double cropping. The growing of two crops in the same year (e.g., winter wheat followed by soybeans) is expedited by no-till since no time is wasted for tillage and seedbed preparation and losses of precious soil moisture associated with conventional tillage are prevented.

The previous crop, the vigor of that crop, and the portion of crop removed from the field during harvest determine the amount of residue on the soil and the protection from erosion. A perennial grass sod or stubble from a high-yielding corn crop or small-grain crop usually provide adequate residue. A crop such as soybeans provides much less residue and may not adequately protect the soil. Winter cover crops such as wheat, rye, or vetch can be seeded after soybeans to provide additional residues. Some weed control can be expected from vigorously growing cover crops and resultant allelopathic chemicals.

Weeds or other vegetation that are growing at crop planting time usually result in greater crop losses than the weeds that emerge with the crop. *As currently practiced, no-till depends totally on herbicide treatments to provide early-season control of these emerged weeds.*

In the early 1980s the standard herbicide program at planting for controlling annual weeds, grass sods, or cover crops in no-till usually consisted of a three-herbicide tank mix. Paraquat, to burn down emerged vegetation, was combined with two residual preemergence herbicides for later germinating weed seeds. One was used for grass control, the other for broadleaves. These herbicides (e.g., alachlor, metolachlor, and pendimethalin for grasses, and metribuzin, atrazine, cyanazine, and linuron for broadleaves) frequently provided some contact foliage activity as well. These herbicides were used without paraquat if weeds were small and easy to control. If large, well-established annual weeds or perennials were present, glyphosate was used in place of paraquat.

In addition to these options, some of the persistent preemergence herbicides (pendimethalin, imazaquin, imazethapyr, chlorimuron, simazine, atrazine) could be applied several weeks ahead of planting and before weed emergence. These "early preplant" treatments reduced the need for burndown at planting. The earlier application dates increased the chances that timely rainfall would leach these herbicides into the soil and result in good weed control.

By the mid-1990s more glyphosate was being used as a burndown at planting due to its reduced price. 2,4-D had been labeled for use ahead of soybeans, and several selective postemergence herbicides provided additional weed control options and made no-till crop production less risky.

The major losses in weed control options in no-till are all the tillage operations and the potential to use preplant incorporated herbicides (Table 14.3). Burn-down treatments, pre- and postemergence herbicides, spot spraying and wipe-on applications, limited hand weeding, crop shading, crop rotation, and allelopathic cover crops are the weed control practices that must make up for the lack of tillage. Of these methods, herbicides provide the only economical and dependable control.

As successful as no-till has been, some important precautions must be taken into consideration when planning and executing a no-till management system.

1. The presence of residues affects the microenvironment at the soil surface. The plant residues shade the soil and reduce the evaporative losses of water, leaving the soil cooler and moister. Cool soil temperatures can delay planting and crop emergence, which can be a problem in the cooler northern regions. The firm, moist soil present in no-till provides ideal conditions for the germination of small-seeded weeds. The plant residue also can interfere with the amount of herbicide that penetrates to the soil surface or to germinating weeds. Bauman and Ross (1983) reported that one-third of the atrazine applied on no-till corn never reached the soil surface.

2. An undisturbed soil surface can have indirect effects on weed control when fertilizers and lime are surface applied. Nitrogen can lower the pH in the top inch of soil, whereas lime can increase it. The efficacy or persistence of herbicides that are affected by soil pH (e.g., the triazines and sulfonylureas) thus may be altered (see Chapter 7).

3. The limited number of weed control options, the relatively restricted management system, the loss of the increment of control normally provided by tillage operations, and the deposition and retention of weed seed on the soil surface (accompanied by lack of mixing with buried seeds) in no-till systems combine to accelerate shifts to different weed species. Changes in dominant species can occur in 1 or 2 yr. A shift from foxtails to crabgrass and fall panicum frequently occurs in no-till corn. Perennials of all kinds (simple, creeping, and woody) increase. Some species rarely encountered as problems in moldboard plow tillage systems may occur. These weeds can include curly dock; dandelion; goldenrod; tall ironweed; and woody species such as brambles, poison ivy, and others. Cool-season annuals such as horseweed (marestail), daisy fleabane, and common groundsel and biennials such as wild carrot sometimes invade no-till. Creeping perennial weeds such as johnsongrass, Canada thistle, quackgrass, and hemp dogbane remain serious

problems when management is changed from a conventional to a no-till system (Table 14.4).

4. Seeds from weeds not controlled remain on the soil surface and provide the next year's flush of germinating weeds. This is in contrast to a conventional system in which plowing mixes and dilutes this season's seeds with the weed seed populations (and genes) from all previous years. Thus, if herbicide resistance appears in newly produced seed and the seed remains on the soil surface, the rate of selection for herbicide-resistant weed populations can increase.

Buildup of simple perennial weeds in no-till systems can be reduced by switching to conventional tillage periodically. Doing so, however, may be inappropriate when the main reason for no-till is to reduce severe erosion potential. Conventional tillage can provide opportunities for the use of intensive control efforts for problem weeds if it is used once every 3 to 4 yr. Tillage at such an infrequent interval, however, is inadequate for controlling infestations of serious creeping perennial weeds.

Many of the more effective treatments for difficult-to-control perennial weeds cannot be used selectively in a growing crop. Efforts should be directed toward working these treatments into the program when crops are not growing. Land in a set-aside program, in which producers are paid to leave the land idle, provides an excellent opportunity to work on serious perennial weeds. In areas where sod crops are to be no-tilled, treatments for perennial weeds can be made the year before a row crop is established. Weeds also can be treated between harvest and crop establishment (between-crop treatments). This treatment, however, requires that the weeds be in an appropriate growth stage. A limited number of herbicides (e.g., 2,4-D and glyphosate) can be used before crop harvest but after the crop has matured.

Chisel/Disk. In this system the moldboard plow is replaced with the chisel plow for primary tillage while retaining the tandem disk for secondary tillage (Table 14.3). It provides weed control options and a seedbed at planting time similar to those provided by the conventional tillage system (Figure 14.1). The critical difference from the conventional system is that properly executed chisel plow operations leave approximately 30 to 80% of the residues on the soil surface after primary tillage. These residues protect against erosion of the soil from the time of the primary tillage (chisel plow) operation until the seedbed is prepared with secondary tillage operations, at which time surface residues are further reduced to 10 to 20% coverage. The period between primary and secondary tillage can be as long as 5 or 6 mo when the chisel plow operation is done in the fall and a summer annual crop is grown.

Chisel/Sweep. This system consists of an initial pass with the chisel plow equipped with points for the initial primary tillage operation, followed by a second pass with sweeps just ahead of planting (Table 14.3). The sweeps lift the soil rather than cut and stir it, so maximum residues (greater than 50%) can be left on the soil surface at planting. In these systems, weed control from tillage is somewhat reduced although not eliminated. One advantage of these systems is that it provides at least one pass of full width tillage for destroying the crowns of simple perennial and biennial weeds. The sweep operation can, however, leave large chunks of intact soil in which many small weeds can continue to grow. Using a foliar-applied herbicide one or more days ahead of the tillage pass and planting should solve this problem. A tool such as a *rolling basket harrow* that will not easily clog with surface residues can be attached to the chisel plow to break down clods and increase control of emerged weeds.

Table 14.4 Effect of Tillage System on Johnsongrass and Canada Thistle Populations[a]

Tillage System[b]	% of Area Covered by Johnsongrass[c]	Number of Canada Thistle Shoots in 10 ft² plots[d]
Moldboard plow/disk	9	6.8
Chisel plow/disk	14	6.4
Chisel plow/sweeps	38	—
No-till	62	28.9

[a]M. A. Ross, unpublished data. Johnsongrass plots located at Madison, Indiana. Canada thistle plots located at Farmland, Indiana.
[b]All plots received herbicide treatments recommended for the tillage system.
[c]Data collected at harvest (1980) the 1st yr in the tillage system.
[d]Data collected (May 1982) after 3 yr in tillage system.

Ridge-Till. In this system, seed is planted into ridges established during the previous year's cropping. Thus, crop residues remain on the soil surface through the winter. In contrast to the chisel/disk and chisel/sweep systems, no separate primary or secondary tillage is conducted (Table 14.3). At planting time, a sweep or other cutting device on the planter is used to scrape off the top of the ridge, and the seed is placed into the band of clean, newly exposed soil. This operation provides a residue-free seedbed and throws any emerged weeds and old crop residues between the ridges. Approximately half of the surface area is covered with crop residues. Once the crop is up and growing, a cultivator is used to rebuild the ridges. The weeds between the ridges are destroyed by the cultivator, and the soil thrown up around the crop buries the small weeds emerging under the crop. Weeds not effectively controlled by these operations can be controlled by herbicides as needed. Volunteer corn, which tends to increase as a problem on most conservation tillage systems, usually can be controlled in the ridge plant system because the plants are thrown between the rows and can be removed with the later cultivation.

Conservation tillage systems may require approaches to weed management that differ from those employed on conventional tillage systems. Weed control in conservation tillage must be approached with a realistic understanding of the problems that can be encountered and the relative effectiveness and availability of practices. *A higher level of management is required on these systems because mistakes can be more costly and fewer alternative corrective practices are available than with conventional tillage.*

Besides the weed control operations used before or at planting, additional treatments during the remainder of the growing season should be anticipated and used when appropriate. Preventive weed control techniques should be employed to the fullest to deter the establishment of new problem weeds. Scattered plants or clumps can and should be spot sprayed before seed or vegetative propagules are produced. The crop itself should be managed to provide maximum shading. When possible, rotations of crops provide greater diversity in weed control options and can alleviate some of the selection pressure toward certain weeds. When several choices are available, rotation of herbicide treatments provides additional diversity. In theory at least, weed species that are controlled effectively should disappear more rapidly with systems in which the soil is not disturbed or is worked only shallowly than with systems in which a new crop of seed is supplied each year by plowing.

Weed Control Practices in Annual Crops

To illustrate some of the variations encountered in weed control practices among annual row crops, we have chosen five crops (corn, soybeans, tomatoes, onions, and small grains) for detailed discussion. A general scheme of the management operations available for these and other annual crops grown under conventional tillage is shown in Figure 14.1. Specific information on the herbicides and weeds controlled in these four crops is given as follows: corn, Tables 14.5 and 14.6; soybeans, Tables 14.7 and 14.8; tomatoes, onions and other annual vegetable crops, Table 14.9; and small grains, Table 14.10.

The information provided in these tables should be used *only* as a starting point for further investigation because specific herbicide recommendations vary from state to state or region to region. Furthermore, the listing is limited because it provides information only on the selectivity of herbicides when used alone. Today it is common practice to combine two or even three herbicides to increase the spectrum of weed kill, minimize crop injury, and otherwise improve performance.

Herbicide combinations offer several benefits. Long-lived herbicides frequently can be used at reduced rates in combination with a short-lived herbicide so that potential residues persisting into the next season can be minimized. Herbicides with differing requirements for rainfall and movement into the soil can provide more consistent control with varying weather conditions. Such combinations also may provide more consistent control on fields with several soil types. Sometimes it is possible to reduce costs by complementing a low rate of an expensive herbicide with a less costly one. Herbicide combinations may be formulated as prepackaged mixes, or they may be tank mixed.

When tank mixes are unable to provide adequate or consistent control, sequential treatments are used instead. Space constraints preclude the listing of the many alternatives for herbicide combinations and sequencing, but such alternatives always should be looked upon as viable options.

Corn

Corn is a warm-season crop that is almost always grown in rows 30 to 40 in. apart. Corn types grown in the United States include field corn for grain or silage, popcorn, and sweet corn. The production of modern corn varieties depends on the production of hybrid seed from inbred lines. Unfortunately, popcorn, sweet corn, and inbreds of all varieties tend to be less vigorous, shorter in height, and more susceptible to herbicide injury than field corn hybrids. Consequently, herbicides labeled for field corn may not be labeled for the other types of corn. In addition, corn for hybrid seed production sometimes is grown in 36- to 40-in. rows (as opposed to standard 30- to 32-in. rows) to facilitate the passage of detasslers. The combination of open space plus less vigorous plants enables weeds to grow longer into the season, be more competitive, and cause greater yield losses.

The large acreages planted with corn have provided a financial incentive to develop a large number of herbicides for weed control in corn. Approximately 30 different parent compounds (examples of parent compounds are 2,4-D acid, dicamba acid, glyphosate, etc.) were available for selective use in corn during the 1997 and 1998 growing seasons. These parent compounds are available in many different formulations. They represent different mode-of-action groups, control different spectra of weeds, and are applied in several different ways (Table 14.5). For example, five herbicides can be used as burndown treatments prior to crop planting or emergence. Of the 16 herbicides that are soil applied, 2 are preplant incorporated only, 9 are used both preplant and preemergence, and 5 are used preemergence only. Of the 14 herbicides providing effective postemergence control, 4 also are soil applied. Another four postemergence herbicides can be used selectively only in herbicide-resistant crops. In addition to individual herbicides, there are over 20 different prepackaged mixtures containing two or more parent compounds (Table 14.5). Some parent compounds (e.g., rimsulfuron and fluthiamide) are used only in mixtures.

In order to put this complex information into perspective, we describe the historical development of corn herbicides. This development parallels that of herbicides in general (see Chapter 5), starting with 2,4-D and evolving to herbicide-resistant crops.

The first herbicide widely used for corn was 2,4-D. This compound was developed in the 1940s and was used primarily as a postemergence treatment for broadleaved weed control. Other auxin-type growth regulator herbicides introduced since then for weed control in corn include dicamba and clopyralid. Dicamba offers postemergence control of numerous broadleaf species

missed by 2,4-D as well as consistent short-lived preemergence activity. Clopyralid is effective particularly for the control of composites (including thistles), smartweeds, and nightshades.

CDAA was introduced in 1957 as a preemergence soil treatment and provided control of annual grasses and a limited number of broadleaved weeds. It was the first of several substituted amide herbicides developed for the corn market.

The introduction of the chloroacetamide herbicide propachlor in 1965 provided a preemergence herbicide less subject to leaching than CDAA. It controlled many annual grasses so that it complemented atrazine in a tank mix. Rates of atrazine could be lowered to reduce carryover, while the decrease in grass control was compensated for by the propachlor. The combination also provided more consistent weed control than either compound alone in areas with variable soil types and rainfall.

Other chloroacetamide herbicides soon were developed. The introduction of alachlor in 1967 provided an effective grass herbicide with fewer handling problems than propachlor and with considerably less skin irritation than CDAA. Alachlor is less water soluble than either CDAA or propachlor, so it is less subject to leaching and tends to provide more consistent weed control on soils with low organic matter. Metolachlor followed alachlor and competed directly with it. More recent chloroacetamide introductions into the corn market include acetochlor (with safener) and dimethenamid. Acetochlor, alachlor, metolachlor, and dimethenamid compete for most of the soil-applied, grass control market in corn. In addition to small-seeded grasses (foxtails, crabgrasses, barnyardgrass, panicums), they control black nightshade and pigweed and suppress common lambsquarters, yellow nutsedge, and shattercane. CDAA is no longer marketed.

The introduction in the late 1950s of preplant incorporated thiocarbamate herbicides such as EPTC provided the first opportunity to control johnsongrass seedlings and shattercane effectively in corn. Unfortunately, crop selectivity was limited. The development of a closely related compound, butylate, provided improved crop tolerance. The discovery that corn could be protected from thiocarbamate injury with certain chemical additives (protectants) resulted in better performance and increased crop safety with both EPTC and butylate. EPTC safened with protectants permitted the use of increased dosages for the improved control of seedlings and suppression of shoots emerging from rhizomes

Table 14.5 Partial List of Herbicides and Herbicide Combinations Labeled for Corn

Common Name	Trade Name(s)	Mode-of-Action/Chemical Group
Burndown prior to crop emergence		
2,4-D	Many trade names	Auxin-type growth regulator/phenoxy acid
Dicamba	BANVEL/CLARITY	Auxin-type growth regulator/benzoic acid
Glyphosate	ROUNDUP/TOUCHDOWN	Aromatic amino acid inhibitor
Paraquat	GRAMOXONE-EXTRA/ CYCLONE CF	Photosystem I energized lipid peroxidation/bipyridilium
Glufosinate	LIBERTY/IGNITE	Glutamate inhibitor-lipid peroxidizer
Preplant incorporated only		
Butylate	SUTAN+	Shoot inhibitor/thiocarbamate
EPTC	ERADICANE	Shoot inhibitor/thiocarbamate
Preplant incorporated or preemergence		
Acetochlor	HARNESS/SURPASS/ TOPNOTCH	Shoot inhibitor/chloroacetamide
Alachlor	LASSO/MICROTECH/ PARTNER	Shoot inhibitor/chloroacetamide
Dimethenamid	FRONTIER	Shoot inhibitor/chloroacetamide
Metolachlor	DUAL	Shoot inhibitor/chloroacetamide
Atrazine	AATREX	Classical photosynthesis inhibitor/ s-triazine
Cyanazine	BLADEX	Classical photosynthesis inhibitor/ s-triazine
Simazine	PRINCEP	Classical photosynthesis inhibitor/ s-triazine
Flumetsulam	PYTHON	Branched-chain amino acid inhibitor/triazolopyrimidine
Isoxaflutole	BALANCE	Carotenoid pigment inhibitor/isoxazole
Preemergence only		
2,4-D	Many trade names	Auxin-type growth regulator/phenoxy acid
Dicamba	BANVEL/CLARITY	Auxin-type growth regulator/benzoic acid
Linuron	LOROX/LINEX	Classical photosynthesis inhibitor/ phenylurea
Pendimethalin	PENTAGON/PROWL	Root inhibitor/dinitroaniline
Propachlor	RAMROD	Shoot inhibitor/chloroacetamide
Early postmergence/postemergence		
2,4-D	Many trade names	Auxin-type growth regulator/phenoxy acid
Clopyralid	STINGER	Auxin-type growth regulator/picolinic acid
Dicamba	BANVEL/CLARITY	Auxin-type growth regulator/benzoic acid
Atrazine	AATREX with crop oil concentrate	Classical photosynthesis inhibitor/ s-triazine
Cyanazine	BLADEX/CY PRO	Classical photosynthesis inhibitor/ s-triazine
Metribuzin	SENCOR	Classical photosynthesis inhibitor/as-triazine)
Bentazon	BASAGRAN	Rapid photosynthesis inhibitor
Bromoxynil	BUCTRIL/MOXY	Rapid photosynthesis inhibitor
Pyridate	TOUGH	Rapid photosynthesis inhibitor
Flumiclorac	RESOURCE	Protox inhibitor/N-phenylphthalimide
Halosulfuron	PERMIT	Branched-chain amino acid inhibitor/sulfonylurea
Nicosulfuron	ACCENT	Branched-chain amino acid inhibitor/sulfonylurea
Primisulfuron	BEACON	Branched-chain amino acid inhibitor/sulfonylurea
Prosulfuron	PEAK	Branched-chain amino acid inhibitor/sulfonylurea
Directed postemergence		
Ametryn	EVIK	Classical photosynthesis inhibitor/ s-triazine
Linuron	LOROX/LINEX	Classical photosynthesis inhibitor/ phenylurea
Paraquat	GRAMOXONE-EXTRA/ CYCLONE CF	Photosystem I energized lipid peroxidation/bipyridilium

Table 14.5 *(Continued)*

Common Name	Trade Name(s)	Mode-of-Action/Chemical Group
Herbicides for resistant corn cultivars		
Glufosinate	LIBERTY	Glutamate inhibitor-lipid peroxidizer
Glyphosate	ROUNDUP ULTRA	Aromatic amino acid inhibitor
Imazethapyr	PURSUIT	Branched-chain amino acid inhibitor/imidazolinone
Sethoxydim	POAST/POAST PLUS	Lipid biosynthesis inhibitor/cyclohexanedione
Prepackaged mixtures		
Grass and broadleaf control		
Acetochlor-EPTC		DOUBLEPLAY
Acetochlor-atrazine		SURPASS 100/HARNESS XTRA/FULTIME
Alachlor-atrazine		BULLET/LARIAT
Atrazine-metolachlor		BICEP II
Dimethenamid-atrazine		GUARDSMAN
Dimethenamid-dicamba		OPTILL
Flumetsulam-metolachlor		BROADSTRIKE+DUAL
Fluthiamide-metribuzin		AXIOM
Nicosulfuron-rimsulfuron-atrazine		BASIS GOLD
Propachlor-atrazine		PROPACHLOR/ATRAZINE
Rimsulfuron-thifensulfuron		BASIS
Broadleaf control		
2,4-D-atrazine		SHOTGUN
Atrazine-cyanazine		EXTRAZINE II/CY-PRO AT
Bentazon-atrazine		LADDOK
Bromoxynil-atrazine		BUCTRIL-ATRAZINE/MOXY-ATRAZINE
Dicamba-atrazine		MARKSMAN
Flumetsulam-clopyralid		HORNET
Flumetsulam-clopyralid-2,4-D		SCORPION III
Primisulfuron-prosulfuron		EXCEED
Imazethapyr-atrazine (imi corn)		CONTOUR
Imazethapyr-dicamba (imi corn)		RESOLVE
Imazethapyr-imazapyr (imi corn)		LIGHTNING
Imazethapyr-pendimethalin (imi corn)		PURSUIT PLUS
Burndown		
Glyphosate-alachlor		BRONCO

of johnsongrass and quackgrass and tubers of yellow nutsedge.

The soil-applied herbicide, simazine, was introduced in the late 1950s for both annual grass and broadleaved weed control. The introduction of atrazine, a triazine herbicide closely related to simazine, soon followed. Although notably weak on crabgrass, fall panicum, and shattercane, atrazine's sixfold increase in water solubility and efficacy as a foliar as well as a soil treatment gave it a definite edge over simazine in terms of consistent performance and flexibility of use. Atrazine became the most used herbicide on corn for three decades or more and has served as the basis for most corn herbicide programs in the Corn Belt states. It was usually combined with a chloroacetamide or thiocarbamate herbicide to provide broad spectrum control. Problem weeds escaping such treatments are controlled with postemergence herbicides as needed. Development of resistant weed populations and detectable quantities of atrazine in ground and surface water have resulted in lowered doses and somewhat restricted use patterns. The long-term availability of atrazine at this time is uncertain.

Another triazine herbicide frequently used on corn was cyanazine. This herbicide had no residual carryover to subsequent crops and was substituted for all or portions of the atrazine when persistence was a problem. It was used frequently in dry climates or where sensitive crops followed corn. This herbicide is being or has been dropped from the herbicide market and suitable alternatives need to be employed.

Other classical photosynthesis inhibitor herbicides used on corn include linuron, which is labeled for use as preemergence and directed postemergence treatments; ametryn, which is labeled directed postemergence; and metribuzin, which is labeled at low doses for postemergence control of annual broadleaf weeds.

The rapidly acting photosynthesis inhibitors bromoxynil, bentazon, and pyridate are labeled postemergence for control of annual broadleaf weeds. Bentazon and pyridate also have activity against yellow nutsedge.

Pendimethalin, the only dinitroaniline herbicide labeled preemergence for corn, is used as a residual surface-applied herbicide that controls seedling johnsongrass and a number of annual grasses. Selectivity depends on the herbicide remaining above the corn roots.

The introduction of the sulfonylurea herbicides nicosulfuron and primisulfuron in 1990 provided selective postemergence control of problem grasses, including johnsongrass and quackgrass. Not all species of annual and perennial grasses, however, are effectively controlled. Neither herbicide is effective for the control of crabgrasses. In addition to seedling johnsongrass and established johnsongrass and quackgrass, the 1998 Accent label (nicosulfuron) claims control for 18 additional grasses and six broadleaves; the 1998 Beacon label (primisulfuron) claims fewer grasses but as many as 30 broadleaves.

Several other branched-chain amino acid inhibitor herbicides have been introduced for use in corn. For example, halosulfuron provides effective control of yellow and purple nutsedge as well as a dozen broadleaved weeds, and flumetsulam controls about 30 broadleaves. Some of these compounds compete with or complement the s-triazine and auxin-type growth regulator herbicides. Should the uses of atrazine be lost or severely restricted, these herbicides could help fill the void.

Individual branched-chain inhibitor compounds tend to provide excellent control of a narrow spectrum of weeds. Thus, each may miss other important species. As a result, they almost always are used or packaged with other compounds of similar or different modes of action. For example, flumetsulam, prosulfuron, rimsulfuron, and thifensulfuron mostly control broadleaved weeds and are available in prepackaged mixtures consisting of two or three components.

Additional chemistries for weed control in corn include the carotenoid pigment inhibitor isoxaflutole, which is labeled as a soil surface application, and the Protox inhibitor flumiclorac, which is used as a postemergence treatment. Isoxaflutole controls several annual grasses and several annual broadleaves, and flumiclorac is labeled for annual broadleaves (particularly pigweeds and velvetleaf).

The most recent advancements in corn weed control are the genetically altered corn cultivars. Glufosinate provides broad spectrum control of annual grasses and broadleaves in glufosinate-resistant corn; glyphosate provides broad spectrum control of annuals plus control or suppression of perennials in glyphosate-resistant corn. Imidazolinone resistance permits the use of the moderately broad spectrum herbicide imazethapyr on corn; sethoxydim resistance permits broad spectrum control of annual and perennial grasses with sethoxydim.

Glufosinate, glyphosate, and sethoxydim do not provide a soil residual or season-long control, so two or more applications of the herbicide or a combination of the herbicide with a soil residual herbicide may be required when multiple weed flushes occur.

Shifts in weed species have been observed with the changing sequence of corn herbicides and tillage practices in the Corn Belt states. These shifts can be attributed to the repeated use of individual herbicides and to the use of corn as the most frequent crop in the rotation. The first shift from annual broadleaved weeds to foxtails in the 1950s was due at least partially to the continuous use of 2,4-D. A shift from foxtails to fall panicum and crabgrass followed the introduction and widespread use of atrazine in the 1960s. Starting in the 1980s, and still continuing, is a shift toward perennial weeds as programs providing good control of annual weeds are adopted, and the tendency toward fewer tillage operations grows. The appearance of amaranth species with tolerance to branched-chain amino acid inhibitors became evident in the late 1990s.

A partial list of the herbicides labeled for corn, their methods of application, and the types of weeds controlled is provided in Table 14.6.

Soybeans

The culture of soybeans is similar to that of corn with the exception that soybeans are grown in rows varying from 6 to 40 in. wide rather than in 30- to 40-in. rows. A definite advantage can be derived from planting soybeans in narrow rows in the northern half of the soybean production area of the United States. In these areas, canopy closure with wide row spacing is less likely to occur because the soybean varieties used tend to be shorter in height, the planting date usually is later, and the growing season is shorter than it is in southern areas. Narrow rows therefore speed canopy closure in northern areas. For example, canopy closure in 7-in. rows versus 30-in. rows in Lafayette, Indiana, starts at approximately 5 and 10 weeks after planting, respectively. Yield and weed control benefits from narrow rows are particularly evident with the late planting dates encountered when soy-

Table 14.6 Partial List of Herbicides Labeled for Corn[a]

	Seedling Grasses						Annual Broadleaves							Established Perennials				
	Foxtails	Large crabgrass	Fall panicum	Shattercane	Woolly cupgrass	Johnsongrass seedlings	Pigweed	Lambsquarters	Jimsonweed	Common ragweed	Velvetleaf	Cocklebur	Black nightshade	Johnsongrass	Quackgrass	Yellow nutsedge	Canada thistle	Field bindweed
Soil applied																		
Acetochlor	X	X	X	S	X	S	X	X	S	S	S	S	X	N	N	X	N	N
Alachlor	X	X	X	S	S	S	X	S	S	S	N	N	X	N	N	X	N	N
Dimethenamid	X	X	X	S	S	S	X	S	S	S	N	N	X	N	N	N	N	N
Metolachlor	X	X	X	S	S	S	X	S	S	S	N	N	X	N	N	N	N	N
Pendimethalin	X	X	X	X	X	X	X	X	S	S	S	S	S	S	N	N	N	N
Atrazine	S	S	N	N	N	N	X	X	X	X	X	X	X	N	S	S	N	N
Simazine	X	S	X	S	N	N	X	X	X	X	S	X	X	N	S	S	N	N
Imazethapyr	S	S	S	S	S	S	X	X	S	S	X	X	X	N	N	S	N	N
Post applied																		
Nicosulfuron	X	S	X	X	X	X	X	S	X	S	S	S	S	X	X	N	N	N
Primisulfuron	S	S	X	X	N	X	S	S	S	X	S	S	S	S	S	S	S	N
Halosulfuron	N	N	N	N	N	N	X	N	X	X	X	X	N	N	N	C	N	S
Bentazon	N	N	N	N	N	N	S	S	X	S	X	X	S	N	N	N	N	N
Bromoxynil	N	N	N	N	N	N	S	X	X	X	X	X	X	N	N	N	S	S
Flumiclorac	N	N	N	N	N	N	S	S	S	S	X	S	N	N	N	N	N	N
Clopyralid	N	N	N	N	N	N	N	N	X	X	N	X	X	N	N	N	X	N
2,4-D	N	N	N	N	N	N	X	X	S	X	X	X	X	N	N	N	S	S
Dicamba	N	N	N	N	N	N	X	X	X	X	X	X	X	N	N	N	S	S
Glufosinate[b]	X	X			X	X	S	X		X	X	X	X	S	S	S	S	S
Glyphosate[b]	X	X	X	X	X	X	X	X	X	X	X	X	X	X	X	S	X	S
Imazethapyr[b]	S	S	S	X	S	X	X	S	X	S	X	X	X	S	N	S	S	N
Sethoxydim[b]	X	X	X	X	X	X	N	N	N	N	N	N	N	N	X	X	N	N

X = good to excellent control; S = suppression; N = no control.
[a]Data compiled from claims on product labels and from weed control ratings from the following universities: Purdue, Ohio State, Michigan State, Kentucky, Illinois, and Iowa State.
[b]Apply only to resistant crop cultivars.

beans are used as a second crop (double crop) after harvest of a winter small grain.

A major shift in management systems for soybeans has been the trend toward no-till from conventional tillage. For example, more than half of the soybean acreage in Indiana was solid seeded in no-till in 1997. This surpasses corn with just a third of the acreage planted in no-till. Good planting equipment, effective burn-down treatments, and selective postemergence herbicides have contributed to this shift from conventional tillage for soybeans. Another advantage of no-till is that the delay in planting caused when tillage is used to prepare the seedbed is avoided. Time saving is particularly important when bad weather or other factors result in late planting dates.

A list of the methods of application, trade names, modes of action, and chemical families shows over 30 parent compounds available for weed control in soybeans for the 1998 growing season (Table 14.7). Two herbicides are used preplant incorporated only, twelve are used both preplant incorporated and preemergence, one is used only preemergence, eighteen are used postemergence (four of these were listed for soil application), and three additional compounds are used for herbicide-tolerant crops and/or for burndown on no-till. Twenty pre-packaged mixtures available to soybean growers with parent compounds are listed as well.

The substituted amide herbicide CDAA was introduced for grass control in soybeans in 1957. CDAA has since been dropped and replaced by the chloroacetamide herbicides (alachlor in 1967, metolachlor in the late 1970s, and dimethenamid in 1994). The chloroacetamides provide effective control of small-seeded grasses (foxtails, crabgrasses, barnyardgrass, and panicums). They also control black nightshade and

Table 14.7 Partial List of Herbicides and Herbicide Combinations Labeled for Soybeans

Common Name	Trade Name(s)	Mode-of-Action/Chemical Group
Burndown prior to crop emergence		
2,4-D	Many trade names	Auxin-type growth regulator/phenoxy acid
(product labels call for delayed planting from 7 to 28 days after application depending on dose)		
Glyphosate	ROUNDUP/TOUCHDOWN	Aromatic amino acid inhibitor
Paraquat	GRAMOXONE-EXTRA/ CYCLONE CF	Photosystem I energized lipid peroxidation/bipyridilium
Metribuzin	LEXONE/SENCOR	Classical photosynthesis inhibitor/*as*-triazine
Linuron	LOROX/LINEX	Classical photosynthesis inhibitor/phenylurea
Imazethapyr	PURSUIT	Branched-chain amino acid inhibitor/ imidazolinone
Flumetsulam	PYTHON	Branched-chain amino acid inhibitor/triazolopyrimidine
Preplant incorporated only, seedling grass control		
Trifluralin	TREFLAN	Root inhibitor/dinitroaniline
Ethalfluralin	SONALAN	Root inhibitor/dinitroaniline
Preplant incorporated and preemergence, seedling grass control		
Pendimethalin	PROWL/PENTAGON	Root inhibitor/dinitroaniline
Alachlor	LASSO/MICROTECH/ PARTNER	Shoot inhibitor/chloroacetamide
Metolachlor	DUAL	Shoot inhibitor/chloroacetamide
Dimethenamid	FRONTIER	Shoot inhibitor/chloroacetamide
Preplant incorporated and preemergence, seedling grass and broadleaved weed control		
Clomazone	COMMAND	Carotenoid pigment inhibitor
Imazethapyr	PURSUIT	Branched-chain amino acid inhibitor/ imidazolinone
Preplant incorporated and preemergence, broadleaf control		
Metribuzin	LEXONE/SENCOR	Classical photosynthesis inhibitor/ *as*-triazine
Imazaquin	SCEPTER	Branched-chain amino acid inhibitor/ imidazolinone
Cloransulam	FIRST RATE	Branched-chain amino acid inhibitor/ triazolopyrimidine sulfonanilide
Flumetsulam	PYTHON	Branched-chain amino acid inhibitor/ triazolopyrimidine sulfonanilide
Sulfentrazone	COVER/AUTHORITY	Protox lipid-peroxidation/ triazolinone
Preemergence, broadleaf control only		
Linuron	LOROX	Classical photosynthesis inhibitor/ phenylurea
Postemergence, grass control		
Fenoxaprop-P	OPTION II	Lipid biosynthesis inhibitor/ aryloxyphenoxy propionate
Fluazifop-P	FUSILADE DX	Lipid biosynthesis inhibitor/ aryloxyphenoxy propionate
Quizalofop-P	ASSURE II	Lipid biosynthesis inhibitor/ aryloxyphenoxy propionate
Clethodim	SELECT/PRISM	Lipid biosynthesis inhibitor cyclohexanedione
Sethoxydim	POAST/POAST PLUS	Lipid biosynthesis inhibitor/ cyclohexanedione
Postemergence, broadleaf control and suppression of some grasses		
Imazethapyr	PURSUIT	Branched-chain amino acid inhibitor/ imidazolinone
Imazamox	RAPTOR	Branched-chain amino acid inhibitor/ imidazolinone
Postemergence, broadleaves only		
Imazaquin	SCEPTER	Branched-chain amino acid inhibitor/ imidazolinone
Thifensulfuron	PINNACLE	Branched-chain amino acid inhibitor/ sulfonylurea

Table 14.7 (*Continued*)

Common Name	Trade Name(s)	Mode-of-Action/Chemical Group
Chloransulam	FIRST RATE	Branched-chain amino acid inhibitor/ triazolopyrimidine
Flumetsulam	PYTHON	Branched-chain amino acid inhibitor/ triazolopyrimidine
Acifluorfen	BLAZER	Protox lipid-peroxidation/ diphenylether
Lactofen	COBRA	Protox lipid-peroxidation/ diphenylether
Fomesafen	REFLEX/FLEXSTAR	Protox lipidperoxidation/ diphenylether
Flumiclorac	RESOURCE	Protox lipid-peroxidation/N-phenylphthalimide
Fluthiacet	ACTION	Protox lipid-peroxidation/oxadiazole
Bentazon	BASAGRAN	Rapid photosynthesis inhibitor
2,4-DB	BUTOXONE/BUTYRAC	Auxin-type growth regulator/phenoxy acid

Herbicides for resistant soybean cultivars

Glufosinate for Liberty Link soybeans	LIBERTY	Glutamate inhibitor-lipid peroxidizer
Glyphosate for Roundup Ready soybeans	ROUNDUP ULTRA	Aromatic amino acid inhibitor

Prepackaged mixtures
Burndown prior to crop emergence

Alachlor-glyphosate	BRONCO

Broad spectrum grass and broadleaf control

Dimethenamid-imazaquin	DETAIL
Fluazifop-P-fomesafen	TORNADO/ TYPHOON
Flumetsulam-metolachlor	BROADSTRIKE+DUAL
Flumetsulam-trifluralin	BROADSTRIKE+TREFLAN
Fluthiamide-metribuzin	AXIOM
Metolachlor-metribuzin	TURBO
Pendimethalin-imazaquin	SQUADRON
Pendimethalin-imazaquin-imazethapyr	STEEL
Pendimethalin-imazethapyr	PURSUIT PLUS
Trifluralin-clomazome	COMMENCE
Trifluralin-imazaquin	TRI-SCEPT

Broadleaf control

Bentazon-acifluorfen	STORM/GALAXY
Chlorimuron-linuron	LOROX PLUS
Chlorimuron-metribuzin	CANOPY
Chlorimuron-sulfentrazone	AUTHORITY BROADLEAF/CANOPY XL
Chlorimuron-thifensulfuron	CONCERT
Lactofen-flumiclorac	STELLAR
Imazaquin-acifluorfen	SCEPTER O.T.

Grass control

Alachlor-trifluralin	FREEDOM
Fluazifop-P-fenoxaprop-P	FUSION

Herbicides for resistant cultivars

Chlorimuron-thifensulfuron	SYNCHRONY STS /RELIANCE for use with sulfonylurea-tolerant soybeans (STS)

pigweed, and suppress common lambsquarters, yellow nutsedge, and shattercane.

Chloramben, a benzoic acid herbicide, was introduced in the early 1960s as a preemergence herbicide. Chloramben provided a broader spectrum of control than the chloracetamides, and for many years it was the herbicide of choice for grass and broadleaved weed control. Like CDAA it has since been dropped from the market.

The dinitroaniline herbicide trifluralin was first available for soybeans in the mid-1960s. Ethalfluralin and pendimethalin were introduced a decade later. The dinitroaniline herbicides control seedling grasses and some broadleaves. Their gradual loss by volatilization and photodegradation coupled with their insolubility in water has limited the use of the dinitroanilines primarily to preplant incorporated treatments. Pendimethalin and trifluralin suppress the development of johnsongrass shoots from rhizomes when used at higher doses combined with sufficient secondary tillage to break rhizomes into short segments. Pendimethalin is labeled for surface application, which makes it available for conservation tillage systems.

Linuron, a classical photosynthesis inhibitor in the phenylurea family of herbicides, was introduced in the mid- to late 1960s for use as preemergence and directed postemergence treatments. Linuron controls seedling broadleaved weeds and suppresses seedling grasses. Strong sorption of linuron to soil colloids limits its soil use to soil surface applications and requires that doses closely match the clay and organic matter content of soil. This makes application of linuron to fields with variable soils difficult. The asymmetrical triazine metribuzin, introduced in 1973, provides a similar spectrum of weed control as linuron but is adsorbed less to soil colloids. Therefore, metribuzin can be used both as preplant incorporated and preemergence treatments and is less sensitive to variable soils. Both herbicides have considerable foliar activity and provide burndown in conservation tillage and directed postemergence applications.

The rapidly acting photosynthesis inhibitor bentazon was labeled in the early 1970s for postemergence control of annual broadleaved weeds (including cocklebur, smartweeds, mustards, and jimsonweed) and yellow nutsedge. Special timing is required to control velvetleaf and giant ragweed effectively. Control of pigweeds, common lambsquarters, morningglories, black nightshade, and common ragweed is marginal or lacking.

With the exception of chloramben, the herbicides available during the 1970s primarily controlled (at doses used in soybeans) either grass species or broadleaved species but not both. Trifluralin and alachlor controlled grasses, whereas linuron, metribuzin, and bentazon primarily controlled annual broadleaved weeds. Consequently, standard programs included a number of tank mixes or sequential treatments. A sequence that provided broad spectrum weed control was trifluralin (preplant incorporated) and linuron (preemergence); combinations included alachlor and linuron (preemergence), trifluralin and metribuzin (preplant incorporated), and alachlor and metribuzin (preplant incorporated or preemergence). Chloramben

(preemergence) alone provided fair broad spectrum control.

By the 1980s these treatments were still being used but were supplemented as needed with postemergence herbicides. The introduction of the postemergence herbicides sethoxydim, fluazifop, and quizalifop in the 1980s (and more recently clethodim and fenoxaprop) provided effective control of perennial grasses including johnsongrass. These herbicides are lipid biosynthesis inhibitors that are specific for grass control. Sethoxydim, fluazifop, and quizalifop, when combined with the herbicides available for broadleaved weed control such as bentazon and acifluorfen, were effective enough to make broad spectrum weed control with postemergence treatments alone a legitimate option.

Some possible treatments by the mid-1980s would have included the combinations of sethoxydim, bentazon, and acifluorfen (early postemergence), or fluazifop and acifluorfen (early postemergence). Most of the herbicide treatments described for the 1970s including trifluralin (preplant incorporated) and bentazon (early post-emergence), or alachlor (preplant incorporated or preemergence) and bentazon and/or acifluorfen (early postemergence) also were available.

Several Protox inhibitor herbicides are available for the control of some of the broadleaved weeds that bentazon misses. Two diphenylether herbicides, acifluorfen and fomesafen, were introduced in the early 1980s. They were followed by lactofen. All are used postemergence. The diphenylethers provide improved control of morningglories, black nightshade, and pigweeds compared to bentazon. Sulfentrazone (a triazolinone), flumiclorac (an N-phenylphthalimide), and fluthiacet (an oxadiazole) were introduced in the late 1990s, also for the control of broadleaves.

Two chemical families of branched-chain amino acid inhibitor herbicides (imidazolinones and sulfonylureas) were introduced to the soybean market in the late 1980s. The imidazolinone imazaquin was first registered in 1986, followed closely by imazethapyr (1989) and a decade later by imazamox. Chlorimuron was the first of the sulfonylurea herbicides to be introduced for weed control in soybeans. It is applied postemergence for control of broadleaved weeds and is mixed with other herbicides for preplant and preemergence soil treatments. Thifensulfuron, which was introduced in 1990, targets common lambsquarters, velvetleaf, pigweed, smartweeds, and mustards.

Another chemical group of branched-chain amino acid inhibitors is the triazolopyrimidines.

Flumetsulam, the first member of this group to be developed as a herbicide, was introduced in 1993 as a component in mixtures with trifluralin and metolachlor to provide broadleaf control. The second member of this family, cloransulam, was introduced to the soybean market in 1998.

As a group, the branched-chain amino acid inhibitors have both soil and foliar activity. They have been widely adopted because of their consistency of performance, efficacy on problem broadleaves, insensitivity to soil colloid interactions, and use at extremely low doses. Individually, these herbicides control a limited spectrum of weeds while missing others, but the broadleaves that they do control are extremely important ones. To increase their spectrum of control, they are almost always used tank mixed or prepackaged with other compounds.

By the 1990s the imidazolinone and sulfonylurea herbicides had become major contributors to the control of broadleaved weeds. To a large extent they replaced linuron and metribuzin, and they compete very well with the older postemergence herbicides such as bentazon and the Protox inhibitors. Tank mixes and sequences continued to be common. For example, combinations of soil-applied dinitroaniline or chloracetamide herbicides with imazaquin or chlorimuron-metribuzin were widely used. Sequential applications of dinitroaniline or chloracetamide herbicides on the soil with postemergence treatments of imazethapyr or chlorimuron were common. On the other hand, imazethapyr was frequently applied alone early postemergence as a single broad spectrum treatment. Grasses and broadleaved weeds escaping this treatment had to be treated with additional postemergence herbicides.

Clomazone, a carotenoid pigment inhibitor, was commercialized in 1985 and was targeted for the control of velvetleaf. Clomazone also controls several other broadleaved weeds and several seedling grasses. Spectacular cases of off-target whitening of sensitive nontarget plants resulted in limitations regarding application. Use of the emusifiable concentrate, which is volatile, on the soil surface is restricted to early-season (early preplant) applications, a time when nontarget species have not yet leafed out. Later treatments must be preplant incorporated. The less volatile microencapsulated formulations are applied preemergence to the soil surface and are not soil incorporated.

Genetically altered soybean cultivars are the most recent advancement in soybean weed control. Glufosinate provides broad-spectrum control of annual grasses and broadleaves in glufosinate-resistant cultivars, while glyphosate provides broad-spectrum control of annuals plus control or suppression of perennials in glyphosate-resistant cultivars. Prepackaged combinations of chlorimuron and thifensulfuron are used with STS (sulfonylurea-tolerant soybean) cultivars. Resistant cultivars permit the use of increased doses of sulfonylurea herbicides for problem weeds without causing damage to the soybean crop and allow the use of glufosinate and glyphosate, which normally kill soybeans.

Glufosinate and glyphosate do not provide control through the soil. The lack of soil activity means that the use of a single application may not provide season-long control, so two or more applications of the herbicide or a combination of the herbicide with a soil residual herbicide may be required when multiple weed flushes occur.

Soybean cultivars being made available to farmers as this publication goes to press will likely have a major impact on weed control practices, the availability of herbicides, the price of herbicides, and the development of future soybean varieties. Glyphosate (when used with glyphosate-resistant soybeans) provides broad spectrum control of annual weeds and control or suppression of perennial weeds. There is no carryover to subsequent crops. A prediction that over 65% of the soybean acreage in the eastern Corn Belt states will consist of glyphosate-resistant soybean cultivars by the year 2000 is probably conservative. Glufosinate, which provides broad spectrum control of annual grasses and broadleaves, could be a second contributor to this market.

Soybeans are shorter than corn, so equipment can be passed between the rows through much of the growing season. This has permitted the successful use of directed postemergence herbicide applications and selective treatments above the crop with glyphosate applied by re-circulating sprayers and wipe-on devices for tall weeds. Although these techniques have been used for controlling problem weeds, they require extra equipment and effort. More expedient and effective treatments such as broadcast applications of postemergence herbicides gradually are replacing the use of such specialized techniques.

The weed species in soybeans are similar to those in corn. Herbicides that can be safely used in soybeans, however, have tended to miss certain annual broadleaved weeds. In the 1970s velvetleaf, jimsonweed, giant ragweed, hemp sesbania, cocklebur, coffeeweed, and annual species of morningglories became more frequent and persistent problems in soybeans than they were in grass crops such as corn. Perennial grasses, especially johnsongrass, were major problems. During the 1980s these same broadleaved weeds persisted as problems. Johnsongrass, however, was effectively controlled with the postemergence herbicides sethoxydim and fluazifop. During the 1990s giant ragweed, common ragweed, common lambsquarters, kochia, pigweeds, water hemps, spotted spurge, and wild poinsettia became more evident. We believe that the extensive use of imidazoline and sulfonylurea herbicides, which did such a good job on velvetleaf, jimsonweed, and the other weeds that were problems earlier, helped select for these weeds. Quackgrass appears to be increasing in some parts of the Midwest due to reduced rates of atrazine in corn.

A partial list of the herbicides labeled for soybeans, their methods of application, and the types of weeds controlled is provided in Table 14.8.

Special Considerations for No-Till Soybeans.
Burn-down treatments to destroy emerged vegetation at the time of planting into no-till initially were made possible with the introduction of paraquat in the early 1960s. Paraquat typically was applied in combination with linuron or metribuzin to provide effective control of annuals and shoot removal of perennials. The later introduction of glyphosate, and the more recent registration of 2,4-D (followed by delayed planting of no-till soybeans), have provided control of belowground perennial structures as well as shoots.

Weed shifts are typical when management is changed. No-till culture has resulted in weed species that are not problems in conventionally tilled soybeans. Some of these include cool-season annuals (horseweed [or marestail], groundsel, and prickly lettuce), biennials (wild carrot and musk thistle), simple perennials (dandelion, pokeweed, and curly dock), less aggressive creeping perennials (goldenrod, asters, tall ironweed, and poison ivy), and a large number of woody species. Burn-down treatments with 2,4-D and/or glyphosate and selective treatments with glyphosate in glyphosate-resistant soybeans effectively control most herbaceous perennial species but are less dependable on woody perennials. Special efforts to eliminate initial scattered woody plants may be merited.

Tomatoes

Two systems for planting tomatoes are employed. Tomatoes for fresh market and those grown in some of the short growing season production areas for mechanical harvest are transplanted. Mechanically harvested tomatoes in the rest of the United States are direct seeded (Figure 14.2). Mechanically harvested tomatoes are used primarily for processing.

Tomatoes are planted after the risk of frost has passed, so weeds encountered tend to be summer annuals similar to those encountered in corn and soybeans. Some of the low growing, less competitive weed species such as galinsoga and common purslane also can be problems.

Weed control practices include tillage, herbicides, good management methods, and hand weeding. Tomatoes are a limited acreage, high-value crop, so the number of labeled herbicides is not great. Selective herbicides available for 1998 include preplant incorporated applications of napropamide prior to direct seeding or transplanting; pebulate, trifluralin, or metribuzin prior to transplanting; postemergence applications of metribuzin or sethoxydim on etablished direct-seeded plants or transplants; and DCPA or trifluralin applied postemergence or directed postemergence on established tomatoes but preemergence to weeds germinating later. Glyphosate or paraquat can be used for nonselective removal of emerged vegetation during the period between crops, on stale seedbeds, and on reduced tillage prior to the emergence of seedlings or before transplanting.

Planting tomatoes in fields with particularly heavy infestations of annual weeds or with creeping perennials should be avoided. Planting dates usually allow the employment of the stale seedbed approach described for soybeans (i.e., the field is prepared for planting and weeds are allowed to germinate). The weeds are then killed prior to direct seeding or transplanting with foliar-active herbicides. Cultivators can be used effectively when the crop plants are small. Weeds not controlled by other methods frequently are removed by hand.

A partial list of the herbicides labeled for tomatoes, their methods of application, and the types of weeds controlled is provided in Table 14.9.

Onions

Onions compete poorly with weeds. Not only are relatively few herbicides available for

Table 14.8 Partial List of Herbicides Labeled for Soybeans.[a]

	Seedling Grasses						Annual Broadleaves							Established Perennials				
	Foxtails	Large crabgrass	Fall panicum	Shattercane	Wooly cupgrass	Johnsongrass seedlings	Pigweed	Lambsquarters	Jimsonweed	Common ragweed	Velvetleaf	Cocklebur	Black nightshade	Johnsongrass	Quackgrass	Yellow nutsedge	Canada thistle	Field bindweed
Soil applied																		
Clomazone	X	X	X	S	X	S	S	X	S	X	X	S	S	S	N	N	N	N
Alachlor	X	X	X	S	S	S	X	S	S	S	N	N	X	N	N	X	N	N
Dimethenamid	X	X	X	S	S	S	X	S	S	S	N	N	X	N	N	X	N	N
Metolachlor	X	X	X	S	S	S	X	S	N	N	N	N	X	N	N	X	N	N
Ethalfluralin	X	X	X	X	X	X	X	X	S	S	N	N	S	S	N	N	N	N
Pendimethalin	X	X	X	X	X	X	X	X	S	S	S	S	S	S	N	N	N	N
Trifluralin	X	X	X	X	X	X	X	X	S	S	N	N	N	S	N	N	N	S
Metribuzin	S	S	S	N	N	N	X	X	X	X	X	S	N	N	N	N	N	N
Imazaquin	X	S	S	S	S	S	X	X	X	X	S	X	X	N	N	S	N	N
Imazethapyr	S	S	S	S	S	S	X	X	S	S	X	X	X	N	N	S	N	N
Post applied																		
Clethodim	X	X	X	X	X	X	N	N	N	N	N	N	N	X	X	N	N	N
Sethoxydim	X	X	X	X	X	X	N	N	N	N	N	N	N	X	X	N	N	N
Fluazifop-P	X	X	X	X	X	X	N	N	N	N	N	N	N	X	X	N	N	N
Quizalofop	X	X	X	X	X	X	N	N	N	N	N	N	N	X	X	N	N	N
Bentazon	N	N	N	N	N	N	S	S	X	S	X	X	S	N	N	X	S	N
Acifluorfen	S	S	S	S	N	S	X	S	N	N	S	S	X	N	N	N	S	S
Flumiclorac	N	N	N	N	N	N	S	S	S	S	X	S	N	N	N	N	N	N
Fomesafen	S	S	S	S	S	S	X	S	X	X	X	X	X	N	N	N	S	S
Lactofen	N	N	N	N	N	N	X	S	X	X	S	X		N	N	N	S	N
Chlorimuron	N	N	N	N	N	N	X	S	X	S	X	X	N	N	N	N	X	S
Thifensulfuron	N	N	N	N	N	N	X	X	S	S	S	S	N	N	N	N	N	N
Imazethapyr	S	S	S	X	S	X	X	S	S	S	S	X	X	S	N	S	S	N
Glufosinate[b]	X	X	X	X	X	X	X	X	X	X	X	X	X	S	S	S	S	S
Glyphosate[b]	X	X	X	X	X	X	X	X	X	X	X	X	X	X	S	X	S	S

X = good to excellent control; S = suppression; N = no control.

[a]Data compiled from claims on product labels and from weed control ratings from the following universities: Purdue, Ohio State, Michigan State, Kentucky, Illinois, and Iowa State.

[b]Apply only to resistant crop cultivars.

Figure 14.2 Bedded direct-seeded tomatoes, a typical horticultural annual row crop, under irrigation in California.

weed control but also onions often are grown in row spacings that are difficult to cultivate. Although expensive, hand weeding frequently is employed. Direct-seeded onions are planted 1/4 to 1/2 in. deep in a finely prepared seedbed. Such seedbeds favor weed seed germination.

Onions are an early planted crop, so they often are infested with both cool- and warm season annual weeds that emerge with the onions. Not much opportunity exists to kill annual weeds prior to early planting. Thus, good weed control in the crops that precede onions is very helpful.

Successful weed control in onions depends on the combination of rotating to fields with

Table 14.9 Partial List of Herbicides Labeled for Vegetable Crops

	Crops															Weeds							
	Beets	Cole crops	Carrots	Cucumbers	Lettuce	Lima beans	Onions	Peas	Peppers (transplant)	Pumpkins	Radishes	Spinach	Tomatoes	Watermelons	Potatoes	Annual grasses	Common purslane	Common ragweed	Black nightshade	Jimsonweed	Lambsquarters	Smartweed	Yellow nutsedge
Preplant incorporated																							
Bensulide		X	X	X			X		X	X				X		X	S	N	N	N	S	N	N
Clomazone									X	X	X					X	X	S	S	X	X	X	N
Cycloate	X											X				X	X	N	S	N	S	S	N
EPTC															X	X	X	S	X	S	S	S	X
Napropamide		X							X				X			X	X	S	N	N	X	S	N
Pebulate													T			X	S	N	X	N	S	S	S
Trifluralin			X	X		X			X				T			X	X	N	S	N	X	S	N
Preemergence or preplant incorporated																							
Alachlor							X									X	X	N	X	S	S	S	S
Bromoxynil								X								N	S	X	X	X	X	X	N
Imazethapyr						X		X								S	X	S	S	S	X	X	N
Metolachlor						X									X	X	X	N	S	N	S	S	X
Metribuzin													T		X	S	S	X	S	X	X	X	N
Naptalam				X										X		S	X	S	S	S	X	S	N
Pendimethalin															X	X	X	N	S	N	X	S	N
Preemergence																							
DCPA		X					X				X					X	X	N	N	S	X	N	N
Ethalfluralin			X											X	X	S	X	N	X	N	S	S	N
Linuron			X											X	X	S	X	S	S	S	X	X	N
Oxyfluorfen	X															S	X	S	X	S	X	X	S
Pronamide					X											X	X	S	S	N	X	X	N
Postemergence																							
Bentazon						X		X								N	X	X	S	X	S	X	S
Bromoxynil							X									N	S	X	X	X	X	X	N
Clethodim							X									X	N	N	N	N	N	N	N
Fluazifop-P			X				X									X	N	N	N	N	N	N	N
Imazethapyr								X								X	X	S	S	X	X	X	X
Linuron		X														S	X	X	X	S	X	X	S
Metribuzin													X		X	S	X	X	S	X	X	S	N
Oxyfluorfen							X									S	X	S	X	X	S	X	N
Phenmediphan												X				S	X	S	X	X	S	X	S
Quizalofop-P				X			X									X	N	N	N	N	N	N	N
Sethoxydim	X		X	X		X	X	X	X				X	X	X	X	N	N	N	N	N	N	N

Crops: X = labeled for crop; T = use on transplants only; Weeds: X = good to excellent control; S = suppression; N = no control.

minimum weed pressure, using available herbicides, cultivating in fields where the crop is grown in wide enough rows, and hand weeding as needed. Onions can be grown flat-planted in rows varying from 14 to 20 in. apart or on beds varying from 32 to 80 in. wide. Two to 8 rows can be planted on 32- to 40-in. beds (Figure 14.3); up to 15 rows are planted on 80-in. beds. Close-spaced rows (less than 16 or 18 in. apart) eliminate the option of cultivation between them. In addition, cultivating onions past the five-leaf stage causes excessive root damage.

A limited number of selective herbicides are available for onions. Selective preplant/preemergence herbicides include bensulide, bromoxynil, and DCPA. Selective postemergence herbicides include bromoxynil, clethodim, fluazifop-P, oxyfluorfen, and sethoxydim. Pendimethalin or trifluralin can be used directed postemergence to established onions for preemergence control of late-germinating weeds. Nonselective postemergence applications of glyphosate and paraquat can be used to control emerged weeds in stale seedbeds and reduced tillage systems prior to crop emergence.

Limited onion acreage coupled with relatively high crop values—and thus high liability—does little to encourage the development of

Figure 14.3 Onions grown in two rows on 32-in. beds.

herbicides or even to maintain labeling for those herbicides that are available. In fact, several have been lost due to liability exceeding profits or products with insufficient market share in other crops to be profitable. Herbicides formerly used for control of weeds in onions and no longer marketed include chloroxuron, chlorpropham, and CDAA.

A partial list of herbicides labeled for onions, their methods of application, and the types of weeds controlled is provided in Table 14.9.

Herbicides used on other major annual vegetable crops also are listed in Table 14.9.

Small Grains

Small grains, including wheat, barley, oats, and rye, historically have been grown in rows ranging from 6 to 14 in. These crops are planted either in the fall or early spring, depending on whether they are winter grains, which require vernalization, or spring grains.

The major weed problems in fall-seeded (winter) small grains tend to be cool-season weeds (winter annuals, simple perennials, and creeping perennials) and, when stands are poor, summer annuals. Specific weeds include winter annuals such as downy brome, cheat, common chickweed, henbit, marestail, and mustards and perennials such as wild garlic, dandelion, and Canada thistle.

Spring-seeded small grains are most often infested with summer annual weeds. Wild oat is a major problem and sometimes so are summer annual grasses including foxtails. Summer annual broadleaved weeds, including common ragweed, sunflower, common lambsquarters, wild buckwheat, kochia, Russian thistle, pigweeds, black nightshade, smartweeds, and velvetleaf, are frequent problems. Under cool conditions, some of the winter annuals may act

as summer annuals and invade spring-planted small grains.

Herbicide-resistant weed populations have appeared in areas in which continuous wheat is the dominant crop.

Crop shading is a major contributor to overall weed control, so good stands, fertilization for good crop vigor, and adequate moisture can have a substantial impact on weeds present at harvest.

As has been the case with other agronomic crops, tillage prior to crop establishment historically has been the major weed control practice. Because they provide more efficient production and erosion control, herbicides have largely replaced tillage operations.

Approximately 20 individual herbicide active ingredients are labeled for weed control in small grains (Table 14.10). Glyphosate and paraquat are used as nonselective postemergence herbicides to obtain dependable burndown of emerged weeds prior to crop establishment. Nonselective treatments also can be used on emerged weeds after harvest to control weeds in place of tillage or for problem weeds.

Selective control of weeds with chemicals is integral to most small grain production. Successful herbicide use depends on finding a weed stage susceptible to the herbicide that coincides with a tolerant stage of the small grain. Product labels describe explicitly the proper timing of application and emphasize avoiding growth stages critical to crop yield. For example, the auxin-type growth regulators (2,4-D, MCPA, and dicamba) can damage small grains when the plants are in the seedling stage or when their meristems are particularly active, such as at tillering, late jointing, and flowering (Figure 14.4). Consequently, they are applied after tillering and prior to jointing or the boot stage, or after the dough stage.

Less reliance has been placed on soil-applied herbicides for control of weeds in small grains than in crops such as corn and soybeans. Soil-applied herbicides are generally restricted to geographic areas and to special weed problems. Preplant incorporated and preemergence herbicides inlude diclofop and triasulfuron for annual bromegrasses; triallate for wild oats; and chlorsulfuron, diuron, linuron, and trifluralin (in restricted geographic areas) for annual grasses and broadleaves.

Several herbicides are used primarily early postemergence and must be applied when annual weeds are small. The two- to three-leaf stage is recommended on all of these product labels; none

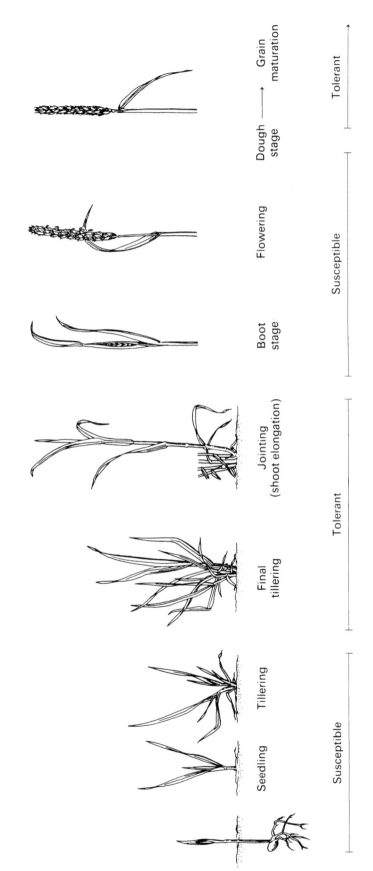

Figure 14.4 Stages of growth of winter wheat susceptible or tolerant to damage from auxin growth regulator herbicides.

exceed the five-leaf stage. Early postemergence herbicides include diclofop, difenzoquat, imazamethabenz, metribuzin, propanil, thifensulfuron, and tribenuron.

Herbicides that can be applied postemergence include fenoxaprop-P for annual grasses and bromoxynil, 2,4-D, dicamba, clopyralid, chlorsulfuron, MCPA, metsulfuron, prosulfuron, and triasulfuron for broadleaves.

Thifensulfuron has been effective for the consistent control of wild garlic in winter wheat. Several other herbicides provide suppression, including triasulfuron, tribenuron, prosulfuron, chlorsulfuron, 2,4-D, and dicamba.

Glyphosate and 2,4-D are labeled for postemergence control of weeds at the crop maturity/preharvest stage.

Some specific problems relating to weed control are associated with the Great Plains and Columbia Basin wheat production regions of the United States. The major crop in these regions is winter wheat, with occasional use of another grain crop such as winter barley or milo. The low rainfall characteristic of these regions leads to a number of problems. Yield potential is limited by lack of moisture, so profits are restricted and the funds available to control weeds may be limited. The lack of vegetation during part or all of the year renders much of the area subject to erosion from wind and water. In fact, it is common practice to crop a field once every 2 yr, allowing it to remain fallow (free from live vegetation) in the alternate year in order to accumulate moisture that weeds and volunteer grains otherwise would use.

Annual weeds may be abundant in these crops at maturity in years when inadequate or variable moisture results in thin stands or limited crop growth. These weeds are usually green. They interfere with harvest and cause spoilage in the harvested grain. A common practice is to apply 2,4-D when the wheat is mature to remove the weeds just prior to harvest.

Annual species of bromegrasses such as downy brome (*Bromus tectorum*), chess (*B. secalinus*), and Japanese chess (*B. japonicus*) also are severe problems in some of these dryland wheat regions. These winter annuals are much more troublesome when wheat is grown year after year than in wheat fallow systems. Deep burial by plowing provides partial control but leaves the soil susceptible to erosion. Several between-crop or fallow herbicide treatments to control these and other weeds can be used to replace tillage operations.

Keeping weeds controlled while maintaining crop residues on the soil surface to prevent erosion presents major management challenges in dryland wheat. Several techniques have been employed successfully. Strip cropping (growing narrow strips of crops perpendicular to the prevailing wind direction), stubble mulch tillage (keeping the wheat stubble on the surface by use of undercutting tillage equipment), and a combination of both have been practiced since the Dust Bowl era. In the stubble mulch system, residues are retained on the soil surface while weeds are controlled with sweep plows, chisel plows, and rod weeders (see Chapter 18).

Historically, several tillage implements have been used in sequence to achieve appropriate soil surface conditions. The introduction of herbicides has provided alternatives to tillage for controlling weeds on these residue management systems. Theoretically at least, chemical fallow could reduce or even eliminate the need for tillage operations.

Limited profit potential obviously limits the amount of inputs affordable for weed control from either tillage or herbicides. In addition, variable rainfall and sporadic germination of weed seeds make the performance of single applications of soil-applied herbicides inconsistent. Current weed management for dryland wheat should combine appropriate tillage, between-crop herbicides, preemergence herbicides, early postemergence herbicides, and postemergence herbicides to match economically the needs of individual situations and weed species.

A partial list of herbicides labeled for small grains, their methods of application, and the types of weeds controlled is provided in Table 14.10.

PERENNIAL CROPS

Perennial crops include a wide array of species grown for diverse purposes. In this section they are grouped into the major categories of herbaceous and woody perennials. Herbaceous perennial crops are further subdivided by type of management system: crops with limited tillage (e.g., asparagus, strawberries, mint, ornamentals); mowed crops (turf for the homeowner and in parks and public areas; forage and hay crops; turf for control of erosion along roads, levees, and airport runways); and crops with minimal management (e.g., ground covers and rangeland). Perennial woody species include tree fruits, nuts, cane fruits, grapes, shrubs, ornamental trees, windbreaks, Christmas trees, and trees for timber and pulpwood.

Perennial crops are maintained from several years to several decades. Crop rotation is a viable

Table 14.10 Partial List of Herbicides Labeled for Small Grains

	Crops					Type of Application								Weeds						
	Spring wheat	Winter wheat	Barley	Oats	Rye	Burndown	Preplant	Preemergence	Early postemergence	Postemergence	Postemergence preharvest	Postharvest for problem weeds	Chemical fallow (between crops)	Winter annual grasses	Winter annual broadleaves	Wild garlic	Summer annual grasses	Wild oats	Summer annual broadleaves	Canada thistle
Glyphosate	X	X	X	X	X	X					X	X	X	X	X		X	X	X	X
Paraquat	X	X	X			X							X	X	X		X	X	X	
Diuron			X	X	X			X	X						X		X		X	
Linuron				X				X									X		X	
Triallate	X	X	X				X	X										X		
Trifluralin	X	X	X				X						X	X			X		X	
Diclofop	X	X	X				X	X	X					X			X	X		
Difenzoquat	X	X	X							X								X		
Imazamethabenz	X	X	X							X					X			X		
Metribuzin	X	X	X							X			X	X	X		X		X	
Propanil	X									X							X		X	
Thifensulfuron	X	X	X	X						X					X	X		X	X	
Tribenuron	X	X	X	X						X					X			S	X	S
2,4-D	X	X	X	X	X				X	X	X	X	X		X	S			X	S
Bromoxynil	X	X	X	X	X				X	X									X	S
Clopyralid	X	X	X	X					X	X		X	X						X	X
Dicamba	X	X	X	X	X		X	X		X		X	X		X				X	S
Fenoxaprop-P	X	X							X	X							X	X		
MCPA	X	X	X	X	X					X					X				X	S
Metsulfuron	X	X	X						X	X			X		X				X	
Prosulfuron	X	X	X	X	X				X	X					X	X			X	S
Triasulfuron	X	X	X				X	X		X				S	X	S			X	S

X = labeled; S = suppression.

option only when perennial crops are maintained for 5 yr or less. In most cases, perennial crops must be left undisturbed to grow. Thus, tillage can be employed only in strips between rows or for occasional renovation. Mowing is an integral part of the management program for many herbaceous perennials.

Management systems for any perennial crop are not diverse. The weeds that persist are well adapted to the crop and management system. Annual, biennial, simple perennial, and creeping perennial weeds all can be found in perennial crops.

Two types of weed situations must be dealt with in raising perennial crops: weeds present at the time of crop establishment and weeds in the established crop. Weeds present at establishment are the same that would occur in any crop planted at that time of year with that particular method of seedbed preparation. Control of weeds during establishment is needed to give the perennial crop a quick start. After establishment, however, the weed population shifts to those species adapted to the crop and its management system.

Establishment of Perennial Crops

The first step in establishment is site preparation, which can vary from extensive tillage and fumigation to no preparation other than opening the soil to plant the crop. The amount of preparation done depends on weeds and other pests present, the crop itself (competitiveness, survivability, value), site location, and planting method.

Perennial crops can be established from transplants or seeds. Normally a transplant becomes established more rapidly and dependably than a plant of the same species grown from seed.

When transplants are involved, the plants must be grown from seeds or cuttings. Growing transplants usually is a separate operation done by someone other than the field grower. Seeds can be started in the greenhouse in flats, pots, or benches using special potting mixes and steam sterilized or fumigated soils. Usually no additional weed control is required at this stage. Cuttings usually are started in special propagation benches in greenhouses with specialized rooting media. Again, usually no additional weed control is required.

Weed control becomes important when the seedlings or cuttings are transferred to outside beds or nurseries or when the seeds initially are planted in beds. Efforts must then be made to eliminate diseases, insects, and other pests as well as weeds. When the volume of soil to grow these plants is limited, the soil can be steam sterilized. Soil fumigants are used if steam is unavailable or the volume of soil is too large to handle. Fumigants provide somewhat less control than steam but fit a wider range of situations. In some cases, entire fields are fumigated. Although some fumigants are applied as drenches (in large volumes of water) and although product labels indicate that water (overhead irrigation) can be used to "seal" the chemical into the soil, best performance is obtained when the treated soil is covered with a tarp. Tarps are removed after a recommended period of time, usually a few days, and the area is planted after the fumigant has dissipated, anywhere from a few days to a few weeks after tarp removal.

Starting perennial plants in beds, nurseries, and specially prepared fields prior to transplanting is a labor-intensive, high-return crop management system. In some instances, the plants produced are transplanted to their final site. In others (e.g., many woody species), plants may be placed in nurseries and grown 1 to several yr prior to final transplanting. In these cases, preparation and management are similar to what is used for annual crops. Selecting herbicides to fit the crop species is the major weed control consideration.

The remaining discussion on establishment relates to perennial crops that are established on the site where they will be used for their long-term purpose.

Removal of troublesome weeds prior to planting is the first critical step in establishment. It should be accomplished either before or during other aspects of site preparation. Undesirable species closely related to or with growth habits similar to the crop should receive special attention. Examples include taprooted simple perennials in forage legumes, perennial grasses in turf, perennial poisonous plants in permanent pastures, low-value woody species in tree plantations, and creeping perennial weeds in nearly all perennial crop situations.

Crops are established on a variety of sites. Examples of typical sites following planting include tilled, bare ground seedbeds; tilled seedbeds with an organic mulch of some kind (straw, peat moss, sphagnum); tilled, bare ground soil in which the perennial crop is planted with a nurse crop; a relatively undisturbed soil with mulch provided by killing the previous crop; relatively undisturbed soil without appreciable cover; and relatively undisturbed soil with growing vegetation.

Establishment of perennial herbaceous species on tilled, bare ground sites by direct seeding starts with tillage operations similar or identical to those used for seedbed preparation of annual crops. Hence, efforts are made to minimize weed competition and maintain adequate moisture for seed germination and seedling survival. This type of site preparation is appropriate if erosion is not a problem. Weed control will be greatly facilitated if selective herbicides for the direct-seeded crop are available.

Mulches can be used on tilled sites to maintain adequate moisture near the soil surface or to reduce soil erosion. On areas of limited size, the mulch can be hauled to the site and spread out. Hydraulic mulchers and seeders are used, for example, in some turf situations. Selective soil-applied herbicides should be applied before the mulch is spread.

Nurse crops (fast growing crops that are planted with the desired crop but that are harvested or die out in a reasonably short period of time) have been used as a means of providing weed control and protection to the crop and soil until the perennial crop is established. Before selective herbicides became available, small-seeded legumes, pasture grasses, and lawn grasses were routinely planted with nurse crops (small grains for legumes and pasture grasses, ryegrasses for cool-season lawn grasses). The nurse crop can, however, slow the development of the desired species. When erosion can be kept in check by other means, the nurse crop can be replaced by the use of herbicides to control weeds, permitting rapid crop establishment as well as increased productivity and increased crop quality. For example, three cuttings of alfalfa hay can be obtained without a nurse crop the first year of growth, in contrast to one cutting if the alfalfa is sown with a small grain. High-quality bluegrass lawns can be obtained in one season with herbicides rather than waiting for perennial ryegrass to disappear over a number of seasons.

Planting into a mulch from a previous crop can provide a sound approach to crop establishment. Tillage can be eliminated if it is not required to shape the terrain or to control weeds. Herbicides such as paraquat and glyphosate can provide dependable control of the vegetation, and planting equipment is available to plant through the herbicide-treated vegetation. Thus, competing vegetation is eliminated and becomes a mulch to conserve soil and moisture. Except for the perennial nature of the crop to be planted, this technique is the same as that described earlier for no-till. Examples include the renovation of turf and legume establishment in pastures. If vegetation to act as a mulch is inadequate, a cover crop can be sown, permitted to grow, and then killed and used as mulch.

Although perennial crop seeds rarely are planted in unprepared sites, transplants, particularly of woody species in low-value sites, can be planted into undisturbed sites. Transplants can be introduced into all those types of planting situations described for seeds.

Some general principles should be adhered to during site preparation and establishment. Control of creeping perennial weeds usually is not evident until several months to a year after control practices are initiated, so sufficient time should be allowed after treatment but prior to crop establishment for meaningful evaluation and retreatment. Retreatment almost always is advisable for perennial weeds that reproduce vegetatively (see Chapter 13). Planting cold-tolerant crops in the late summer or fall permits summer annual weeds to be killed by frosts. Unfortunately, winter annuals or cold-tolerant perennial weed species can be a problem with this technique. Selective early postemergence herbicides should be applied according to crop stage and weeds present. Many grass crops should be past the juvenile stage before post-emergence herbicides are applied. Mowing can be used to reduce competition from tall weeds, particularly in herbaceous perennial crops with low growing points such as turf or forages.

Established Perennial Crops

Herbaceous Perennial Crops

Crops with Limited Tillage. These crops include food and ornamental crops that are maintained in beds or rows, such as asparagus, mint, strawberries, and perennial flowerbeds. Tillage operations are used to help control weeds and diseases, stimulate stand vigor, maintain rows, and facilitate harvest. Three crops, asparagus, strawberries, and mint, are described as examples of this type of management system. A partial list of herbicides labeled for herbaceous perennial food crops is provided in Table 14.11. A partial list of herbicides labeled for perennial flowerbeds (herbaceous ornamentals) is provided in Table 14.16.

Asparagus. This crop is grown in beds maintained for as many as 12 to 15 yr. Shallow tillage over the tops of the beds and cultivation between the rows have been standard practices. The beds are worked with shallow tillage in the spring before spears emerge to provide weed control and an acceptable surface for harvest. The spears are harvested over a period of 4 to 12 wk. If weeds become troublesome during harvest, the beds can be reworked or treated with 2,4-D amine, or dicamba (in western states). After harvest, cultivation is used between the beds. The tops of the beds are lightly tilled, and the ferns then are allowed to grow until fall to replenish the root system (Figure 14.5).

Several herbicides are available that supplement or replace some of the tillage operations. Selective soil-applied herbicides include napropamide, norflurazon, trifluralin, diuron, linuron, metribuzin, and terbacil. These herbicides should provide residual control of germinating weed seeds. Selective postemergence treatments include 2,4-D and dicamba for broadleaved weeds and fluazifop-P and sethoxydim for grass weeds.

Glyphosate and paraquat are used for the nonselective removal of emerged weeds prior to spear emergence or after harvest. Asparagus is more tolerant of glyphosate than most herbaceous crops, so it can be used directed postemergence once the ferns have attained adequate size.

Strawberries. Strawberries are established as transplants either in sheet mulches or in tillage systems where they are maintained in rows or beds with cultivation. Fumigation of the soil followed by covering with plastic tarps is a common practice just before establishment.

Table 14.11 Partial List of Herbicides Labeled for Herbaceous Perennial Food Crops[a]

	Asparagus	Mint	Strawberries	Pineapple	Sugarcane
Soil applied					
DCPA			X		
Napropamide	X	X	X		
Norflurazon	X				
Pendimethalin					X
Simazine			X		
Trifluralin	X				X
Soil or post-applied					
Atrazine					X
Bromacil				X	
Diuron	X	X		X	X
Hexazinone				X	X
Linuron	X				
Metribuzin	X				X
Oxyfluorfen		X			
Terbacil	X	X	X		X
Post-applied					
Ametryn				X	X
Asulam					X
Bentazon		X			
Bromoxynil		X			
Clethodim			X		
2,4-D	X		X		X
Dicamba	X				X
Fluazifop-P	X		X	X	
Sethoxydim	X	X	X		
Nonselective post					
Glyphosate	X		X	X	X
Paraquat	X	X	X	X	X

[a]Some herbicides may be limited to geographic areas or to nonbearing crops only. Check current labels and state recommendations.

Fields heavily infested with serious perennial weeds are not suitable for strawberries and should be avoided. Such fields should be planted in crops that result in practices that effectively control problem perennial weeds for a year or more before planting strawberries.

If strawberries are produced using only cultivation and hand weeding, weed control costs may be as high as several hundred dollars per acre per year. These costs can be substantially reduced with herbicides. DCPA, napropamide, simazine, and terbacil are soil applied for control of annual weeds. DCPA and napropamide can be used at any time, whereas simazine and terbacil are more restricted (see labels). Paraquat can be applied as a shielded directed spray between rows. 2,4-D is used to control established broadleaved weeds after harvest.

Mint (Spearmint and Peppermint). Mints are creeping perennials with stolons. They are harvested like a hay crop in that the tops are cut and chopped in the field. The shoot material then is steam distilled to extract the mint oil. Major growing regions are the upper Midwest and the Pacific Northwest. Weeds compete with mint to reduce yield and contribute off-flavors to the mint oil at harvest, resulting in lowered oil quality.

In the Midwest, mint fields are plowed shallowly following the first killing frost to protect the stolons from winterkill and to control a rust

Figure 14.5 Well-maintained and weed-free rows of asparagus in the fern stage.

disease. Plowing should not bury the stolons more than 4 1/2 to 5 in. The mint beds are prepared in the spring by harrowing or some other form of shallow tillage to level the beds and control any annual weeds present.

A limited number of herbicides are available for use in peppermint and spearmint. Paraquat is used to remove emerged weeds before mint shoots emerge. Diuron (preemergence after the last cultivation in the spring in the Northwest), napropamide, oxyfluorfen, and terbacil are used preemergence to the weeds. Terbacil, bentazon, and bromoxynil are used postemergence to control broadleaved weeds, and in some states sethoxydim is used postemergence to control annual and perennial grasses.

In the Pacific Northwest, flaming or applications of contact herbicides are used rather than winter tillage, so the management for mint in this region more closely resembles no-till production systems or those used in hay production. Efforts to use less tillage also have been made in the Midwest.

Mowed Perennial Crops. These crops include lawn-quality turf, roadside grassses, well-managed pastures, and hay crops. Perennial weeds tend to dominate, although biennials can become problems in less frequently mowed sites. Summer annuals that survive mowing also can be troublesome. Winter annuals frequently develop during periods of winter dormancy and pose problems in the late spring. Weeds tend to establish themselves when stands or canopies are opened up by mowing, disease, insect damage, mechanical damage, drought, or winter stress.

Weeds in mowed perennials include winter annuals such as chickweed, many mustards, winter annual bromegrasses, and speedwell; biennials such as bull thistle, musk thistle, and wild carrot; simple perennials such as dandelion, plantains, black medic, and curled dock; and creeping perennials such as quackgrass, yellow nutsedge, bermudagrass, pasture grasses, and white clover; and some summer annuals such as crabgrass, green foxtail, goosegrass, barnyardgrass, and knotweed.

Lawn-Quality Turf. This type of turf is expected to provide a pleasant, uniform-appearing cover for homes, businesses, schools, and parks and yet take a reasonable amount of foot traffic. A good-quality turf is uniform in color, texture, plant density, and growth rate. Standard management practices include weekly mowings; fertilization one or more times a year; control of insects, weeds, and diseases when needed; irrigation when rainfall is lacking; and reseeding or resodding when bare spots occur.

Turfgrasses are divided into cool-season and warm-season species. The major cool-season turfgrasses are Kentucky bluegrass, red and chewings fescue, creeping bentgrass, the perennial ryegrasses, tall fescue, and rough bluegrass. The major warm-season grasses include Bahiagrass, bermudagrass, centipede grass, kikuyugrass, St. Augustine grass, and zoysia grass.

We define four aspects related to management of weeds in lawns: site preparation, turfgrass establishment, management practices to maintain a competitive established turf, and control of weeds invading the established lawn.

Site Preparation. The first step in achieving the desired quality of turf is to properly prepare the site prior to planting. Special efforts should be made to eliminate undesirable established perennial grasses such as pasture grasses, which include tall fescue, smooth bromegrass, redtop, and orchardgrass, and weedy perennial grasses such as quackgrass and bermudagrass. These wide-leaved grasses tend to grow faster than the narrow-leaved grasses used for lawns, thus accentuating an already undesirable difference in surface texture. These perennial grasses are difficult if not impossible to selectively remove once turf is established.

Two or more treatments frequently are required to eliminate perennial grasses. Control efforts may be sequential herbicide treatments, a herbicide treatment followed by clean tillage, or several repeated tillage operations. Use of soil fumigants (e.g., dazomet) may be warranted where very high value limited areas are involved. Some of the herbicides used for site preparation include diquat, cacodylic acid, glufosinate, glyphosate, and pelargonic acid. Of this group, glyphosate is most likely to translocate and control well-established perennials.

The site should be free from compacted soil layers or areas and large hard objects such as chunks of buried concrete, rocks, sheets of dry wall and insulation, boards, other building materials, and similar objects that limit development of turfgrass roots and the movement of water.

The site should be appropriately graded so that areas that impound water or that are sharply elevated and tend to dry out are eliminated. After grading, the seedbed must be prepared properly, and fertilizers such as lime, phosphorus, potassium, and other immobile soil amendments should be mixed into the soil prior to planting. If topsoil, peatmoss, or manure is hauled in, care should be taken to avoid the introduction of perennial grasses, sedges, and other problem weeds.

Turfgrass Establishment. Establishment of a vigorous turf using high-quality seed or sod is the second step. The idea is to establish a healthy, thriving turf that can resist invasion by weeds. A good-quality turf is initiated at planting time by investing in quality varieties and by selecting species and varieties that tolerate disease, climate extremes, and other conditions typical of the geographic area. Low expenditures at planting can be costly later if a poor lawn is the result.

Because the goals of individual property owners can vary, including the amount of management they are willing to use, the type of turfgrass planted should be chosen carefully. Is the goal to obtain a picture-perfect lawn that has uniformity of leaf size (width), color, and growth rate? And is the owner willing to expend time and money to maintain such a lawn? If so, an improved Kentucky bluegrass variety or a mixture of varieties might be the best choice. Is the goal to have a durable lawn, one that is green from spring through fall, establishes from seed rather quickly, keeps children and household pets out of the mud, withstands some dry periods, and tolerates summer heat and humidity without special attention? If so, mixtures of fine fescues, perennial ryegrass, and Kentucky bluegrass might be the answer. Is the goal to have a green ground cover that is fertilized sparingly and rarely weeded? If so, a lawn-quality tall fescue might be appropriate. The time to decide on goals is at establishment. It should be noted that some new dwarf or slow-growing varieties of turfgrass, introduced to reduce the number of mowings, may allow excessive weed invasion and not resist wear.

Turfgrass germination, emergence, and establishment are maximized by maintaining moisture at the soil surface. Irrigation and mulching help achieve these objectives. Seed germination that is rapid and prolific results in good stands and quick canopy closure.

Weeds that appear in an emerging turf must be controlled as soon as practical. One of the major causes of seedling failure is competition from weeds. Weeds will be the same as those germinating in an annual crop planted at the same time of year. Sometimes adjusting the planting time can reduce the amount of weeds that compete. For example, cool-season turfgrasses can be planted in late summer or early fall to minimize competition from summer annual weeds.

Warm-season annual weeds will cease to grow as the weather cools. Winter annuals, however, sometimes can cause problems.

Turfgrasses that are seeded in the spring can be treated with herbicides to reduce competition from summer annual weeds. Siduron can be soil applied at planting to selectively control foxtails, crabgrasses, and barnyardgrass. Bromoxynil can be applied on new seedings for broadleaved weeds but is restricted to non-home sites. Fenarimol, a fungicide, is used to suppress *Poa annua* in newly seeded bentgrass. Several herbicides can be used once seedling turfgrasses are tall enough to be mowed. Examples include fenoxyprop-P for seedling grasses and low doses of 2,4-D for annual broadleaves.

Mowing can be initiated once the grass is emerged and rooted. The main concern of early mowing is damage to the grass from the wheels of the mowing equipment. Mowing, however, can reduce competition from weeds that canopy over the young grass.

Management Practices to Maintain a Competitive Established Turf. Good management practices are the major contributors to weed control.

An integral part of management is proper mowing. Mowing height and frequency are major contributors to the competitive ability of lawn-quality turf. Higher cutting (e.g., heights of 1.5 to 3 in.) allows good shoot growth, which shades out weed seedlings and encourages the development of extensive root systems, which increases the tolerance to dry weather. Turf management experts suggest that mowing remove only 1/3 of the shoot.

A lawn should be mowed at least weekly. Mowing intervals that are too long result in shock to the grass, which delays recovery, creates stress, and leaves long grass clippings that can clump and smother the lawn and are unsightly. Frequency of mowing has more impact on appearance than height of mowing.

Fertilization and watering should be done properly to favor a healthy, competitive turf that discourages weed development. Turfgrasses have relatively high nitrogen requirements. Fall fertilization with a balanced fertilizer to promote root development is the foundation of a good fertility program. Additional fertilizer applications should be made as needed to maintain optimal growth. Irrigation on established turf should result in soaking of the root zone and be done at infrequent intervals. Frequent light waterings coupled with close mowing is the recipe for ensuring the invasion of crabgrasses and other annual weeds.

The presence of certain weed species usually is a manifestation of a management problem. Goosegrass, prostrate spurge, and knotweed grow in areas with excessive traffic and soil compaction. Barnyardgrass and nutsedge indicate periodic standing water. *Poa annua, Poa trivalis*, violets, moss, and algae are favored by moisture and shade. Carpetweed and sheep sorrel frequently indicate infertile, dry, acid soil conditions. Sandbur is best adapted to infertile, dry, sandy sites. Application of a herbicide for these situations provides only a temporary solution. Modification of the site offers a much more permanent solution to the weed problem.

Preventive techniques should be employed if weed seeds are being introduced from the outside into weed-free areas. For example, mowing equipment that might have seeds of annual grasses clinging to it should be cleaned. Manures and top dressings free of weed seeds and tubers should be chosen, and wind-dispersed weeds such as dandelions and thistles adjacent to the clean area should be controlled (Figure 14.6).

When bare areas appear, special efforts should be made to employ adequate weed control measures and to encourage turf growth and reestablishment.

Additional chemical weed control can compensate for a less vigorous turf. Examples include turf that gets heavy wear, turf that is treated with growth regulators to reduce height and thus provides less shading, or turf on which the use of nitrogen fertilizer has been reduced.

Control of Weeds in Established Lawns. Weed control should not have to be done on a frequent basis. Most well-maintained turfgrass should not require chemical weed control every year. Once a lawn is well established, weed control should be used on an as-needed basis rather than routinely.

Weeds that are major problems in lawn-quality turf are those that can withstand frequent close mowings. These include summer annual broadleaves such as knotweed, carpetweed, yellow wood sorrel, speedwell, lambsquarters, prostrate spurge, shepherdspurse, kochia, and purslane. Winter annual broadleaved weeds include several mustards, common chickweed, and henbit. Simple perennial broadleaved weeds include dandelion, buckhorn plantain, common plantain, curly dock, and black medic. Ground ivy, field bindweed, Canada thistle, red sorrel, and white clover are creeping perennial broadleaves that are found in turf.

Broadleaved weeds in established turf usually are controlled with postemergence applications of the auxin-type growth regulators 2,4-D, dicamba, mecoprop, 2,4-DP, MCPA, triclopyr, and clopyralid. Formulations available to homeowners typically contain 2,4-D alone, or mixtures of 2,4-D plus mecoprop, 2,4-D plus dicamba, and 2,4-D plus mecoprop plus dicamba. Weed species vary considerably in their response to these herbicides (Table 14.12) and dictate the individual herbicide or mixture to be used. Special efforts should be made to avoid applications of dicamba and clopyralid over the roots of desirable woody ornamentals. Bromoxynil, a nitrile herbicide, can be used early postemergence to control annual broadleaves.

Although the auxin-type growth regulators can be used to treat all life cycle groups of broadleaves, the creeping perennials most likely

Figure 14.6 Dandelion-infested lawn on the left is a potential source of seed to the weed-free lawn on the right.

Table 14.12 Response of Broadleaved Weed Species to Postemergence Herbicides for Bluegrass Turf[a]

Weed Species	2,4-D	Mecoprop	Dicamba
Black medic	N	F	G
Carpetweed	G	F	G
Chickweeds	N	F–G	G
Curly dock	F	P	G
Dandelion	G	F–G	G
Ground ivy	P	F	F–G
Henbit	F	F	G
Knotweed	N	F	G
Plantains (buckhorn, common)	G	P	N
Purslane	F	N	G
Red sorrel	N	N	G
Speedwell	F	F	F
Spurge (prostrate)	P	F	F–G
Thistles	F–G	F	G
White clover	F	G	G
Wild carrot	G	F–G	G
Wild garlic and onion	F	N	F–G
Yarrow	F	P	G
Yellow woodsorrel	F	F	F

[a]Data from Street et al. 1977
N = none; P = poor; F = fair; G = good.

will require two or more treatments for complete control.

Broadleaves in established turf also can be controlled with preemergence applications of isoxaben. As with most soil-applied treatments, applications must be made before germination of the weeds and must be watered or rained into turf for good performance. Several of the herbicides for preemergence grass control also control germinating annual broadleaved weeds.

Summer annual grasses most frequently encountered in established turf are hairy crabgrass, smooth crabgrass, goosegrass, and foxtails.

Several herbicides are used preemergence (pregermination to the weeds) to selectively control annual grasses in established turf. Often called "crabgrass preventers," they include pendimethalin, benefin, bensulide, DCPA, napropamide, trifluralin, dithiopyr, oryzalin, and prodiamine. Most also control several germinating annual broadleaved weeds. Selective herbicides with postemergence activity on grasses include fenoxaprop-P for warm-season annuals and established johnsongrass and bermudagrass and dithiopyr for crabgrass. Asulam and MSMA also are labeled for selective postemergence use on annual grasses.

The perennial grasses are the most difficult group of weeds to remove selectively from established turf. Examples include tall fescue, quackgrass, nimblewill, bermudagrass, dallisgrass, *Poa annua* (which can be either annual or perennial), bentgrass, Bahia grass, kikuyugrass, rescuegrass, Italian ryegrass, and velvetgrass. They can be removed by hand or, if they occur in limited patches or as scattered plants, by spot treatments with effective symplastically translocated herbicides.

Total renovation of the turfgrass is needed when too much area is infested with perennial grasses (Figure 14.7). Glyphosate is the herbicide most likely to be applied to kill all vegetation for site preparation or renovation of poor-quality turf. Cacodylic acid also is labeled for this use. If the treated area already is smooth and on the desired grade, the residue from the killed sod can be used as mulch for the seeding of desired species. The seed, however, must be planted through the grass residue and placed so that it comes in good contact with the soil.

Certain perennial grasses can be controlled partially by management practices that favor the competing turfgrasses while providing less than ideal conditions for the weedy species. Vigorous power raking combined with a relatively long mowing height and care to avoid overwatering can favor Kentucky bluegrass over nimblewill and bentgrass. Asulam, MSMA, and DSMA claim selective control of a limited number of perennial weedy grasses. Chlorsulfuron is labeled for selective postemergence control of tall fescue, perennial ryegrass, and several other weeds in established turf.

Grasslike perennial weeds include yellow nutsedge and wild garlic. Selective control of yellow nutsedge in established turf is claimed on labels for herbicides containing halosulfuron, bentazon, imazaquin, MSMA, and DSMA. Herbicides with the potential for selective control of wild garlic include 2,4-D, dicamba, and imazaquin.

Figure 14.7 Comparison of well-maintained, weed-free bluegrass lawn (left) with one infested with tall fescue (right). The infested lawn likely will have to be totally renovated in order to reestablish a good-quality turf.

Emerged weeds can be treated when the turf is dormant with diquat and pelargonic acid.

A partial list of herbicides labeled for use in established turfgrass is provided in Table 14.13.

Crops Mowed One or More Times per Year. Roadsides, grass roadways, sodded waterways, sod strips in orchards, well-managed pastures, and hay crops are typical of this management level. Except for roadsides, these crops are fertilized. When severe stand problems develop, the crop stand should be reestablished.

Roadsides. Roadsides are planted with grass species to provide erosion control and a pleasing appearance. The vegetation must be short enough to provide good visibility at intersections and interchanges and to allow easy sighting of traffic signs, obstructions, and so on. Mowing two or more times a year usually is enough to keep the grass at an acceptable height and to control most tall weeds and brush species. The greater the frequency of mowing, the more effective the weed control will be. Infrequent mowing (one to three times) allows growth of tall perennial herbaceous species such as Canada thistle, common milkweed, hemp dogbane, johnsongrass, sweet clover, and Russian knapweed. Some biennials such as wild carrot, poison hemlock, common mullein, and several biennial thistles can flower under these circumstances. In some instances, particularly in arid areas, annuals such as sunflower, kochia, downy brome, and Russian thistle can be problems when mowing is infrequent.

Herbicide applications to roadsides can be made every 1 or 2 yr for problem broadleaved weeds. These treatments often are handled by custom applicators or state or county personnel. The treatments usually involve relatively narrow strips of vegetation over long distances, so drift-control techniques and equipment often are employed to prevent injury to nontarget plants. Herbicide applications also are appropriate in low-lying areas or ditches where water collects and mowing equipment cannot be used. The design of borrow ditches along interstate highways has resulted in ideal habitats for cattails and other plants that impede water flow and sometimes obstruct visibility.

In areas too steep to mow, ground covers such as sedums and crownvetch often are used in place of grass. Once established, these plantings require little or no attention.

Established Forage Crops. Forage crops include alfalfa, clovers, alfalfa-grass and legume-grass mixtures, and grasses for hay and improved pasture. These crops are relatively competitive during the growing season so that weeds tend to invade during the winter dormant period and immediately after cutting. Hay crops generally are harvested from two to four times per season depending on climatic conditions and species present. Major exceptions are mountain meadow hay, which is usually harvested once, and the irrigated alfalfa hay produced in some of the desert states of the Southwest, in which six or more cuttings are obtained.

Pure stands of legumes can be established using herbicides for weed control, which results in rapid development of the legume crop. Selective herbicides are available for soil-applied and early postemergence treatments during establishment. Benefin, EPTC, imazethapyr, bromoxynil, 2,4-DB, and sethoxydim are the choices currently available.

Table 14.13 Partial List of Herbicides Labeled for Turfgrasses

	Warm-Season Turfgrasses					Cool-Season Turfgrasses					Weeds							
	Bermudagrass	Centipedegrass	St. Augustine grass	Zoysia grass	Overseeded ryegrass	Kentucky bluegrass	Bentgrass	Fine-leaf fescues	Tall fescue	Perennial ryegrass	Annual grasses	Annual broadleaves	Chickweeds	Established dandelions	Established legumes	Perennial grasses	Nutsedges	Wild garlic
Selective at turf seeding																		
Bromoxynil			X			X	X	X	X	X		X	X					
Siduron						X	X	X	X	X	X							
Selective for germinating weed seedlings on established turf																		
Benefin	X	X	X	X		X		X	X	X	X	X						
Bensulide	X	X	X	X		X	X	X	X	X	X	X						
DCPA	X	X	X	X	X	X	X	X	X	X	X	X						
Dithiopyr	X	X	X	X		X	X	X	X	X	X	X	X					
Imazaquin	X	X	X	X								X	X			X	X	X
Isoxaben	X	X	X	X		X	X	X	X	X		X	X					
Napropamide	X	X	X								X	X	X					
Oryzalin	X	X	X	X					X		X	X						
Oxadiazon	X		X	X		X	X		X	X	X	X						
Pendimethalin	X	X	X	X		X		X	X	X	X	X	X					
Prodiamine	X	X	X	X		X	?	X	X	X	X	X	X					
Siduron						X	X	X	X	X	X							
Simazine	X	X	X	X							X	X	X					
Trifluralin (with benefin)	X	X	X	X		X		X	X	X	X	X	X					
Selective postemergence on established turf																		
2,4-D	X	X		X	X	X	X	X	X	X		X	S	X	S			X
2,4-DP (dichlorprop)	X	X	X	X	X	X	X	X	X	X		X	X	X	X			X
Asulam	X		X								X					X		
Bentazon	X	X	X			X	X	X	X	X		X					S	
Bromoxynil	X		X	X		X	X	X	X	X		X	X					
Chlorsulfuron	X					X	X	X				X	X			X	X	X
Clopyralid (with triclopyr)	X	X		X		X	?	X	X	X		X	X	X	X			
Dicamba	X	X	X	X	X	X	X	X	X	X		X	X	X	X			X
Dithiopyr	X	X	X	X		X	X	X	X	X	S							
Ethofumesate	X		X	X		X	X		X	X	X	S	X				S	
Fenoxaprop-P				X		X	X	X	X	X	X					S		
Fluazifop-P				X					X		X					X		
Glyphosate	X										X	X	X		S	S		
Halosulfuron	X	X	X	X		X	X	X	X	X		X					X	
Imazaquin	X	X	X	X							X	X	S	X	S	X	X	X
MCPP (mecoprop)	X		X			X	X	X				X	X	X	X			S
Metribuzin	X										S	X						
MSMA	X			X		X					S	S	S			S	S	
Sethoxydim		X							X		X					X		
Triclopyr	?		?	?		X	?		X	X		X		X	X			

Turfgrasses: X = labeled; ? = check label for turf species and variety tolerance; Weeds: X = control; S = suppression.

Once the crop is well established, late fall treatments are applied mostly to control winter annuals. These residual herbicides are applied after the crop is partially dormant and before the soil is frozen. Small annual weeds are controlled by many of these treatments, which include di- uron, pronamide, metribuzin, terbacil, and hexazinone. Quackgrass can be removed from established alfalfa with sethoxydim or pronamide. Problems in selecting herbicides are encountered when a legume is mixed with a perennial grass because herbicides that control weedy grasses also injure pasture grasses. Some postemergence treatments can be used on forage crops during dormancy (e.g., paraquat, 2,4-DB). MCPA is used when alfalfa is dormant to control mustards.

Well-managed pastures are harvested by grazing and mowing. Typically, a period of overproduction in late spring permits the cutting of vegetation for hay. Mowing helps to maintain a more palatable mixture of forages and discourages the growth of taller undesirable species such as brush. Improved pastures usually consist of a mixture of grass and legume species, so herbicides must be selected and their use must be judiciously timed to provide minimum injury to the legumes. Herbicides selective for legumes, spot treatments, selective placement, and other techniques that minimize injury can be used for limited weed infestations. When pastures are renovated (e.g., legume reestablishment once every 3 to 5 yr), broadcast herbicide treatments should be used 6 mo or more prior to legume planting to control broadleaves and brush.

A partial list of herbicides labeled for use on forage legumes is provided in Table 14.14.

Perennial Crops Generally Not Mowed.

These crops include rangeland, ground covers, and sods to prevent soil erosion, and unimproved pastures (Figure 14.8). With the exception of ground covers, these crops consist primarily of grass species.

Management of grazing animals and control of weed species that are particularly objectionable are the major management practices in rangelands. Grazing animals select the most palatable species, so overgrazing often provides the less palatable species with a competitive advantage. Brush, poisonous plants, and noxious weeds frequently occur in overgrazed rangelands. Burning, chaining, occasional mowing, and spot treatment with herbicides are practices typically employed when weed control is needed. Wipe-on applications also have potential for these types of areas. Renovation is used only when severe stand problems occur.

Ground covers for erosion control include a variety of broadleaved species that spread by trailing stems and stolons. Crownvetch in the eastern United States and sedum in the arid parts of the United States are the prevalent species. Ornamental ground covers include

Table 14.14   Partial List of Herbicides Labeled for Forage Legumes

	Crops				Type of Application						Weeds								
												Winter annuals			Summer annuals		Simple perennials		Creeping perennials
	Alfalfa	Birdsfoot trefoil	Clover	Others	Preplant	Preemergence	Early postemergence	Postemergence	Dormant	Between crop	Annual bromes	Mustards	Chickweeds	Mints	Grasses	Broadleaves	Dandelion	Curly dock	Quackgrass
Before establishment																			
Glyphosate											X	X	X	X	X	X	X	X	X
During establishment																			
Benefin	X	X	X		X							X			X	X			
EPTC	X	X	X	X	X							X			X	X			X
Bromoxynil	X						X					X				X			
2,4-DB	X	X	X				X					X				X		X	
Sethoxydim	X	X		X			X	X							X				X
Imazenthapyr	X					X	X					X			X				
Established stands																			
Diuron	X	X	X							X		X	X		X				
Pronamide	X	X	X	X		X	X				X	X	X		X	X		X	X
MCPA	X		X					X	X			X							
Metribuzin	X			X					X		X	X	X	X		X			
Paraquat	X								X	X					X				
Sethoxydim	X	X		X			X	X							X				X
Imazethapyr	X					X	X	X				X			X	X			
Terbacil	X									X	X	X	X	X	X	X			
Hexazinone	X									X	X	X	X	X	X				

Figure 14.8 Yucca, an unpalatable weed, in an overgrazed pasture in western Kansas.

ajuga and vinca, whereas dichondra is used in lawns. When well established, ground covers require little management. Herbicides are labeled for use on the occasional weeds that come through or in stands in which the ground cover is only sparsely established.

A partial list of the herbicides labeled for pastures, unimproved sods, and rangeland is provided in Table 14.15. A partial list of the herbicides labeled for ground covers is provided in Table 14.16.

Perennial Woody Plants

Woody plantings usually are long term and involve continuous maintenance of the crop for several decades. Examples include forests, forest plantations, nut crops, fruit orchards, vineyards, ornamental plantings, and windbreaks. Some shorter-term (typically 1 to 10 yr) plantings include nurseries and Christmas tree plantations.

Although woody species have numerous uses, the principles involved in their establishment and the practices used for weed control generally are similar for all species.

Except for natural reseeding in some forest and woodlot situations, most perennial woody plants are started from seeds, cuttings, or grafts in greenhouses, plant beds, or nurseries. The growing plants are then transplanted to the permanent site. Planting configurations are either in rows or randomly spaced. When access for management operations is required, the plants are placed in rows. In some cases, the plants are placed in check rows (equidistant plantings) that permit management operations to be conducted in both directions (Figure 14.9). Randomly spaced plants in close proximity to one another greatly limit the number of operations that can be done with large equipment.

Competing weeds have the greatest impact when woody plants are small and soon after

transplanting (Figure 14.10). The maximum effect of interference is observed the year of transplanting. Somewhat less effect can be noted the second year, and by the third or fourth season, interference may be minimal (Bey et al. 1975). Obviously, the growth rates and heights of both the woody crop plant and weeds will influence the impact of the interfering vegetation. Short plants or those that initiate growth slowly will be adversely affected by weeds for a longer period.

Special consideration should be given to perennial weeds prior to crop establishment. Two or more herbicide applications may be required starting a year or more ahead of planting to control species such as Canada thistle, bindweeds, mugwort, horsenettle, and woody species.

Once the plants have been transplanted, the soil around them can be managed in one of several ways: as bare soil, as solid stands of mowed grass, as bare soil next to the woody plant rows with strips of mowed sod in between, as mulch, and as unattended understory.

Bare ground culture provides the opportunity to select from the maximum number of weed control methods. Tillage, herbicides, hand weeding, or any combination of these methods can be used when appropriate (Figure 14.11). Disadvantages associated with bare soil culture include soil erosion on sloping land or soils subject to blowing. Problems also can be encountered when equipment must be used for spraying, pruning, or harvesting and the bare soil or tilled bare soil is wet. Wet sod supports ground equipment much better than wet bare soil.

Solid stands of mowed sod can be maintained beneath plantings of woody species, but the sod can compete with tree roots for nutrients and water as well as harbor mites, insects, and rodents that attack the trees. Mowed strips of sod with bare soil maintained next to the trees is a frequently used compromise that reduces the shortcomings of completely bare soil or a solid sod cover system (Figure 14.12). The bare soil strips are maintained routinely with directed sprays of herbicides.

Mulches are appropriate when their expense is justified by the value of the planting. They are used frequently in ornamental plantings. Mulches also have potential for use on new plantings in nurseries, orchards, and vineyards.

In forests, woodlots, older plantations, and some ornamental plantings, the surviving understory is allowed to grow. If done at all, weeding consists only of removing particularly objectionable species.

Table 14.15 Partial List of Herbicides Labeled for Pastures, Rangelands, and Unimproved Sods

| | Crops | | | | | | Weeds | | | | |
| | Pasture | | Unimproved Sods | | | | | | | | |
	Cool	Warm	Cool	Warm	Rangeland	CRP[a]	Annual Grasses	Annual Broadleaves	Perennial Grasses	Perennial Broadleaves	Brush
Atrazine		X		X	X	X	X	X	X	X	X
Bromoxynil			X	X		X		X			
Chlorsulfuron			X				X	X	X		
Clopyralid	X	X	X	X	X	X		X		X	X
2,4-D	X	X	X	X	X	X		X		X	X
Dicamba	X	X	X	X	X	X		X		X	X
2,4-DP	X	X	X	X	X			X		X	X
Fenoxaprop-P											
Glyphosate	X	X	X	X	X	X	X	X	X	X	X
Hexazinone		X		X	X	X	X	X	X	X	X
MCPA	X	X	X	X	X			X		X	
Metsulfuron	X	X	X	X				X	X	X	X
Pendimethalin			X	X		X	X	X			
Prodiamine			X	X			X	X			
Triclopyr	X	X	X	X	X					X	X

[a]CRP = Conservation Reserve Programs.

294

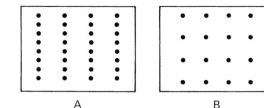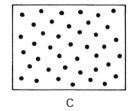

Figure 14.9 Typical woody plant spacings: rows (A), check rows (B), and random spacings (C). Check-row spacing permits greatest flexibility in management operations.

Use of mechanized weeding equipment can cause considerable damage to established woody species. Tillage equipment can result in root pruning (just as it can in large herbaceous crops). Another serious problem is encountered when tillage and mowing equipment, used close to the plants, damage the lower stems of trees. In addition to the problem of physical damage, wounding opens the plants up to invasion by disease organisms.

Some herbicides have good selectivity when applied directly over the top of small woody species. Clopyralid on conifers; hexazinone on pines; and sethoxydim, clethodim, fluazifop-P, and fenoxaprop on nearly all woody species are examples.

Woody crop plants already are rooted at transplanting and shoot growth continues to add to the height of the plant, so that differential plant height and differential placement can be used to enhance selectivity. Herbicides with activity through the foliage are routinely applied as directed sprays to emerged weeds. Paraquat, glyphosate, amitrol, 2,4-D amine, sulfometuron, and glufosinate have labels for such treatments in at least some established woody species. Water-soluble formulations of these compounds can be allowed to contact the mature bark of the crop stems but not the leaves or green stems. If selectivity is limited, the herbicides can be directed under the foliage.

Herbicides that provide weed control primarily through soil activity can be applied over the top of short crop plants if foliage is not damaged (sometimes formulation solvents and adjuvants can cause contact injury). Once in the soil, selectivity depends either on the woody crop plant having a true tolerance for the compound or on the positioning and maintenance of the herbicide in the soil in a zone above the roots of the woody species. Oryzalin and diclobenil are examples of herbicides with minimum soil movement. Some soil-applied herbicides affect only germinating seedlings so that the established woody plants avoid injury.

Soil-applied herbicides that must be incorporated into the soil create special problems in established woody crops because incorporation equipment can be difficult to use. In some cases, overhead irrigation or rainfall can provide reliable incorporation into the soil. Because volatility and photodegradation can cause surface losses, granular formulations can extend the life of soil-applied herbicides.

Figure 14.10 Herbaceous weeds (giant foxtail, ragweed) overtopping 1-yr-old Christmas tree plantings.

Figure 14.11 Bare soil culture in a California walnut orchard.

Figure 14.12 Strips of sod between apple tree rows reduces competition and help support equipment.

A herbicide with soil activity and one with foliar activity can be tank mixed or used in sequence to control weeds developing from seed in the soil as well as those that already have emerged. A partial list of the herbicides used in woody crops is given in Table 14.16.

Control of Unwanted Woody Species

Shrubs (woody perennials with multiple shoots and limited height) are termed brush when encountered as problem species. Tall growing plants with single stems are considered trees.

Woody species can be problems in grasslands (rangeland, pastures), home grounds, on utility rights-of-way, borders of water courses, fencerows, maturing forests and woodlots, new forest plantings, orchards, nurseries, ornamental plantings, and previously forested sites that have been harvested and are to be replanted. Unwanted plants can vary from one or two isolated specimens to a small patch to a population of general overstory. *Timber stand improvement* is a term used to describe the selective removal of unwanted trees from forests and woodlots to enhance productivity.

Some specialized techniques and terminology are associated with the control of woody plants. Mechanical control methods include mowing, chaining, shearing, bulldozing, girdling, burning, and felling (cutting with a saw). Herbicide applications include cutting and spraying the cut stumps, broadcast foliar spraying, soil applications of leachable soil-active herbicides, frill and stem injections, and basal spray treatments. Foresters sometimes use the term *silvicide* in place of *herbicide*.

Shearing and bulldozing are used to prepare sites previously populated with woody plants. The unwanted plants are cleared from the sites with a V-blade that shears off small-diameter trees and a bulldozer that uproots them. The tops and roots usually are piled and burned. The land may be further prepared for replanting with heavy disks.

Girdling consists of removing all the bark in a horizontal strip around the lower part of the trunk so that food from the leaves cannot reach the roots (Figure 14.13). The roots starve, and the tree eventually dies.

Burning has been used to keep brush species from invading grasslands and also is used to remove understory and trash in mature forests.

Large trees frequently are removed by cutting with a saw (felling). Sprouting of new shoots from the cut stump can be prevented by spraying the freshly cut stumps with herbicide. This technique is called *cut and spray*.

If solid stands of undesirable plants are a problem, broadcast spraying can be done. In many cases tall plants and rough terrain dictate the use of aerial applications. Herbicides such as 2,4-D, triclopyr, and glyphosate can be used in site preparation prior to transplanting. Broadcast treatments frequently are made to mixed stands of hardwood and conifers to inhibit growth of the hardwoods and favor the growth of conifers. This process is called *conifer release*. Herbicides used for this purpose include imazapyr, hexazinone, and clopyralid.

Woody species are most susceptible to foliar applications of the phenoxy herbicides from full leaf expansion until the foliage starts to turn color. Effectiveness decreases as plants approach fall maturation, and the efficacy of phenoxy application past mid-August in most parts of the country is questionable. Glyphosate and fosamine, on the other hand, are best applied to some species from mid-August to fall color.

Individual woody plants frequently are treated with applications of relatively leachable herbicides to the soil. Such treatments are made on an individual plant or clump of plants. Herbicides can be applied as pellets, granulars, or spray solution over the individual root zone and allowed to leach into the roots. These treatments usually are nonselective, so care must be taken to avoid using them when desirable woody species are in the immediate vicinity.

Frill treatments consist of making ax cuts around the trunk of a tree and then applying herbicide into the cuts. Special injectors are available to make the cuts and dispense herbicide at the same time. Herbicide is used at full strength or diluted somewhat with water and applied at low volume. Injectors include a hollow tube with a cutting bit and an injection handle. The hollow

Table 14.16 Partial List of Herbicides Labeled for Ornamentals, Ground Covers, and Woody Plants

	Stage of Crop					Weeds				Placement			Crop								
	Site preparation	New seedlings	New transplants	Established plants	Container grown	Germinating seedlings	Emerged weeds	Perennial grasses	Perennial broadleaves	Soil applied	Post-applied	Directed post	Herbaceous ornamentals	Ground covers	Deciduous shrubs	Conifers	Deciduous trees	Cane fruits	Tree fruits	Citrus fruits	Nuts
2,4-D	X			X			X		X	X	X						X		X		X
Alachlor			X	X	X	X				X					X	X					
Asulam				X			X				X					X					
Atrazine				X	X	X	X	X		X						X					
Bentazon				X			X			X	X		X	X	X	X	X		NB		
Clethodim		X	X	X			X	X		X	X		X	X	X	X	X		NB		
Clopyralid				X			X		X	X						X					
Dazomet	X									X											
DCPA		X	X	X		X				X					X	X	X				
Dichlobenil				X		X		X	X	X						X	X	X	X		X
Diquat	X			X			X					X				X	X				
Diuron				X			X			X								X	X	X	X
Fenoxaprop-P		X	X	X			X	X		X			X	X	X	X					
Fluazifop-P		X	X	X			X	X		X	X		X	X	X	X	X	NB	NB	NB	NB
Glufosinate	X			X			X				X				X	X	X		X		X
Glyphosate	X			X			X	X	X		X				X	X	X		X	X	X
Hexazinone	X			X		X	X	X	X	X	X					X					
Imazaquin				X		X				X				X	X						
Isoxaben			X	X	X	X				X				X	X	X	X				
Napropamide			X	X	X	X				X				X	X	X	X		X	X	X
Naptalam			X	X		X				X					X	X					
Norflurazon			X	X		X				X						X	X		X	X	X
Oryzalin			X	X	X	X				X				X	X	X	X	NB	NB	NB	NB
Oxadiazon			X	X	X	X				X				X	X	X	X				
Oxyfluorfen		X	X	X	X	X	X			X	X					X					
Paraquat	X			X			X				X	X					X	X	X	X	X
Pelargonic acid	X			X	X		X					X		X					NB		NB
Pendimethalin				X	X	X				X				X	X	X	X		NB		NB
Prodiamine			X	X		X				X				X	X	X	X				
Pronamide				X			X	X	X	X	X	X			X	X	X	X	X		
Sethoxydim		X	X	X			X	X			X		X	X	X	X	X	X	X	X	X
Simazine				X		X				X						X			X	X	X
Terbacil				X		X	X					X						X	X		X
Trifluralin			X	X		X				X				X	X	X	X		X	X	X

NB = non-fruit-bearing trees or canes.

tube serves as a reservoir for herbicide and as a handle for the implement. This barlike device is used to make cuts at the root collar (Figure 14.14). Another implement is the Hypo-Hatchet Tree Injector, a specially designed hatchet that feeds herbicide from a reservoir worn on the user's belt into the cuts made by the hatchet (Figure 14.15). Spacing of cuts vary from touching to several inches apart, depending on trunk diameter and susceptibility of the individual tree species to the herbicide. Injection provides rapid control of most woody species and limits basal sprouting. It may be impractical in stands of dense brush with small-diameter stems.

Basal sprays consist of saturating the lower portion of the woody stem with herbicide. The objective is to wet individual stems to run off from a height of approximately 18 in. to the ground line and including the root collar zone. *Dormant cane* is the term sometimes used to describe basal treatments applied during the dormant season to brush. Herbicides applied as basal treatments usually are mixed in an oil or oil–water carrier to enhance penetration of the bark. They can be used at any time during the year.

Thinline basal bark applications are made to stems no greater than 6 in. wide with a thin

Figure 14.13 A girdled tree truck. (H. A. Holt, Purdue University)

Figure 14.15 A Hypo-Hatchet Tree Injector. (H. A. Holt, Purdue University)

stream of undiluted herbicide about 6 in. above the base of the plant.

Herbicides that are effective as foliar applications for woody weed species include 2,4-D, 2,4-DP, clopyralid, dicamba, imazapyr, metsulfuron, picloram, fosamine, triclopyr, glyphosate, and amitrole. Herbicides for soil application include picloram, tebuthiuron, metsulfuron, bromacil, and hexazinone.

BARE SOIL SITES

Total vegetation control is used in certain areas where weeds are unsightly, harbor vermin, collect moisture, present a fire hazard, block visibility, or otherwise interfere with the function or overall appearance of the site. It also may be required by regulations to ensure worker safety in industrial sites. Examples of sites where permanent elimination of vegetation is the goal include parking lots, fencerows, industrial sites, around buildings, under paved areas, highway shoulders, guardrails and traffic markers, railroad beds (Figure 14.16), railroad yards, tank farms for petroleum storage, oil pumping and

well stations, substations and structures for electrical power, and lumber storage areas.

Usually no crop or desirable species is located in such areas, so it is possible theoretically to use any type of control measure that removes plant shoots. Possibilities include mechanical methods such as hand weeding, shallow tillage, smothering with a barrier mulch, or repeated flaming. Herbicide treatments include repeated use of contact herbicides (e.g., diquat, paraquat, pelargonic acid, glufosinate), periodic use of symplastically translocated herbicides (e.g., glyphosate, 2,4-D, dicamba), or use of soil-applied herbicides including both persistent (e.g., diuron, simazine) and highly persistent herbicides (e.g., picloram, tebuthiuron). Soil pH, dosage, and weather can result in the shift of some herbicides from the persistent to the highly persistent category (Table 14.17). The situation, size of the area, surrounding vegetation, terrain, weeds present, and economics determine the choice of programs.

In most cases the use of highly persistent soil-applied herbicides provides the most practical solution to bare ground situations. Initial

Figure 14.14 A tree injector for herbicide application. (H. A. Holt, Purdue University)

Figure 14.16 Weed-free railroad beds treated with a long residual soil-applied herbicide.

Table 14.17 Partial List of Herbicides Labeled for Bare Soil Sites

	Annual Grasses	Annual Broadleaves	Perennial Grasses	Perennial Broadleaves	Brush	Highly Persistent?
Ametryne	X	X				Depends on dose
Asulam	X		X			No
Bromacil	X	X	X	X	X	Depends on dose
Bromoxynil		X				No
Chlorsulfuron	X	X	X	X	X	Depends on soil pH
Clethodim		X		X		No
Clopyralid		X		X	X	Depends on dose
Dicamba		X		X	X	No
2,4-D		X		X	X	No
Dichlobenil	X	X	X	X		Depends on dose
Diquat	X	X				No
Diuron	X	X	X	X		Depends on dose
2,4-DP		X		X	X	No
DSMA	X	X	X	X		No
Fluazifop-P	X		X			No
Glufosinate	X	X	S	S	S	No
Glyphosate	X	X	X	X	X	No
Hexazinone	X	X	X	X	X	Depends on dose
Imazapyr	X	X	X	X	X	Yes
Isoxaben		X				Yes
Metsulfuron	X	X	X	X	X	Depends on soil pH
MSMA	X	X	X	X		No
Norflurazon	X	X				Yes
Oryzalin	X	X				Yes
Paraquat	X	X				No
Pelargonic acid	X	X				No
Pendimethalin	X	X				No
Picloram		X		X	X	Yes
Prodiamine	X	X				Yes
Prometone	X	X	X	X		Yes
Sethoxydim	X		X			No
Simazine	X	X				Depends on dose
Sulfometuron	X	X	X	X	X	Depends on soil pH
Tebuthiuron					X	Yes
Triclopyr		X		X	X	No

X = control; S = suppression.

control can be included as part of the site preparation and may consist of mechanical soil disturbance, use of foliar-applied translocated herbicides, use of highly persistent herbicides, or a combination of several methods. It is important that the type of vegetation on the site be identified correctly at the outset so these treatments as well as subsequent ones can provide adequate control.

Uninjured perennial root systems can send shoots through mulches and even asphalt if not properly treated prior to covering the area. The plants should be treated either under good growing conditions with effective herbicides or with an effective soil residual applied sufficiently ahead of time to leach into the root zone. If this is not done, a herbicide applied on the soil prior to a barrier mulch or asphalt cover may have little or no effect because it cannot move through the soil into the root zone. We have observed bigroot morningglory emerging through 6 in. of asphalt that was underlain with a blanket of herbicide at the time of paving. Perennials routinely penetrate through plastic film.

Much of the activity of highly persistent herbicides is through the soil rather than through the foliage, even when these compounds also possess foliar activity. Normally they persist more than one season, so the most economical way to use them is to apply the amount needed to provide season-long control initially and then to reapply just enough in subsequent seasons to bring the total amount to the original level.

Several of the highly persistent herbicides are mobile in the soil. This characteristic plus the fact that they are used at rates that control all vegetation and persist a season or more means that the potential for injury to nontarget vegetation and movement to nontarget areas is maximized. Problems consist of drift, lateral movement, leaching into shallow groundwater, uptake by the roots of adjacent established plants, and the presence of residues in the soil if the use of the area should change. Drift control techniques should be used with spray applications. Dust-free granules or pellets also can be used in areas where potential damage from wind-carried herbicide exists. These herbicides *should not* be applied to frozen ground because such applications provide maximum potential for lateral movement. Care also should be exercised in making applications to sloping sites since both soil erosion and lateral movement with rapid runoff of water are likely to occur.

A frequent problem encountered with highly persistent soil-applied herbicides occurs when trees outside the treated area pick up the herbicide because their roots extend into the treated soil. Roots of large trees can extend more than 100 ft from the tree itself. Applications close to established trees or shrubs can cause injury; thus, herbicide choice and rate of application are important considerations.

Special care should be exercised where mobile, persistent compounds can leach into sources of groundwater that are likely to be used for irrigation or domestic purposes.

Soil type and vegetation present are important aspects to be considered when making applications of highly persistent soil-applied herbicides. When high organic matter soils, charcoal, cinders, and other adsorbants are present, some of the tightly bound herbicides may not perform adequately. In contrast, leaching may be a problem on sites with little clay or organic matter. If vegetation is present at the time of application, the treatment may need to include a herbicide (e.g., paraquat, diquat, a phytotoxic oil) to provide immediate top kill in addition to the soil activity. When shallow-germinating annuals are present, a herbicide with limited soil mobility (e.g., diuron, simazine) can be adequate; deep-rooted species require a soil mobile compound (e.g., bromacil, hexazinone, picloram, tebuthiuron). Two or more herbicides frequently are needed when several species are involved, particularly if some weeds are shallow rooted and others are deep rooted.

A partial list of herbicides labeled for bare soil sites is provided in Table 14.17 (several of the herbicides, such as diuron and simazine, are selective at low application rates and are used in cropland).

MAJOR CONCEPTS IN THIS CHAPTER

Appropriate weed control practices for any given situation are determined by the management system, weeds present, crop vigor, economics, and available technology.

The more restricted the number of operations in a management system, the more fully the practices that remain must be utilized for weed control.

Crop competition for weed control should be exploited fully whenever possible. It is very cost effective.

Most cropping systems provide periods when full sunlight reaches the soil and favors weed development. Supplemental weed control is needed most at these times.

Integrating effective weed control practices into the crop management system is the key to attaining a successful weed control program.

Annual crops are solid seeded (drilled) or planted in rows (to facilitate passage of spray equipment, between-row cultivation, or harvest). Solid-seeded crops depend on shading to offset the contribution lost from between-row herbicide applications and cultivation.

Agronomic row crops return fewer dollars per acre, require less special care, are grown on more acres, provide more shading, and generally are more easily harvested than horticultural crops.

The high value (high liability) of horticultural crops coupled with limited acreage (small market potential) results in high risks for companies interested in developing and selling herbicides for horticultural crops. Consequently, fewer herbicides are available.

Examples of annual crops are agronomic crops such as corn, soybeans, and wheat and horticultural crops such as tomatoes and onions.

Weeds appear during the establishment and production of annual crops at three critical times: (1) prior to crop planting, (2) as the crop emerges, and (3) as the crop is growing. The timing of weed control practices focuses on these three important periods of weed presence and is determined by the crop, the crop management system, and the weeds that are present.

Control practices for weeds established prior to planting or crop emergence are accomplished with primary and secondary tillage, early preplant herbicides, and burn-down herbicides.

Selective control of germinating weeds emerging with the crop depends on preplant or preemergence herbicides, allelopathic residues, sheet mulches, and shallow selective tillage.

Selective control of emerged or late-emerging weeds in an emerged crop is achieved with shading from the crop canopy; early postemergence, postemergence, and directed postemergence herbicides; wipe-on treatments; spot treatments; preharvest mature crop treatments with herbicides; cultivation between rows; and hand weeding.

When tillage is used, the typical weed control program for an annual crop in rows would likely include seedbed preparation, two or three herbicides (either tank mixed or applied in sequence), and at least one cultivation. On a high value poorly competing crop, additional cultivations, herbicide treatments, or hand weeding would be used.

Fields with especially troublesome or dense weed populations may not be appropriate sites for crops that compete poorly.

Herbicides have made planting into seedbeds with little or no tillage (conservation tillage) a viable option.

Conservation tillage systems (to reduce soil erosion) leave soil residues to protect the soil surface. This protective mulch generally is obtained by eliminating some or all of the tillage operations. The weed control contribution lost (including use of preplant incorporated herbicides) must be replaced (usually with herbicides) if such systems are to be employed as continuous standard management practices.

A high level of management is required on conservation tillage systems because mistakes can be more costly and fewer alternative corrective practices are available than is the case on systems using extensive tillage. Serious weed problems must be anticipated and adequate control measures must be provided with the limited number of options available.

Rapid shifts of weed species are encountered frequently on no-till systems.

Rotation of weed control practices (particularly herbicides) provides a means of preventing rapid buildup of adapted weeds in reduced tillage systems.

During establishment, perennial crops usually have the same weed control problems as annual crops. Major differences occur after the crop has been established 1 to several yr.

Perennial crops offer little or no opportunity for crop rotation. Management systems tend to lack diversity, and persistent weeds tend to be well adapted to the crop and its management system.

Special emphasis on site preparation prior to crop establishment to eliminate problem species usually is warranted for perennial crops.

Established perennial crops offer an opportunity to obtain chemical selectivity between the weeds and the crop by using age difference, directed sprays, and depth protection due to differential root position and herbicide movement.

Asparagus, strawberries, and mint are examples of perennial herbaceous crops with a limited amount of tillage sometime during the year.

Mowed perennial crops include lawn-quality turf, turf mowed one to six times per year (turf on roadsides, well-managed pastures), and hay crops.

Rangeland, unimproved pastures, and various grass and broadleaved ground covers to reduce soil erosion rarely if ever are mowed.

Shading is the major weed control practice for most herbaceous perennial crops. Weeds are problems in these crops primarily when stands are thinned or weakened.

Management to maintain a good stand of vigorous plants is the first step in controlling weeds in perennial herbaceous crops.

The more frequently a herbaceous perennial sod crop is mowed the more likely surviving weeds will be limited to those with growing points close to the soil surface.

Management of grazing animals is critical to the quality of permanent pastures and rangeland.

Woody species rely on shading for control of weeds. When shading cannot be attained, control is obtained with tillage, differentially placed herbicides, mowing, and mulch (in high-value plantings).

Methods for controlling woody plants include mowing, chaining, bulldozing, girdling, felling (and spraying the stumps), broadcast foliar sprays, soil applications with leachable herbicides, frill and stem herbicide injections, and basal spray treatments.

Total vegetation control (bare ground sites) typically is achieved and maintained with herbicides. Highly persistent soil-applied herbicides are used to maintain a weed-free soil surface and then are supplemented with other practices and herbicides as needed.

The life of long residual soil-applied herbicides coupled with their general broad spectrum activity and high mobility in the soil (in some cases) dictates that care be exercised to prevent their moving off-site laterally or leaching downward into root zones of valuable trees and shrubs or into water sources.

Weed control efforts that are cost effective under the wide diversity of management systems and weed species normally encountered are tillage, herbicide use, and sometimes mowing.

TERMS INTRODUCED IN THIS CHAPTER

Basal spray
Chaining
Check row
Chisel/disk tillage system
Chisel/sweep tillage system
Conifer release
Conservation tillage
Conventional tillage system
Cool-season turf species
Cut and spray
Direct seeded
Dormant cane treatment
Felling
Frill injection
Girdling
No-till system
Nurse crop
Renovation
Ridge-till system
Solid seeded
Thinline basal bark application
Warm-season turf species

SELECTED REFERENCES ON WEED CONTROL SITUATIONS

Fermanian, T. W., M. C. Shurtleff, R. Randell, H. T. Wilkinson, and P. L. Nixon. 1997. Controlling Turfgrass Pests, 2nd ed. Prentice-Hall, Upper Saddle River, New Jersey.

Maynard, D. N. and G. J. Hochmuth. 1997. Knott's Handbook for Vegetable Growers. John Wiley and Sons, New York.

Nyland, R. D. 1996. Silviculture: Concepts and Applications. McGraw-Hill, New York.

Siebert, A. C. and J. J. Vorst. 1996. Crop Production. Stipes Publishing L.L.C., Champaign, Illinois.

Walstad, J. D. and P. J. Kuch. 1987. Forest Vegetation Management for Conifer Production. John Wiley and Sons, New York.

Also check recommendations from appropriate state universities, the Cooperative Extension Service, and USDA/ARS.

AQUATIC PLANT MANAGEMENT

Aquatic plants are found in almost every body of water. In natural systems such as lakes, rivers, and ponds, moderate growths serve useful purposes by producing oxygen, food, and cover for fish and other aquatic organisms. In overabundance, however, some species become "weedy" as they crowd out desirable plants, adversely affect other aquatic life, and interfere with human uses of water. Aquatic weeds are major management problems in such diverse sites as lakes, farm ponds, irrigation canals, rivers, home aquariums, and swimming pools. Their control is essential if the quantity, quality, and availability of water for drinking, agriculture, recreation, wildlife, and transportation are to be maintained.

The economic impact of aquatic weeds in terms of dollar loss is difficult to estimate. Aquatic sites are basically noncropland sites, so crop yields and losses cannot be measured. Some of the monetary losses due to aquatic weeds have been estimated by calculating the amount spent for their control. In 1996, $15 million was allocated by the state of Florida for the control of primarily three weed species in public waters and wetlands: hydrilla ($12 million), water hyacinth and other free floating weeds ($2 million), and melaleuca ($1 million).[1] This value does not include the amount spent in Florida on other aquatic weed species or by the federal govern-

ment or private sectors. An earlier survey dating from 1987 (Steward 1989) suggested that for every dollar spent by a state agency for aquatic weed control in Florida, $2.50 was spent by the private sector. This would amount to an additional $37.5 million over and above the $15 million spent by the state in 1996. These values can be multiplied manyfold, since aquatic weeds infest waters throughout the United States (including Alaska) and Canada as well as throughout the world. Specific problems caused by aquatic weeds include the following:

1. Aquatic weeds restrict or prevent recreational activities such as swimming, fishing, water skiing, and boating (Figure 15.1).

2. The presence of aquatic weeds can be hazardous to swimmers and water skiers because of the danger of becoming entangled in weeds and drowning. Ladders, rocks, and rafts in swimming pools and other recreational areas become slippery and hazardous when covered with algal growths.

3. Aquatic weeds cause foul tastes, unpleasant odors, and discoloration of potable water supplies stored in surface impoundments such as reservoirs and ponds. Even large lakes such as Lake Michigan, which provides drinking water for parts of the Chicago area, are impacted periodically with foul taste-causing organisms. The offenders, primarily microscopic algae, release aromatic oils and other compounds into the water that can cause fishy, grassy, or rotten fruit odors. Treatment with copper sulfate and

[1]Quotation from D. W. Doggett, in *Aquatics,* Summer 1996, Vol. 18, No. 2, pg. 3.

Figure 15.1 An infestation of filamentous algae prevents the recreational use of this shallow lake.

passage through sand or charcoal filters to kill and remove odor-causing algae are standard water quality improvement procedures that must be used in many municipal and private water supply reservoirs. An additional taste-related problem is that of off-flavors in fish flesh caused by blue-green algae.

4. Fish populations and fish harvests are affected adversely by excessive weed growth. An overabundance of plants can lead to overpopulation and stunting of fish, and death and decomposition of vegetation can cause fish kills. The topic of fish kills and stunting is described in more detail in the section on Diagnosis and Prevention of Fish-Related Problems. The harvest of fish, both by netting and by hook and line, can be greatly impeded by the presence of weeds. The invasion of water hyacinth into Lake Victoria and other large lakes in Africa has led to the displacement of human communities that previously depended on fishing for their livelihoods.

5. Aquatic weeds block the flow of water in irrigation and drainage systems (Figure 15.2). According to Bureau of Reclamation figures

Figure 15.2 Elodea infestation in a drainage canal greatly impedes the flow of water.

from 1976, more than 47,000 miles of irrigation waterways in 14 western states were seriously infested with aquatic weeds. There is no reason to believe that these figures have changed significantly over the years. Just a single species, parrotfeather (*Myriophyllum aquaticum*), infested more than 600 miles of waterways in California alone in 1985 (Anderson 1989). In Asia, water flow in an irrigation system designed to provide water to 1.4 million acres of land was reduced 80% by submersed weed growth in the first 5 yr of use (Holm et al.1969). Weeds also obstruct water flow in irrigation canals by clogging control gates, filters and screens, distribution lines, intake pipes, sprinkler heads, and other associated equipment. In drainage systems, aquatic weeds cause the backup of water and flooding of fields and highway culverts.

6. A major cause of water loss is by evaporation from the leaf surfaces of free-floating or shoreline plants. Water hyacinth plants, for example, release three to four times more water from their leaf surfaces than what is lost by evaporation from a free water surface alone (Holm et al. 1969). Certain shoreline plants called phreatophytes not only transpire surface water but also draw from groundwater supplies. Many of the tree species of phreatophytes such as salt cedar, cottonwood, and willow actually cause groundwater levels to drop measurably during the day while they are actively transpiring. Timmons (1960) reported that phreatophytes infested approximately 14 million acres of stream channels, canals, reservoirs, and floodplains in the western states and were responsible for the loss of over 20 million acre-ft of water annually.

7. Aquatic weeds catch debris and sediment, causing waterways to fill in at accelerated rates. The accumulation of plant materials and sediments in ponds, lakes, and canals can eventually result in the formation of marshes or dry land if the process is permitted to continue without correction (Figure 15.3).

8. Navigation of rivers and waterways is difficult when they are clogged with aquatic weeds. Weeds impede boat movement, become entangled in boat propellers, and plug the cooling systems of motors, causing extensive damage. Water hyacinth has prevented boat traffic from using major waterways in Florida (Figure 15.4), California, and parts of the Nile River in Africa. Aquatic weeds, if left unchecked, have the potential to cause similar problems in portions of the Panama Canal.

9. Free-floating vegetation and submersed plants that become uprooted can disrupt hydro-

Figure 15.3 A pond gradually being taken over by cattails.

electric and drinking water intake systems in large reservoirs. This has been a serious problem in South America and Africa. The mats of vegetation can clog the turbines or intakes, shutting down the ability of the system to generate electricity or deliver water. Large screens or barriers are used to prevent vegetation from entering the turbines and intakes, but these structures must be cleaned periodically in order to prevent their collapse from the weight of the vegetation. In the Weija Reservoir, which provides drinking water for the capitol city of Accra, Ghana, the weight of aquatic vegetation helped to cause the collapse of the dam.

10. Aquatic weeds provide habitat for noxious insects such as mosquitos. Weed growth prevents normal wind and wave agitation of the water surface, providing undisturbed sites for the development of mosquito larvae. In tropical parts of the world, such habitat promotes the development of mosquitos that cause malaria. The larvae and pupae of *Mansonia* mos-

Figure 15.4 Water hyacinth in the St. Johns River, Florida. (Aquatic Plant Control Operational Support Center, Jacksonville [FL] district, U.S. Army Corps of Engineers)

quitos, which transmit rural filariasis and encephalitis, actually bore into aquatic plants in order to obtain sufficient oxygen for development. Other undesirable organisms sometimes associated with aquatic vegetation include leeches and the snails that act as hosts for the microscopic organisms that cause swimmer's itch. Snails that inhabit dense vegetation at the edges of water bodies are the vectors for the organism that causes schistosomiasis, one of the most important public health problems in the tropics and subtropics (Gangstad and Cardarelli 1989).

11. Water infested with certain types of microscopic blue-green algae blooms can become toxic to animals that drink the water. The most frequent outbreaks of toxicity have occurred in the central states and provinces (e.g., Arkansas, Iowa, Minnesota, Alberta). Little is known about the circumstances under which algal toxins are produced, so it is impossible to predict when or where an outbreak will take place. Although the situation is uncommon, when it does occur the effects on livestock, pets, and wildlife are rapid and usually fatal.

12. The growth of unsightly, foul-smelling weeds in a body of water greatly lowers the aesthetic appeal of waterfront properties, with a resultant decline in property values. Aquatic weeds detract from the appearance of ponds on golf course grounds, ornamental and reflecting pools in shopping malls and landscaped areas, swimming pools, and home aquariums. The recent emphasis on the construction of ponds and lakes as water retention sites and major landscaping features for housing developments, apartment complexes, and office and industrial parks has greatly increased the frequency of aquatic weed problems. These bodies of water often are shallow and surrounded by well-fertilized fields, lawns, and greens, so controlling aquatic weeds becomes a major maintenance consideration.

MAJOR AQUATIC PLANT GROUPS AND HABITATS

People who find aquatic weeds tangled on their fishing lines or boat props usually call these plants "moss." Technically very few aquatic plants are true mosses. Most fall into two major botanical groups: algae and flowering plants.

Algae are simple plants without true roots, leaves, or flowers. They reproduce by cell division, by plant fragmentation, or by spores. They are found either free-floating in the water or attached to other plants, bottom sediments, rocks, or other solid substrates (Figure 15.5). Three major types of algae are microscopic algae (also called phytoplankton when suspended in the water), mat-forming filamentous algae, and the Chara/Nitella group. Extremely heavy growths of phytoplankton that color the water or form surface scums are called algal blooms.

Aquatic flowering plants have stems, roots, leaves, and flowers (although these structures are sometimes highly reduced or modified) and reproduce by seeds or vegetatively by creeping perennial structures such as rhizomes, tubers, roots, or stolons. Like algae, many aquatic flowering plants can spread by fragmentation (plant parts that break off and reroot). True mosses and other groups of plants such as aquatic ferns can be weedy, but they are encountered less frequently as problems in temperate climates than are algae and flowering plants. *Salvinia*, a free-floating fern that is a serious problem in Asia and Africa, is listed as one of the world's worst weeds (Holm et al. 1977).

Aquatic flowering plants are organized into four distinct groups based on where they are found in a body of water (Figure 15.6).

1. *Submersed plants* live beneath the water surface and usually are rooted in the bottom sediments. These plants are found at water depths of less than 1 ft to more than 20 ft (depending on water clarity). Their leaves are thin and often are dissected into narrow segments to facilitate gas exchange. Although the majority of the plant body is underwater, the flowers or flower clusters in most species extend above the water surface. Examples of submersed plants include the watermilfoils, hydrilla, the naiads, elodea, coontail, and many pondweeds.

2. *Free-floating plants* live unattached on or just below the water surface. Most have roots that extend into the water for nutrient uptake. The bulk of the abovewater portions of the plant consists of flattened or boat-shaped clusters of leaves with the stem reduced to a short segment. Examples of free-floaters include water hyacinth, water lettuce, duckweed, and watermeal.

3. *Rooted floating plants* are rooted in the bottom sediments in shallow water at depths of roughly 1 to 5 ft. The plants consist of floating or erect leaves that extend from an underground rhizome. The flowers are often quite conspicuous and showy. Examples include water lily, spatterdock, watershield, and the floating-leaved pondweeds such as American pondweed.

4. *Emergent plants* extend above the water surface and are rooted in sediments at depths less than 2 to 3 ft. These plants also are referred to as marginal or shoreline plants. Examples include herbaceous species such as cattails, bulrushes (tules), common reed (*Phragmites*), Reed's canary grass, arrowheads, and water-willow and woody species such as willow, salt cedar, and melaleuca.

INVASIVE SPECIES

Over the past 100 years, a number of exotic aquatic plant species have been introduced into North America and have displaced native aquatic

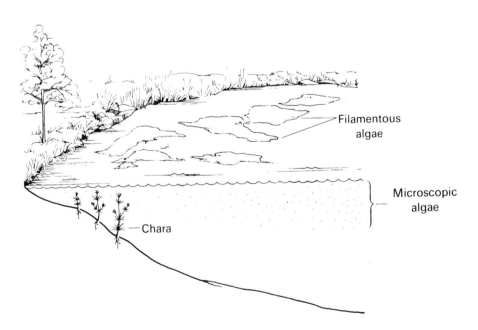

Figure 15.5   Common habitats for algae.

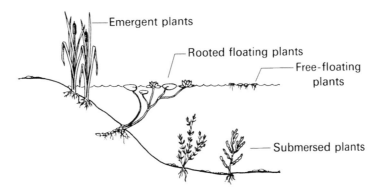

Figure 15.6 Common habitats for aquatic flowering plants.

plant populations. The result is a monoculture of the invading plant that reduces species diversity and can degrade the habitat for fish and other wildlife. It is estimated that over $100 million per year is spent to control exotic aquatic plant pests in the United States (Army Corps of Engineers Fact Sheet, 1995).

Submersed and Free-Floating Invaders

Three invasive species that have become well established in North America are the free-floating plant, water hyacinth (*Eichhornia crassipes*), and two submersed species, hydrilla (*Hydrilla verticillata*) and Eurasian watermilfoil (*Myriophyllum spicatum*).

Water hyacinth was introduced from the Amazon Basin in 1884 as an ornamental plant because of its attractive spikes of large purple flowers (Figure 15.7). It spread so rapidly that control measures were begun in Florida in the 1890s. Its current distribution is in Florida, southern Georgia, along the Gulf Coast, and into some parts of California. It also is an extremely serious weed in Africa, Asia, and South America and is ranked as one of the world's worst weeds (Holm et al. 1977). Its stoloniferous growth (Figure 15.8) produces dense mats of free-floating vegetation (Figure 15.4).

Hydrilla was introduced into Florida in the 1960s, probably from Central Africa by way of South America, through the aquarium trade. The plant spreads extremely rapidly. A 1994 survey of public waterways in Florida showed more than 97,000 acres choked with hydrilla, almost double the acreage infested in 1992.[2] Hydrilla

[2] 1994 Florida Aquatic Plant Survey Report, available from Bureau of Aquatic Plant Management, Florida Department of Environmental Protection, 2051 East Dirac Drive, Tallahassee, FL 32310-3760.

also is found in other southeastern states, in Texas, along the Atlantic coast as far north as Connecticut, and in very small, confined areas in California and Washington state where every effort is being made to eradicate it. It has not yet spread to the central and northern portions of the United States.

Eurasian watermilfoil, which probably was introduced in the 1940s from Europe and Asia, is distributed throughout the United States and into Canada. Its most recent invasions have been into the natural lakes system of Minnesota and the Okanagan and Columbia River

Figure 15.7 Water hyacinth flowers.

Figure 15.8 A water hyacinth plant. A new plant has sprouted from the end of the stolon.

systems of British Columbia and Washington state, respectively.

A key characteristic of the invasive submersed species is their tendency to form canopies over the water surface (Figure 15.9). For example, both Eurasian watermilfoil and hydrilla can grow from the sediment to the surface in waters as deep as 15 ft. As the plant nears the surface, it begins to

Figure 15.9 Invasion of an aquatic site by Eurasian watermilfoil: (Top) A few stems extend to the surface and begin to form a canopy over lower-growing native species; (Bottom) Eurasian watermilfoil has completely crowded out the native species and has formed an extensive canopy of foliage over the water surface.

branch and forms a canopy of vegetation over the surface. This growth habit shades out lower-growing native species, resulting in domination by the invading species. Canopy formation also can adversely affect fish because it prevents the normal movement of air into water and causes an imbalance in oxygen content in the water column (high at the surface where the photo-synthesizing stems and leaves are located but low under the canopy where plant growth is being shaded).

Although not rooted, the same canopy-type effect is produced by water hyacinth, which can completely cover the surface of the water and shade out submersed plant growth.

Canopy formation not only is deleterious to the plant and animal life in a body of water, but it also has a negative impact on human uses of bodies of water, primarily because we tend to use the uppermost portion of the water column for our activities, such as fishing, swimming, and boating.

All three invaders are well adapted for spread by vegetative means. Hydrilla reproduces by frag-mentation, stolons above the sediment surface, rhizomes below it, tubers, and turions (short, cocklebur-like stem segments produced in the axils of the leaves) (see Figure 15.26). As many as 500 tubers/sq ft of sediment have been recorded. In Rodman Reservoir in Florida, an infestation of 15 acres in 1973 spread to 3000 acres by 1975.

Eurasian watermilfoil, which develops from an underground root crown, readily fragments, and some of its stems can lie prostrate along the sediment surface and root at the nodes to pro-duce new plants. In the Chesapeake Bay area, an infestation of 100 acres in 1965 spread to 8000 acres within 3 yr.

Stolon production is the only means of vege-tative reproduction in water hyacinth. Even so, the plant has an extremely rapid growth rate and is able to double its coverage every two weeks.

None of these plants has significant repro-duction from seed. In fact, the reason water hyacinth has not invaded the central or northern states is because it has no overwintering mecha-nism such as the production of dormant seeds. The vegetation, which is confined to the water surface, is killed by heavy frosts and ice cover. Hydrilla and Eurasian watermilfoil, on the other hand, are protected from extreme cold by virtue of vegetative structures in the sediments and by being located underwater below the ice layer. Although hydrilla mostly has a southern distri-bution, its presence in northern China indicates that it could grow and survive in the northern states of the United States.

Emergent Invaders

Three invasive emergent plant species that have received much recent attention are purple loosestrife (*Lythrum salicaria*), melaleuca (*Melaleuca quinquenervia*), and salt cedar (*Tamarix* spp.). Purple loosestrife is herbaceous whereas melaleuca and salt cedar are woody species. Unlike the free-floaters and submersed species, all three of these emergent species are prolific seed producers, and the dense stands of plants that originate from these seeds crowd out native species such as cattails, willows, and bulrushes.

Purple loosestrife (Figure 15.10) was introduced to northeastern North America in the mid-1800s and has gradually been spreading southward. It is well established in Canada and in the northeastern and midwestern United States, with scattered populations in Alabama and Texas. It is currently present in 26 states and infests over 400,000 acres (NAPIS database at www//ceris.purdue.edu/napis/). It is a problem primarily in wetland areas where its prolific growth and seed production (a single plant can produce up to 2 million seeds a year) allow it to crowd out native species and reduce wildlife diversity. In addition to seed production, the plant can spread vegetatively and overwinter from underground structures. Purple loosestrife has been designated a noxious weed by several states (e.g., North Dakota, South Dakota).

Melaleuca is currently limited to Florida, but its destructive effect on the Everglades and other natural areas and its rapid spread make this plant particularly noxious. It currently infests 1.5 million acres and, of this, approximately 43,500 acres consists of pure melaleuca stands (Bodle et al. 1994). Melaleuca was introduced from Australia in 1906. Because of its extremely high evapotranspiration rate, one of its original uses was to dry up and drain wetland areas for farming. Its success at doing this is one of the major reasons it is considered a serious pest in areas where preservation of wetlands is now the goal. The trees can be up to 80 ft tall, and dense, tangled stands of up to 4000 trees per acre make access impossible and obliterate both plant and animal diversity.

Salt cedar was purposely introduced into the southwestern United States in the late 1800s to serve as windbreaks and for bank stabilization. Its major disruptive property is its ability to evapotranspirate water from the water table at rates as high as 9 acre-ft of water per acre of vegetation per year (Kerpez and Smith 1987). Since salt cedar also is drought resistant, it can continue to survive even though it may have caused the water table to be greatly lowered, whereas native species cannot tolerate such water loss. Most bird species find the seeds inedible, and the dense thickets prevent access of larger animals to the riverbank.

GOALS FOR AQUATIC PLANT MANAGERS

A diverse community of native aquatic plant and wetland species is much more desirable than a monoculture because it has less of a negative impact on human uses and provides habitat for a diverse community of animals. Native submersed species tend to be low growing, provide oxygen throughout the water column, and aid in sediment stabilization. Native submersed species, however, can pose weed problems when growing in shallow water (e.g., ponds and shallow portions of lakes and reservoirs) where they can top out at the surface.

Aquatic vegetation harbors many organisms including tiny invertebrates such as *Daphnia* and copepods, larger aquatic insects, crayfish, and small fish and amphibians that provide food

Figure 15.10 Purple loosestrife. This plant is about 5 ft tall.

for larger fish and for waterfowl. Fishermen know that weed beds are excellent places to fish because the shade, cover, and food organisms they provide attract large fish. The "ideal" amount of vegetation in a body of water needed to support a healthy fish population is somewhat controversial, but 20% coverage is commonly recommended.

From this discussion it should be obvious that the goal of an aquatic plant manager is not to eliminate all vegetation but to promote the growth of native species, when feasible, and to suppress or control invasive species and other plant populations that are excessive or prevent essential human activities.

CONDITIONS FOR AQUATIC WEED GROWTH

Two of the major factors that regulate aquatic plant growth are sunlight and nutrients. In aquatic habitats, the depth of light penetration dictates the depth to which underwater plants will grow. The amount of nutrients determines the quantity of vegetation that can be produced. Other important factors include adequate growing temperatures and, for rooted plants, a stable substrate and protection from wave action. The factors that regulate aquatic plant growth most often are optimal in shallow water, so many aquatic weed problems also are limited to "shallow" sites. The potential for a body of water to develop an aquatic weed infestation can be estimated by evaluating the availability of each of these growth-regulating factors.

Light

Submersed plants do not grow in water that is so turbid or deep that all light is blocked; however, they can grow at *very* low light intensities, even as low as 1% that of surface light intensity on a midsummer day. The portion of a body of water in which enough light can penetrate to support aquatic plant growth is termed the photic zone. The bottom of the photic zone generally is defined as the depth at which the light intensity is equivalent to 1% of full sunlight. In silty ponds, the depth of the photic zone may be only a few inches, whereas in extremely clear bodies of water such as Crater Lake, Oregon, the photic zone can extend to a depth of 300 ft. The shallower the body of water, however, the more likely it is that the photic zone will extend to the sediment bottom where rooted weeds can grow.

The depth of the photic zone, and therefore the potential of a body of water to support aquatic

weeds, can be estimated using a Secchi disk (Figure 15.11), a disk-shaped piece of metal painted with white and black sections that hangs horizontally in the water. The Secchi disk is lowered into the water on a chain or rope marked in feet. The depth at which the disk disappears is noted. The disk is then slowly raised and the depth at which it becomes visible again is noted. The average of these two depths is the Secchi disk depth. Multiplying this value by three gives a rough estimate of the depth of the photic zone (Figure 15.12).

Factors that reduce water clarity include suspended sediments and phytoplankton growth. Another factor that influences water clarity is water hardness. Most people recognize hard water as that which requires large amounts of soap to lather or that, on evaporation, forms a deposit of salts on the container. Technically, water hardness is a measure of calcium and magnesium. These ions can bind with and cause the precipitation of suspended colloidal particles such as clays and organic matter that normally reduce the penetration of light. Soft, acid waters with low calcium and magnesium concentrations often are stained with organic substances that prevent good light penetration. Hard waters (hardness values greater than 60–75 ppm $CaCO_3$) tend to be clearer and therefore weedier than soft waters. Although hydrilla grows prolifically in many of the clear water lakes and waterways of Florida, it seldom is found in the waters of northern Florida, many of which have high concentrations of tannic (brown-staining) acids. Water hardness can be measured easily with inexpensive test kits.

Nutrients

Bodies of water that drain fertile watersheds suitable for agriculture tend to support heavier

Figure 15.11   A Secchi disk.

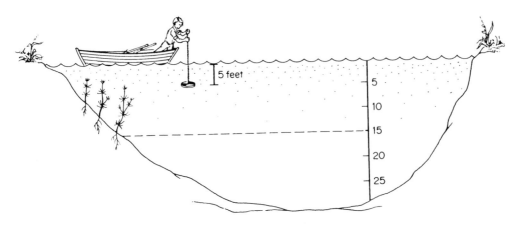

Figure 15.12 The Secchi disk depth is 5 ft. Therefore, the depth of the photic zone in this pond is approximately 15 ft. This depth is the lowest depth at which rooted plants can grow.

growths of aquatic weeds than waters located in poorly developed soils low in organic matter and nutrients. Lakes and ponds formed in the poorly developed soils of granitic basins and volcanic areas (e.g., portions of the western United States) seldom are weedy unless directly contaminated by human activities.

Besides being released from the underlying materials, nutrients can enter aquatic systems from precipitation, streams, groundwater, and runoff from urban and rural areas. In general, the larger the watershed (area draining into a body of water), the more productive the body of water will be in terms of plant growth.

Specific nutrient contributions from urban watersheds include sewage effluents, storm sewer drainage, and septic field seepage. A major agricultural source is runoff from fertilized fields, feedlots, and nearby pastured areas. Runoff is a major contributor to nutrient enrichment, so it is not unusual to see the most prolific weed growths occurring in shallow shoreline areas where the runoff is first received.

The nutrients most often regulating aquatic plant growth are carbon, nitrogen, and phosphorus. Of the three, phosphorus generally is conceded to be the most important. The nitrogen and carbon content of most natural fresh waters exceeds that of phosphorus by an order of magnitude or more, so phosphorus is most likely to be the first nutrient to become limiting to plant growth. Furthermore, the addition of small amounts of phosphorus to phosphorus-depleted waters results in inordinately large increases in plant biomass. Calculations based on the elemental ratios of phosphorus, nitrogen, and carbon in plant tissues indicate that the addition of 1 lb of phosphorus will generate 500 lb of plant biomass whereas 1 lb of nitrogen or carbon will only generate 71 and 12 lb, respectively.

The direct addition of phosphorus or nitrogen to water is most important in the growth of plants that obtain their nutrients *directly from the water* (i.e., phytoplankton, filamentous algae, and free-floating flowering plants). Rooted plants, on the other hand, obtain most of their phosphorus and other nutrients *from the sediments* through root uptake (Carignan and Kalff 1980, Barko and Smart 1980). Sediment fertility appears to be a more critical factor than water fertility in regulating rooted plant growth and probably is one reason dense stands of submersed weeds sometimes can be found growing in clear, clean waters that do not have a high nutrient content. Continuous inputs of nutrients to the water, however, still can have an important impact on rooted plant growth since these nutrients eventually become incorporated into the sediments.

Temperature

Most aquatic plants undergo optimum growth under conditions of warming waters in late spring and early summer and reach their maximum biomass in midsummer. Shallow water tends to warm up more quickly than deeper water, so these areas usually are the first to show growths of aquatic weeds. Shallow waters provide aquatic plants with long growing periods, even in the northern states where the growing season normally is short.

Temperature is a major determining factor in the life cycle and geographic distribution of aquatic plants. In temperate zones, the onset of cold water temperatures in the fall causes most aquatic plants to die back to the sediments. Some underwater plants that live in deep water can survive through the winter under an ice cover with little loss in biomass. Examples include largeleaf pondweed, Robbins pondweed, and Eurasian watermilfoil.

Other plants grow best in late summer and early spring. Examples include curlyleaf pondweed and certain filamentous algae such as *Spirogyra*. In natural lakes in the Midwest, curlyleaf pondweed appears in late summer, persists over the winter, and reaches its maximum biomass in late spring. The plant dies down in midsummer to be supplanted by other submersed plants such as Eurasian watermilfoil and then begins new growth again in the fall.

Substrate

A stable substrate is required for the attachment of rooted aquatic plants. Sand tends to shift and therefore is a poor substrate in flowing waters or along shorelines of large lakes exposed to strong wind and wave action. In smaller bodies of water and in protected areas of streams and lakes, sand as well as silt and clay interspersed with some organic matter provides an excellent rooting medium for most aquatic plants. Rock and large gravel substrates are not conducive to the growth of rooted flowering plants because they provide little in the way of fertile sediments for nutrient uptake.

Runoff and erosion of terrestrial sediments lead to a buildup of soil along shorelines and at the mouth of inflowing streams. Shallow areas with gradually sloping bottoms are prime locations for weed infestations. Conversely, bodies of water that are deep and have steep sides provide few sites for plant attachment.

AQUATIC WEED CONTROL METHODS

Of the available control methods, chemicals are the most frequently used because they are effective and work relatively rapidly. Measures that might alleviate a severe aquatic weed problem over the long term (for example, elimination of nutrient inputs) should not be ignored, however. In assessing aquatic weed problems, all traditional methods of control (preventive, mechanical, biological, and chemical) should be evaluated for short- and long-term effectiveness, applicability to the management situation, level of plant control desired, and cost. Another method of control with direct applicability to the aquatic situation is habitat alteration.

Preventive Weed Control

Preventive aquatic weed control consists of two major objectives: prevention of weed spread and reduction of shallow areas where plants can root.

Prevention of Aquatic Weed Spread

Aquatic plants seem to appear in new ponds or lakes quickly, even though the nearest body of water may be miles away. Algal spores can be carried by wind, and spores, seeds, and plant fragments can be carried on the feet and feathers of waterfowl or the fur of other animals. Humans also are primary movers of aquatic weeds, frequently spreading them by moving boats and boat trailers without proper cleaning. Using aquatic plants to provide packing or shade for bait minnows or worms and disposing of aquarium plants directly into bodies of water or through sewer systems are common methods of spread. Hydrilla probably was introduced into Washington state when it was used as packing material for ornamental waterlilies. Brazilian elodea (*Egeria densa*), a native of South America, is spreading in the eastern and Gulf states because it is one of the most widely sold aquarium plants in the world (sold as anacharis, or waterweed). Fanwort (*Cabomba caroliniana*), a native of Florida, is now present in many northern states because of its widespread use as an aquarium plant.

It is nearly impossible to prevent the spread of aquatic weeds by animals, wind, or water, but any human activity associated with weed spread should be curtailed. The rate of spread can be reduced by education, inspections and quarantines, and eradication of initial infestations.

Education. Education mostly takes the form of written materials such as pamphlets (Figure 15.13) and posters. Such educational materials and warnings often are posted at public boat docks, and new aquatic weed invasions receive considerable coverage by the print and television media. An example is the extensive media coverage given to the first appearance of Eurasian watermilfoil in lakes in the Minneapolis/St. Paul area in the 1980s. Similar publicity has been given to the invasion of hydrilla in the Potomac River in the Washington, D.C., area and the introduction and spread of Eurasian watermilfoil in Lake Tahoe, California. The problems caused by melaleuca are well publicized throughout Florida.

Inspections and Quarantines. Inspections of boats and boat trailers for the presence of Eurasian watermilfoil fragments at boat docks (e.g., in Minnesota) and at the borders of states or provinces (e.g., upon entering British Columbia) may be slowing the spread of this plant. Several states (e.g., Indiana) prohibit the sale of

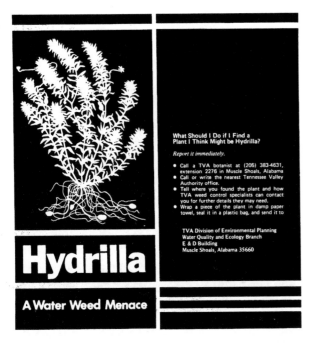

Figure 15.13 Portions of a pamphlet from the Tennessee Valley Authority describing hydrilla and the problems it causes.

ornamental purple loosestrife cultivars because they can interbreed with the weedy cultivar, thereby increasing its spread. At the present time no federal quarantines are in effect for aquatic weed species. Several aquatic weed species are on the Federal Noxious Weed List, including hydrilla and melaleuca. This listing means that permits are required to move these species into or through the United States (see Chapter 4). In 1994 approximately 2300 noxious plant specimens (both terrestrial and aquatic) were intercepted at ports of entry (NAPIS database at www//ceris.purdue.edu/napis/); however, it is probably safe to say that federal inspections prevent only a small number of exotic weed species, including aquatic ones, from entering the United States. For example, an average of four new aquatic pests (both plant and animal) are introduced into San Francisco Bay every year. Florida provides an ideal avenue for the introduction of tropical and subtropical plant species.

Eradication of Initial Infestations. Eliminating initial infestations has been the major strategy in California's effort to prevent hydrilla from gaining a foothold in its extensive irrigation, river, and water-holding systems. Hydrilla is classified as a "Type A" pest, which means that the plant (like the Mediterranean fruit fly) is to be eradicated whenever possible and that its sale, intrastate movement, and importation are illegal. The procedures that the California De-

partment of Food and Agriculture have employed are drastic but effective. The strategy is to treat the infested area with an aquatic herbicide (mostly fluridone), drain the site, scoop out and dispose of the soil, treat the remaining base materials with a soil fumigant, and then refill with water after the appropriate waiting period.

Although these methods are expensive and not feasible for all situations, a program of hand pulling or spot treatments with herbicides of new infestations could significantly reduce the rate of aquatic weed spread. If successful, the initial expense of eradication is extremely cost effective compared to the expense of long-term control. Controlling dissemination is the only method by which an extremely aggressive plant like hydrilla can be prevented from extending its range.

Elimination of Shallow Areas

New ponds and lakes should be constructed to avoid extensive shallow areas less than 3 ft deep. Shoreline edges should be deepened to at least 3 ft to reduce sites for rapid plant establishment. The only exception should be swimming areas in which sharp drop-offs may be hazardous to young children. Swimming areas can be maintained relatively free of submersed vegetation by laying down black plastic and covering the area with sand.

Reduction of shallow areas in existing ponds or lakes can be accomplished with dredges or draglines but usually at high cost. One advantage to removing sediments is that nutrients and plants also are removed. A dumping site for the spoils must be available, however.

Habitat Alteration

A control methodology unique to aquatic sites involves the alteration of the environment in which the weeds are found. The major targets for alteration are the factors that enhance aquatic plant growth: light, nutrients, temperature, and substrate.

Light Alteration

Techniques that shade submersed plants can be effective control strategies. These include the use of benthic barriers and of dyes.

Benthic Barriers. Covering bottom sediments with black plastic and other commercial products (also called substrate liners) to shade emerging submersed plants can be useful on a small scale for controlling submersed weeds. Areas amenable to benthic barriers include boat docks, swimming beaches, and small water gardens. The barrier must be anchored into the sediments and must be porous enough to allow gases that build up in the sediment to be released. Otherwise, gas buildup will cause the barrier to balloon up off the sediments.

Dyes. Nontoxic dyes, which act as light screens, can be used to inhibit submersed plant growth. An example is the blue dye Aquashadow, which absorbs light that otherwise would be used for photosynthesis. Aquashadow can be applied easily (Figure 15.14), disperses readily in a body of water, and reduces growth of plants at depths greater than 2 ft. The dye concentration must be maintained throughout the growing season, so its use is limited to ponds without outflow. Also, it must be applied before or as the weeds emerge in the spring; once weeds reach the water surface, the dye has little effect. In the Midwest, an initial application plus a midseason application are suggested.

Another shading technique, but one that is *not* recommended for the typical lake property or pond owner, is fertilization. Fertilizers stimulate the bloom of microscopic algae, which then shade out the underwater rooted plants. The technique is used primarily in the southeastern United States in catfish farming operations. Once initiated, fertilization must be continued over the growing season to maintain the bloom. If fertilization is stopped and the water is allowed to clear, heavy infestations of submersed weeds can be expected because the sediments eventually receive much of the added fertilizer. The technique requires constant monitoring and the means to correct imbalances in algal populations and oxygen content. For most bodies of water, efforts should be made to reduce nutrient inflow, not increase it.

Figure 15.14 Aquashadow can be poured directly from the container into the water.

Nutrient Control

A major effort in recent years, particularly in streams and larger bodies of water such as lakes and reservoirs, has been the control of essential plant nutrients such as phosphorus, nitrogen, and carbon. Most of the effort has centered on phosphorus because of its importance in regulating aquatic plant growth, and because it is the easiest of the three elements to control. Sufficient quantities of carbon to support extensive plant growth can enter water from the atmosphere (CO_2) or are available from dissolved bicarbonate (HCO_3^-) compounds. Nitrogen is available from many sources including rain, lightning, groundwater, and the fixation of atmospheric nitrogen by blue-green algae. For many lakes and streams, however, the major inputs of phosphorus are from point sources such as sewage treatment plants, septic tanks, storm sewers, and feedlots. Most of these sources can be identified, so appropriate control measures can and should be initiated; for example, phosphorus can be removed through tertiary treatment at sewage treatment facilities, drainage from feedlots can be directed away from the water body, sewer systems can be installed in place of inadequate septic systems, and laws restricting the use of phosphate detergents can be passed, as they already have been in many states.

The results of phosphorus limitation through the control of point sources has been spectacular in some cases. An excellent example is that of Lake Washington in Seattle, Washington, which was plagued with heavy blooms of microscopic algae. The installation of a diversion system to direct sewage effluent away from the lake resulted in a decrease in the algal populations and the improvement of water clarity (as measured by a Secchi disk) from 3 ft to 10 ft over a period of 7 yr (Edmundson 1970). Unfortunately, water clarity improved to the point that submersed plants, notably Eurasian watermilfoil, invaded Lake Washington and now cause major problems at boat docks and in shallow areas.

Not all bodies of water, however, obtain nutrients from point sources. When nutrient input is from diffuse sources such as surrounding agricultural fields, control is much more difficult. Some of the methods that should be considered for reducing nutrient inflow include the following:

1. Installing grass filter strips along drainage areas and around receiving waters to prevent runoff and to absorb nutrients.

2. Discontinuing the fertilization of grass around the body of water. Fertilization of high maintenance lawns and golf courses can be a major source of nutrients in ponds and lakes.

3. Preventing livestock from entering the body of water. Animals not only fertilize the water but tear down banks and increase soil erosion. A pipe and trough can be installed to allow stock watering nearby. The presence of excessive numbers of waterfowl, such as Canada geese, also should be discouraged.

4. Practicing conservation tillage methods in areas that are subject to erosion.

5. Constructing a settling or retention pond or wetland area to receive nutrients from an inflowing stream or drainage site (Figure 15.15) before the flow reaches the main body of water.

6. Avoiding the addition of fertilizers to a body of water. In fact, except in commercial fish operations, fertilizers *never* should be added to a body of water.

7. Checking for hidden sources of nutrients such as septic fields and drainage tiles. Septic systems can be checked using dyes available from state or local health boards.

Once nutrients are present in a body of water, usually stockpiled in the sediments, certain methods can be used to prevent them from being resuspended in the water where they can be used to support algal growth. These methods include nutrient precipitation with alum, the removal of nutrient-rich sediments by dredging, and aeration.

The principle behind aeration is that oxygen-rich waters tend to cause phosphorus normally suspended in the water to be precipitated to the sediments. Another advantage to aeration is to provide constant oxygen levels for fish. Aerating

equipment for lakes works on one of two principles. It either injects air into anaerobic bottom water or it pumps bottom water to the surface where it is mixed with air and reintroduced to the bottom of the lake. Simpler aeration devices and fountains agitate the surface waters but seldom are effective in maintaining necessary dissolved oxygen levels for fish. Aeration is beneficial for deep lakes with anaerobic bottom waters because it opens up new water areas for fish life and prevents the lake from going completely anaerobic in the summer or winter.

Methods that either reduce nutrient inputs or cause the precipitation of nutrients into the sediments (such as alum and aeration) are most effective for controlling organisms such as phytoplankton that receive their nutrient supply from the surrounding water. Even though every effort should be made to reduce the nutrient content of water, there is no substantial evidence to suggest that these efforts will result in an immediate decrease in established rooted plant growth. Reducing nutrient inputs as a control technique for rooted plants probably is most effective in newly established bodies of water in which the sediments have not yet become heavily nutrient laden.

Temperature Alteration

Winter drawdowns are effective for controlling many submersed and rooted floating weeds. This technique involves removing water to expose the shallow areas to drying and freezing conditions. Drawdown can be achieved with water flow control structures built into the pond, lake, or ditch; the installation of siphoning systems to lower the water level; or naturally as a result of receding shorelines during periods of low rainfall. One of the benefits of a partial drawdown is to concentrate the fish in a small, deep area away from the shallow weed zone. Concentration enables more effective predation of small fish by large fish, resulting in an overall improvement in fish quality. Another benefit is the drying and compaction of the sediments, which deepens the water body.

Substrate Alteration

The shoreline can be lined with rocks or other types of riprap to prevent both erosion and aquatic weed establishment. Rooted aquatic plants need a sediment in order to anchor and obtain nutrients; rock or concrete blocks can reduce this growth effectively.

Figure 15.15 Drainage from fields flows through a cattail wetland before entering a pond (background).

Mechanical Weed Control

Methods to remove existing stands of aquatic weeds mechanically include hand pulling, raking, the use of various handheld cutting instruments, and the use of various types of mechanized equipment. Hand-removal (Figure 15.16) can be temporarily effective, but it is extremely time consuming and laborious. Regrowth from seeds and underground plant parts can be expected. Hand-removal is a major method of aquatic weed control in developing countries.

Hand pulling can be effective in restricting the spread of invasive species. For example, the first plant of an invasive emergent such as purple loosestrife that appears at a site should be hand pulled (both underground and aboveground portions) before it can set seed. Once the infestation has spread to large areas, hand-removal becomes less of a viable option and other methods, such as chemical control, must be used.

Mechanized equipment includes draglines, which are commonly used to remove vegetation and sediments from irrigation and drainage ditches. Draglining is an effective weed control method, but it is expensive and usually must be repeated every 3 to 4 yr. Mechanical weed harvesters are used primarily on large lakes or rivers. These machines (Figure 15.17) cut off underwater rooted vegetation 4 to 5 ft below the water surface and collect the plants for transport to shore. Weed cutters without a harvesting capability are not recommended because the cut plant fragments can easily spread and reroot in uninfested areas.

One of the benefits of mechanical harvesting is that plant material is not left in the water to decompose; removal reduces the chances for fish kills. Removing plant material also aids in removing nutrients. Only a small portion, 2–3% in some cases, of a lake's nutrient content is con-

Figure 15.17 A mechanical aquatic weed harvester. (Aquamarine Corporation, Waukesha, WI)

tained in the aquatic weeds, however. If continued over a long period of time, harvesting may lower the nutrient content of a body of water significantly, but only if new inputs of nutrients are prevented from entering the lake and nutrients are not recycled from the sediments.

A number of factors should be considered before investing in a mechanical harvester. High equipment and maintenance costs can make these machines, at least initially, an expensive form of weed control. Mechanical harvesting is like mowing a lawn. The plants continue to grow from the uncut portions, so harvesting must be done several times in a season to maintain open water. Mechanical harvesters are unsuitable for removing vegetation in water less than 3 to 4 ft deep. Another consideration is vegetation disposal. A dumping area from which vegetation cannot wash back into the lake should be available.

Harvested aquatic weeds can be used as mulches or fertilizers in gardens and fields. Research has been conducted to determine other potential uses of the wet vegetation, such as livestock feed or as a source of methane gas. So far, the availability of feed and fuel from more nutritious or cheaper sources has prevented adoption of harvested aquatic vegetation for these uses. Another problem is the fact that aquatic plants typically contain high percentages of water. As much as 85 to 95% of the fresh weight of an aquatic plant can be water; in addition, the presence of water on the outside of the plant adds to the overall wet condition. This is in contrast to a 70 to 90% water content in terrestrial plants. Before harvested aquatic plants can be transported or processed into livestock feed, they must be dried. The large amount of energy required for drying (particularly in humid or cool areas where air drying is not feasible) has prevented the economical use of this material.

Figure 15.16 Hand raking aquatic weeds. (Photo courtesy of Clark Throssell, Purdue University)

Biological Weed Control

Some of the most spectacular and promising results with biological controls to date have been achieved on aquatic weeds. The aquatic environment presents a habitat roughly analogous to rangeland in which biocontrols also have been used successfully. In both cases, the environment is relatively undisturbed, does not usually support high value crops, and can be so extensive in area that the costs of chemical and mechanical control operations may be prohibitive.

Of the three major strategies for biological control (Chapter 4), the classical approach and grazing animals are the most widely used in aquatic situations. Although many fungi have been tested for efficacy as biocontrol agents, such manipulated agents have not yet been developed successfully for the aquatic plant control market.

The Classical Approach

Successful biocontrol agents for aquatic weeds, like their terrestrial counterparts, must be adaptable and voracious. Selectivity, particularly among organisms feeding on emergent plants, is desirable; organisms must not show any tendency toward a shift in food preference to valuable terrestrial species.

Many of the classical biocontrol agents currently being used are specific to weeds found in the southeastern states, such as water hyacinth and alligatorweed (*Alternanthera philoxeroides*). This is the geographic area in which much of the early research on biological controls for aquatic weeds was conducted.

One of the most spectacular examples of a successful biocontrol is the use of insects for the control of alligatorweed, a plant that forms floating mats over the surface of the water. The release of the South American flea beetle (*Agasicles hygrophila*) in 1964 and two other insects over the next 7 yr has reduced alligatorweed in waterways from Virginia to Texas to the point that it is no longer a problem. Damage by the insects either kills the plant outright or weakens it, making it vulnerable to disease, wind or wave action, or low doses of herbicides.

Two weevils (*Neochetina eichhorniae* and *N. bruchi*) and a moth (*Sameodes albiguttalis*) have been introduced in Florida for water hyacinth control. The widespread distribution of *N. eichhorniae*, in particular, and signs of feeding on water hyacinth in central and south Florida, indicate that these insects are well established. Their activity, along with a vigorous program of 2,4-D treatments, probably accounts in part for the reduction of water hyacinth coverage from a high of 125,000 acres of public waterways in 1959 to a coverage of 1650 acres in 1994. However, the insects by themselves, although weakening the plant, do not control water hyacinth. In areas where no mechanical or chemical control has been attempted for years, feeding damage is visible but water hyacinth coverages are as high as 100%.

The leaf-mining fly *Hydrellia pakistanae* was released in the southeastern United States in 1987 as a potential biological control for hydrilla. Although *H. pakistanae* is now widespread throughout the region, its potential to cause hydrilla declines either by itself or in tandem with herbicide control has not been determined.

An initial screen of organisms that feed on melaleuca in its native habitat in Australia led to the listing of over 400 species. Of these, the Australian weevil (*Oxyops vitiosa*) appears to have the greatest potential and is being tested in Florida (Purcell and Balciunas 1994).

Research on potential biocontrol agents for more northerly distributed aquatic weed species also is being conducted. In 1987 a study was initiated to screen 120 insects associated with purple loosestrife in Europe (Malecki et al. 1993). Of these, five insects have been released in the United States. They provide a multidimensional approach for control since they target different parts of the plant. Two leaf-mining beetles (*Galerucella calmariensis* and *G. pusilla*) defoliate the plant, a root-mining weevil (*Hylobius transverovittatus*) attacks the main storage tissue, and two flower-feeding beetles (*Nanophyes marmoraturs* and *N. brevis*) reduce seed production. The latter is particularly important to prevent the spread and reestablishment of the plant.

The occurrence of natural die-offs of Eurasian watermilfoil in Vermont led to the discovery of a milfoil-feeding weevil (*Euhrychiopsis lecontei*) that was causing significant damage to established stands. Studies suggest that this insect may have potential to control Eurasian watermilfoil when introduced into infested waters.

Grazing Animals

Grazing animals include waterfowl and herbivorous fish.

Waterfowl. White Chinese geese, muscovy ducks, and cygnet swans have been used successfully in small ponds for removing floating and submersed plants rooted in shallow water. One pair of swans is reported to maintain an acre weed-free, and three to eight geese or ducks per acre may reduce duckweed populations (Holm and Yeo l981). Major problems with waterfowl are that they are extremely aggressive during certain parts of the year, they must be protected from predators, their droppings litter the shoreline, and they contribute to shoreline erosion and water fertility.

Herbivorous Fish. A great deal of attention has been given to herbivorous fish. Fish species native to the United States normally do not feed on vegetation, so species that are strictly vegetarian have been introduced from other parts of the world. The two most notable introductions have been the tilapia (*Tilapia* or *Sarotherodon*) and the grass carp or white amur (*Ctenopharyngodon idella*).

Tilapia. The Congo tilapia (*Tilapia melanopleura*) feeds on phytoplankton and unicellular algae when young and feeds on larger plants as an adult. Tilapias, which are cultivated for food widely throughout the warm regions of the world, are extremely efficient at converting plant tissue into fish biomass. Their rapid reproduction, however, sometimes results in large populations of undersized fish that are not marketable. This trait also can cause problems for native fish species if the tilapia is permitted to escape into our waters. The spread of this fish thus far has been checked by its intolerance to cold temperatures. This intolerance limits its potential as a biological control agent in this country, primarily to Florida. For example, the Java tilapia (*T. mossambicus*) dies at water temperatures of 46 to 48°F and cannot survive winters in ponds in central and southern Alabama. *Tilapia zillii* was introduced in irrigation canals in the Imperial Valley of southern California for vegetation control, but the canals had to be restocked each year.

Grass Carp. Survival in cold water is not a problem for the grass carp or white amur (Figure 15.18). This fish is a native of the rivers of China, Manchuria, and Siberia, so it tolerates cold water. For example, it can easily survive the winter in ice-covered ponds and lakes in Indiana and Iowa. The grass carp is a strict vegetarian, feeding on some filamentous algae, Chara, and most submersed weeds. Although it has been known to eat vegetation just above the shoreline, it is not considered an effective control for emergent vegetation such as cattails or large free-floating weeds such as water hyacinth.

Grass carp eat their weight in vegetation every day, increase in weight by 1 to 10 lb per year, and may weigh as much as 30 to 50 lb in large lakes. Recommendations for the stocking

Figure 15.18 The grass carp (*Ctenopharyngodon idella*).

rate of fish vary from state to state. In some cases the stocking rate is based on the severity of the weed infestation. In others, the stocking rate is based on the level of control desired. For example, stocking rates in Indiana are 15 to 30 fish per vegetated acre. The lower stocking rate is intended to prevent complete removal of vegetation; the higher stocking rate is used in waters where no vegetation can be tolerated, such as irrigation canals, aquaculture ponds, and fire-control ponds. The fish should be 8 to 12 in. long in order to prevent loss from fish predators such as largemouth bass. Grass carp can provide long-term control, for they are believed to have a life span of about 16 yr.

Although the grass carp is in the same family of fishes as the common carp, it does not have the undesirable attributes of the common carp. For example, it does not roil the sediments and muddy the water as does the common carp. The grass carp should not be confused with the Israeli or the mirror carps, which are strains of common carps bred for improved food quality.

The grass carp does not appear to reproduce naturally in static bodies of water such as ponds and lakes, but some reproduction has been reported in the large river systems in this country. All commercial breeding of this fish is conducted at fish hatcheries, where spawning is induced with hormone injections and the fertilized eggs are cultured under specially controlled conditions. In addition, the eggs can be subjected to certain stress conditions in order to induce triploidy. Triploid organisms have three sets of chromosomes rather than the normal two sets of chromosomes characteristic of most organisms, which are diploid. The triploid condition renders the fish sterile so that they cannot reproduce naturally. Triploid grass carp therefore cannot multiply and crowd out our native game species, and it is the only form of the fish that is approved for stocking by many states. Not all states approve the importation and sale of grass carp; *consult your state fisheries biologists or university extension personnel to determine if purchasing grass carp is legal in your state.*

Although the grass carp has provided good control of aquatic vegetation in many situations, it is not the solution for all bodies of water. Since its effects on vegetation may not be noticed for a year or more, it is difficult to determine if enough fish are still present in the water to be effective. If results are not observed within 3 yr, restocking is probably necessary because the fish may have died, escaped, or been removed by predators. The fish may move away from target areas (e.g., around boat docks) to feed in parts of the lake where there may be less disturbance. In addition, the grass carp prefers certain plant species over others. For example, it will consume native species such as pondweeds before it will feed on truly troublesome weeds such as Eurasian watermilfoil.

If retention of native species is a goal, the grass carp is not the best weed control choice. In fact, the grass carp should not be introduced into natural bodies of water (such as natural lakes, wetlands, rivers, etc.) where it could destroy native vegetation and remove habitat for fish and other animals. Some of the states that have banned grass carp importation (e.g., Michigan, Wisconsin, Minnesota, and others) are concerned that the grass carp could destroy the diverse aquatic flora that is essential to maintaining a healthy fishery.

Chemical Weed Control

The most frequently used method of aquatic weed control in the United States is the application of aquatic herbicides. Herbicides are relatively easy to apply, provide effective and rapid control, and can be used in a wide variety of situations. They can be used selectively to remove certain plant species and leave others. In fact, herbicides do such an excellent job of controlling aquatic weeds quickly that people tend to overlook other methods of control, particularly preventive and habitat alteration methods that reduce or eliminate the actual cause of aquatic weed growth (e.g., nutrient inputs and shallow water). A sound management program incorporates the use of aquatic herbicides when necessary but also includes steps that provide a long-term solution in addition to short-term, cosmetic ones.

Once the decision to use an aquatic herbicide has been made, at least seven important considerations are essential in planning a successful program: (1) proper identification of the weed or weeds, (2) uses of treated water and environmental impact of the herbicide, (3) choice of the most appropriate herbicide to fit the specific situation, (4) dosage and amount of herbicide to apply, (5) timing of the treatment and water temperature, (6) method of application, and (7) probability of retreatment, even within the same year.

Aquatic Weed Identification

Weed identification is essential because most aquatic herbicides are relatively selective. Help in identifying aquatic weeds can be obtained from cooperative extension services and state fisheries biologists or through publications available from state and federal agencies and industry. If a weed sample is to be mailed for identification, it should be removed from the water, allowed to drip drain for a few minutes, and placed in a plastic bag or wrapping for shipment. Water never should be added to the sample, for it will increase the rate at which the specimen deteriorates. A dry plant that can be rewet is preferable to an overly wet one that may decompose during shipment. Exceptions are phytoplankton blooms and scums, which can be collected only in water. These minute algae are extremely delicate and susceptible to rapid deterioration. Water samples should be collected and immediately shipped or hand delivered to the appropriate agency for identification. Brief descriptions of the major aquatic weed species follow.

Algae. The major categories of algae are microscopic algae, filamentous algae, and the Chara/Nitella group.

Microscopic Algae. Blooms of microscopic algae impart color to the water and on calm midsummer days can rise to the water surface to cause greenish or red-streaked scums (Figure 15.19). Although these organisms are a major source of oxygen in bodies of water, the sudden die-off of dense blooms can cause oxygen depletion and fish kills. Examples of bloom-forming algae (Figure 15.20) are the blue-green algae (also called the cyanobacteria) *Microcystis, Anabaena,* and *Aphanizomenon,* which turn water a pea soup green or yellowish green, and the euglenoid algae *Trachelomonas* and *Euglena sanguinea,* which can turn water red-brown or sometimes a brilliant red. These red color-causing algae are not toxic to fish and animals. The red tide algae, dinoflagellates, which occur

Figure 15.19 Blue-green algal blooms often cause scums and streaks on the water surface.

in the sea, are toxic to fish and cause shellfish poisoning in mammals.

Filamentous Algae. These plants form dense free-floating or attached mats (Figure 15.1) consisting of green hairlike strands. The growths are often the first of the aquatic weeds to appear in the spring, usually around the edges (Figure 15.21) or along the bottoms of ponds and shallow portions of lakes. As the algae photosynthesize, oxygen bubbles become trapped in the mats, causing them to rise to the surface of the water where they become visible as floating clumps of vegetation. Several of the most common filamentous green algae (Figure 15.20) can be identified by their texture. *Spirogyra* is bright grass green and slimy to the touch. *Cladophora* is referred to as the cotton mat alga because of its texture, and *Pithophora*, which is very coarse, is called the horsehair alga. *Hydrodictyon* forms a soft, green, netlike growth. Some blue-green algae (e.g. *Lyngbya*) also form thick, free-floating mats. Another blue-green, *Oscillatoria*, normally grows as a coating on the bottom of a pond or lake. If the sediments are disturbed by heavy winds or rain, the slimy dark greenish-blue or black clumps can detach from the bottom and float to the surface.

Chara and Nitella. Chara (*Chara* spp.) and Nitella (*Nitella* spp.) resemble flowering plants in that their "leaves" are needle-like and arranged in whorls along a "stem" (Figure 15.22). The plants are submersed and attached to the sediments by rootlike rhizoids. They can easily be confused with flowering plants (Figure 15.23); identifying them as algae is essential to applying proper control measures. Common names for Chara are stonewort (because of the calcified, brittle

texture of the plant) and muskweed (for its peculiar skunklike odor). Chara typically is found in hard waters and therefore is abundant through the Midwest and the central portions of the Southeast. Nitella is more characteristic of soft waters and can cause problems in the northeastern states. Low-growing Chara and Nitella species provide good habitat for fish.

Flowering Plants. Flowering plants include submersed plants, free-floating plants, rooted floating plants, and emergent plants.

Submersed Plants. Most submersed plants are perennial and can reproduce from rhizomes, stolons, tubers, or turions. Propagation from seed tends to be minimal. An exception to this is the naiad group, which reproduces from seed only.

Submersed plants are identified on the basis of leaf arrangement on the stem (alternate, opposite, whorled, or basal) and leaf shape.

Plants with an alternate leaf arrangement have one leaf per node. The best known group of submersed aquatic plants with alternately arranged leaves are the pondweeds (*Potamogeton*). *Potamogeton* contains more species (60) of aquatic plants than any other aquatic genus. Several species are quite weedy; for example, curlyleaf pondweed (*P. crispus*), sago pondweed (*P. pectinatus*), and leafy pondweed (*P. foliosus*) (Figure 15.24). In some situations, *Potamogeton* species can be beneficial. The seeds and tubers of sago pondweed are important sources of food for many aquatic waterfowl species, and the plant is often introduced into managed waterfowl areas. Other species such as largeleaf pondweed (*P. amplifolious*) provide excellent habitat for game fish in large lakes and reservoirs and should be left undisturbed if not present in nuisance proportions.

Plants with an opposite leaf arrangement have two leaves attached opposite to one another at a node. The group of aquatic plants best illustrating this characteristic is the naiad genus (*Najas*). Examples include southern (*N. guadalupensis*) and brittle naiad (*N. minor*) (Figure 15.25).

Plants with a whorled leaf arrangement have three or more leaves attached at a node. Two of the most invasive aquatic weeds, Eurasian watermilfoil and hydrilla, belong in this group. Native species such as coontail (*Ceratophyllum demersum*) and American elodea (*Elodea canadensis*) also can infest shallow bodies of water and crowd out other species. The leaf shapes of these important plants are illustrated in Figure 15.26.

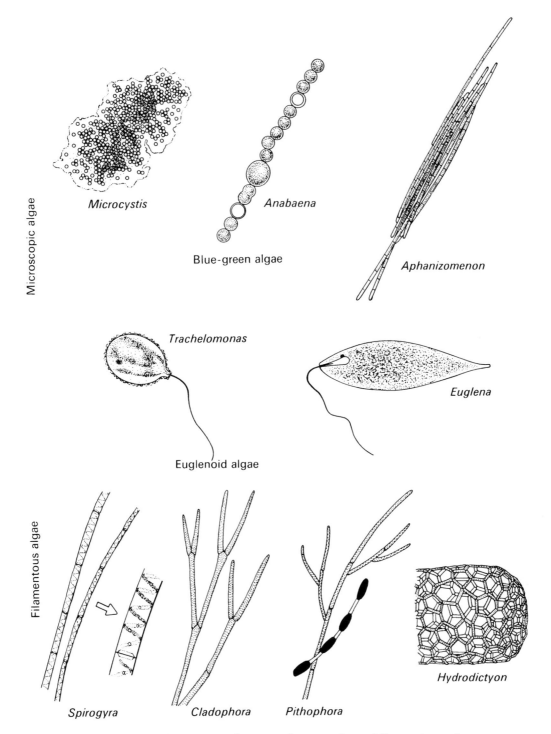

Figure 15.20 Common nuisance microscopic and filamentous algae.

Plants with a basal leaf arrangement have a very short stem that is buried in or emerges a few cm from the sediment (Figure 15.27). The leaves are long and linear and extend from the stem upward through the water. Examples of plants with basal leaves are tapegrass, also called eelgrass (*Vallisneria americana*), and strap-leaf sagittaria (*Sagittaria kurziana*). These plants provide excellent cover for fish but can be weedy in shallow waters.

Free-Floating Plants. Free-floaters vary considerably in size and shape, ranging from large plants with well-developed root systems, such as water hyacinth (Figures 15.4, 15.7, 15.8) and water lettuce (*Pistia stratiotes*), to tiny plants with few or no roots, such as the duckweeds.

The most common members of the duckweed family are duckweed (*Lemna*), giant duckweed (*Spirodela*), and watermeal (*Wolffia*) (Figure

Figure 15.21 Early spring growth of filamentous algae often begins in shallow portions of a pond.

15.28). Individual plants of giant duckweed are about 1/4 in. in diameter and contain many roots per plant. Duckweed is characterized by one root per plant, whereas watermeal has no roots. Watermeal is the smallest of all the flowering plants; it appears as fine green grains floating on the water surface (Figure 15.29). It often is mistaken for seeds of other plants. Growth of the duckweeds may be so prolific that a blanket of plants 1 to 2 in. thick can completely cover the water surface, causing bottom water to become anaerobic and thus unsuitable for fish life. The duckweeds require nutrient-enriched water because they are unable to obtain their nutrients from the soil. Sewage treatment lagoons, highly fertile ponds, and stagnant portions of lakes are ideal sites for duckweed growth.

Unlike water hyacinth, the duckweeds have an overwintering mechanism in waters subject to ice cover in winter. Starch accumulation in

Chara

Figure 15.22 Chara or muskweed.

Figure 15.23 The overall appearance of Chara is like that of flowering plants. Care must be taken to identify this plant correctly.

the late summer and fall causes the duckweeds to become heavy and sink to the bottom. The plants remain on the bottom sediments over the winter. Increasing water temperatures and light in the spring promote photosynthesis and the production of oxygen, which causes the plants to become buoyant, rise to the surface, and begin multiplying. The doubling time for reproduction of watermeal is one of the most rapid of any organism: 1 day. New plants "bud off" vegetatively from an internal pouch in the parent plant.

Rooted Floating Plants. These plants are identified to genus by leaf shape and petiole attachment (Figure 15.30). The petiole is attached at the center of the leaf in water lotus (*Nelumbo*) and watershield (*Brasenia*) and at the cleft or edge of the leaf in waterlily (*Nymphaea*), spatterdock (*Nuphar*), and floating leaved pondweeds such as American pondweed (*Potamogeton nodosus*). Leaves vary in shape from circular (water lotus, waterlily) to heart-shaped (spatterdock) and can lie on the water surface (waterlily, watershield) or extend above it (water lotus, spatterdock). The leaves of all these plants emerge from underground rhizomes. The massive rhizomes of spatterdock are branched and can be 6 to 7 in. in diameter (Figure 15.31).

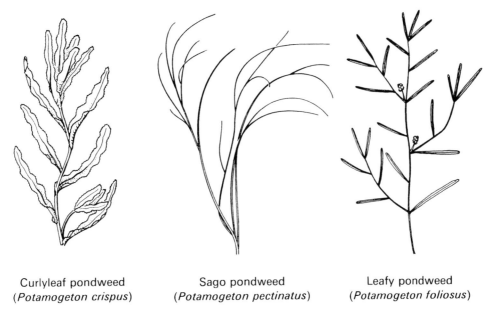

Curlyleaf pondweed
(*Potamogeton crispus*)

Sago pondweed
(*Potamogeton pectinatus*)

Leafy pondweed
(*Potamogeton foliosus*)

Figure 15.24   Common *Potamogeton* species. The leaf arrangement is alternate.

Infestations of these plants in shallow water hamper recreational activities and impede water flow. On the other hand, these plants provide valuable waterfowl habitat and their control in natural bodies of water is not permitted in some states.

Emergent Plants. Emergent plants are divided into herbaceous and woody species. Herbaceous plants are further divided into grasses (and grasslike plants) and broadleaves.

One of the most common of the grasslike plants is the cattail (*Typha*) (Figure 15.32). Although not in the grass family, these plants re-semble giant grasses and form a brown flowering spike in midsummer. They reproduce by seed and rhizomes and are notorious for rapidly filling in shallow areas of ponds and ditches. Their rhizome growth not only weakens dams but attracts muskrats that can cause further damage to the dam structure. Other grasslike plants include a number of sedges such as the bulrushes or tules (*Scirpus* sp.) and the spike rushes (*Eleocharis* sp.). Many true grasses also are found along waterways and in shallow water. These include giant reed (*Phragmites communis*), Reed's canary grass (*Phalaris arundinacea*), cutgrass (*Leersia hexandra*), and torpedograss (*Panicum repens*).

Emergent aquatic plants with "broad" leaves include arrowheads (*Sagittaria* sp.), creeping water primrose (*Jussiaea repens*), water smartweed (*Polygonum* sp.), and invasive species such as purple loosestrife and parrotfeather (*Myriophyllum aquaticum*). Like sago pondweed, the smartweeds are highly valued for waterfowl because they produce edible seed and provide habitat protection. On the other hand, excessive growths can become undesirable. Many terrestrial smartweeds are somewhat tolerant of wet ground and often grow up through the middle of a newly excavated and water-filled pond. Most species die out in deep water after a year or two.

Woody species commonly found along the water's edge include willows (*Salix* sp.) and button bush (*Cephalanthus occidentalis*). These plants increase water loss through transpiration and also make access to a body of water difficult. Invasive woody species include salt cedar and melaleuca.

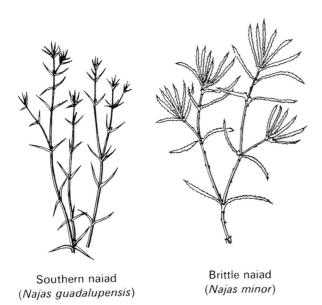

Southern naiad
(*Najas guadalupensis*)

Brittle naiad
(*Najas minor*)

Figure 15.25   Common *Najas* species. The leaf arrangement is opposite.

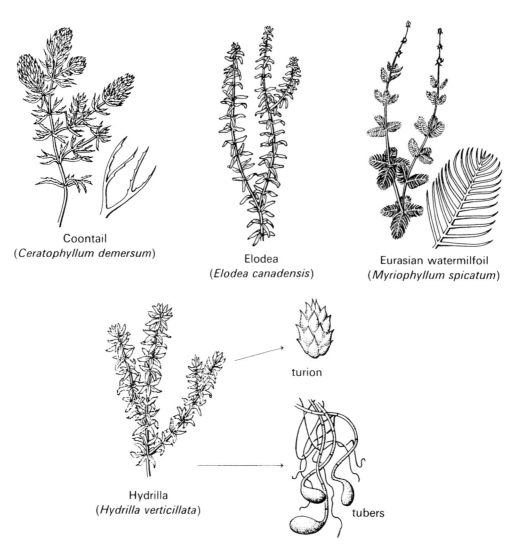

Coontail
(*Ceratophyllum demersum*)

Elodea
(*Elodea canadensis*)

Eurasian watermilfoil
(*Myriophyllum spicatum*)

turion

Hydrilla
(*Hydrilla verticillata*)

tubers

Figure 15.26   Aquatic weeds with whorled leaf arrangements.

Water Use and Environmental Impact

The introduction of most aquatic herbicides into the aquatic environment requires the restriction of the water use until the herbicide has been degraded, inactivated, or dissipated. Consequently, determining the present *and* potential uses of a body of water is one of the most critical steps in choosing an aquatic herbicide. Restrictions usually involve the use of water for drinking (potable water), swimming, livestock watering, and irrigation and the use of fish for consumption (Table 15.1). The period of restriction, which varies from herbicide to herbicide, depends on potential toxicity to nontarget organisms, the concentration of the herbicide in the water immediately after application, and the persistence of the compound in the water.

Toxicity to Nontarget Organisms. The potential for an aquatic herbicide to be toxic to an-imals is related to its mode of action in plants. Herbicides that are site-specific for a metabolic process that occurs in plants but does not occur in animals tend to have relatively low toxicity to animals. Fluridone and glyphosate inhibit the synthesis of carotenoids and aromatic amino acids, respectively, neither of which is synthesized by animals. 2,4-D competes with the activity of auxin, a hormone that is not produced by animals. The oral LD_{50}s of these three aquatic herbicides ($>$10,000, 5600, and 764 mg/kg, respectively; see Table 5.4) places them in the moderate to low toxicity category (Table 5.3).

Although the modes of action of copper sulfate and endothall are not well understood, they and diquat are probably general membrane and cell disruptors. Therefore, they have the potential to destroy animal cells, which is reflected in rather low LD_{50}s of 470, 206, and 230 mg/kg, respectively (Table 5.4). These compounds are in the high toxicity category (Table 5.3), and yet

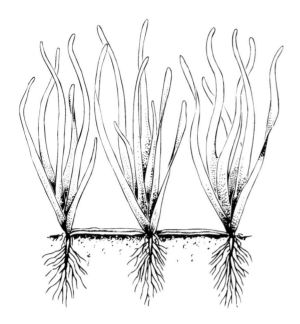

Figure 15.27 Eelgrass, a plant with a basal leaf arrangement.

Figure 15.29 A handful of watermeal plants from the surface of a pond.

they are the most commonly used of the aquatic herbicides. Why do they not cause toxicity problems to humans and other organisms? Part of the answer is that these products are used at extremely low dosages. Therefore, when they are applied to water, the concentration of aquatic herbicides is reduced to levels that are nontoxic to most nontarget organisms. For example, a comparison of typical concentrations in treated water immediately after application with the 96 hr LC_{50}s to bluegills (Table 15.2) indicates that there is a wide safety margin for most aquatic herbicides. For example, the amount of diquat present in the water immediately after application is approximately 116 times less than the concentration that kills 50% of the test bluegill population.

An exception to the wide safety margin is the dimethylamine salt of endothall (Hydrothol). The herbicide concentration in water (0.05 to 1.0 ppm) overlaps the 96 hr LC_{50} (0.94 ppm). This herbicide can be toxic to fish even after dilution. It should be applied in such a way that

fish have an opportunity to move to untreated water as soon as they detect the compound in the water. Interestingly, the potassium salt of endothall (Aquathol) has a much wider margin of safety; its concentration in treated water is as much as 100 to 600 times less than its LC_{50}.

All of the herbicides just listed (except for copper, which has not been submitted) have successfully undergone the reregistration process with the Environmental Protection Agency (see Chapter 5). Most are in the E carcinogenic category (noncarcinogenic; see Table 5.5). 2, 4-D is category D. The extensive literature on copper, particularly its effects on human health, indicate that it is noncarcinogenic. These data plus the fact that copper is a required mineral in the human diet suggest that the EPA is unlikely to add any new restrictions to copper, at least in the near future.

Another factor that protects nontarget organisms from toxicity is the relatively short persistence of aquatic herbicides in the water.

Persistence. Factors that affect the persistence of herbicides in water include herbicide chemistry (water solubility, affinity for soil colloids, stability to degradation) and environmental factors (e.g., temperature, microbial populations, water flow, presence of soil colloids and ions). Diquat and copper are adsorbed to suspended soil colloids, which then settle to the bottom. Although copper is not

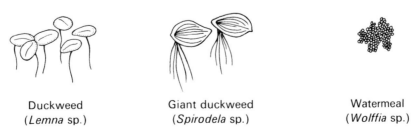

| Duckweed (*Lemna* sp.) | Giant duckweed (*Spirodela* sp.) | Watermeal (*Wolffia* sp.) |

Figure 15.28 Members of the duckweed family.

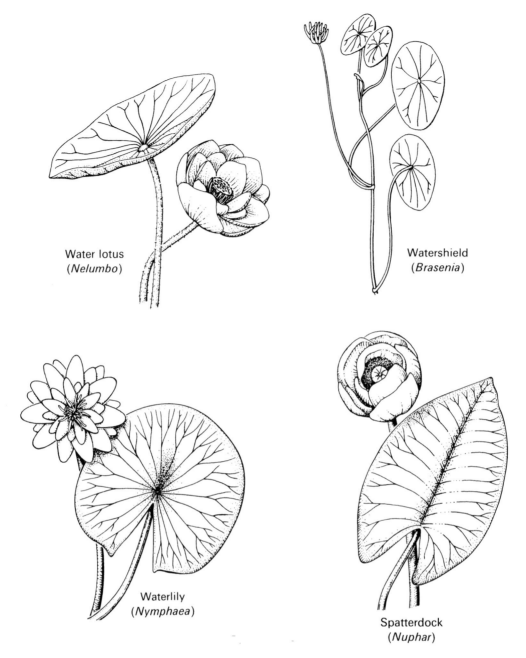

Figure 15.30   Rooted floating plants.

Water lotus
(*Nelumbo*)

Watershield
(*Brasenia*)

Waterlily
(*Nymphaea*)

Spatterdock
(*Nuphar*)

broken down, and diquat is only slowly degraded by microbes, the bonding between the herbicide and the soil colloid is so strong that these compounds are not released back into the water. Glyphosate is also adsorbed to colloids, but it is rapidly broken down by microbes. Endothall and 2,4-D are microbially degraded, and fluridone is photodegraded.

The rapidity with which these herbicides are degraded or removed from water via adsorption is reflected in the water use restrictions, which extend for only a few days to no more than a month for most of the herbicides (Table 15.1). The most severe restrictions are on 2,4-D, which cannot be used in water intended for drinking, livestock watering, and irrigation. These restrictions are more a reflection of lack of data development by industry than any negative characteristic of the herbicide itself. 2,4-D is such an inexpensive herbicide that the financial incentive needed to justify spending the money to obtain the necessary data to remove these restrictions is lacking.

Proper state or local authorities should be consulted regarding special restrictions that may not be noted on the herbicide label. In some states, permits are required before application can be made. Restrictions almost always are imposed for streams and public or multiple use lakes and reservoirs.

Figure 15.31 Rhizome of a spatterdock plant.

Choice of Herbicide

The checklist on page 329 on weed identification and the uses, characteristics, and dimensions of the body of water summarizes the information required for proper selection of an aquatic herbicide. Other considerations that enter into the selection process include cost, efficacy under different environmental conditions, suitability to application equipment, and need for additives.

A list of the major aquatic weeds and the herbicides labeled for their control is provided in Table 15.3 and is summarized here.

Herbicide Recommendations. Herbicide labels and state recommendations and use restrictions always should be consulted for detailed informa-

Figure 15.32 Cattails choking an irrigation canal in California.

tion. For example, 2,4-D is packaged in a number of formulations, most of which are not cleared for use in water or can only be used for emergent vegetation control. *Always* use the package that is labeled for aquatics; only that label has the appropriate use information. Applying unlabeled products violates federal regulations.

Herbicides (Algicides) for Microscopic Algae. Copper sulfate and copper chelates are used for microscopic algae control. Early in bloom development the algae may be dispersed throughout the water, whereas later the bloom may form a surface scum. Low concentrations of these chemicals can be used to thin the algal populations during development. After scum formation, treating only the upper 2 ft of the water with a copper compound is required. The amount of chemical is calculated for the upper 2 ft and is applied by spraying the water surface or injecting the chemical just beneath the surface. The rapid kill of heavy blooms of microscopic algae can cause oxygen depletion; therefore, early treatment is preferable.

Herbicides (Algicides) for Filamentous Algae. Copper sulfate, copper chelates, diquat, and the dimethylamine salt of endothall are labeled for filamentous algae. The copper compounds are by far the most widely used algicides for algae control. They should be dispersed in the water if the growth is underwater. If the mats are floating, at least some of the copper should be dissolved in water and sprayed directly on the mats. This increased contact is essential because copper must come in direct contact with the algal cells to be effective. Treatment with copper should be done on bright sunny days to enhance activity.

Herbicides (Algicides) for Chara. Copper sulfate, copper chelates, and the dimethylamine salt of endothall are used for Chara. This alga can be difficult to kill, particularly in hard water where the plants accumulate a covering of calcium carbonate (limestone) that may prevent adequate algicide penetration. Two or more treatments may be necessary. Best results are obtained when the plants are young and uncalcified.

Herbicides for Submersed Flowering Plants. Diquat, the dimethylamine and potassium salts of endothall, fluridone, and granular formulations of 2,4-D are labeled for the control of submersed flowering plants. The naiads and elodea are most susceptible to diquat; watermilfoil, to 2,4-D; and the pondweeds, to the potassium salt of endothall and diquat. Fluridone tends to be nonselective. Difficult-to-control weeds include

Table 15.1 Use Restrictions on Water Treated with Aquatic Herbicides (in Number of Days)

	Human				Irrigation		
	Drinking	Swimming	Fish Consumption	Livestock Watering	Turf	Forage	Food Crops
Copper-based compounds	0	0[a]	0	0	0	0	0
Diquat (Reward)	1–3	0[a]	0	1	1–3	1–3	5
Potassium salt of endothall							
(Aquathol Granular)	7	0[a]	3	0	7	7	7
(Aquathol K)	7–25	0[a]	3	7–25	7–25[d]	7–25	7–25
Dimethylamine salt of endothall							
(Hydrothol 191 Granular)	7–25	0[a]	3	7–25	7–25	7–25	7–25
(Hydrothol 191)	7–25	0[a]	3	7–25	7–25	7–25	7–25
Fluridone (Sonar)	0[b]	0[a]	0	0	7–30	7–30	7–30
Glyphosate (Rodeo)	0[b]	0[a]	0	0	0	0	0
2,4-D	[c]	0[a]	0	[c]	[c]	[c]	[c]

[a]Although this compound has no waiting period for swimming, it is always advisable to wait 24 hr before permitting swimming in the area of treatment.
[b]See label for distance allowed from potable water intake.
[c]Do not use for this purpose.
[d]May be used for sprinkling bent grass immediately.

eelgrass, cabomba, brittle naiad, and hydrilla. Some success with hydrilla control has been obtained with copper chelates and copper–diquat mixtures, particularly when applied as invert emulsions.

Herbicides for Free-Floating Flowering Plants. Diquat, fluridone, glyphosate, and 2,4-D are used for free-floating plants. Diquat, glyphosate, and liquid formulations of 2,4-D are used on water hyacinth, whereas diquat and fluridone are used on the duckweeds. Spraying to provide good plant coverage with 2,4-D and diquat is essential. Effective control of the duckweeds with diquat depends on contact with as many plants as possible, including those washed up above the shoreline. Duckweed control with diquat can be inconsistent because the compound may be inactivated by the organic matter that is usually associated with these plants, and regrowth can be initiated by dormant forms on the sediment bottom or plants that were not killed. Fluridone (AS formulation only), which is applied through the water, is a more consistent, albeit expensive,

alternative and can provide more than one-season control of duckweed.

Herbicides for Rooted Floating Flowering Plants. Diquat, the endothall salts, and fluridone are labeled for the floating-leaved pondweeds (e.g., American pondweed, Illinois pondweed). Glyphosate and 2,4-D are the most commonly used herbicides for plants such as waterlilies and spatterdock. Glyphosate is applied as a spray to the leaf surface. 2,4-D granules are used for sediment application, but their effect can be inconsistent.

Herbicides for Emergent Flowering Plants. Glyphosate and liquid formulations of 2,4-D are used for controlling emergent plants. Fluridone provides partial control only; its effects are most noticeable as a temporary whitening of emergent vegetation when the compound is used for submersed weed control. Control with glyphosate is dependent on adequate spray coverage and the addition of a wetting agent to Rodeo (the formulation cleared for water) to aid in cuticle penetration. Since

Table 15.2 Comparison of Aquatic Herbicide Concentrations in Treated Water Immediately After Application and Toxicity Data (96 Hr LC$_{50}$) for Bluegills

Herbicide	Herbicide Concentration in Treated Water (ppm)	96 hr LC$_{50}$ (ppm)[a]
Copper sulfate	1.0	44
Diquat	1.5[b]	175
Endothall (dimethylamine salt)	0.05–1.0	0.94
Endothall (potassium salt)	0.5–3.0	308
Fluridone	0.01–0.15	14
2,4-D (amine salt)	0.1[b]	524
Glyphosate	0.4[b]	120

[a]Data from *Herbicide Handbook* of the Weed Science Society of America (1994). See also Table 5.6.
[b]Calculated for 4 acre-ft.

CHECKLIST FOR AQUATIC WEED CONTROL PROGRAM

AQUATIC WEED IDENTIFICATION

Type of plant:
    Algae _____

    Flowering plant _____

Habitat:
    Free-floating _____

    Submersed (underwater) _____

Attachment:
    Rooted _____

    Nonrooted _____

    Rooted floating _____

    Emergent _____

Plant species _____

WATER USES

      Fishing _____   Drinking (potable) Water _____

      Swimming _____

      Irrigation: Livestock:

          Crop _____   Dairy _____   Beef _____

          Turf _____   Other _____

CHARACTERISTICS OF THE WATER BODY AND WATERSHED

      Pond _____   Lake _____   Water outflow: _____

      Reservoir _____   Canal _____   Seasonal? _____

      Stream _____   A lot or a little? _____

      Water inflow: Constant? _____

          Source _____   Downstream uses _____

          Can it be regulated? _____   Can it be regulated? _____

Obvious nutrient sources (feed lots,
    fertilized fields, septic systems) _____

Weeds throughout _____   Weeds along edge _____

Potential for nontarget affects: Previous treatment history:
    Trees and desirable shoreline species? _____   What was used? _____

    Sensitive fish species (e.g., trout)? _____   What was the result? _____

                                                        _____

                                                         _____

DIMENSIONS

Area _____   Average Depth _____   Volume (acre-feet) _____

Table 15.3 Aquatic Herbicides Labeled for the Control of Commonly Occurring Aquatic Weeds

Aquatic Weed	Copper chelate	Copper sulfate	Diquat (Reward)	Amine salt of endothal (Hydrothal 191)	Potassium salt of endothall (Aquathol)	Fluridone (Sonar)	Glyphosate (Rodeo)	2,4-D
ALGAE								
Microscopic algae	X	X						
Filamentous algae	X	X	X	X				
Chara	X	X		X				
FLOWERING PLANTS								
Submersed								
American elodea	X		X	X		X		
Bladderwort			X			X		+[b]
Brazilian elodea						X		
Brittle naiad			X	X	X	X		
Cabomba						X		
Richardson's pondweed						X		
Coontail			X	X	X	X		+[b]
Curlyleaf pondweed			X	X	X	X		
Eelgrass				X		X		
Eurasian watermilfoil			X	X	X	X		X[b]
Flatstem pondweed			X	X	X	X		

Table 15.3 (Continued)

Aquatic Weed	Copper chelate	Copper sulfate	Diquat (Reward)	Amine salt of endothal (Hydrothal 191)	Potassium salt of endothall (Aquathol)	Fluridone (Sonar)	Glyphosate (Rodeo)	2,4-D
Horned pondweed				X	X			
Hydrilla	X		X	X	X	X		
Leafy or small pondweed			X	X	X	X		
Sago pondweed			X	X	X	X		
Small pondweed			X	X	X	X		
Southern naiad	X		X	X	X	X		
Watermilfoil			X	X	X	X		X[b]
Waterstargrass					X			X[b]
Free-floating								
Common duckweed			X			X[a]		
Water hyacinth			X				X	X[c]
Watermeal						+[a]		
Rooted floating								
American lotus						+	X	
American pondweed			X	X	X	X		
Floatingleaf pondweed			X	X	X	X		
Illinois pondweed			X	X		+		
Largeleaf pondweed			X	X	X	X		
Spatterdock						X[a]	X	+[b]

Table 15.3 (Continued)

Aquatic Weed	Copper chelate	Copper sulfate	Diquat (Reward)	Amine salt of endothal (Hydrothal 191)	Potassium salt of endothall (Aquathol)	Fluridone (Sonar)	Glyphosate (Rodeo)	2,4-D
Water lily						X[a]		+[b,c]
Watershield						+		+[b]
Waterthread pondweed			X	X	X	X		
Emergent								
Alligatorweed						+	+	X[c]
Arrowhead								X[c]
Bulrush (tules)							X	X[c]
Cattail			X			+	X	
Creeping water primrose						+	X	X[c]
Phragmites							+	
Purple loosestrife							X	
Spikerush						+		
Water smartweed						+	X	+[c]
Willow							X	X[c]

+Labeled for partial control only.
[a]AS formulation only.
[b]2,4-D granules only. Check label to make sure product is cleared for use in or near water.
[c]2,4-D liquids only. Check label to make sure product is cleared for use in or near water.

glyphosate translocates to underground parts, it provides longer-lasting control than contact materials such as diquat (which is labeled for cattail control). Herbaceous and woody broadleaves can be treated selectively with 2,4-D in order to prevent damage to grass species that provide habitat and erosion control.

Aquatic Herbicides. Some general principles that apply to the effectiveness of aquatic herbicides are as follows:

1. Herbicides that do not translocate (contact herbicides) include the copper compounds, diquat, and endothall. These compounds enter the plant from the water.

2. Filamentous algae, particularly those that produce thick mats (e.g., *Cladophora, Pithophora,* and *Lyngbya*), may require a copper treatment once every few weeks during the growing season. The thick mats appear to prevent adequate penetration of the copper to all cells. The mats may look burned and may sink to the bottom, but they typically will reappear at the surface within a week or two and resume growth.

3. Most submersed, rooted floating, and emergent flowering plant species are perennial and develop from underground structures. Therefore, contact herbicides such as diquat and endothall kill the shoot tissue but seldom provide complete plant kill. These compounds usually have to be applied each year and sometimes more than once within a single year.

4. Contact herbicides work quickly, often within a few days or less. They are effective if rapid results are desired. Another advantage of their rapid activity is that contact herbicides can be used as spot or partial treatments. By the time they are diluted to nontoxic doses as they move into untreated water, they have already affected the target plants. Thus, contact herbicides are excellent when it is desirable to control plants in some areas of bodies of water but not others.

5. Fluridone moves into plants through the water and the sediment and is symplastically translocated to the growing tips. Therefore, it can provide excellent long-term control of submersed plants that develop from underground structures, such as hydrilla and Eurasian watermilfoil. Most flowering plants are sensitive to its mode of action (carotenoid inhibition), so it tends to be nonselective at normal use rates. Although it is excellent in controlling invasive species, its ability to leave native species unharmed at normal use rates is questionable at this time.

6. Fluridone is slow acting; the plants must be in contact with the compound for at least several weeks. Therefore, fluridone cannot be used as a spot or partial treatment but only as a whole pond treatment or on areas in large lakes that are greater than 5 acres.

7. Most submersed native aquatic plants are monocots. Eurasian watermilfoil, along with other milfoils and coontail, are exceptions. They are dicots. Therefore, 2,4-D, which is a broadleaf weed killer, can be used selectively to remove Eurasian watermilfoil from stands of native plant species.

8. Glyphosate has no activity in water. Its herbicidal action is limited to emergent plants, some rooted floating plants, and large free-floaters such as water hyacinth.

A more detailed description of the algicides and herbicides that are effective in controlling aquatic weeds is provided here.

Copper Sulfate. Copper salts were among the earliest known herbicides for both terrestrial and aquatic weed control. Copper sulfate (bluestone, blue vitriol) was first used in 1904 for water treatments; its use today is widespread. It is one of the few herbicides that can be used for algae control in potable water supplies.

Copper sulfate is used strictly for algae (including Chara) control. It has little or no effect on flowering plants at the normal use rates. Most copper sulfate products are formulated as copper sulfate pentahydrate (CSP). It can be used at a maximum dosage of 4 ppm CSP (1 ppm copper ion), but the normal use rate is 1 ppm CSP (0.25 ppm copper ion) or 2.7 lb of CSP/acre-ft of water. There are no restrictions on the use of the water following treatment (Table 15.1); however, copper sulfate should not be used in trout-bearing waters. Drinking, swimming, or fishing activities can be initiated immediately after treatment, although it is desirable to wait 24 hr to let the metallic smell of the copper in the water dissipate.

The activity of copper sulfate is affected greatly by the carbonate alkalinity of the water. Copper sulfate dissociates in water, releasing the copper ion (Cu^{2+}), which is the active ingredient. Copper combines with negatively charged ions, such as carbonates and other ions in water with a high carbonate alkalinity (greater than 250 ppm $CaCO_3$). These copper complexes precipitate out of the water column and prevent the copper from entering algal cells. Consequently, higher concentrations of CSP usually are required to control algae in alkaline ("hard") waters than in softer, acid waters. On the other hand, when carbonate alkalinity is low (less than 50 ppm $CaCO_3$), most of the copper goes into solution in the water rather than precipitating.

Although this may be excellent for algae kill, fish kills also can occur. For example, the LC_{50} to bluegills in hard water is 44 ppm of copper sulfate but only 0.884 ppm in soft water. Trout, on the other hand, are extremely sensitive to copper in both hard and soft waters. Fortunately, the moderate amounts of alkalinity (50 to 250 ppm $CaCO_3$) characteristic of a large part of U.S. waters acts to protect most fish species from lethal doses of copper so that fish kills are relatively rare. Most fish kills following copper sulfate treatments appear to result from heavy algae kill and accompanying oxygen depletion rather than from the algicide itself.

Copper sulfate is a contact herbicide. Therefore, direct exposure of the algae to the compound is required, and good distribution in the water where the plants are growing is essential. A standard method of applying copper sulfate is to place the crystals or granules in a burlap sack and dissolve them in the water by dragging the sack behind a boat. If free-floating mats of filamentous algae are present, powdered copper sulfate can be dissolved in water and sprayed directly on the surface mats as well as injected into the water. Copper sulfate solution is extremely corrosive to metals, and equipment used for spraying should be washed thoroughly after application. Probably the least preferred method of application for free-floating algae is to toss the crystals or granules into the body of water since they fall to the bottom and the copper may adsorb to sediment particles rather than disperse in the water. This type of treatment, however, may be suitable for mats lying on the bottom and for Chara control. Treatment should be made on sunny days for light appears to enhance the activity of the compound.

Sometimes, when filamentous algae mats are treated with copper, the algae is killed but the brown mats remain on the water surface. In this case it will take a rain or wind event to cause the mats to dissipate and sink.

Copper sulfate has a fairly short period of phytotoxicity. It is quickly adsorbed to soil, inorganic ions, and organic materials suspended in the water. These particles fall to the bottom sediments, removing the chemical from the water column. This rapid inactivation plus the low dosages required to kill algae and the low mammalian toxicity of the compound after it is dissolved in water account for the lack of restrictions on water use after treatment.

Copper Chelates. The copper chelates (copper held in an organic molecule) are formulations that prevent the precipitation of copper from solution.

They should provide somewhat longer-lasting results than copper sulfate, particularly in alkaline water. Another advantage of chelated copper compounds is that they are slightly less toxic to fish than copper sulfate. Manufacturers of these products, however, still caution against their use in waters of low alkalinity and in waters containing trout. Chelated forms of copper can be formulated as liquids as well as granules and thus are more adaptable to various types of application equipment than is copper sulfate. They also are less corrosive than copper sulfate to metals and equipment.

The copper chelates are used primarily for algae control; some submersed flowering plants also are susceptible. The copper ethylenediamine complex (Komeen) is labeled for controlling hydrilla, southern naiad, and elodea; the copper ethanolamine complex (Cutrine Plus) is labeled for hydrilla control.

Like copper sulfate, the copper chelates are contact materials, move into the sediments, and have no water use restrictions other than toxicity to trout. They are considerably more expensive than copper sulfate per unit of copper. Other than the advantage of having different formulations to work with, it is unclear at this time whether their effectiveness justifies their increased cost.

Diquat. Diquat (Reward) is used primarily for controlling submersed and free-floating flowering plants. This contact herbicide must be well distributed in the water or on the plant for optimal activity. It also provides a quick temporary burndown of filamentous algae and emergent plants such as cattails, but the effect is seldom long lasting. Diquat kills the leaves of cattail but does not translocate to the underground rhizomes.

Diquat is formulated as a bromine salt. When the formulation is added to water, however, diquat dissociates from the bromine to form the cation (positively charged ion). This property makes the compound extremely susceptible to adsorption by negatively charged clay and organic matter particles. Thus, diquat should never be applied to water that is turbid from suspended soil particles or to plants covered with mud deposits.

Diquat is removed from the water column by adsorption to suspended sediment and plant material that then settles to the bottom.

Diquat is available only as a liquid. It often is used as a tank mix with the liquid forms of copper chelates and the dipotassium salt of endothall. This tank mix provides a broad spectrum of control for aquatic weeds plus the convenience of working with liquid formulations.

Endothall. Two salts of endothall are used in aquatic weed control. One is the mono(N,N-dimethylalkylamine) salt of endothall (Hydrothol l91) and the other is the dipotassium salt (Aquathol). Even though the parent compound is the same, the two salt formulations differ appreciably in their characteristics.

The dimethylamine salt is labeled for controlling both algae (filamentous and Chara) and submersed flowering plants. Thus, its spectrum of control is broad. The dipotassium salt, on the other hand, is effective on submersed flowering plants (particularly *Potamogeton* species) but has little if any effect on algae. Fish also respond differently to these two compounds (Table 15.2). Fish can be extremely sensitive to the dimethylamine salt (particularly the liquid formulation) at normal use rates but not to the potassium salt. The potential of the former compound to cause fish kills has greatly limited the use of an otherwise effective herbicide. The dimethylamine salt should be applied only as a partial or shoreline treatment so as to leave untreated water to which fish can escape.

Both of the salts are contact herbicides and require reasonably good distribution to be effective. They are subject to rapid microbial breakdown, and the typical half-life is only 7 days. The restrictions on the use of treated water range from 7 to 25 days for irrigation on food crops, domestic uses, and livestock watering, depending on the dosage (Table 15.1). Both forms of endothall are available as liquid and granule formulations.

Fluridone. Fluridone (Sonar) is a broad spectrum herbicide that is available as a sprayable aqueous suspension (AS) or as a slow-release pellet (SRP). It is primarily used for the control of submersed flowering plants but also has activity on duckweed and watermeal (the AS formulation only) and a few emergents. Emergents are seldom targeted for fluridone treatment because it provides only partial control; glyphosate is the better choice for emergent plant control. Fluridone does not control algae.

Fluridone is the only compound that can control weeds for more than a year. For best results, it should be applied during the early stages of rapid weed growth. Complete weed removal is slow (unlike the contact herbicides diquat and endothall) and can take as long as 30 to 90 days, even though the typical whitening symptoms may become visible within 7 to 10 days. The benefit of this is that fish kills may be minimized because the slow action of the herbicide reduces oxygen depletion problems.

Because its activity is slow and it has the potential to be diluted out, fluridone cannot be used as a spot treatment. It must be applied to the entire pond surface or to an area no smaller than 5 acres in larger bodies of water. The half-life of fluridone ranges from 5 to 60 days and averages 20 days under summer water conditions. It disappears by photodegradation and to a lesser degree by plant and microbial metabolism. Since plants can take up fluridone from the sediments, desirable trees and shrubs growing in the water or near the shoreline with their roots extending into the water may be injured.

Fluridone has no restrictions on swimming or fishing after application (Table 15.1); it should not be applied within 1/4 mile of a potable water intake. Treated water should not be used for irrigation for 7 to 30 days after application, depending on the crop.

Glyphosate. Glyphosate is the primary herbicide for the control of rooted floating and emergent plants. It must be applied only to the aerial parts of the plant. Glyphosate has no activity if applied to the water because it is diluted and possibly inactivated by soil particles suspended in the water. Therefore, glyphosate does not control plants that are completely submersed or have a majority of their foliage underwater. The herbicide is translocated symplastically to underground structures and can provide long-term control.

The formulation labeled for aquatics (Rodeo) does not contain surfactant. Surfactant *must* be added separately to ensure cuticular penetration and adequate control.

The only restriction on the use of water that may have been inadvertently sprayed with glyphosate is that it not be used within a half mile upstream of a potable water intake. Extremely cool or cloudy weather following treatment can slow the activity of glyphosate and delay visual effects of control. In general, control is slow and may not be visible for several weeks after application.

The foliage must not be wet to the point of runoff. If too much solution runs off the plant, not enough glyphosate will remain on the plant to be effective. Washoff of sprayed foliage by boat backwash or by rainfall within 6 hr of application must be avoided. Glyphosate treatments appear to be most effective for cattail control when applied in late summer or fall, as these plants move storage products to the rhizomes.

2,4-D. In general, 2,4-D is not as effective on underwater plants as it is on terrestrial plants. An exception is the granular ester formulation

(Aquakleen, Navigate) of 2,4-D, which is particularly effective on watermilfoil and coontail.

The liquid formulations of 2,4-D are limited primarily to the treatment of plants with aerial parts; for example, water hyacinth, rooted floating plants, and ditch-bank growths of cattails, bulrushes (tules), and many broadleaved species. The labels of these liquid formulations usually state "application should be made when leaves are fully developed *above* the water line and plants are actively growing." This kind of wording eliminates the use of liquid formulations for submersed weed control.

Restrictions on 2,4-D vary, but in general the herbicide should not be used in water intended for irrigation, domestic use, or watering dairy herds.

2,4-D is available as amine salts or esters. Esters, although more effective for plant kill than the amines, tend to be more toxic to fish (Table 5.6). The degree of toxicity, however, appears to be associated with how the ester is formulated. Some liquid ester formulations are not cleared at all for aquatic use. Other liquids are cleared, and the granular esters have very low fish toxicity. *Always consult the label to determine if the product has been cleared for aquatic use.*

Acrolein. This compound is used to control submersed plants and algae in irrigation ditches primarily in the western states. According to Klingman and Ashton (1982), hand weeding and dredging costs of $80 to $400/mile were reduced to $10 to $40/mile by applying aromatic solvents such as acrolein.

Acrolein is metered into the water from pressurized containers and eliminates plants up to 25 miles downstream. Acrolein is a potent irritant and lachrymator (tear inducer) that must be used with special equipment and precautions.

Acrolein can be used on water intended for irrigation because it dissipates within 24 to 48 hr under normal conditions. Acrolein is extremely toxic to fish and other aquatic organisms.

Dosage and Amount to Apply

To calculate the amount of herbicide to apply, three basic determinations must be made: (1) the area or volume to be treated, (2) the dosage required to control the problem weed or weeds, and (3) the amount of formulated material required to give the proper dosage.

Determining the Area or Volume to Be Treated. Herbicide treatments usually are limited to the area of water containing the weeds. Therefore, application may be partial (marginal or spot treatments) or complete. In large bodies of water, it usually is necessary to treat only shoreline areas or boat channels, whereas small ponds often require a complete treatment. Dosages usually are increased for marginal or spot treatments to overcome the effects of dilution. These treatments are best accomplished with quick-acting contact herbicides (copper, diquat, endothall). When the total volume of a body of water is being treated, the actual application of the material should be concentrated in the specific areas where the plants are growing.

Herbicide applications are made either on a surface-area or volume basis, depending on the habitat of the weed. For algae and submersed plants, the amount of herbicide usually is calculated for the water volume. In some cases in which the vegetation is growing along the bottom and special injection equipment is being used, only the bottom acre-ft of water need be treated. In the case of rooted floating and emergent plants as well as drawdown treatments, calculations are made on an area basis in a manner identical to that for terrestrial weeds. The following discussion concentrates on treatments based on water volume.

To determine the volume of a body of water, two parameters must be known: surface area in acres and average depth in feet.

In many cases, the surface area of a body of water is known from the time of lake or pond construction. In other cases, the area has to be determined, usually by measuring the dimensions of the pond or area to be treated. The formula used depends on the shape of the area.

Rectangular area

acres = length (ft) $\times$ width (ft) / 43,560 ft^2

Circular area

acres = 3.1416 $\times$ radius (ft)^2 / 43,560 ft^2

Triangular area

acres = 1/2 base (ft) $\times$ height (ft) / 43,560 ft^2

If the area to be treated is not a regular shape, it should be divided into sections of standard geometric shapes. The area of each section should then be calculated and the calculations added together.

Average depth can be estimated as roughly half the maximum depth of the pond, lake, or portion to be treated. A more reliable method of calculating average depth is to take depth soundings at intervals along a transect or transects, depending on the size of the body of water (Figure 15.33). Measuring can be done with a

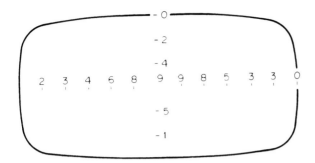

Figure 15.33 Transects of a water area should be taken to determine average depth.

long pole, a chain, or a weighted rope marked off in feet. The measurements (plus a zero) are then added and the sum is divided by the number of readings. For example, the average depth of a pond with depth readings of 2, 5, 8, 5, 2 and 0 ft is 22/6 or 3.66 ft.

The most common mistake people make is to think they know the maximum or average depth without taking actual measurements. Often they assume that the maximum depth is the same as when the impoundment was first constructed. With sediment inflow and vegetation dieback over the years, however, depth gradually is lost in most bodies of water. What may have been 15 ft deep at the time of construction may only be 10 ft at the time of treatment. Using the 15-ft depth to estimate average depth would lead to an overdose of chemical, a possible fish kill, and a waste of money. Measurements always should be taken before treatment, even if the only measurement taken is maximum depth. On man-made ponds or lakes, the deepest portion usually is located near the dam or outlet end.

The volume of the water is expressed in acre-feet and is calculated using the following formula:

surface area (acres) × average depth (ft)
    = acre-feet

EXAMPLE: A man-made pond (Figure 15.33) is roughly rectangular in outline with dimensions of 240 × 120 ft. Depth measurements taken at 20-ft intervals along two transects across the lake are 2, 3, 4, 6, 8, 9, 9, 8, 5, 3, 3, and 0 ft and 2, 4, 9, 5, 1, and 0 ft. What is the volume of the pond in acre-feet?

Area: $\dfrac{240 \text{ ft} \times 120 \text{ ft}}{43{,}560 \text{ ft}^2} = 0.66$ acre

Average depth:

2 + 3 + 4 + 6 + 8 + 9 + 9 + 8 + 5 + 3 + 3 + 0
    = 60/12 = 5 ft

2 + 4 + 9 + 5 + 1 + 0 = 21/6 = 3.5 ft

5 ft + 3.5 ft = 8.5/2 = 4.25 ft

Volume: 0.66 acres × 4.25 ft = 2.8 acre-ft

Determining Dosage. The next step is to check the herbicide label for the dosage recommended for the weed species. Dosage usually is given in parts per million (ppm). Most labels also give dosage recommendations for spot or partial treatments.

Determining Amount Required. The correct amount of formulated herbicide to use per acre-foot of water to give the required ppm usually is provided in a table on the herbicide label. The amount also can be calculated easily if the following relationship is used: Since an acre-foot of water weighs approximately 2.7 million (2,718,144) lb, 2.7 lb of any material dissolved in 1 acre-ft of water will equal 1 ppm (by weight). Therefore,

$$\frac{2.7 \text{ lb}}{\text{acre-ft}} \times \text{ppm desired} \times \frac{\text{acre-ft}}{\text{treatment}} = \frac{\text{lb required}}{\text{treatment}}$$

EXAMPLE: How much copper sulfate pentahydrate (CSP) is required to treat 2.8 acre-ft of water at a recommended dosage of 2 ppm?

$$\frac{2.7 \text{ lb}}{\text{acre-ft}} \times 2 \text{ ppm} \times \frac{2.8 \text{ acre-ft}}{\text{treatment}} = \frac{15.1 \text{ lb CSP}}{\text{treatment}}$$

EXAMPLE: If the recommendation calls for the treatment of 2.8 acre-ft with copper sulfate pentahydrate at a rate of 1 ppm Cu (25% of CSP is elemental copper, Cu, which is the active ingredient), how much copper sulfate pentahydrate is required to achieve a final concentration of 1 ppm Cu in water?

$$\frac{2.7 \text{ lbs}}{\text{acre-ft}} \times 1.0 \text{ ppm} \times \frac{2.8 \text{ acre-ft.}}{\text{treatment}} =$$

$$\frac{7.56 \text{ lb Cu required}}{\text{treatment}}$$

$$\frac{7.56 \text{ lb Cu}}{\text{treatment}} \times \frac{1 \text{ lb CSP}}{.25 \text{ lb Cu}} = \frac{30.2 \text{ lb CSP}}{\text{treatment}}$$

EXAMPLE: Aquathol K liquid is labeled at a rate of 2 ppm a.i. for coontail control. Aquathol K contains 4.23 lb of active ingredient (dipotassium endothall) per gallon. How many gallons of Aquathol K are required to treat 2.8 acre-ft at a rate of 2 ppm a.i.?

$$\frac{2.7 \text{ lb}}{\text{acre-ft}} \times 2 \text{ ppm} \times \frac{2.8 \text{ acre-ft.}}{\text{treatment}} =$$

$$\frac{15.1 \text{ lb of a.i. required}}{\text{treatment}}$$

$$\frac{15.1 \text{ lb a.i.}}{\text{treatment}} \times \frac{1 \text{ gal Aquathol K}}{4.23 \text{ lb. a.i.}} =$$

$$\frac{3.57 \text{ gal Aquathol K}}{\text{treatment}}$$

State or federal recommendations or guidelines may vary as to whether the information is given as total formulated material or active ingredient. In all cases, the label must be consulted for proper dosage information.

Timing of Treatment and Water Temperature

Most aquatic herbicides should be applied in mid- to late spring or early summer when the plants are young and growing vigorously and before they have gone to seed. Not only are the weeds more susceptible at this stage, but also there is less biomass to treat than later in the season. Treatment of a heavily infested area in June, July, or August can result in massive plant decomposition accompanied by oxygen depletion and fish kill. If vegetation must be treated at these times, only one-third to one-half of the area should be treated at any one time, followed by a second application 2 to 3 wk later.

Another reason for early treatment is that cold water contains more oxygen than warm water, providing a greater margin of safety for fish. For example, oxygen-saturated water at 65°F contains 9.2 ppm oxygen, whereas water at 85°F contains only 7.5 ppm. Treatment, however, should not be made when the water is too cold (less than 60°F). Although plants grow at these temperatures, they may not be metabolizing rapidly enough to take in sufficient quantities of herbicide. In addition, treatment should be delayed when there is a chance of heavy rains washing the herbicide downstream.

With the possible exception of fluridone, plant parts should be visible at the time of treatment. There is no substantial evidence that fall treatments of submersed weeds or algae are particularly effective. By then, seeds or algal spores have been produced to provide for growth the following season.

Application Methods

Aquatic herbicides are formulated as liquids and granules. The liquids are applied in a water carrier, whereas granules are applied directly to the water or exposed sediment surfaces. Invert emulsion systems for submersed weed control are also used to improve target placement. Large bodies of water may require aerial spraying. The various types of treatments are as follows:

Volume Treatments. Liquid treatments to a water volume can be accomplished in several ways. The least desirable method is to pour the herbicide directly from the container. This method does not ensure mixing. Liquids almost always are diluted in water to facilitate even distribution. Dilution ratios are given on the label. When a water carrier is used, the solution is sprayed on the surface and allowed to sink to the bottom or is injected into the water a foot or deeper below the water surface. Injection can be accomplished by siphoning the solution into the water with a boat bailer or by pumping the chemical into trailing hoses attached to a boom (Figure 15.34). For aquatic applications with a boat, motor-powered pumps are often adapted with a "Y" intake to pump water from the lake and chemical from the container simultaneously. A hose fitted with a handgun, boom, or subsurface injector is used on the discharge side of the pump.

Bottom Acre-Foot Treatments. Herbicides can be injected into bottom water by connecting weighted brass pipes to 15- to 30-ft hoses extending from a boom on the spray boat. The herbicide is released into the water through small holes bored about 6 in. from the end of the pipe. For best results, pipes should be no more than 3 ft apart. Application rates are based on the lower 1 or 2 ft of water. The technique is particularly effective early in the growing season when submersed plants are still short. It works best in lake or static water sites with firm sandy bottoms; it is not recommended for flowing water or muddy bottoms.

Bottom treatments also can be made using weighted invert emulsions or polymers (Figure 15. 35). The consistency of invert emulsions is like mayonnaise. When weighted with a copper compound and injected into the water, the in-

Figure 15.34 Injection system showing chemical tank, boom, and trailing hoses.

Figure 15.35 Trailing hoses dispersing a mixture of Aquathol, Hydrothol 191, Cide-Kick (adjuvant), and Nalquatic (polymer). (Gordon Baker, South Florida Water Management District, West Palm Beach)

vert sinks to the bottom, where it sticks to the plant (Figure 15.36). Polymers are added to an herbicide–water mix to give a sticky, viscous material that also sinks to the bottom. Polymers can be used through conventional equipment, whereas inverts require special equipment to ensure proper mixing and formation of the invert. These systems have been used successfully for hydrilla control in Florida and have potential for use in other parts of the country, particularly by commercial applicators, for submersed weed control.

Granular Treatments. Granular herbicides usually are applied with a granule spreader or by hand (Figure 15.37). These preparations sink to the bottom and release the herbicide into the water. Granules have the advantage of being easy to apply in a uniform manner. Disadvantages are that they usually are more expensive than liquid herbicides, and their bulk takes up what is usually limited space on the applicator boat.

Figure 15.36 Invert particles consisting of Cutrine Plus and Ivod as the inverting oil adhering to Chara plants. (Gordon Baker, South Florida Water Management District, West Palm Beach)

Figure 15.37 Granular applicator being used to disperse herbicides into a lake.

Emergent Plant Treatments. Plants with the majority of their leaf area above the water surface (emergents, free-floaters, and rooted floaters) usually are sprayed with aqueous solutions of herbicides (Figure 15.38). A wetting agent is required for cuticle penetration. Dosage is calculated on a surface-area basis.

Treatment of Water Conveyance Systems. Flowing water in ditches and canals requires control techniques different from those used in static systems. Herbicide solutions usually are injected or allowed to drip into the water, and the water is used to disperse the herbicide. In drip systems, constant-flow metering devices slowly drip the chemical into the water. The dosage rate on the herbicide label usually is given as the amount of herbicide per hour per cubic feet per second (C.F.S.). C.F.S. can be determined using the following formula:

average width (ft) × average depth (ft) × velocity (ft/sec) × 0.9 = C.F.S.

Large-volume applications of a dry concentrate are used to "slug" or introduce high

Figure 15.38 Spraying cattails and other emergent plants in a drainage culvert along an interstate highway.

concentrations of a chemical to the water. The concentration gradually decreases downstream, providing control and permitting the safe use of the water. This procedure commonly is used with copper for algae control.

Retreatment

Most aquatic herbicides have relatively short persistence in the environment. This characteristic is desirable from the standpoint that restriction times on the use of water can be fairly short, but it is undesirable in that vegetation not killed by the treatment, or vegetation that comes back from seeds, spores, or underground structures, can reinfest an area soon after treatment. This problem is particularly true of algae. Algae often must be retreated several to many times per season, whereas flowering plants usually require only a touch-up treatment to catch missed or late-sprouting plants. Unfortunately, new weed species that previously were not problems sometimes appear within the season of treatment. A common weed shift is the appearance of Chara after submersed flowering plants are controlled. Another is the bloom of microscopic algae after submersed plants are killed or vice versa.

Retreatment in subsequent years almost always is required as weeds return from seeds, spores, or vegetative propagules, or as new species appear. A successful chemical weed control program should be regarded as a long-term program requiring continuing treatments over time.

DIAGNOSIS AND PREVENTION OF FISH-RELATED PROBLEMS

Some plant growth is beneficial for fish. Microscopic algae serve as the base of the food chain, and larger plants provide habitat and protection for fish food organisms and fish. Both of these types of plants are major sources of oxygen in water. An overabundance of plants, however, can have severe adverse effects on fish populations.

Fish Stunting

Stunting of fish is a result of overpopulation and lack of predation. Rooted weeds provide such excellent cover and protection for young fish that predation by larger fish is reduced, resulting in an overpopulation of small fish. Food is rapidly depleted, and combined with human fishing pressure on the larger fish, the end result is a pond or lake filled with stunted fish. Stunting also can occur in waters turbid with algal blooms because feeding is prevented by poor visibility.

The only cure for a severely stunted population, besides controlling the aquatic vegetation, is to poison the site (usually with rotenone) and restock with the correct ratio of predator to forage fish. An imbalance in fish populations can be prevented by removing forage fish such as bluegills when they are caught rather than returning them to the water. Another good practice is to initiate a limit on the size and number of predaceous fish such as bass and trout that can be removed in order to maintain a proper predator-to-forage-fish ratio.

Fish Kills

Weed-associated fish kills typically result from oxygen depletion in the water. Normally, oxygen levels from plant photosynthesis are highest during midday or early afternoon. Plants cease photosynthesizing at night but continue to utilize oxygen for respiration throughout the nighttime hours. Thus, oxygen levels are lowest just before sunrise. The presence of an algal bloom or a heavy infestation of weeds can cause a significant drain on the oxygen content of the water. The effect on fish will be most noticeable during the early morning hours, when they come to the surface "gasping" for air. This temporary loss of oxygen may not result in a direct fish kill, but it can stress the fish, making them more susceptible to disease, stunting, or the presence of external toxicants.

A second major cause of oxygen depletion, plant death and decomposition, can result in kills of a significant proportion of the fish population. There are two reasons for this. First, plant photosynthesis is shut off, and if the water is not sufficiently aerated by wind (as is often the case on calm midsummer days or during periods of ice cover in winter), oxygen cannot be replenished. Second, the bacteria and fungi decomposing the plant material remove oxygen from the water by respiration.

Unfortunately, the causes of natural vegetation die-offs are not well understood. All aquatic plants probably are susceptible to die-offs, but the best known offenders are microscopic algae (particularly blue-green algae), which become buoyant and form thick yellowish-green surface scums during periods of calm, clear, warm weather. In some cases the die-off may be triggered by a change in the weather to cloudy, cool, or rainy conditions. The death and rapid decomposition of these algae have been noted to cause a drop in dissolved oxygen content from 8 to 0 ppm within 24 hr (Boyd et al. 1975). The oxygen-depleted condition persisted for 5 days until a new algal bloom became established. Such prolonged exposure to oxygen concentrations of 1 ppm or less can result in a major fish kill.

Even without causing an immediate fish kill, surface layers of microscopic algae, free-floating plants such as duckweed and water hyacinth, or canopy-forming submersed plants can aggravate oxygen depletion problems by preventing light penetration to the deep waters. Without light, other plant life cannot photosynthesize and produce the oxygen needed to support fish.

Herbicide treatment of large amounts of aquatic plant material also can cause oxygen depletion if the vegetation dies and decomposes rapidly. Prevention of massive vegetation die-offs is one of the primary reasons aquatic herbicides are applied in the spring or early summer, when the amount of vegetation is minimal. Applications also can be staggered so that only a part of the body of water is treated at any one time.

A good aquatic weed control program is the best prevention of weed-related fish kills. Some controversy about whether all weeds should be removed exists, however. Fishermen know that some of the best fishing spots lie along the edges of weed banks. Certainly in large multiple-use lakes, weed banks in backwater areas can be highly desirable as fish habitats. Small, shallow bodies of water such as farm ponds and small lakes (< 20 acres), on the other hand, should be kept relatively weed free (no more than 20% coverage). The volume of water is so low and weeds can spread so rapidly that deterioration of fish populations and fish kills can occur with only partial weed infestations. If weeds are completely removed, fish habitat can be replaced by sinking tree limbs, tires, or other structures into the water.

Once a fish kill is in progress, little can be done to stop it. Aeration should be provided, but running a boat motor usually is not sufficient to reduce mortality. Lake-installed aerators are more useful, but sometimes even they cannot overcome the oxygen deficit. The best way to prevent a fish kill is to control the vegetation before a fish kill can occur.

Winter Fish Kills

Winter fish kills can be just as frequent and are based on the same principles as summer fish kills. The formation of an ice cover followed by an accumulation of snow on the ice blocks light penetration to the water beneath. Once again, photosynthesis ceases while plant and animal respiration continues to draw oxygen out of the water. The situation can be aggravated further if the body of water was heavily weed infested the summer before because decomposition of plant material during the winter adds to the oxygen loss. Winter fish kills may not be evident at the time they occur because the dead fish may decompose or wash downstream before or as the ice cover breaks. It may not be until fishing season starts that a drastic reduction in catchable fish is apparent.

Winter fish kills can be prevented by maintaining a good weed control program during the summer, by eliminating sources of nutrients that stimulate algal growth, by keeping portions of the ice free of snow, and, if necessary, by providing agitation to maintain an open water area.

Diagnosis

The diagnosis of the cause of a fish kill is extremely difficult, even for experts. If water color changes from an obvious green color to a dark brown or gray prior to the kill, then plant decomposition and oxygen depletion should be suspected. The behavior of the fish may include crowding in shallow water, gasping at the surface, and crowding into streams or inlets. The fish will be particularly stressed at night and during the early morning hours. In an oxygen-related fish kill, the largest fish typically die first. Sensitive species such as trout, bass, and shad die more readily than carp, bullheads, and other hardy types.

Not all fish kills are caused by weed die-offs. Runoff of ammonia from feedlots or septic fields during heavy rainfall or spring thaws can cause fish kills. Fish are very susceptible to insecticides sprayed aerially on nearby fields that may drift or soil-applied insecticides washed in by heavy rains. Fish suffering from insecticide poisoning often show erratic movements. The young fish usually die first and often are contorted when pulled from the water. Outbreaks of disease are not uncommon and can kill large numbers of fish. Usually one or several related species are affected.

The major problem in diagnosing fish kills is that water quality conditions can change rapidly, making it difficult to determine events after the fact. A person encountering a fish kill should note the following: time of day, weather, location, color and smell of the water, unusual appearance and behavior of the live fish, and species and sizes affected. If toxicants are suspected, samples of water and fish should be frozen and sent to an appropriate agency or analytical laboratory for determination of the cause of death. Many pesticides degrade within a few days, so the immediate collection of samples is essential. Disease diagnosis can be made only on live fish or those preserved live in a pickling agent such as 10% formalin.

Oxygen concentrations in water can be tested using field kits, but again, the measurements must be taken shortly after the kill in order to determine whether oxygen depletion was the cause.

MAJOR CONCEPTS IN THIS CHAPTER

Aquatic weeds interfere with recreational uses of water, cause foul tastes, hinder the flow of water in irrigation canals and ditches, cause large water losses through transpiration, and make water transportation difficult.

The major groups of aquatic weeds are the algae and flowering plants. Algae consist of microscopic forms, mat-forming filaments, and Chara. The four major groups of flowering plants are submersed, free-floating, rooted floating, and emergent plants.

Invasive species, such as the submersed plants hydrilla and Eurasian watermilfoil, and the free-floating plant water hyacinth, are particularly troublesome because they crowd out desirable native aquatic plant species, degrade the habitat, and prevent human uses of the water. These species persist and spread primarily by producing vegetative reproductive structures.

Invasive emergent plants, such as purple loosestrife, melaleuca, and salt cedar, are prolific seed producers. They also degrade the environment and provide little valuable habitat for wildfowl and other animals.

Environmental factors regulating aquatic plant growth are light, nutrients (particularly phosphorus and nitrogen), temperature, and the type of substrate. These factors most often are optimal in shallow water, so many aquatic weed problems also are found in shallow sites. Aquatic plants can grow at light intensities as low as 1% of surface light intensity.

Preventive aquatic weed control means preventing weed spread and reducing shallow areas where plants can root. Weed spread can be slowed by education, inspections, and elimination of initial infestations.

Habitat alteration can be an effective method of aquatic weed control. Techniques include those that reduce light (dyes) and nutrients (watershed management) or alter temperature (drawdowns) and substrate (lining the shoreline with riprap).

Mechanical control methods include hand removal, draglining, and machine harvesting. Harvested weeds can be used as mulches or fertilizers. Weed harvesting is not suitable for removing plants from water less than 3 to 4 ft deep.

Biological weed control has considerable potential in aquatic sites. Insects and herbivorous fish such as the grass carp are being used with success in certain parts of the country.

Important considerations in using herbicides for aquatic weed control are (1) weed identification, (2) uses of the water to be treated and environmental impact, (3) choice of an appropriate herbicide, (4) dosage and amount of herbicide to apply, (5) timing of treatment and water temperature, (6) method of application, and (7) probability of the need for retreatment.

Copper sulfate is effective only for algae control. Other types of herbicides must be used to control flowering aquatic plants.

Applying the proper dosage involves accurate calculation of the surface area or volume of the body of water to be treated. Volume can be calculated as average depth (ft) $\times$ surface area (acres) and is expressed as acre-feet. Herbicide dosages are given in ppm. Labels usually provide the amount of chemical to be applied per acre-foot in order to give the appropriate ppm of active ingredient.

Most aquatic herbicides should be applied in mid- to late spring or early summer when plants are growing vigorously and there is less biomass to treat. Treatments of heavily infested areas later in the season invite fish kills because of the oxygen depletion that accompanies plant decomposition.

Weed shifts after herbicide treatment are not uncommon in aquatic sites. A common weed shift is the appearance of Chara after submersed flowering plants are controlled.

Major fish problems caused by excessive weed growth include fish stunting and fish kills.

Diagnosis of fish kills is extremely difficult. Besides plant decomposition and oxygen depletion, other causes of fish kills can be ammonia runoff, insecticides, or disease. Water and fish samples should be collected immediately and frozen for later analysis if a pesticide is suspected as the cause of a fish kill.

TERMS INTRODUCED IN THIS CHAPTER

Acre-foot
Aeration
Algae
Algal bloom
Alkalinity
Alum
Benthic barrier
Blue-green algae
Canopy formation
Drawdown
Emergent plant
Filamentous algae
Fish kill
Fish stunting
Free-floating plant
Grass carp
Habitat alteration
Invert emulsion
Microscopic algae
Photic zone
Phreatophyte
Phytoplankton
Riprap
Rooted floating plant
Secchi disk
Submersed plant
Tilapia
Triploid grass carp
Turion
Water hardness
Weed harvester

SELECTED REFERENCES ON AQUATIC PLANT IDENTIFICATION

Listed in Appendix A

SELECTED REFERENCES ON AQUATIC PLANT BIOLOGY AND ECOLOGY

Goldman, C. R. and A. J. Horne. 1983. Limnology. McGraw-Hill, New York.
Hutchinson, G. E. 1975. A Treatise on Limnology. Vol. 3. Limnological Botany. John Wiley and Sons, New York.
Sculthorpe, C. D. 1971. The Biology of Aquatic Vascular Plants. Edward Arnold, London.

SELECTED REFERENCES ON WATER ANALYSIS

Boyd, C. E. 1979. Water Quality in Warmwater Fish Ponds. Agricultural Experiment Station, Auburn University, Auburn, Alabama.
Lind, O. T. 1979. Handbook of Common Methods in Limnology. The C. V. Mosby Co., St. Louis, Missouri.
Wetzel, R. G. and G. E. Likens. 1979. Limnological Analyses. W. B. Saunders, Philadelphia, Pennsylvania.
Literature and water analysis equipment are available from Hach Chemical Company, Box 907, Ames, IA 50010.

SELECTED REFERENCES ON AQUATIC WEED CONTROL

Aquatic Weeds, Their Identification and Methods of Control. Fishery Bulletin No. 4. Department of Conservation, Springfield, Illinois.
Gallagher, J. E. and W. T. Haller. 1990. History and development of aquatic weed control in the United States. Rev. Weed Sci. 5:115–191.
Hoyer, M. V. and D. E. Canfield, Jr., Eds. 1997. Aquatic Plant Management in Lakes and Reservoirs. Center for Aquatic Plants, University of Florida. Available online at http://aquat1.ifas.ufl.edu/hoyerapm.html
Making Aquatic Weeds Useful. 1976. National Academy of Sciences-National Research Council, Washington, D.C.
Pieterse, A. H. and K. J. Murphy, Eds. 1989. Aquatic Weeds. The Ecology and Management of Nuisance Aquatic Vegetation. Oxford University Press, Oxford.

HERBICIDE FORMULATIONS AND PACKAGING, CARRIERS, SPRAY ADDITIVES, AND TANK MIXES

In order to understand herbicides and how they work one must have an understanding of the forms (formulations) in which they are provided by the manufacturer, the various additives (adjuvants) that are added by the applicator to herbicides to enhance their handling and weed control properties, and the carriers used to dilute or suspend herbicides during application. These topics plus the implications from the standard practice of tank mixing herbicides are addressed in this chapter.

HERBICIDE FORMULATIONS AND PACKAGING

The term *formulation* has two related meanings in reference to chemical weed control:

A *formulation* is a herbicide preparation supplied by the manufacturer for practical use. The formulation includes all contents inside the container: active ingredient (actual herbicide) plus inert ingredients (everything else) such as solvents, diluents, encapsulating materials, and various adjuvants.

Formulation also is the process, carried out by the manufacturer, of preparing herbicides for practical use. Through the formulation process, the manufacturer provides the user with a herbicide in a form that is convenient to handle and which, if used correctly, can be applied accurately, at the correct dose, uniformly, and with safety to the applicator.

Most herbicides are formulated so they can be applied in a convenient and suitable carrier. Most commonly, sprayable preparations are formulated to be diluted with and applied in water, fertilizer solutions, or diesel consistency oils. Some herbicides are formulated to be applied as dry granular or pelleted preparations directly from the container. Dusts usually are not used for herbicides because they present problems with handling and because they drift to nontarget vegetation.

The way in which a herbicide has been formulated often can be recognized because an abbreviation of the formulation type may be added to the trade name or listed very close to the trade name. For example, Aatrex 4L is a 4 lb/gal liquid (suspension concentrate) formulation, Prowl 3.3 EC is a 3.3 lb/gal emulsifiable concentrate formulation, Sencor DF is a 75% dry flowable (water dispersible granule) formulation, and Command 3 ME is a 3 lb/gal microencapsulated formulation.

Labeling, however, must be interpreted with caution. In some cases, letters after the trade name do not signify formulation. For example, the STS in Synchrony STS stands for sulfonylurea-tolerant soybeans. However, a separate product, Synchrony STS DF is a dry flowable formulation of the herbicide. The B in Conclude B stands for broadleaf control (the product contains acifluorfen and bentazon, which control broadleaved weeds) while the G in Conclude G stands for grass control since the product contains acifluorfen and sethoxydim, which control grasses. Treflan HFP

344

indicates an emulsifiable concentrate of Treflan with a high flash point.

Unfortunately, too many herbicide labels fail to provide any description of the type of formulation used. Only after reading the mixing instructions does it become evident what the probable formulation might be.

The designations and descriptions of the various herbicide formulations can vary considerably from label to label, thus making label interpretation complex and confusing. We have atttempted to sort through this confusion by compiling a list of formulations and their abbreviations found on herbicide labels and have combined this information with designations and descriptions provided by the *Herbicide Handbook* (Weed Science Society of America, 1994) and a book entitled *Pesticide Formulation and Adjuvant Technology* (1996). We first provide an outline for orientation, and this is followed by descriptions of each formulation. The formulations are grouped under the broad categories of sprayables and dry applications.

I. Sprayable formulations
   A. Water-soluble formulations
      1. Soluble liquids (SL)
      2. Soluble powders (SP); see also soluble packets (SP)
      3. Soluble granules (SG)
   B. Emulsifiable formulations
      1. Emulsifiable concentrates (E or EC)
      2. Gels (GL)
   C. Liquid suspensions (L or F) to be dispersed in water
      1. Suspension concentrates (SC) and aqueous suspensions (AS)
      2. Emulsions of a water-dissolved herbicide in oil (EO) and emulsions of an oil-dissolved herbicide in water (EW)
      3. Microencapsulated formulations (ME) or capsule suspensions (CS)
   D. Dry solids to be suspended in water
      1. Wettable powders (W or WP)
      2. Water-dispersible granules (WDG, WG, DG) or dry flowables (DF)
II. Dry applications
   A. Granules (G)
   B. Matrix granules (G)
   C. Pellets (P) or tablets (TB)

General Considerations

In the previous list, the soluble liquids (a type of water-soluble formulation), the emulsifiable formulations, and the liquid suspensions are formulated and packaged as liquids. The soluble powders and soluble granules (also types of water-soluble formulations), the dry solids, and the materials used in dry application are formulated and packaged as dry products. The three sizes of dry products are powders, granules, and pellets or tablets. Powders are small enough so that the individual particles cannot be seen with the naked eye. Examples are finely ground flour and confectioners' sugar. Granules are large enough so that the individual particles can be seen with the naked eye. However, the particle size is still small, roughly no more than 2.5 mm in diameter. Common substances in this size range include ground coffee, table sugar, and table salt. Pellets and tablets are larger than granules, generally in the size range of aspirin tablets (approximately 10 mm in diameter) or larger.

The groupings and individual types of formulations are described here.

Sprayable Formulations

A sprayable is any formulation that can be moved in a liquid carrier through a sprayer. A sprayable may actually be packaged in a dry form (for example, as a soluble granule or a wettable powder), but it will be added to the liquid carrier and then dispersed through a sprayer.

Water-Soluble Formulations. Water-soluble formulations mix completely in water. *Soluble liquids* mix into water with minimum agitation and, once dissolved, require no additional agitation. These formulations form clear spray solutions (although they may have color). They require wetting agents for maximum foliar activity. Some products contain wetting agents in the formulation; for other products, wetting agents must be added to the spray tank. Most of these formulations contain 2 to 4 lb of active ingredient per gallon.

Soluble powders are dry, finely divided solids, while *soluble granules* are of a larger particle size. Most soluble powders and granules are soluble salts of various compounds. Considerable stirring or agitation may be needed to dissolve these herbicides, but once in solution they remain in that state indefinitely. They form clear solutions in the sprayer tank and require a surfactant for maximum foliar activity. The surfactant may need to be added to the sprayer tank. Typical formulations contain 40 to 95% active ingredient.

Emulsifiable Formulations. Emulsifiable formulations are nonpolar (oily) liquids containing emulsifiers. Once added to the spray tank, the oil-soluble fraction containing the herbicide will be dispersed in the water. Spray tank emulsions are milky and require some agitation to prevent the components from separating. Normal flow through the bypass line on the sprayer generally is adequate to prevent separation. These formulations rarely require additional surfactant for foliar applications. The most widely used emulsifiable formulation (and the most widely used liquid formulation of any type) is the *emulsifiable concentrate*. Most EC formulations contain 2 to 6 lb of active ingredient per gallon but some have 7 or 8 lb. EC formulations are soluble in oil.

Gels are relatively new products that are thickened emulsifiable concentrates packed in water-soluble bags. Gels can be formulated so they resist leaking from pinhole size tears in the bags. The bags are premeasured so the user knows exactly how much herbicide is being added to the spray tank. For example, Buctril Gel is packaged in 2 1/2 pint soluble bags. The contaminated water-soluble bag is dissolved by the water in the spray tank, thus removing the environmental hazard posed by empty container disposal.

Liquid Suspensions to Be Dispersed in Water. These formulations are all water-dispersible liquids and are frequently designated as liquids or flowables (L or F). As a general rule, they are packaged as finely ground solids suspended in a liquid system. Water-dispersible liquids readily disperse when they are added to the tank, and the particles become suspended in the water carrier.

The size of the particles in a liquid suspension are the same or smaller than those of a wettable powder (a dry formulation described later), and yet because the particles are suspended in liquid, they tend to have advantages over wettable powders. For example, there is less need to make a slurry with a liquid suspension than with a wettable powder. In addition, liquid suspensions provide dust-free handling and mixing.

Because the herbicide particles of a liquid suspension are not soluble in water but rather are suspended in water, the particles will settle out if the spray mixture is allowed to stand without agitation. These formulations form opaque milky liquids in the spray tank. Four pounds of active ingredient per gallon is a common concentration for these herbicides.

There are several types of liquid suspensions. The herbicide may be labeled L or F (for example, Aatrex 4L), but it will be one of the specific types, each of which carries its own abbreviation. Aatrex 4L is actually a suspension concentrate and could be designated Aatrex SC.

Suspension concentrates are finely divided solids suspended in a liquid such as water or oil. If suspended in water, they also can be called *aqueous suspensions*. They are added to water in the spray tank for dispersal. Suspension concentrates contain surfactants to keep the particles dispersed and tend to have smaller amounts of organic solvents than emulsifiable concentrates. They consist of ingredients having a low water solubility and can settle out during storage.

Emulsions of a water-dissolved herbicide in oil consist of fine globules of the herbicide in water dispersed within a continuous liquid oil phase. An *emulsion of an oil-dissolved herbicide in water* is just the opposite; in other words, it consists of a herbicide in fine globules of oil dispersed within a continuous water phase. Although these formulations have many of the same characteristics as other liquid suspensions, they are packaged as liquid droplets in a liquid phase rather than as finely ground solids in a liquid phase. In contrast to emulsifiable concentrates, which consist of a single oil-based phase, these emulsions are packaged in two phases, water and oil. The advantage is the ability to mix two unlike or incompatible formulations in the same package. For example, Squadron, which is a combination of pendimethalin, which is oil soluble, and imazaquin, which is water soluble, is an EW formulation.

Microencapsulated formulations are small particles consisting of a herbicide core surrounded by a barrier layer, usually made up of a polymer shell. They also are referred to as *capsule suspensions* because the capsules are suspended in a liquid medium. In contrast to emulsifiable concentrates, which require a large amount of solvent to ensure suspension, microencapsulation greatly reduces the amount of solvent needed. Thus, microencapsulation reduces the cost of formulation and the amount of solvent (which can sometimes be toxic) that will be dispersed into the environment. The polymer shell protects the user from contact with the herbicide, shields the herbicide from degradative agents such as light, and reduces the rate and amount of herbicide loss due to volatization and leaching.

Dry Solids to Be Suspended in Water. Solids are dry formulations that are added to the liquid carrier for dispersal through a sprayer. In contrast to soluble powders or granules, which completely dissolve in water, these formulations are present as particles that are suspended in the water of the spray tank. In contrast to some liquid formulations, dry solids are typically free of organic solvents, are easy to pick up in case of spillage, and are stable at below freezing temperatures.

Wettable powders are finely ground solids that can be readily suspended in water. The usual recommendation is that they be mixed with a small quantity of water to form a light paste (slurry) prior to introduction into the tank. Continuous moderate to vigorous agitation is needed to keep these materials suspended in the spray tank; a solid precipitate forms at the bottom of the tank as soon as agitation stops. This much agitation usually requires a mechanical or hydraulic agitator in addition to the return line. The spray mixture is opaque. These formulations normally contain 50 to 80% active ingredient. Most wettable powders are used as soil treatments; however, they sometimes are used on foliage. When foliage applied, these formulations frequently require an additional wetting agent.

Water-dispersible granules also are called dry flowables and dry granules. They are dry formulations of granular dimensions. The granules themselves are made up of very finely ground solids combined into discrete particles with suspending and dispersing agents. They can be introduced into the sprayer tank without making a slurry first; they disperse without clumping. Water-dispersible granules pour more cleanly from the container than do suspension concentrates or wettable powders, and they produce less dust than wettable powders. Uniform density makes constant metering by volume theoretically possible. The higher cost of producing water-dispersible granules is partially offset by less bulk, resulting in lower shipping and storage costs than for wettable powders. Their other properties appear to be similar to liquid suspensions.

Dry Application Formulations

This group of herbicides is applied directly to the target in a dry form. Because these formulations are applied directly from the package to the field without dilution, the herbicide is supplied at relatively low concentrations on the dry carrier.

Granules are dry formulations of herbicides and other components in discrete particles generally less than 10 mm³ in size. The herbicide is coated on the exterior of the particle. A variety of dry substances have been utilized as granule components including clay minerals, dry fertilizers, and ground plant residues. Herbicide concentrations are typically 2 to 20%.

In general, granular herbicides require slightly more rainfall to leach into the soil than sprayable formulations do. The rate of herbicide loss to volatilization can be retarded if formulated as a granule. Recently, *matrix granules* made of starch, in which the herbicide is trapped within the particle, have been used to slow the release of certain herbicides into the soil; however, slow-release technology has not yet been widely adopted by herbicide manufacturers.

Pellets are dry formulations of herbicides and other components in discrete particles usually larger than 100 mm³. *Tablets* are in the form of small, flat plates. Pellets and tablets frequently are used for spot applications. Herbicide concentrations typically are 5 to 20%.

Trends in Formulation and Packaging Technology

Developers and users continue to search for herbicide products that are convenient to ship, store, and handle; that provide accurate, convenient application and consistent weed control with no crop injury; that are safe to the applicator; and that are environmentally safe, all at economical costs. Recent developments in formulation technology, such as gels, microencapsulated formulations, and water-dispersible granules, along with new types of packaging such as returnable containers, packaging that dissolves in the sprayer tank, and closed-handling systems all help meet these objectives.

One of the major incentives for the development of new technology has been the regulatory pressure placed on industry to reduce the use of volatile organic solvents in pesticide formulations. Organic vapors are implicated in causing poor air quality, and organic solvents are flammable, thus producing a fire hazard. Organic solvents readily penetrate skin, maximizing the oral and dermal toxicity of the active herbicide ingredient. As a matter of fact, many organic solvents are more toxic to humans than the herbicide ingredients dissolved or suspended in them. Of all of the formulations, the emulsifiable concentrates tend to contain the greatest amounts of organic solvents, and they are being replaced gradually by microencapsulated formulations and water-dispersible granules, both of which contain low to no organic solvents in their formulation.

In addition to vapors and toxicity of organic solvents, industry researchers are addressing other safety considerations in the development of formulation technology, all in an effort to prevent contaminating handlers of the product. Dusts from wettable powders are particularly troublesome, and the splash and spill from liquids also can be a problem. Gels, encapsulated formulations, and water-dispersible granules are almost dust free, and water-dispersible granules are relatively easy to clean up if spilled. The barrier surrounding the encapsulated formulations prevents skin contact with the active herbicidal ingredient, and also reduces volatilization, leaching, and degradation of the herbicide. Water-soluble packaging also eliminates skin contact.

Another consideration is efficiency of shipping and storage. The tendency is to produce high density (high weight per volume) products in easily transportable units. For example, wettable powders, which tend to be very powdery and fluffy, take up more volume per unit weight than do water-dispersible granules, which are basically wettable powders condensed into discrete granules. An analogy would be that of confectioners' sugar (a soluble powder) and granulated sugar (a water-soluble granule).

Any of the dry formulations are less likely to be adversely affected by freezing temperatures than are the liquids.

Packaging technology has been directed toward the development of packages that are easy to handle, prevent contamination to the user, and can be disposed of so as to prevent environmental contamination.

Disposable containers must be emptied into the spray tank and then triple-rinsed with water also put into the spray tank. Some containers are more amenable to rinsing, such as containers with wide tops and rounded corners. A formulation that readily flows out of the container, such as a water-dispersible granule, leaving little residue also is helpful in completely emptying the package.

A relatively new strategy in disposal of empty single-use (one-way) containers is the use of water-soluble packaging that dissolves in the spray tank. This is part of the technology behind gel formulations. Other water-soluble packages are designated *soluble packets* (SP) and contain dry products such as wettable powders and water-dispersible granules. Water-soluble packets contain premeasured quantities of herbicide and are most amenable for products that are applied at ounces per acre. For example, one packet of Beacon Accu-Pak treats two acres at the standard rate of 0.76 oz/ acre. Water-soluble packages are themselves packaged in a waterproof bag, which is uncontaminated and easily disposed of. Special care must be taken in how the water-soluble bags are added to and dissolved in the spray tank.

The use of returnable (two-way) containers is a strategy that avoids the need to dispose of packaging. This approach is usually accompanied by a closed handling system. *Closed handling systems* allow the user to transfer herbicide from the container to the spray tank without operator exposure. For example, when alachlor, which is a restricted-use herbicide, is applied to over 300 acres in a year, a closed handling system is required for its transfer to the spray tank in order to reduce human exposure. Closed handling systems for large quantities of herbicide are combined with two-way mini-bulk tanks. A typical unit is a 110-gal container that can be lifted with a forklift and loaded onto a small farm truck. The units are equipped with a device that allows the herbicide to be metered into the spray tank without exposure to the user. Mini-bulk tanks with closed handling systems provide the convenience of not having a lot of containers to handle. They work very well for liquids, but dry products are more difficult to meter. Bayer Corporation has devised a pneumatic (air blast) system that allows transfer and measurement of water-dispersible granules into the spray tank.

HERBICIDE CARRIERS

A carrier is a gas, liquid, or solid substance used to dilute or suspend a herbicide during application.

Liquid Carriers

Liquid carriers for spray applications include water (most widely used), liquid fertilizers, and diesel and similar viscosity oils. Water is abundant, relatively cheap, and a good carrier. Problems can be encountered with hard water and dirty water. Hard water (water with excessive amounts of dissolved calcium and magnesium salts) can result in Ca and Mg ion antagonism with the herbicide or even cause herbicides with anionic charges to precipitate (e.g., 2,4-D formulated as a salt or acid). Water with sand, silt, or other grit can cause wear on sprayer parts and can clog screens and nozzles. Algae and other plant growth or residues also can plug the system.

Liquid fertilizer solutions frequently serve as carriers for soil applied herbicides. Combining fertilizer and herbicide applications is the major advantage although additional burndown activity can be obtained on emerged vegetation. Such applications should be made before the crop has emerged to avoid crop injury.

Oils usually serve as carriers for special uses. Diesel and other mineral oils are used routinely for dormant applications of herbicides to woody species. Oils have been investigated as carriers for selective postemergence applications. Such a use requires oils that are not phytotoxic. Both mineral oils and vegetable oils have potential for use at lower-volume applications. Oils provide the advantage of improved cuticular wetting and penetration when compared to aqueous carriers.

The choice of a specific liquid carrier can affect the performance of foliar-applied herbicides; for example, the use of oil carriers results in better retention by plant foliage than does the use of water carriers. The type or volume of carrier has little or no effect on the performance of soil-applied herbicides (assuming that the carrier and herbicide can be evenly and accurately applied at the correct dosage).

Dry Carriers

Dry carriers are used to apply herbicides without further dilution and are the major components of granules and pellets. They include attapulgite, kaolinite, and vermiculite, as well as dry fertilizers, polymers of starch, and many other substances.

Dry fertilizers can serve as good carriers for herbicides. Turf herbicides have been applied in this manner for years. Granular bulk fertilizers have been impregnated with soil-applied herbicides and successfully applied to large acreages. The herbicide usually is mixed with dry fertilizer by a farm chemical retailer and then custom applied.

Impregnating dry bulk fertilizers with herbicides is done simply by using a rotary drum (or similar) mixer equipped with a spray nozzle positioned inside the mixer to provide uniform spray coverage of the tumbling fertilizer. Information and restrictions for mixing are provided on the herbicide label. Labels from some herbicides recommend the addition of an absorptive powder to the prepared pesticide–fertilizer mixture to ensure that the two components stick together and to provide a dry, free-flowing mixture.

Uniform and accurate spreading of the fertilizer–herbicide mixture to soil should result in weed control comparable to that obtained by using these same herbicides applied with other carriers.

Gas Carriers

Gas carriers, unlike liquid and dry carriers, are not used for suspension but as propellents. Compressed air, carbon dioxide, and nitrogen are used to pressurize small handheld or backpack sprayers. For large-scale field applications, air is used along with the dry or liquid carrier to deliver the pesticide to the intended target. To date, such field devices have been used primarily for applying fungicides and insecticides (orchard sprayers, power dusters, some fan-propelled ground sprayers). This principle could be applied to herbicides. Air-driven types of equipment are tested from time to time. Air delivery granular applicators, which have been developed and commercialized recently, are an advance in granular application technology.

SPRAY ADDITIVES

Herbicides make up 65% of the pesticides sold in the United States. Most require adjuvants either in the formulation or spray tank or both. An *adjuvant* is any additive used with a herbicide that enhances the performance or handling of the herbicide. There are two sources of adjuvants: formulation adjuvants and application adjuvants. A *formulation adjuvant* is added to the product during formulation by the manufacturer and requires no action on the part of the applicator. An *application adjuvant* is purchased and added to the spray tank by the applicator. Such adjuvants can greatly enhance the activity of foliar-applied herbicides. The term *adjuvant* often is restricted to application adjuvants.

There are many different types and functions of adjuvants. Foy (1996), for example, listed over 25 functions of adjuvants, ranging from buffering to improved leaf wetting. In general, the proper adjuvant will about double the foliar activity of a herbicide over the activity of that herbicide when used by itself. It can be difficult, however, for a field specialist to predict the impact of unfamiliar adjuvants. Many components interact when spray solutions are applied to weeds: each of the herbicides that is added to the tank; the adjuvant(s), solvents, and other materials in the herbicide formulation; the adjuvants (and their components) that are added; and the plant cuticle. Each component partitions into each of the others, depending on individual solubilities and charges. Therefore, the effect of a new adjuvant must be tested for each adjuvant, herbicide, and weed combination. This is the same process that is used to evaluate herbicide performance, and it is risky to do any less for individual adjuvants. Unless willing to do such testing, the best bet for users in the field is to select adjuvants suggested by the herbicide manufacturer.

Adjuvants are generally classified on the basis of their use rather than their chemistry. Chemical properties do, however, determine the suitability of compounds for individual uses. Commonly used adjuvants that, as a group, improve handling, wetting, retention, and entry of the herbicide into the plant shoot will be described briefly here. These include surfactants (emulsifiers and wetting agents), oils, and fertilizer salts. Additional types of adjuvants and their uses are listed in Table 16.1.

Table 16.1 Different Types of Spray Additives and Their Uses[a]

Acidifier: a material that is added to spray mixtures to lower the pH.

Activator: a material that increases the effectiveness of the herbicide.

Adjuvant: any additive used with a herbicide that enhances the performance or handling of the herbicide.

Buffer: a compound or mixture that, when contained in solution, causes the solution to resist change in pH. Each buffer has a characteristic limited range of pH over which it is effective.

Compatibility agent: a material that allows the mixing or improves the suspension of two or more formulations when applied together as a tank mix. They are used most frequently when a liquid fertilizer is the carrier solution for a herbicide.

Crop oil concentrate: oil-based material that enhances herbicide penetration through the leaf cuticle.

Defoamer: a material that eliminates or suppresses foam in the spray tank so that pumps and nozzles can operate correctly.

Drift control agent: a material used in liquid spray mixtures to reduce spray drift.

Emulsifier: a material that promotes the suspension of one liquid in another.

Fertilizer: certain fertilizers added to the spray tank can enhance penetration of the herbicide into the leaf.

Humectant: a material that increases the water content of a spray deposit and thereby slows the drying of the deposit on the leaf surface. Theoretically this should increase the amount of time that the herbicide has to penetrate into the leaf.

Penetrant: a material that enhances the ability of a herbicide to penetrate a surface, such as the leaf cuticle.

Safener: A material that reduces toxicity of herbicides to crop plants by a physiological mechanism.

Spreader: a material that increases the area that a droplet of a given volume of spray mixture will cover on a target. Same as a wetting agent.

Sticker: a material that causes the spray droplet to adhere or stick to the target. It reduces spray runoff during application and washoff by rain.

Surfactant: a material that improves the emulsifying, dispersing, spreading, wetting, or other surface-modifying properties of liquids.

Wetting agent: a material that reduces interfacial tensions between water droplets and the leaf cuticle.

[a]Adapted from Foy and Pritchard (1996).

Surfactants

Surfactants (a coined word derived from *surface-active-agent*) are materials that favor or improve the emulsifying, dispersing, spreading, wetting, or other surface-modifying properties of liquids. The individual molecules of all surfactants consist of a lipophilic (fat-loving) end and a hydrophilic (water-loving) end. Two of the major types of surfactants are emulsifiers and wetting agents.

The common names and chemistries of 19 different surfactants are provided in the *Herbicide Handbook* (WSSA, 1994). Rather than listing them separately, their general characteristics as emulsifiers and wetting agents are summarized here. Two additional classes of surfactants that will be described are the organosilicones and the *N*-alkyl pyrrolidones.

Emulsifiers. *Emulsifiers* are substances that promote the suspension of one liquid in another. They are most commonly used to disperse oil in water. Emulsifiers by themselves are rarely added to the spray tank. Rather, they are present in products that are added to the tank, such as emulsifiable concentrate and water-dispersible formulations of herbicides and crop oil concentrates. Emulsifiers act as surface active agents by partitioning their lipophilic ends into the oil particles and their hydrophilic ends into water, thus linking the two (Figure 16.1). This action prevents the oil droplets from combining to form larger droplets that would otherwise rapidly separate out from the water carrier.

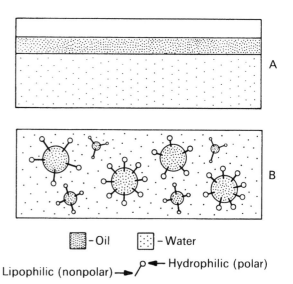

Figure 16.1 (A) Water and oil without emulsifier. (B) Emulsifiers link oil and water particles, enabling oil droplets to become suspended in water.

Wetting Agents.

Wetting agents reduce interfacial tensions between surfaces that normally would repel one another, such as water droplets on a waxy leaf cuticle. Molecules of the wetting agent orient themselves so that the lipophilic ends partition into the waxy cuticle of leaves and the hydrophilic ends partition into the water droplets. This effectively reduces water tension and enables the liquid to spread and make better contact with the surface of the plant. Wetting agents frequently are added to the spray tank at concentrations of 1/4 to 1/2% of the total spray volume.

Four groups of surfactants can be used as wetting agents: anionic, cationic, nonionic, and amphoteric. These various forms of ionization occur on the hydrophilic ends of the surfactant molecules. Anionic and cationic surfactants have electrical charges in water (negative and positive, respectively). Nonionic surfactants do not form an overall electrical charge. Amphoteric surfactants have various amounts of positive and negative charges depending on the pH of the solution.

Anionic surfactants are excellent wetting agents. Both anionic and cationic surfactants, however, have the disadvantage of reacting with other ions, possibly even the herbicide itself. Cationic surfactants can be toxic to plants and are not often used.

Nonionic surfactants are the type usually available for addition to spray tank solutions. They are good dispersing agents, are soluble and stable in cold water, and have low toxicity to both plants and animals. Since they do not ionize in aqueous solutions, nonionics are compatible with most herbicides and are unaffected by hard water. In other words, they do not form insoluble salts with calcium and magnesium ions and precipitate out. Anionic surfactants often are blended with nonionics to improve the wetting and emulsifying properties of the herbicide formulation.

Just because a product lowers the surface tension of water or increases the wettability of a spray solution does not mean it can be used as a wetting agent. In some cases, soaps and liquid household detergents have been used, frequently with disappointing results. Soaps can combine with hard water to form precipitates in the spray tank. In addition, many liquid detergents and soaps make too much foam for use in a spray tank. Household detergents may have relatively low concentrations of surfactant (10 to 20%); agricultural surfactants contain 50 to 90%. The ingredients in detergents also may not have Environmental Protection Agency (EPA) approval for use in agricultural sprays.

Organosilicones.

Organosilicones are silica-based surfactant molecules that reduce the surface tension of water droplets to a much greater degree than other surfactants. Their addition to a water–herbicide solution results in greater spreading of the spray drop on the leaf surface than even that predicted by the reduction in surface tension. This property allows for better adherence of spray drops to leaves that are highly water repellent. It also may allow for penetration of the solution through stomates, although cuticular penetration is probably more important.

In theory, the spreading out of a spray drop to this degree can increase the evaporative surface and cause the droplet to dry out before adequate penetration into the leaf surface can be achieved; however, the organosilicones have significant humectant properties that slow drying.

Organosilicones are not effective for all herbicides; for example, some are antagonistic to the activity of glyphosate.

N-alkyl Pyrrolidones.

This class of surfactant has unusually good physical properties such as low vapor pressure, good thermal and chemical stability, low phytotoxicity, and rapid biodegradability. The major advantage of this class is that it provides benefits usually obtained by a combination of several other surfactants; that is, it provides the properties of a spreader, sticker, wetter, and penetrant.

Oil-Based Adjuvants

These materials are mixtures of surfactants and nonphytotoxic light oils derived from either petroleum or oil seed crops. The surfactants enable these oils to emulsify in water. *Crop oils* and *crop oil concentrates* are added to aqueous solutions of certain herbicides to enhance their effectiveness as foliar treatments. They act mostly to increase absorption of the herbicide through the cuticle of the leaves. Postemergence applications of herbicides such as atrazine, bentazon, and the aryloxyphenoxy propionates provide more consistent weed control when applied with crop oil concentrates than with standard wetting agents.

Crop oils contain 1 to 2% surfactant and are used at concentrations of 1 gal in 20 to 25 gal of spray solution/acre. Crop oil concentrates contain 15 to 20% surfactant and are used at about 1 to 2 pts per 20 to 25 gal of spray/acre. Crop oil concentrates have replaced crop oils in popularity because of the reduction in the volume of adjuvant required. Crop oil concentrates are available as mineral oils plus surfactant, seed oils plus surfactant, and methylated seed oils plus surfactant.

Fertilizer Salts

Ammonium sulfate (AMS), 28% nitrogen (mixtures of ammonium nitrate and urea), and 10-34-0 fertilizer solutions are *fertilizer salts* that are added to spray tanks at a concentration of 2 to 5% to enhance herbicidal activity. The effect appears to be enhanced absorption of the herbicide into the plant, and a faster burndown of the vegetation usually is observed. Herbicide performance on annuals and some simple perennials can be improved. For example, the control of velvetleaf is almost always increased. Control of regrowth (long-term kill) of creeping perennials, however, may not be improved since rapid top kill appears to decrease translocation to vegetative propagules. Fertilizer salt solutions also can improve herbicide performance under conditions of low humidity and other stress conditions that tend to reduce herbicide penetration and efficacy.

AMS also has been shown to overcome decreased herbicidal activity due to antagonism from dissolved salts such as calcium and magnesium in the spray solution. Herbicides that are susceptible to antagonism from salts include 2,4-D, bentazon, acifluorfen, glyphosate, and nicosulfuron. Sometimes antagonism between herbicides (for example, bentazon plus sethoxydim) can be reduced by adding fertilizer solutions to the spray solution.

TANK MIXES INVOLVING HERBICIDES

There can be many advantages to mixing individual chemicals together in the same sprayer tank. By combining applications of liquid fertilizer, herbicide, and possibly an insecticide, the number of application trips over the field can be significantly reduced. In addition, herbicide mixtures can control a broad spectrum of weed species, provide more consistent control, and help prevent the development of herbicide-resistant populations. Mixtures are an important part of the trend toward prescription recommendations for mixed weed populations. Mixtures containing herbicides that are used at lower doses or have less of a negative environmental impact can substitute for single herbicide treatments that are typically used at larger doses or are more persistent.

Several factors must be taken into account when chemicals are tank mixed. Two such factors are the potential for physical incompatibility in the tank and the potential for altered plant responses, particularly when the solution is applied to the foliage.

When two chemicals are used in a tank mix (or applied in sequence), it is possible to obtain three degrees of plant response. An *additive* response occurs when the effect of the two chemicals is equal to the predicted effect based on the activity of each chemical applied separately. This frequently is the case with most herbicides when they are tank mixed.

A *synergistic* response occurs when the effect of the two chemicals is greater than the predicted effect based on the activity of each chemical applied separately. A well-known synergistic effect is an adverse one: crop injury increases when primisulfuron or nicosulfuron are applied with organophosphate insecticides. A positive example of synergism was the addition of tridiphane to mixtures of atrazine and cyanazine. Although tridiphane is a potent meristematic inhibitor, its use alone resulted in very little herbicidal activity. However, when combined with atrazine, cyanazine, or an atrazine–cyanazine mix, tridiphane inhibited the activity of the enzyme (glutathione-S-transferase) that is responsible for detoxifying triazine herbicides. Thus, the herbicidal effect was much greater than applying either tridiphane or the triazine compounds alone. Tridiphane was effective in concept; however, marketing considerations resulted in its being removed from sale.

Antagonism occurs when the effect of the two chemicals is less than the predicted effect based on the activity of each chemical applied separately. Antagonism has been noted with mixtures of 2,4-D, acifluorfen, or bentazon with the selective postemergence grass herbicides (aryloxyphenoxy propionates and cyclohexanediones; see Table 6.1).

Most herbicide applications to the soil can be made in combination with other chemicals, providing that all components are kept uniformly dispersed in the spray solution. Mixtures that can be successfully applied selectively to crop foliage are more limited and must be chosen with care in order to avoid crop damage or reduced control.

Tank Mixes Applied to Soil

Soil applications can sometimes consist of as many as two or three pesticide formulations mixed in a liquid fertilizer carrier solution. Thus, it is possible to have finely ground solids, water-soluble liquids, ions (both inorganic and organic), high concentrations of salts, oils or oil-soluble compounds, and miscellaneous other compounds all mixed in the same sprayer tank.

In general, the individual herbicides can be expected to maintain their chemical integrity. Since the physical characteristics of each of these different formulations vary considerably, however, mixing them can cause their components to separate into layers or to coagulate and precipitate in the spray tank. When this occurs, the mixture is termed incompatible. Problems are encountered most frequently among herbicide combinations when flowables, dry flowables, or wettable powders are mixed with emulsifiable concentrates. Liquid fertilizers appear to accentuate this problem.

Most labels for herbicides applied to the soil contain specific directions for mixing and agitating the chemicals when they are to be used in combination with other herbicides and fertilizer solutions. For example, labels will indicate the sequence in which the chemicals should be added to the spray tank. These labels also describe procedures for predicting the compatibility of compounds in the spray tank.

Procedures for mixing a herbicide with a liquid fertilizer (taken from the label of an emulsifiable concentrate herbicide) recommend that a small-scale test (in a 1-qt widemouthed glass jar) be conducted using the components to be tank mixed in the proportions and order to be used in the spayer. The suggested steps, once compatability has been established, are as follows:

1. Fill the spray tank at least two-thirds full of fertilizer solution.
2. Start moderate agitation and keep it going.
3. Add, mix, and disperse wettable powders or water-dispersible granules.
4. Add flowables (liquid). Allow thorough mixing.
5. Add emulsifiables. Disperse them thoroughly.
6. Finish by adding water solubles and their recommended companion surfactants.
7. Maintain agitation until the tank is emptied.

The addition of a compatibility agent may aid in keeping mixtures suspended, but thorough and constant agitation in the sprayer tank usually is required. In most cases, agitation must be initiated during mixing and continued until the solution is sprayed onto the soil. It is not uncommon for a mixture of chemicals in the tank to leave a sticky precipitate on the bottom of the sprayer when agitation is stopped. In many cases these precipitates cannot be resuspended. As a result, the chemicals are wasted, the sprayer requires tedious cleaning, a suitable disposal site must be found for the precipitate, and valuable and irreplaceable time is lost.

Tank Mixes Applied to Foliage

Foliar applications do not lend themselves particularly well to unlimited tank mixing, especially when maintaining crop selectivity is important. Liquid fertilizer solutions tend to increase foliage burn. Several cases of serious crop injury have occurred when warnings not to apply triazine or chloroacetamide herbicides in liquid nitrogen solutions following corn emergence have been ignored. Increased injury also can occur with combinations of several herbicides, particularly when an emulsifiable concentrate or an adjuvant is added. Addition of 2,4-D and dicamba to early postemergence treatments of atrazine and crop oil can cause severe injury to corn.

On the other hand, tank mixes of some postemergence herbicides can result in decreased herbicide activity (antagonism). If a specific tank mix combination is not labeled or has not been suggested by a responsible advisory agency and backed by sound research, it should *not* be used for selective postemergence applications.

MAJOR CONCEPTS IN THIS CHAPTER

By formulating herbicides, the manufacturer provides the user with the herbicide in a form that is suitable for practical use.

Sprayable formulations include the water-soluble liquids (SL), water-soluble powders (SP), water-soluble granules (SG), emulsifiable concentrates (E or EC), gels (GL), suspension concentrates (SC) and aqueous suspensions (AS), emulsions of a water-dissolved herbicide in oil (EO), emulsions of an oil-dissolved herbicide in water (EW), microencapsulated formulations (ME), wettable powders (W or WP), and water-dispersible granules or dry flowables (WDG or DF).

Dry formulations for direct application are most commonly granules (G). Pellets (P) and tablets (TB) sometimes are used for special purposes.

Packaging technology has been directed toward development of packages that are easy to handle, prevent exposure to the user, and can be disposed of so as to prevent environmental contamination. New strategies to accomplish these goals include water-soluble packets (SP), in which the packaging material dissolves in the spray tank, the use of returnable containers, and closed handling systems, which allow the user to transfer the herbicide from the container to the spray tank without exposure.

The most commonly used carrier for spray applications of herbicides is water. Other carriers include liquid fertilizer solutions and diesel and mineral oils. Granular fertilizers can be used as carriers for dry applications.

Choice of carriers should have minimal influence on soil-applied herbicides but can substantially alter the performance and selectivity of foliar applications.

An adjuvant is any substance that enhances the performance or handling of the herbicide. A formulation adjuvant is added to the product during formulation by the manufacturer. An application adjuvant is added to the spray tank by the applicator.

Spray adjuvants include surfactants (emulsifiers, wetting agents), crop oils, crop oil concentrates, fertilizer salts, compatibility agents, drift control agents, defoamers, and stickers.

Tank mixing of herbicides with other herbicides, with liquid fertilizers, or with other pesticides can reduce the number of application trips over the field, provide control of a broader spectrum of weed species, and prevent the development of herbicide-resistant populations. On the other hand, tank mixes can result in physical incompatibility problems in the spray tank and can cause poor application or altered plant responses.

The effect of two chemicals mixed together can be additive, synergistic, or antagonistic.

Spray tank mixes can contain water, dissolved salts from fertilizers, suspended solids from herbicides, water soluble herbicides, oils from emulsifiable concentrates or crop oil concentrates, and other additives. Incompatibility in the tank is most likely when solids and oils both are present.

Mixing procedures on labels should be followed. A typical procedure is to fill the tank two-thirds full with the carrier, start agitation, and then add components in the following order: dry solids, solids suspended in liquid, oil solubles, and water solubles. Each component should be thoroughly mixed before the next component is added. Agitation should be stopped only when the sprayer tank is empty.

When used as a carrier, liquid fertilizer can cause extensive crop injury if the crop has emerged.

TERMS INTRODUCED IN THIS CHAPTER

Additive effect
Adjuvant
Antagonism
Application adjuvant
Closed handling system
Compatibility agent
Crop oil concentrate
Defoamers
Dry carrier
Dry flowable
Emulsifiable concentrate
Emulsifier
Flowable
Formulation
Formulation adjuvant
Gas carrier
Gel
Granule
Liquid carrier
Microencapsulated formulation
Mini-bulk tank
Organic solvent
Organosilicone
Pellet
Returnable (two-way) container
Soluble packet
Spreader
Sticker
Surfactant (surface-active-agent)
Suspension concentrate
Synergistic response
Tablet
Water-dispersible granule
Water-soluble granule
Water-soluble liquid
Water soluble powder
Wettable powder
Wetting agent

SELECTED REFERENCES ON FORMULATIONS, CARRIERS, ADJUVANTS, AND TANK MIXES

Adjuvants for Herbicides. 1982. Weed Science Society of America, Lawrence, Kansas.

Foy, C. L. and D. W. Pritchard, Eds. 1996. Pesticide Formulation and Adjuvant Technology. CRC Press, Boca Raton, Florida.

Herbicide Handbook, 7th ed. 1994. Weed Science Society of America, Lawrence, Kansas.

HERBICIDE APPLICATION

Herbicides must be properly placed in relation to the crop and the weed and be adequately and evenly distributed if they are to be effective weed control tools. Learning the principles and techniques involved in the actual application of herbicides is just as important as choosing the correct herbicide for a specific weed control situation.

The most serious problems encountered with herbicide use often stem from improper application technique. These problems include injury to off-target vegetation, nonperformance in the control of weeds, injury to crop plants present, and injury to subsequent crops. While improper diagnosis, incorrect herbicide selection, and adverse weather conditions can account for a portion of these problems, the majority are due to mistakes made during application. These mistakes could be eliminated through the proper use of present-day application technology.

The need for careful attention to application details is reinforced by a survey conducted by University of Nebraska agricultural engineers who spot-checked pesticide application equipment as it was being operated in the field. The equipment included 95 sprayers and 38 granular applicators. Application accuracy was deemed acceptable if the amount of chemical applied was within ±10% of the desired dosage. *The study showed that three out of every five pesticide applications made were outside this range.* Even worse, these misapplications deviated an average of ±30% from the desired dosage.

Errors were attributed to inaccurate calibration, incorrect mixing, worn and improperly operated equipment, and failure to read the product label. Specifically, 46% of the applicators made mistakes calibrating their sprayers, 5% made mistakes in mixing chemicals in the spray tank, and another 12% made mistakes in *both* sprayer calibration and mixing. A major cause of mixing errors was not knowing the exact volume of liquid in the sprayer tank. This problem was due to incorrect tank markings (or lack of them altogether) as well as to not having the tank level when making volume readings.

Worn and improperly maintained sprayer components, particularly nozzles, were observed frequently. Nozzle-related problems included the presence of several types of nozzles on the same boom, badly worn nozzle tips, variation in delivery between nozzles (even of the same type), damaged nozzles due to improper nozzle cleaning with wires or knives, plugged nozzles, and nozzles without screens. Broken or inaccurate pressure gauges were also common.

Granular equipment was found to be operated with about the same degree of accuracy as the sprayers. Variation in granule output between boxes at the same setting was a major contributor to uneven distribution. Poorly maintained equipment such as worn hopper agitators, bent or partially plugged delivery tubes, and partially sealed or rusted hopper outlets were frequently observed.

The major conclusion of this study was that the primary problem with agricultural chemicals is not the chemicals themselves but the people who apply them. Misapplication wastes an estimated $1 billion annually, which includes money spent for wasted chemicals and increased losses due to pests and crop injury. Furthermore, instances of inaccurate application invariably invite criticism from environmental groups along with increased interference and regulation from governmental agencies.

Weed management practitioners must understand several interrelated subjects in order to apply herbicides properly and obtain the maximum benefit from them. These subjects include conventional herbicide application equipment, special techniques and equipment to keep herbicides on target, methods for incorporating soil-applied herbicides, and calibration of application equipment.

APPLICATION EQUIPMENT

Effective and accurate herbicide use requires that the application equipment be correctly operated and maintained. The most frequently employed device for herbicide application is the sprayer.

Conventional Agricultural Sprayers

A typical sprayer for the application of herbicides consists of the following components (Figure 17.1): a *tank* to hold the spray solution; a

pump to deliver the spray solution through the system and maintain pressure; a *pressure regulator* to maintain a constant pressure in the system; a *pressure gauge* to provide pressure readings; *nozzles* on a *boom* or *handgun* to deliver the spray to the target effectively; a *shutoff* or *selector valve* to provide instantaneous delivery or shutoff of spray solution to the nozzles; an *agitator* (either mechanical or hydraulic) to keep the spray mixture evenly dispersed in the sprayer tank; a *bypass* line to return excess spray solution to the tank so the pressure regulator can function; *strainers* or *screens* (suction, in-line, nozzle) to keep large particles such as rust and grit from clogging the system and particularly the nozzles; a *suction line* to bring the spray solution from the tank to the pump; and various hoses, fittings, clamps and other parts to make the system function.

Self-propelled sprayer units, multipurpose trailer-type sprayers, and other sprayer systems can be purchased assembled and ready-to-use. Sprayer systems also can be constructed or modified to meet specific needs. Individual components can be purchased and assembled by anyone who understands how sprayers operate, has some mechanical ability, and knows about the specific type of herbicide application needed.

Pumps

The sprayer pump is the heart of the system. Most farm herbicide sprayers are designed to spray at relatively low pressure (30 to 60 lb per square in. [psi]) and are equipped with inexpensive pumps such as the centrifugal pump,

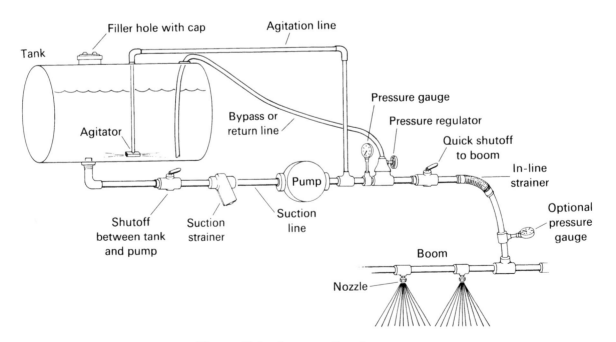

Figure 17.1   A conventional sprayer.

turbine pump, or roller pump. Sprayers used to apply insecticides and fungicides at high pressures of 100 to 500 psi usually are equipped with the more elaborate piston pumps. Gear pumps, flexible impeller pumps, and diaphragm pumps are used less frequently.

The *centrifugal pump* is popular on sprayers used for herbicides because of its reasonable cost and good serviceability. It can handle with ease nearly all sprayable formulations, including coarse or abrasive materials. Pressure develops as a result of centrifugal force (Figure 17.2A). The spray solution enters the pump through the center of a rotating impeller (rotor). The impeller forces the liquid to the outer edge of the housing and releases it to the delivery system through an outlet. The maximum pressure that can be developed with these pumps is about 75 psi. At 30 psi, centrifugal pumps should deliver 50 to 70 gal/min. Most sprayers equipped with centrifugal pumps can be used without a pressure regulator because shutting off the delivery system does not result in the drastic changes in pressure common with some other types of pumps.

The speed of operation of the impeller in these pumps is 1200 to 3500 revolutions per minute (rpm). Since the standard operating speed for tractor power takeoffs is only 540 rpm on standard and 1000 rpm for high speed, the centrifugal pump requires an extra belt and pulley system or a gear system for attaining proper operating speed. Some are run with hydraulic motors connected to the hydraulic system of the tractor. Equipment to increase rotor speed makes centrifugal pumps somewhat heavier and larger than roller pumps.

The *turbine pump* is used with herbicides and fertilizer solutions. It is similar to a centrifugal pump except that the spray solution is both picked up and delivered at the periphery of the impeller (Figure 17.2B). In contrast, the centrifugal pump picks up the spray solution through the center of the impeller and then delivers it to the periphery. The turbine pump has the high-volume and low-pressure characteristics of the centrifugal pump but works efficiently at slower speeds. It will operate directly from a 1000 rpm power takeoff and can be used either direct drive or with a step-up drive from a 540 rpm power takeoff. Pump capacity, however, is reduced at 540 rpm. The turbine pump delivers a slightly lower volume and is more sensitive to increasing pressures than the centrifugal pump.

The *roller pump* frequently is used when an inexpensive low-pressure delivery system is needed. Although the sales literature indicates that the roller pump is capable of generating pressures up to 300 psi, the pressure drops rapidly with wear. Most of these pumps are operated at pressures below 100 psi. Gritty water

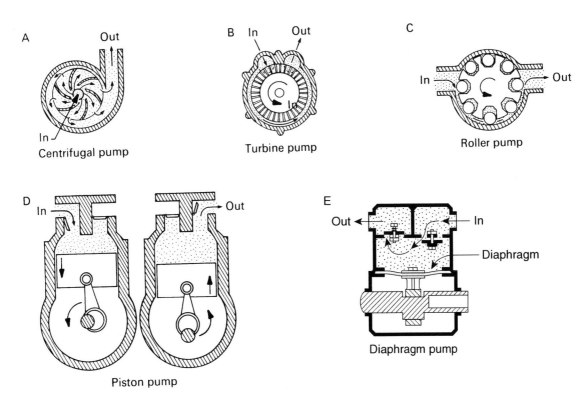

Figure 17.2   Sprayer pumps.

or abrasive wettable powders can cause rapid wear of the rollers in these pumps. Although the initial cost is low, the relatively short life of the roller pump may make the centrifugal pump a better buy.

The roller pump has loose rollers fitted into slots in an impeller that is mounted off-center in the pump housing (Figure 17.2C). The rotating impeller forces the rollers to follow the pump housing. The liquid is propelled ahead of the rollers and through the outlet into the spraying system. Delivery is semiquantitative since liquid is trapped between the rollers.

The *piston pump* is the most versatile of the pumps. It is designed to handle all sprayable formulations and to develop pressures up to 800 psi. Piston pumps are positive displacement pumps that deliver an exact volume of spray each time the piston makes a cycle. They work on the principle that a receding piston pulls liquid into the chamber through a valve that opens under negative pressure and then forces the liquid out another valve that opens as the piston advances (Figure 17.2D). A typical piston pump has two synchronized pistons. The piston pump requires that the sprayer system be fitted with an air chamber (surge tank) to dampen the pulsating action of the pump and to deliver the liquid as a steady flow.

Piston pumps can be either motor driven or ground wheel driven. When attached to ground-driven wheels, they can be used as metering devices to ensure that the amount of herbicide delivered per acre is constant even though the ground speed of the rig may vary considerably. Variations in speed can occur when a rig is driven up and down hills, over soils of differing firmness, or anytime the driver has difficulty maintaining constant ground speed.

Piston pumps are expensive. Unless such a pump is needed for metering, most herbicide sprayer rigs are overequipped when piston pumps are used.

The *diaphragm pump* has a flexible diaphragm that moves up and down in a chamber, creating suction and pressure. Intake and discharge valves open and close to force liquid through the pump (Figure 17.2E). These valves are separated from the diaphragm mechanism.

Pumps should be selected to deliver the volume needed to provide sufficient spray on the target, agitation, some return through the bypass, and compensation for frictional losses. The order of providing adequate volume among commonly available sprayer pumps is, from most to least, the centrifugal, turbine, roller, piston, and diaphragm pumps.

Pumps should be mounted on sprayer equipment so that they fill with spray solution when spray solution is in the tank. Having the pump full of liquid when it starts minimizes wear. Positive displacement pumps (roller and piston) are self-priming and will pick up liquid even when started dry. Nonpositive displacement pumps (centrifugal and turbine) must be full of liquid (primed) in order to function. The diaphragm pump has semi-positive displacement. Certain spray solution characteristics, such as an abundance of foaming, and wear can prevent the unit from self-priming.

Nozzles

Nozzles help determine the amount of spray delivered, the spray pattern, and the distribution of the herbicide on the target. Several different types of nozzles commonly used for herbicide application are discussed here. Those used specifically for drift control are discussed in the section on keeping herbicides on target.

A typical nozzle consists of a nozzle body, a strainer, a nozzle tip, and a cap (Figure 17.3). The cap holds the strainer and nozzle tip in place.

Commonly used materials for nozzle bodies include brass, nylon, stainless steel, and aluminum. Nozzle tips are made either from these materials or from hardened stainless steel. Initial cost and resistance to abrasion, corrosion, solvents, breakage, and thread stripping all need to be considered when choosing nozzles and nozzle tips.

Most nozzles are manufactured in standard dimensions so that strainers and tips can be changed to meet individual spraying needs. The three most commonly used nozzle types are the flat fan, flood (deflector), and hollow cone nozzles (Figure 17.4).

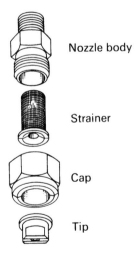

Nozzle body

Strainer

Cap

Tip

Figure 17.3   **Parts of a sprayer nozzle.**

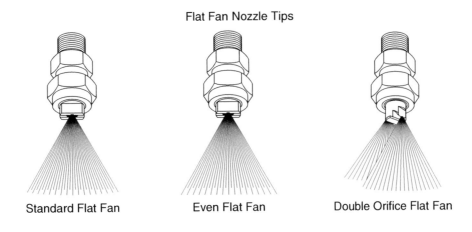

Flat Fan Nozzle Tips

Standard Flat Fan   Even Flat Fan   Double Orifice Flat Fan

Flood (Deflector) Nozzle Tips

(side)

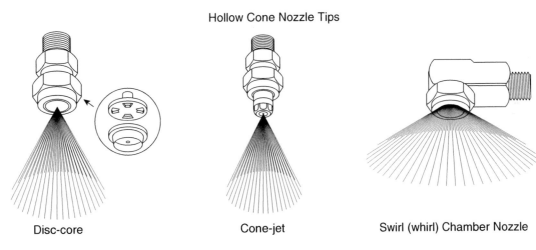

Hollow Cone Nozzle Tips

Disc-core   Cone-jet   Swirl (whirl) Chamber Nozzle

Figure 17.4 Nozzles and nozzle tips and spray patterns. (L. E. Bode and B. J. Butler, University of Illinois, Urbana, IL)

Flat Fan Nozzles. The standard *flat fan* tips produce a fan-shaped pattern and are used for broadcast spraying. The spray volume tapers off at the edge of the spray pattern, so these nozzles should be overlapped to produce a uniform delivery pattern across the entire width of the boom. Sprayers equipped primarily for herbicide spraying frequently are fitted with flat fan nozzle tips. They work well for uniform broadcast applications of soil-applied herbicides and other pesticides and for foliage applications with symplastically translocated herbicides or when the amount of foliage is not excessive.

Even flat fan (even-edge) tips are used for banding and for postemergence directed spray applications. The spray pattern of this tip retains full volume to the outside edge of the pattern. These nozzle tips are unsuitable for solid boom spraying.

Extended-range flat fans are flat fan nozzle tips that have been machined to deliver the same spray pattern at both high and low pressures.

Double-orifice flat fan tips are used to provide a more thorough coverage of foliage and crop residue than that obtained with the standard flat fan. The spray is delivered from two openings in the nozzle tip, thereby hitting the target at two angles instead of one.

Flood (Deflector) Nozzles. Deflectors produce a fan-shaped pattern and are used for broadcast applications of fertilizers, soil-applied herbicides, and soil-applied insecticides. Most commercial applicators equip their sprayers with flood nozzles. They provide a tapered pattern, but the pattern is much less precise than the tapered delivery of the regular flat fan. Flood nozzles and tips require a 100% overlap of the spray pattern for uniform delivery across the boom width.

Hollow Cones. Hollow cone nozzles and nozzle tips produce a pattern that is somewhat uneven, but they provide a satisfactory broadcast pattern when the nozzles are tilted at an angle away from the ground. Depending on the nozzle specifications and the manufacturer's recommendations, the angle of tilt should be set somewhere between 10 and 90 degrees.

Several types of hollow cone nozzles are used for herbicide applications: the cone-jet, the swirl or whirl chamber nozzle, and the Raindrop nozzle. Another hollow cone nozzle is the *disc-core nozzle*, which is used on aircraft and ground equipment for insecticide and fungicide applications.

Cone-jet tips are used when good coverage of foliage is essential. They are used for post-emergence applications of contact herbicides, defoliants, desiccants, and insecticides. They also can be used for band or directed herbicide spraying. These tips frequently are operated at somewhat higher pressures than others used for herbicide application. They are used when small droplet size and thorough coverage are desired. Because of the small droplet size, increased herbicide drift can occur with these tips.

Swirl chamber (or *whirl chamber*) *nozzles* are constructed so that the spray enters at right angles to the nozzle chamber (other nozzles feed the liquid straight through). The deflected spray then enters the nozzle chamber and passes through the tip to form a wide-angle hollow cone pattern. These nozzles are used for soil applications ahead of tillage implements and also can be mounted on cultivators and other appropriate equipment for directed postemergence applications.

The Raindrop and other drift-control nozzles are discussed in the section on keeping herbicides on target.

Sprayer Tanks

Sprayer tanks can be constructed of any of several materials, including stainless steel, aluminum, aluminized steel, fiberglass, galvanized steel, and various high-impact plastics. The nature of the spray solution to be used, the sprayer setup, and the length of service needed are factors to consider in selecting a tank. Stainless steel tanks probably are the most adaptable to all situations, but they usually cost the most initially. Tanks made of soft or galvanized steel should be avoided because they tend to react with some herbicide and other spray solutions. Soft or galvanized steel provides a continuing supply of rust flakes that clog the spraying system. Some spray solutions also may react with aluminum and coated steel. Fiberglass and plastic are resistant to chemical reactions and do not rust, but tanks made of these materials are more subject to breakage than metal tanks are.

Anyone who constructs and maintains spray equipment should keep a file of information on currently available sprayers and sprayer parts. This and other technical information on application equipment and techniques can be provided by reputable equipment manufacturers and local dealers.

Devices for Above-Crop Placement

Devices for placing herbicides on weeds taller than the crop have been on the market since the late 1970s. These specialized pieces of equipment are designed to supplement standard weed control practices. They provide the grower with another technique for attacking weeds at a stage of development for which there are not many satisfactory techniques, that is, after the weeds have emerged above the crop. In addition, they provide very economical, automatic spot treatments. Herbicide solution is used only when the weed to be controlled is present. Thus, herbicide cost is minimized to the point that an otherwise expensive herbicide can be applied for only a few dollars per acre.

There are two basic types of above-crop applicators: recirculating sprayers (RCS) and wipe-on devices.

Recirculating Sprayers

As the name of this sprayer implies, the spray solution from the RCS that is not intercepted by target vegetation is collected and returned to the sprayer tank to be used again. In contrast, spray solution from conventional sprayers not intercepted by the weeds falls to the ground or onto the crop, where it is lost or can cause crop injury. The recirculating feature of the RCS not only conserves herbicide but the spray pattern can be directed in such a way as to provide control of certain weeds without causing damage to the crop (Figure 17.5). Selectivity with the RCS is determined by the differential height between weeds and crop. Weeds at least 6 in. taller than the crop can be controlled by placing the nozzles in a horizontal position and spraying above the top of the crop. The unused spray is intercepted by the collector box or pad and trough and returned to the sprayer tank by siphons. Properly operated, the spray nozzles produce both a very narrow pattern of spray (in fact, almost a straight stream between collectors) and a minimum of "fines" (fog- and mist-sized particles) that are susceptible to drift. Thus, spray contact with the crop plants is minimized while return of the spray to the collectors is maximized.

Herbicides that can be expected to work in recirculating sprayers are those that translocate symplastically in plants, for example, glyphosate and 2,4-D. These compounds are particularly suited to the RCS because they require contact with only a small portion of the foliage in order to kill weeds effectively. The RCS safely permits the use of nonselective herbicides when crops are present.

Although the RCS has been used primarily for applications of glyphosate in soybeans and cotton, it has potential in many situations in which tall weeds are growing in a relatively short crop. Vegetable crops, turf, pastures, and grain sorghum are all likely crop candidates. Johnsongrass, shattercane, giant ragweed, volunteer corn, sunflowers, and Jerusalem arti-

choke are some of the important weeds that attain heights sufficient to make them suitable for control with the RCS.

Wipe-On Devices

Wipe-on devices designed to provide selective placement of herbicides to weeds above the crop more economically than the RCS are available. Commonly used devices include the wedge wick (Figure 17.6), rope wick (Figure 17.7), and sponge wick.

The initial cost of wipe-on applicators is less than a recirculating sprayer, but the delivery of the herbicide solution is limited by the wicking action. As a result, the speed of travel is limited, and two passes frequently are required. Wipe-on devices, however, are much preferred by farmers over the RCS because of their overall simplicity, ease of operation, and reduction of required maintenance. This preference serves to illustrate the advantage of equipment that is simple to operate and maintain. Wipe-on devices have even been modified for handheld and between-row applications.

RCS or wipe-on units should be mounted on the tractor (or high-clearance equipment) so

Figure 17.6   A wedge wick selective applicator.

Figure 17.5   Recirculating sprayer. (Monsanto Corporation, St. Louis, MO)

Figure 17.7 A rope wick being used on johnsongrass in soybeans. (Monsanto Corporation, St. Louis, MO)

that good crop clearance, good operator visibility, and rapid height adjustment can be achieved with ease ahead of the power unit. The units should be mounted in front on a hydraulic lift system.

Compressed Gas Powered Sprayers

Sprayers used for herbicide applications to small areas (lawns, gardens, experimental plots, and spot applications) frequently are powered by compressed air, carbon dioxide, or nitrogen (Figure 17.8). These sprayers allow an experienced operator to apply herbicides with a great deal of precision. The total unit, including the 1/2- to 3-gal spray tanks, is light enough to be carried by one person. Such systems can be mounted on bicycle wheels or small motorized vehicles when large amounts of spraying need to be done.

Booms up to 10 ft wide can be used with these sprayers to facilitate broadcast application. Regulators are required to deliver constant volume at constant pressure and to provide accurate delivery. The regulator is mounted on the propellant tank when the propellant tank is separate from the spray solution tank. If the same tank is used for both the propellant (usually air) and the spray solution, the regulator is mounted in the delivery line to the boom and nozzles. Since the spray solution passes through this regulator, it must be resistant to solvents.

When sprayers do not have a regulator, the spray is mixed to a given volume (e.g., 1% volume/volume) and then applied to a certain wet-

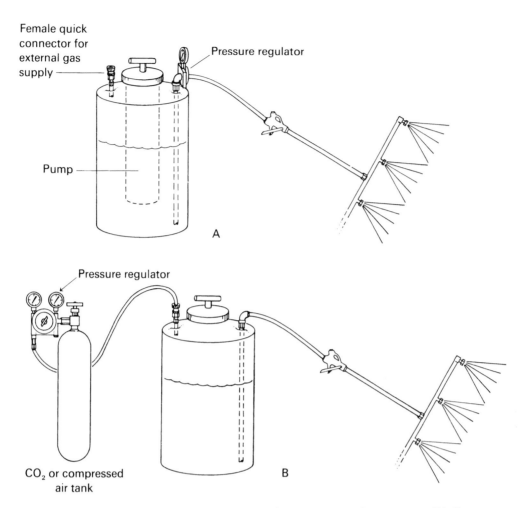

Figure 17.8 Two types of compressed gas powered sprayers: (A) Sprayer containing all the compressed gas (from either an internal pump or external gas supply) needed to expel the spray volume. Pressure regulator is present to maintain constant delivery. (B) Sprayer in which constant pressure is supplied from an external CO₂ or compressed air tank.

ness (e.g., until the leaves are shiny, until they are wet, or until the solution runs off).

KEEPING HERBICIDES ON TARGET

Spray drift is the movement of herbicides off target following their release from application equipment so that the chemicals never reach the intended target. Spray drift is caused by the presence of small-diameter spray droplets that fall slowly and are easily moved off target by air currents (Table 17.1). In addition, water can evaporate from the spray droplets, leaving solid herbicide particles suspended in the air (aerosols) and susceptible to air movement.

The number of fine spray droplets must be reduced if herbicide drift is to be reduced. The typical spray nozzle produces many different particle sizes ranging from very small to very large droplets. The range and average size of these particles depends on the pressure of the spray system, the nozzle shape, the volume of liquid dispersed (which depends on the size of the nozzle orifice), the angle of the spray pattern, the viscosity of the spray solution, and the density of the carrier solution. If all of these factors but one are kept constant, droplet size will increase as (1) the pressure is lowered, (2) orifice size and spray volume increase, (3) the angle the spray leaves the nozzle decreases, or (4) the viscosity of the spray solution increases. Table 17.2 shows the effect of spray angle and nozzle pressure on droplet size.

The average spray droplet size is frequently expressed as the VMD (volume mean diameter) or the droplet size at which one-half the spray volume consists of large droplets and one-half consists of small droplets. Since it takes many more small droplets to make up one-half the spray volume, there always will be more small droplets than large droplets present in a typical

Table 17.1 Effect of Spray Droplet Size on Spray Drift[a]

Droplet Diameter (μm)	Type of Droplet	No. Droplets/sq in. From 1 gal Spray/A	Time Required to Fall 10 ft in Still Air	Distance (ft) Droplet Travels in Falling 10 ft	
				1 mph Wind	5 mph Wind
1	↑		28 hr		
5	Fog	9,000,000	66 min	*5810	*29,050
10	(aerosol)	1,151,000	17 min	*1496	*7,480
20		144,000	4 min	*352	*1,760
25		81,000			
50	↓	9,000	40 sec	*59	*295
100	Mist	1,160	11 sec	15.4	77
200	Fine spray	144	4 sec	5.9	30
240	Medium spray	78			
300	↓	43			
400	Coarse spray	18	2 sec	3.0	15
600		5.4	1.7 sec	2.5	12
800		2.3			
1000	↓	1.1	1 sec	1.5	7

[a]Data from Klingman and Ashton (1982), Matthews (1979), and Warren (1976).
*These droplets could easily evaporate before reaching the target.

spray pattern (Figure 17.9). It is *the fog- and mist-sized droplets (those less than 100 μm in diameter) that are most susceptible to drift* (Table 17.1).

On the other hand, as spray droplet size increases, the number of drops per unit area decreases (Figure 17.10), resulting in poorer distribution and poorer coverage of leaf and soil surfaces (Table 17.1). There is therefore a need to maintain droplet sizes large enough to minimize drift and yet maintain sufficient numbers of drops to ensure good coverage.

In theory, maximum coverage without drift can be attained in a pattern in which all the droplets have diameters between 200 and 300 μm (Figure 17.11). Unfortunately, such uniform droplet sizes are difficult to achieve because most application equipment produces a wide range of droplet sizes. In fact, most devices that eliminate droplet sizes smaller than 150 μm do so by producing much larger droplets.

A number of factors other than sprayer pressure, nozzle shape, and liquid viscosity influ-

Table 17.2 Effect of Spray Angle and Nozzle Pressure on Average Diameters of Spray Droplets (μm)[a]

Nozzle Type[b]	Spray Angle	Nozzle Pressure		
		15 psi	40 psi	60 psi
4005	40°	900	810	780
6505	65°	600	550	530
8005	80°	530	470	450
11005	110°	410	380	360

[a]Adapted from Spraying Systems Co. Catalog 43A. Spraying Systems Co., P.O. Box 7900, Wheaton, IL 60189-7900.
[b]Flat fan nozzles. All deliver 0.5 gal/min at 40 psi.

ence drift. These include distance of the boom above the target, relative humidity, and wind velocity. Some factors relate specifically to application with aircraft; for example, shear forces on the spray patterns caused by wind, turbulence associated with the aircraft, positioning of nozzles on the aircraft, angle of the aircraft to the target during spray release, and atmospheric stratification.

The height of the spray boom is important because the closer it is to the target the less time the spray droplet will be suspended in the air and therefore susceptible to drift and evaporation. Low relative humidities result in more rapid evaporation of droplets and increased drift, while increased wind velocity increases the distance of drift. One exception is when herbicides are applied aerially to still air during atmospheric inversion conditions. The herbicide droplets will hang suspended at the inversion interface and eventually will move great distances before settling out and causing damage to vegetation. If there is normal air mixing, the fine droplets and vapors can either settle immediately or be carried into the upper atmosphere where the herbicides are dissipated and degraded.

Herbicide sprays released from aircraft at relatively low volumes and high speeds are subject to wind-induced shear forces that increase the breakup of spray streams, cones, and sheets into finer particles. Turbulence caused by helicopter rotors and the tips of fixed-wing aircraft increases the lifting and suspension of fine droplets into the air and increases the number of small droplets.

Reducing Spray Drift

Drift can be reduced with standard spraying equipment by keeping pressures down, using large spray volumes, keeping the boom close to the target, and avoiding wind. Problems with spray drift also can be reduced by making applications at times when particularly sensitive plant species are not present (e.g., in spring before sensitive plant species emerge or leaf out, in fall after sensitive crops mature, or even in winter when no crops are present) and by avoiding applications near overly sensitive crops. It also helps to avoid using herbicides that cause injury at low doses and yet require relatively large doses for weed control.

Besides the techniques and precautions described earlier, a number of special techniques and equipment can result in the outstanding reduction of driftable particles. Every technique may not apply to all situations and herbicide formulations, but it should be possible to find an economical and effective drift control technique for most herbicide applications.

Special drift control techniques include modifications of nozzle/nozzle tip operation and design, alterations in spray consistency or viscosity with additives, specialized low-pressure applicators, shielded sprayers, selective hand placement, and subsurface injection into water and soil.

Modified Nozzle Operation and Design

Most conventional nozzles and nozzle tips (flat fan, even flat fan, double-orifice flat fan, cone-jet, disc-core) are manufactured to operate properly at pressures of 25 to 60 psi. These pressures are too high to reduce the number of fine droplets. Although reducing the pressure can increase the number of large droplets, often the pattern will be distorted if nozzles are operated at pressures of 5 to 15 psi. Thus, lowering the pressure on the system and raising the boom height may not provide the desired results.

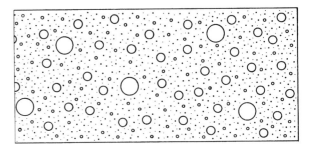

Figure 17.9 A typical spray pattern consists of droplets with diameters ranging from 50 to 1000 µm. The approximate volume mean diameter is 400 µm. Note the greater number of small droplets in relation to large droplets.

Some standard nozzles/nozzle tips maintain their spray patterns reasonably well when operated at lower pressures, even though the delivery volume decreases. These include the extended-range flat fans, some deflector (flood) nozzles, and the swirl (whirl) chambers.

A number of equipment modifications for drift control result in nozzles and nozzle tips that easily replace standard nozzles and nozzle tips. These nozzles and tips are designed to produce large droplets at low pressures while still maintaining desired spray patterns. They include the TeeJet LP (low-pressure) flat fan, the TG Full Cone, the Drift Guard flat fan tips (DG TeeJet), the Turbo FloodJet, and AI TeeJet flat fan tips, all manufactured by Spraying Systems, Inc.; the Raindrop and Raindrop RA nozzles manufactured by Delevan, Inc.; and the TurboDrop injector nozzle manufactured by Greenleaf Technologies.

The TeeJet LP (low-pressure) flat fan is a regular tip machined to provide (at half the pressure) the same volume and overall pattern as the standard flat fan tip does. The other devices, however, are more complex multiple component systems. The TG Full Cone and Drift Guard flat fan tips (DG TeeJet) have inserts or distributors as components of the tips.

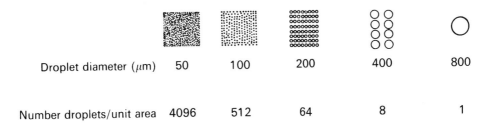

Droplet diameter (µm)	50	100	200	400	800
Number droplets/unit area	4096	512	64	8	1

Figure 17.10 Relationship between spray droplet diameter and number of droplets per unit area. As droplet diameter increases by a factor of two, droplet number decreases by one-eighth.

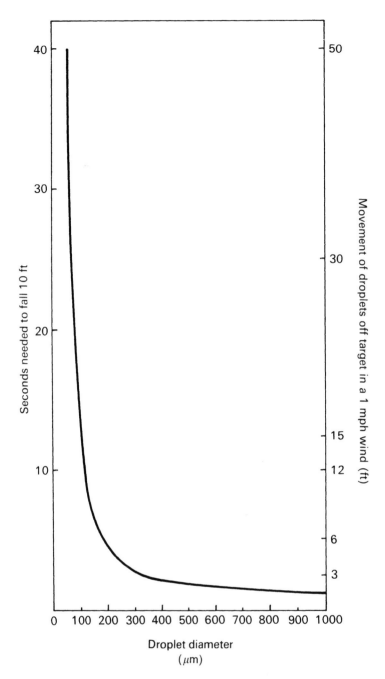

Figure 17.11 Effect of spray droplet diameter on time needed to fall and distance of drift when released 10 ft above target. Note that minimum droplet diameter needed to reduce drift is in the 200 to 300 μm range. Increasing the droplet diameter beyond 300 μm provides little benefit.

The remaining devices have metering and distribution orifices separated by a mixing chamber, into which air enters the nozzle. Air trapped within the liquid theoretically increases the volume of the spray solution, thus providing better coverage. These devices provide a reduction in pressure as the solution passes through the nozzle. Standard sprayer system pressures of 40 to 50 psi are used to deliver the spray solution through a small initial orifice of the nozzle and into a mixing chamber. Air is siphoned (aspirated) into the chamber from the outside through the distribution orifice or through ports in the nozzle tip or nozzle body. The air in the mixing chamber, in conjunction with a large delivery orifice (shaped to give the desired delivery pattern), produces a sizeable pressure drop and results in the delivery of large droplets of spray. Thus, the delivery is metered internally through a small initial orifice, and the spray solution is spread and the pattern determined at the second, larger terminal orifice. This arrangement contrasts with the tips of conventional nozzles, which both meter (determine the amount of solution) and determine the pattern.

The Raindrop double-chambered drift control nozzle consists of a disc-core hollow cone nozzle (a standard type used for foliar applications of insecticides and fungicides) to which a special cap with a much larger orifice has been added (Figure 17.12A). The disc-core nozzle meters the spray solution into the chamber formed by the additional cap. The large outside delivery orifice (rather than ports) allows air to enter the chamber so that the pressure and velocity of the liquid is reduced as it passes through the cap.

A second raindrop nozzle, called the Raindrop RA nozzle, is a modified swirl chamber nozzle with multiple chambers (Figure 17.12B). Both of the Raindrop nozzles can be used with all common sprayable herbicide formulations on aircraft and ground equipment. The large spray droplets are formed and discharged around the edge of the orifice in a hollow cone pattern (Figure 17.12C).

The TurboDrop injector nozzle (Figure 17.12D) works on the same concept as the previous nozzles. Unlike the Raindrops, air enters the mixing chamber through intake ports; the solution is then distributed through standard tips (flat fan, hollow cone, flood) that are larger than the metering oriface. The AI TeeJet is a flat fan tip that also uses ports to mix air into the system before the spray solution leaves the tip.

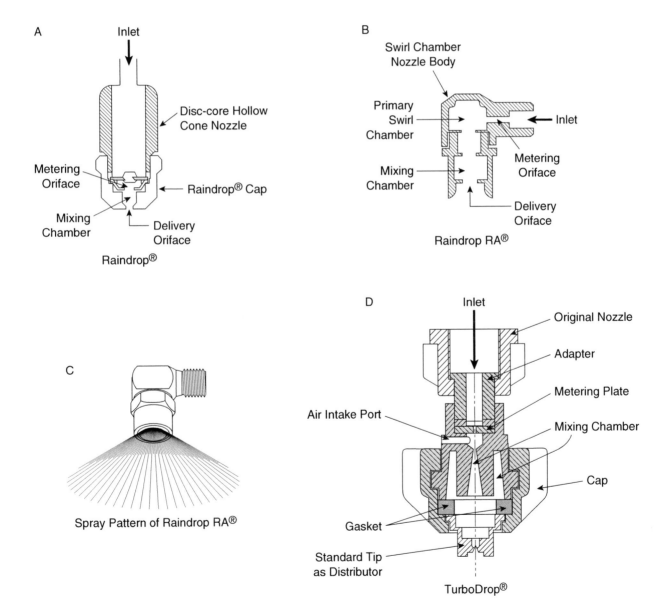

Figure 17.12 Special nozzles to reduce spray drift.

Alteration of Spray Viscosity

Alterations of spray viscosity have been obtained by using invert emulsions, thickening and particulating agents, and polyvinyl polymers. All these materials alter the characteristics of spray solutions and reduce 80 to 90% of the drift caused by small droplets.

Invert emulsions are high-viscosity mixtures of oil, water, and emulsifier (Figure 17.13). Oil, water, and emulsifier also are the components for standard emulsions, which usually consist of 1 part of oil to 10 to 100 parts of water. Standard emulsions have a consistency similar to water and also produce spray patterns similar to water. On the other hand, if the amount of water in the mixture is reduced (for example, 1 part oil in 6 to l0 parts water plus emulsifier) and outside energy is supplied, the resultant emulsion forms a viscous mayonnaise-like mixture in which the water is suspended (or dispersed) in oil. This high-viscosity liquid or invert forms large drops of spray that substantially reduce drift. Unfortunately, inverts are difficult to force through sprayer equipment, and the large droplet size results in relatively poor coverage of the target area (Figure 17.14A). Inverts require special application equipment, increase the spraying time, and are limited to herbicides that translocate well in the plant. The oils used to make invert emulsions are less subject to evaporation than water. As a consequence, drops of invert reach the target with less loss and remain moist on the leaf surface longer than do aqueous sprays.

Invert emulsions can be made by placing the oil, emulsifier, and herbicide together in a single mixing container and vigorously stirring the mixture as the water is added to form the mayonnaise-like solution. The mixture then can be transferred to a suitable sprayer system for application.

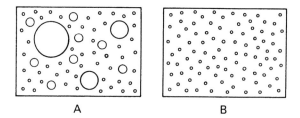

Figure 17.14 (A) Spray from Raindrop nozzles, invert emulsions, or polyvinyl polymers consists of droplets with varying diameters greater than 100 µm and as large as 1600 µm. Drift of fines is eliminated, but foliage coverage is poor. (B) Spray from a CDA is delivered as droplets of relatively uniform diameter of 200 µm. Excellent foliage coverage as well as reduced drift is attained.

A two-liquid system of making invert emulsions eliminates the premixing step. It consists of placing the appropriate components into separate tanks and creating the invert by forcing the two liquid streams together. The energy of the two combining streams creates the invert emulsion. Such bifluid systems can be fitted to either helicopter or ground applicators.

Invert emulsions have been used extensively for herbicide applications to rights-of-way from helicopters. The cost of oil puts invert emulsions at an economic disadvantage, particularly if other drift control methods can be substituted.

Water-thickening additives such as swellable polymers that produce a solution of gelatinous particles were developed and used in the 1970s. They have been discontinued primarily because of cost, special equipment requirements, and handling problems.

Polyvinyl (polyacrylamide) polymers increase the cohesiveness of spray solutions and frequently are added to the spray tank to reduce drift. They change the properties of the liquid in the spray stream in proportion to the amount added. Surfactant must be present in the spray solution for these polymers to function properly. If inadequate amounts of surfactant are present in the herbicide formulation itself, additional surfactant must be added to the sprayer tank.

One advantage that polymers have over invert emulsions and thickeners is that they can be used with conventional sprayers, pumps, and nozzles. The additive, however, may alter the output and delivery pattern so that recalibration and changes in nozzle tips, boom height, or nozzle spacing may be needed to provide uniform patterns.

These polymers may not work with sprayable formulations containing solids, such as wettable powders and flowables.

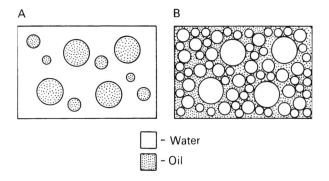

☐ - Water

▨ - Oil

Figure 17.13 (A) Standard emulsion. (B) Invert emulsions result in the suspension of water droplets in oil.

Specialized Low-Pressure Applicators

Various types of specialized low-pressure devices have been evaluated for reducing spray drift. They include solid streams of liquid delivered with spinning or vibrating tubes, rotary atomizers, and airfoil needle and small-diameter tube orifice booms.

Solid Streams Delivered with Tubes in Spinning or Vibrating Devices. Liquid delivered through a tube exits as a stream that breaks into relatively large droplets upon leaving the orifice. Combining the solid stream with spinning or vibrating devices provides additional breakup of the stream and delivers a swath of spray solution on target with a minimum of fine droplets.

The principle upon which this technique is based can be demonstrated using a garden hose with an adjustable bayonet-type nozzle. The change in droplet size can be observed as the angle of discharge changes as the orifice is shut down: the pattern of water flow changes from a solid stream to a cone made up of relatively fine droplets.

Hydraulically fed spinning devices with tubes providing streams of droplets are used for application with ground equipment along rights-of-way and on boats. An example is the Directa-Spray (Figure 17.15). A spinning nozzle system with spokelike tubes radiating from a cylindrical center rotates at speeds of 46 to 70 rpm and is powered by an electric motor. The spray solution is supplied to the nozzle at pressures of 25 to 50 psi produced by a conventional pump. The swath width varies from 15 to 30 ft, depending on the pressure on the system and the delivery arc, which can be varied from 90 to 360 degrees. The device gives relatively good drift control with conventional aqueous spray solutions and also can be used with invert emulsions or thickened spray solutions. This nozzle will handle any of the sprayable formulations.

Vibration can be used in place of the spinning motion to break up solid streams as they exit from multiple orifices. These devices provide results comparable to those obtained with spinning devices.

Other spinning devices have been developed for single-row band application behind a planter unit, for gravity-fed applicators for lawns, and for helicopter applications.

Rotary Atomizers. With this approach, herbicide is metered under pressure on to the center of an open revolving cone or disk. The centrifugally propelled liquid migrates outward along ridges on the upper surface of the cone and is forced off a series of teeth along the outer edge of the cone (Figure 17.16), which causes the spray to be broken into droplets of relatively uniform size. Spray distribution and droplet size are determined by the speed of rotation and the resultant centrifugal force of these machines.

The devices that work on this principle are called controlled droplet applicators (CDA). They are powered with either an electric or a hydraulic motor that is hooked into the power unit, typically a tractor. Each unit has the capability of projecting a spray width up to 78 in. The promotional literature claims that at the low (2000 rpm) operating speed, droplet size should be in the range of 175–250 μm if the flow rate is 0.5 to 1 qt/min (Figure 17.14B). At the high (5000 rpm) operating speed, droplet size is supposed to be 75 to 100 μm in diameter if the flow rate is 4 liquid oz/min. The low operating speed is suggested for soil-applied herbicides and for symplastically

Figure 17.15 A Directa-Spray unit showing spray pattern. (H. A. Holt, Purdue University, West Lafayette, IN)

Figure 17.16 Cone of a controlled droplet applicator. (T. N. Jordan, Purdue University, West Lafayette, IN)

translocated foliage herbicides. The higher operating speed is suggested for contact herbicides, desiccants, fungicides, and insecticides in areas where extra coverage is needed. Difficulty in obtaining proper patterns may be encountered with formulations containing solids.

Another rotary atomizer, the Herbi, is a handheld unit powered with batteries. It consists of a smaller disk that yields a smaller droplet size than the CDA (Figure 17.17).

Equipment that provides maximum coverage for the volume of spray used and without driftable particles would be expected to be widely adopted, but this has not happened. Pattern displacement caused by crosswind frequently is encountered with the uniform droplet applicator. Since all the droplets are the same size, the entire spray swath shifts uniformly rather than having droplets of different sizes moving at varying distances. Skips and overlaps result. Making applications parallel with the wind direction or making applications when there is uniform wind speed helps, but such conditions are not always attainable. Poor spray droplet penetration into or through the crop canopy presents a problem. In addition, the equipment has been expensive and cumbersome to use.

Airfoil Needle and Small-Diameter Tube Orifice Booms. The Microfoil Boom (developed by Amchem, now absorbed by Rhone-Poulenc Co.) illustrates the principle of an airfoil-shaped boom. This device applies conventional aqueous solutions directly from helicopters. The nozzles deliver spray through narrow diameter tubes (or hollow needles) into a zone of minimum air turbulence (Figure 17.18). The resulting streams of spray break into sheets consisting of uniform drops large enough to resist drifting.

Figure 17.18 Spray stream from a Microfoil Boom. (Union Carbide Agricultural Products Co., Research Triangle Park, NC)

Special precautions to keep the spray solutions free from dirt, grit, and gum have to be made. The original microfoil did not handle wettable powder or water-dispersible liquids because the solid residues clogged the nozzles. Even with soluble and emulsifiable sprayables, considerable routine cleaning of the needlelike orifices is needed.

Shielded or Hooded Sprayers

Shielded or hooded sprayers help keep wind away from the spray pattern as it leaves the nozzle (Figure 17.19). Drift is reduced by nearly 50%, and movement of the spray to the target is improved. Although the primary purpose of such equipment is crop safety, reduced release into adjacent nontarget areas also can be expected. Shielded sprayers are used for directed applications in row crops and woody species grown in rows.

Figure 17.17 A Herbi.

Figure 17.19 Sprayer shields reduce herbicide drift to nontarget vegetation. (T. N. Jordan, Purdue University, West Lafayette, IN)

Selective Hand Placement

Selective hand placement of herbicide sprays offers considerable drift reduction opportunities. Most hand applications are spot treatments and result in application to individual plants or patches of plants on a small percentage of the total area. The amount of herbicide solution required in comparison to broadcast spray applications can be reduced proportionally. Some hand-placement operations, such as trunk hatchet injection and root collar injections for controllling unwanted trees, place the herbicide directly in the target plant. Virtually no environmental contamination occurs. Appropriate drift control nozzles, additives, and techniques previously discussed can be used with handheld equipment to yield additional reduction in off-target movement.

Subsurface Injection

Subsurface injection is limited to a few soil treatments and to aquatic situations. When herbicides can be released into the soil or under the water surface and still perform satisfactorily, their drift potential from fines or vapors is virtually eliminated. Mixing herbicides into the soil also minimizes lateral movement of soil-applied herbicides with runoff.

HERBICIDE INCORPORATION TECHNIQUES AND EQUIPMENT

The effectiveness of soil-applied herbicides to control weeds depends on three major conditions: (1) the herbicide must move or be moved into the soil; (2) it must come in contact with weeds that are germinating and emerging, usually from the upper l to 3 in. of soil; and (3) moisture must be present in order for the herbicide to move into the plant. *Herbicides that are applied to a dry soil surface and remain in that position during seed germination and emergence or during the growing stages of established plants do not provide adequate weed control.*

A herbicide moves into the soil in two major ways. One is through movement with water provided either by rainfall or overhead irrigation. The other is by mechanical incorporation (i.e., the mixing of soil-applied herbicides into the soil using mechanical devices). In some parts of the country, the amount and frequency of rainfall usually is sufficient to move the herbicide into the soil. In areas where rainfall cannot be depended on, growers must use methods they can manipulate, such as overhead irrigation or mechanical incorporation.

Rain and Irrigation Incorporation

The amount of water required for rain or irrigation incorporation depends on several soil-, herbicide-, and plant-related factors. Increasing the clay content or organic matter, increasing the depth of developing weeds, increasing the affinity of the herbicide for soil colloid, and decreasing herbicide solubility all increase the amount of water needed to leach the herbicide into the soil. The amount of water required for movement typically ranges from 1/2 to 1 in., depending on the factors mentioned ealier.

Several soil-applied herbicides are labeled for application through irrigation systems. This method is commonly known as herbigation. Advantages of this incorporation technique include the reduction in herbicide application time and not having to run motorized applicators or incorporation equipment over the field. With irrigation, the use of herbicides requiring incorporation is possible theoretically for reduced tillage systems and management systems without tillage. Herbicides can be delivered broadcast and in a uniform manner if the irrigation system itself is capable of delivering uniform amounts of irrigation water over the area to be treated.

The amount of water needed to deliver the herbicide varies somewhat with the herbicide being used and the soil texture of the field, but 1/2 to 1 in. of overhead irrigation generally is recommended. In pressurized systems (center pivot and other sprinkler systems), the herbicide is injected into the irrigation water. This method requires a suitable injection pump that is adjusted to deliver the correct amount of herbicide into the water. If the herbicide is to be diluted before injection, agitation of the solution usually is required. The injection pump should be connected to the system so that it shuts off when the well ceases pumping. The system should be equipped with an anti-siphoning device to prevent herbicide from being drawn into the well if pumping should stop.

Commercial formulations of several thiocarbamate herbicides, alachlor, metolachlor, and some combinations of these are among the herbicides labeled for center pivot irrigation. Some of the thiocarbamate herbicides also are labeled for application with furrow or border irrigation water. In this case, the herbicide is gravity fed into the irrigation water.

Advantages of Mechanical Incorporation

When dependable movement of herbicides can be obtained with rainfall or overhead irrigation, mechanical incorporation may not be required.

Even in areas of relatively high rainfall, however, dry periods coinciding with and following herbicide application to the soil can lead to herbicide failure. Since rainfall cannot be manipulated and timed, mechanical incorporation can present an attractive option. In addition, incorporation can reduce herbicide loss due to photodegradation or volatilization from the soil surface; provide better placement of the herbicide for the control of weeds developing from large, deep-germinating seeds and vegetative propagules; and provide an opportunity for the grower to combine herbicide application with other management operations.

Herbicides that work either as soil surface or incorporated treatments show less variability in performance when incorporated into the soil than when surface applied. Even though the average of the results of the two methods over time is nearly identical, incorporation reduces the extreme fluctuations in weed control observed with surface-applied herbicides. Thus, over a period of years, incorporated herbicides provide more consistent weed control results, whereas surface-applied herbicides, which are dependent on rainfall to move into the soil, can provide excellent results one year and poor results the next.

In general, soil-applied herbicides can be grouped according to their suitability for mechanical incorporation: (1) herbicides that work reasonably well whether left on the soil surface (providing precipitation is adequate) or incorporated; (2) herbicides that provide inadequate weed control or result in excessive crop injury when incorporated into the soil; and (3) herbicides that *must* be mechanically incorporated, either to prevent their loss from the soil surface or because they require special placement for consistent weed control performance.

The triazine, chloracetamide, imidazolinone, and sulfonylurea herbicides are in the first group and can be used either as surface-applied or incorporated herbicides. The second group of compounds includes those that bind tightly to soil colloids and lose their activity when mixed into soil. Examples are linuron and oxyfluorfen. In contrast, propachlor, naptalam, and dicamba are subject to excessive leaching when incorporated into the soil. They can move out of the zone where the weeds are germinating and into the crop root zone, where they can cause crop damage.

Those herbicides requiring incorporation because they are subject to loss from the soil surface include the thiocarbamates (lost through volatility), most of the dinitroanilines (lost through volatility and photodegradation), and clomazone (tends to cause vapor injury to nearby vegetation). Mechanical incorporation also is helpful for the control of plants developing from deep-germinating seeds, tubers, rhizomes, or roots. Herbicides used for this purpose include chloroacetamides on yellow nutsedge, dinitroanilines and thiocarbamates on rhizome johnsongrass, and triazines on bur cucumber.

One of the major incentives for incorporation is the opportunity for the grower to combine herbicide application with other over-the-field operations. For example, insecticides or fertilizers can be applied after primary tillage and before planting. If these chemicals require incorporation (and they frequently do), the herbicide can be mixed with one or both of the materials. All the components are then applied together and mixed into the soil with the same tillage operation. Extra trips to either apply or incorporate the herbicide thus are eliminated. Much of the initial use of preplant incorporated atrazine in the Corn Belt resulted because chemical dealers provided atrazine at or near dealer cost in liquid nitrogen solutions with no additional charge for the herbicide application.

Another incorporation option is to apply the herbicide during a secondary tillage operation rather than as a separate application following planting.

When using a persistent herbicide such as trifluralin, growers can apply and incorporate the herbicide at their convenience, even several weeks before planting, rather than at the critical and hectic planting period when there are many other problems to worry about and time is precious.

Disadvantages of Mechanical Incorporation

Disadvantages associated with the incorporation of soil-applied herbicides include the purchase and maintenance of additional equipment, extra trips over the field, loss of critical time during planting season, loss of soil moisture, soil compaction on wet soils, limited utility in reduced-tillage systems, and instances of increased crop injury.

If equipment is used primarily for herbicide incorporation, the purchase price, storage, and maintenance of the equipment represent added production costs to the farmer. In most cases the purchase price is several thousand dollars.

One or two extra trips over the field usually are required to provide adequate incorporation. Theoretically, in a conventional tillage sequence it is feasible to make one pass with a primary tillage implement, one pass with a secondary tillage tool, and then plant the crop. This amount of tillage not only is inadequate for soil herbicide incorporation but usually is considered the minimum starting point for the incorporation process. Thus, at least one and sometimes two additional trips over the field are needed. Extra trips, however, are not particularly unusual even in standard tillage sequences. In those programs in which secondary tillage is essential for reducing weed populations before planting, a second or even a third pass may be required to ensure that weeds unearthed by the first pass do not become reestablished in the soil.

Extra trips for herbicide incorporation may compete with time needed for timely planting. This loss can be extremely important in areas where planting is limited to those relatively few days when the soil is suitable for working and planting.

Soil moisture is lost when the soil is tilled. One or two secondary tillage passes during herbicide incorporation can be expected to dry out the soil to below the moisture level required for seed germination. Thus, crop seed germination will be delayed until moisture can be replenished.

Soil compaction can occur if incorporation is attempted on wet soils. If soils are sufficiently dry for proper incorporation, compaction should not be a major problem. Unfortunately, incorporation on soils that are too wet frequently is attempted and results in both disappointing weed control performance and damage to the soil.

Crop injury may increase when selectivity depends on keeping the herbicide above the root zone of the developing crop. For example, pendimethalin can be used effectively preemergence but not preplant incorporated in corn.

Mechanical mixing to incorporate soil-applied herbicides generally is incompatible with cropping systems in which secondary tillage is limited to one or two passes (except for powered tillers or combination tillage tools) or in which secondary tillage is not included in the system (no-till).

Incorporation Techniques

The number of passes needed for herbicide incorporation is an important consideration for the grower, particularly since the number may not be the same as the number needed for secondary tillage. Herbicide incorporation can be achieved dependably only after the seedbed has been sufficiently smoothed and chunks have been broken down. At least one pass with a secondary tillage implement following primary tillage usually is required *before* mechanical incorporation is initi-

ated because primary tillage usually leaves the soil surface in relatively large chunks and almost always leaves it uneven. The normal sequence for herbicide incorporation is primary tillage first, followed by one pass with a secondary tillage tool for leveling and pulverization, herbicide application to the soil surface, and then two passes with an incorporation implement. Some types of equipment such as power-driven rototillers, the Roterra, and certain bed conditioners will incorporate the herbicide in a single pass.

The type of equipment and practices for secondary tillage normally used by the grower will dictate whether additional trips over the field will be required for incorporation. If only one or two secondary tillage operations normally are used, then incorporation probably will require extra trips. If three normally are used, then the herbicide probably can be applied and incorporated during the normal operations.

The condition of the soil at the time of herbicide application and incorporation frequently is more important than the choice of implement used for incorporation. *Herbicides applied to cloddy, wet, or uneven soils; to heavy, unevenly distributed residues; or to extensive weed growth should not be expected to provide effective weed control even when mechanically incorporated with the prescribed number of tillage operations.* Incorporation equipment usually does not work well under such conditions. In addition, the herbicide will be unevenly distributed and concentrated in hot spots or streaks within the field (Figure 17.20).

Weather conditions following application may have tremendous influence on an incorporated herbicide. If the herbicide is incorporated into dry soil and no subsequent moisture is received, poor performance may result. Even well-placed, mechanically incorporated herbicides require some moisture to work. In contrast, rainfall can compensate for inadequate incorporation, provided the herbicide has not been lost, streaked, or placed too deeply.

Implements Used for Incorporation

Many types of tillage equipment are used for herbicide incorporation. Detailed descriptions of this equipment are provided in Chapter 18. Following is a discussion of the use of these implements as incorporation devices.

The tillage equipment used for soil incorporation of herbicides includes power-driven tillers, disk harrows, rolling spoked wheeled implements, field cultivators, sweeps and blades, and tools that combine two or more of these components.

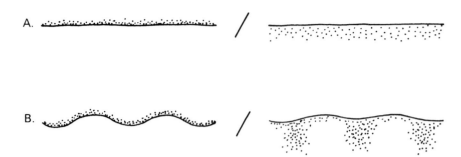

Before incorporation   After incorporation

A.

B.

Figure 17.20   Application and incorporation of a herbicide to uneven soils will result in areas with too much herbicide and others with too little. (A) Even soil surface. (B) Uneven soil surface.

Power-Driven Tillers

Equipment powered directly off the tractor includes the rototiller (rotovator), the Roterra, and power-driven harrows such as the Vicon. The rototiller provides precision incorporation over a depth of several inches and gives the most flexibility of the available tools. When equipped with L-shaped knives, one pass with the rototiller provides uniform mixing to the depth of cut. When equipped with spikes, this type of tiller provides uniform but more shallow incorporation. It may leave the soil in better condition than the rototiller with knives.

Power-driven vertical-tine implements such as the Roterra and power-driven harrows provide a horizontal stirring action and give uniform but shallow incorporation. If the herbicide is applied to the soil prior to the passage of the Roterra, incorporation less than 1 in. deep can be expected. If the spray is directed into the rotating teeth (and flowing soil), incorporation to 2 in. deep can be obtained. Other implements with similar action (such as the Vicon) should provide similar results.

The relatively slow speed, high purchase costs, and high power requirements of the rototiller and vertical-tine implements have limited their acceptance for large acreages of commercial crop production.

Disks

The disk (tandem disk harrow) is a commonly available and employed implement for incorporating soil applied herbicides. Disks normally mix the herbicide to approximately one-half the depth of cut. A disk running 5 in. deep therefore would be expected to incorporate herbicide to a depth of 2 1/2 in. Making two passes at different angles is suggested almost universally. One pass with a disk usually will distribute the chemical in hot spots. The second pass provides a redistribution and evening out of the pattern. Attaching a harrow behind the disk aids in the incorporation process. Finishing disks with small-diameter blades (20–22 in.) and 7-in. spacings between the blades are the best choice. Those with 9-in. spacings can be made to work on coarse- and medium-textured soils if properly operated. Those implements with 11-in. spacings between disk blades (e.g., offset disks) are not suitable for incorporation; they are more of a primary tillage tool.

One attempt at eliminating an extra tillage pass for herbicide incorporation has been to place the spray nozzles between the front and rear gangs on the tandem disks so that the front gang levels and pulverizes (acts as the first secondary tillage operation) while the back gang mixes the herbicides into the soil. All of this is accomplished on the first pass. A second pass completes the incorporation procedure.

Field Cultivators

Field cultivators also are used to provide incorporation. These implements vary in design from rigid shank models with large sweeps to those with flexible Danish tines. Field cultivators used for incorporation usually consist of two or three rows of spring steel shanks with spacings of 6 to 9 in. with sweeps attached. Incorporation to 3 in. can be achieved with these implements if the equipment is properly operated. Two passes at different angles ensure success, as does a relatively high speed (6 mph) of operation. The field cultivator should be operated with the front and back level to prevent the back shovels from running deeper and causing streaking.

Growers have exhibited a great deal of interest in recent years in one-pass incorporation (also referred to as surface blending, stroking in, or limited-tillage incorporation). In most cases, the equipment suggested is of the field cultivator type with one or two additional secondary tillage implements either attached as an integral portion of the unit or pulled behind it. Soil conditions, however, can have more impact on success than the incorporation device. The soil must be friable and dry enough to flow well through the tillage implements. If improper soil conditions are encountered during incorporation, results can be disappointing. More failures can be expected using one-pass rather than two-pass incorporation.

Rolling Cultivators and Treaders

Mulch treaders and rolling cultivators are implements with curved spoked wheels that run at an angle to the direction of travel and push soil sideways like a disk blade. There are open spaces in the wheel, however, so inversion is not as thorough and soil penetration is not as deep as with a disk blade. These implements provide incorporation to a depth of 1 1/2 to 2 in. and work well on coarse- and medium-textured soils. Most herbicide labels suggest two passes. The rolling cultivator can be adjusted to provide incorporation on a raised bed. The mulch treader has two gangs of wheels on a single axle similar to a tandem disk. Considerable firming of the seedbed can be obtained with the mulch treader.

Rotary Hoe

As an incorporation device, the rotary hoe is limited to shallow mixing. It can be used after planting a large-seeded crop. The rotary hoe generally is considered inadequate for herbicides that require mixing to depths of 1/2 in. or more. On the other hand, it can be used without significant dilution and loss of herbicide activity and may help decrease injury from layers of herbicide concentrated on the soil surface by dispersing them into the upper 1/4 to 1/2 in. of soil. The rotary hoe can be used for weeding or breaking of the soil crust without significant reduction of herbicide performance.

Combination Seedbed Conditioners

Bed conditioners are implements with more rows of tools than a field cultivator so that the amount of stirring provided approaches that of two passes with a standard field cultivator. A one-pass herbicide incorporation can be obtained with these implements.

Some units consist of the components of three or more individual tillage tools coupled into one machine. Often called do-alls, these machines consist of either field cultivator or spring tooth harrow teeth preceded by a disk gang or followed by a chopping reel and several sets of harrow teeth. A smoothing or leveling device follows behind.

Subsurface Layering Devices

Blades or sweeps are used for subsurface layering of herbicides into the soil. Layering can be used effectively with highly volatile herbicides such as the thiocarbamates. These herbicides are layered 3 to 4 in. below the soil surface and then, because of their volatility, diffuse upward into the germinating weed zone, where they can kill the seedlings. In other cases, the herbicide is layered into the soil but above the sprouting plant or plant part and acts as a contact barrier by preventing weed emergence. Herbicides in this category include trifluralin for field bindweed control and dichlobenil for perennial weed control in orchards. Subsurface layering can be accomplished by fitting a straight blade or individual sweeps with nozzles that are directed to spray the soil layer that has been exposed by the cut. Another type of sweep developed by agricultural scientists at the University of Nebraska includes upward-projecting, pressurized, close-spaced spray streams to provide hydraulic incorporation into the soil above the blade. This device is useful for mixing herbicide into upper soil under crop residues that must be maintained on the soil surface in reduced tillage systems such as the stubble mulch tillage system.

EQUIPMENT CALIBRATION AND ASSOCIATED PROCEDURES

Accurate herbicide application at the proper dosage is a critical aspect of chemical weed control technology. The two major steps prior to the actual application of the herbicide are:

1. Determination of sprayer delivery, technically referred to as calibration.
2. Determination of the prescribed dosage and the amount of herbicide formulation to add to the sprayer tank to provide the prescribed dosage.

Step 1: Sprayer Calibration

Sprayer calibration procedures include preparing the sprayer for proper use and then calibrating or calculating the delivery rate of the sprayer under operating conditions.

Preparing the Sprayer

A precalibration sprayer check is needed before the calibration process is initiated to ensure that the sprayer components operate properly.

Flush and clean the sprayer to remove the dirt, flakes, and grit that inevitably accumulate. Then add clean water to the tank and check individual components for proper functioning. Specifically, check the pump, hoses, pressure regulator, pressure gauges, shutoff valves, agitator, strainers, and nozzle tips. Pressure gauges and nozzle tips require extra attention because they are especially susceptible to malfunction. Repair or replace malfunctioning parts.

Select and install correct nozzles or nozzle tips for the spraying job. Clean all strainers and then check the nozzles for delivery rates, pattern, and angle. All of these factors should be the same for each nozzle on the boom. Replace defective, inaccurate, or mismatched nozzle components.

Calibrating the Sprayer

The delivery rate of the sprayer should be determined under normal operating conditions. The desired constant speed, constant pressure, herbicide carrier, type of terrain to be sprayed, and effect of other equipment associated with the application on the speed of the power unit must be established. Differing soil surfaces, slip-page, attached equipment, and slope variation all can result in a change in travel speed and can alter sprayer delivery.

The factors that determine the volume of spray delivered per unit area include travel speed (once established, travel speed should be held constant); swath (effective boom) width; volume of spray solution through the nozzles as determined by pressure, orifice size, number of nozzles, and nozzle spacing; and viscosity of the spray solution. Viscosity becomes a major consideration when fertilizer solutions or oils are used as carrier solutions or when drift-control additives are employed. Changing any of these factors alters sprayer delivery and creates the need for recalibration.

The methods for calibrating sprayers are numerous and varied. Properly executed, all are correct. We think the most practical approach to calibration is to divide the component processes into simple steps, assign units to each component, and then make the appropriate calculations.

We also think that you should learn calibration from scratch so you obtain a thorough understanding of the factors and computations. Once you have mastered these concepts, understanding other calibration methods will be easy. There are two basic methods of calibration.

Calibration Based on Spraying a Measured Area. An area is sprayed at a constant operating speed and pressure. The area sprayed is determined by multiplying the swath (effective boom) width by the distance traveled (Figure 17.21A) during the calibration run. This figure is divided by the number of square feet in an acre (43,560) to yield the total number of acres sprayed during

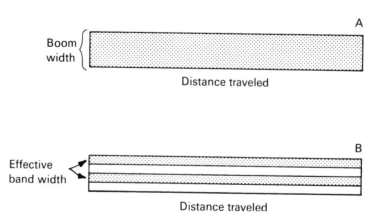

Figure 17.21 Area sprayed is determined by multiplying the effective swath width by the distance traveled. With broadcast applications, the effective swath width is the full width of the boom (A); with band applications, it is that fraction of the width that is actually treated (B).

the calibration run. The amount of spray solution used to spray that area is then determined by refilling or taking measurements in the tank. Combining this figure with the area sprayed results in the delivery rate (gal/acre). The equations for making these calculations are as follows:

(17.1) swath width (ft) $\times \dfrac{\text{ft}}{\text{calibration run}}$

$\times \dfrac{1 \text{ acre}}{43,560 \text{ ft}^2} = \text{acres/calibration run}$

(17.2) The amount of spray solution used during the calibration run is measured and equals gal/calibration run.

Equations (17.1) and (17.2) are combined to give the number of gallons that are sprayed per acre.

(17.3) $\dfrac{1 \text{ calibration run}}{\text{acres}} \times \dfrac{\text{gal}}{\text{calibration run}}$

$= \text{gal/acre}$

A critical step in this procedure is determining the amount of spray solution used. Measuring the spray solution in the sprayer tank often is the major source of error in calibration and mixing. If possible, the sprayer should be parked in the same spot when the tank is initially filled and after the calibration run has been made. The tank can then be refilled with a volumetric measure to determine the amount of solution used. Graduated scales attached to the sprayer or portable rulers may be more convenient, but they must be accurately marked to be of value. Care should be taken to ensure that the sprayer is level when volume readings are taken.

Calibration Based on Speed and Nozzle Output. In this method, the volume of nozzle output per unit time (gal/h or gal/min) is combined with the area traveled per unit time (acres/h or acres/min) to determine sprayer delivery. This is done in three basic steps:

First, nozzle output per unit time (gal/h) is determined by catching the spray solution from a nozzle or nozzles over a given period of time.

(17.4) number of nozzles $\times \dfrac{\text{gal}}{\text{nozzle min}}$

$\times \dfrac{60 \text{ min}}{\text{h}} = \text{gal/h}$

Second, the area traveled per unit time (acres/h) is calculated by determining the spray vehicle speed (miles/h) and the swath width.

(17.5) $\dfrac{\text{mi}}{\text{h}} \times \dfrac{5280 \text{ ft}}{\text{mi}} \times$ swath width (ft)

$\times \dfrac{1 \text{ acre}}{43,560 \text{ ft}^2} = \text{acres/h}$

Finally, the nozzle output per unit time and the area per unit time are combined to yield the delivery rate in gal/acre.

(17.6) $\dfrac{\text{gal/h}}{\text{acres/h}} = \text{gal/acre}$

Another calculation that is essential in eventually determining the amount of herbicide to add to the spray tank is the acres that can be sprayed with one tank load. Two pieces of information are required: the volume of the spray tank (gal/tank) and the gallons sprayed per acre as determined during the calibration run (from equation [17.3] or [17.6]).

(17. 7) $\dfrac{\text{gal}}{\text{tank}} \times \dfrac{\text{acres}}{\text{gal}} = \text{acre/tank}$

Step 2: Determination of Dosage and Amount of Herbicide to Add to the Sprayer Tank

Determining the prescribed dosage of a herbicide depends on diagnosing the situation properly (weeds, soils, crop, growth stage, temperature, relative humidity, etc.) and carefully matching the situation with suggestions from official recommendations and product label information. Prescribed dosages are given in lb/acre, kg/ha, oz/1000 ft^2, qt/acre, l/ha, or other appropriate units of mass or volume per unit of area.

Suggested dosages may be expressed either as the amount of formulation or as the amount of active ingredient per unit area. Herbicide product labels express dosages as amount of formulation (herbicide preparation as it comes from the package). State and federal agencies and research reports, however, frequently express dosages in terms of active ingredient. The purchaser or applicator needs to be able to convert the amount of active ingredient to the amount of formulation.

A single herbicide may be available in two or more different formulations. For example, atrazine is available as a 90% water-dispersible granule (90WDG) and a 4 lb/gal water-dispersible liquid suspension (4L). A prescribed dosage from state university recommendations may be given as 1.5 lb active ingredient (a.i.) of atrazine per acre. One-and-one-half pounds of active ingredient are contained in 1.67 lb of 90WDG and 1.5 qt of 4L. How are these calculations made?

For a dry formulation, convert the percentage of the active ingredient in the formulation to a whole number and divide into the pounds of active ingredient needed per acre.

Example: 1.5 lb/acre of active ingredient is prescribed. The available formulation is a 90% WDG (i.e., 1 lb of formulation contains 0.90 lb a.i.).

(17. 8) $\dfrac{1 \text{ lb formulation}}{0.90 \text{ lb a.i.}} \times \dfrac{1.5 \text{ lb a.i.}}{\text{acre}}$
= 1.67 lb formulation/acre

Therefore, the dosage in terms of formulation to be applied is 1.67 lb formulation/acre. If the amount of formulation is to be expressed in grams, use the following conversion: lb $\times$ 454 g/lb = g.

For a liquid formulation, similar calculations are made except that the active ingredient per gallon is inserted in the equation.

Example: 1.5 lb/acre of active ingredient is prescribed. A 4 lb a.i./gal formulation is available.

(17.9) $\dfrac{1 \text{ gal formulation}}{4 \text{ lb a.i.}} \times \dfrac{1.5 \text{ lb a.i.}}{\text{acre}}$
= 0.375 gal formulation/acre

Gallons can easily be converted to other units. For example, 0.375 gal is equal to

Quarts: 0.375 gal $\times$ 4 qts/gal = 1.5 qt

Pints: 0.375 gal $\times$ 8 pts/gal = 3 pt

Liquid ounces: 0.375 gal $\times$ 128 oz/gal = 48 oz

Liters: 0.375 gal $\times$ 3.785 *l*/gal = 1.42 *l*

The amount of herbicide formulation to be added to the spray tank is determined by multiplying the acres/tank (from equation [17.7]) with the formulation/acre (from equation [17.8] or [17.9]).

(17. 10) $\dfrac{\text{acres}}{\text{tank}} \times \dfrac{\text{amount formulation}}{\text{acre}}$
= amount formulation/tank

Example Problems

Broadcast Application (Problem 1). A sprayer is equipped with a 1000-gal tank and a 40-ft boom. You wish to apply 1.5 lb of trifluralin/acre. The formulation to be used is Treflan EC (a 4-lb/gal emulsifiable). In a calibration run of 544.5 ft, 12.5 gal of solution are applied. Determine the amount of Treflan EC needed for each tank load.

Step 1 (equation [17.1]). Sprayer calibration (based on spraying a measured area)

40 ft swath $\times \dfrac{544.5 \text{ ft}}{\text{calibration run}} \times \dfrac{1 \text{ acre}}{43,560 \text{ ft}^2}$
= 0.5 acres/calibration run

Step 2 (equation [17.2]). The amount of spray solution used during the calibration run equals 12.5 gal/calibration run.

Step 3 (equation [17.3]).

$\dfrac{1 \text{ calibration run}}{0.5 \text{ acres}} \times \dfrac{12.5 \text{ gal}}{\text{calibration run}} = 25 \text{ gal/acre}$

Step 4 (equation [17.7]).

$\dfrac{1000 \text{ gal}}{\text{tank}} \times \dfrac{1 \text{ acre}}{25 \text{ gal}} = 40 \text{ acres/tank}$

Step 5 (equation [17.9]).

$\dfrac{1 \text{ gal Treflan EC}}{4 \text{ lb a.i.}} \times \dfrac{1.5 \text{ lb a.i.}}{\text{acre}}$
= 0.375 gal Treflan EC/acre

Step 6 (equation [17.10]).

$\dfrac{40 \text{ acres}}{\text{tank}} \times \dfrac{0.375 \text{ gal Treflan EC}}{\text{acre}}$
= 15 gal Treflan EC/tank

NOTE: Steps 5 and 6 can be combined easily and are done so in the following problems.

$\dfrac{1 \text{ gal Treflan EC}}{4 \text{ lb a.i.}} \times \dfrac{1.5 \text{ lb a.i.}}{\text{acre}} \times \dfrac{40 \text{ acres}}{\text{tank}}$
= 15 gal Treflan EC/tank

Broadcast Application (Problem 2). A broadcast sprayer is equipped with a 1500-gal tank and a 33-ft boom. Twenty nozzles deliver 0.6 gal/min each. The herbicide applications are to be made at 7.5 mph. Sulfometuron will be applied at a dose of 3 oz a.i./acre. The available formulation is Oust (75% WDG).

Step 1 (equation [17.4]). Sprayer calibration (based on speed and nozzle output)

20 nozzles $\times \dfrac{0.6}{\text{nozzle min}} \times \dfrac{60 \text{ min}}{\text{h}} = 720 \text{ gal/h}$

Step 2 (equation [17.5]).

$\dfrac{7.5 \text{ mi}}{\text{h}} \times \dfrac{5280 \text{ ft}}{\text{mi}} \times 33 \text{ ft} \times \dfrac{1 \text{ acre}}{43,560 \text{ ft}^2}$
= 30 acres/h

Step 3 (equation [17.6]).

$\dfrac{720 \text{ gal/h}}{30 \text{ acres/h}} = 24 \text{ gal/acre}$

Step 4 (equation [17.7]).

$\dfrac{1500 \text{ gal}}{\text{tank}} \times \dfrac{1 \text{ acre}}{24 \text{ gal}} = 62.5 \text{ acres/tank}$

Step 5 (combined equations [17.8] and [17.10]).

$$\frac{1 \text{ oz Oust}}{0.75 \text{ oz a.i.}} \times \frac{3 \text{ oz a.i.}}{\text{acre}} \times \frac{62.5 \text{ acres}}{\text{tank}}$$
$$= 250 \text{ oz Oust/tank}$$

Band Application. An eight-row planter set on 30-in. rows is equipped to spray a 15-in. band over each row. The recommendations prescribe a broadcast dose of metribuzin at 0.3 lb a.i./acre. Lexone 75DF is the source of metribuzin. In a 544.5 ft calibration pass, 4 gal of solution are used. How much Lexone 75DF do you add to the 500-gal spray tank to deliver the correct metribuzin dosage?

Effective swath width must first be determined (*Note:* Only a fraction of the planter width is treated; see Figure 17.21B). Therefore,

$$8 \text{ rows planted} \times \frac{30 \text{ in}}{\text{row}} \times \frac{15 \text{ in. sprayed}}{30 \text{ in. planted}} \times \frac{1 \text{ ft}}{12 \text{ in.}}$$
$$= 10 \text{ ft sprayed} = \text{effective swath width}$$

Step 1. Sprayer calibration (based on spraying a measured area)

$$10 \text{ ft} \times \frac{544.5 \text{ ft}}{\text{calibration run}} \times \frac{1 \text{ acre}}{43,560 \text{ ft}^2}$$
$$= 0.125 \text{ acres/calibration run}$$

Step 2. The amount of spray solution used during the calibration run is measured and equals 4 gal/calibration run.

Step 3.

$$\frac{1 \text{ calibration run}}{0.125 \text{ acres}} \times \frac{4 \text{ gal}}{\text{calibration run}} = 32 \text{ gal/acre}$$

Step 4.

$$\frac{500 \text{ gal}}{\text{tank}} \times \frac{\text{acre}}{32 \text{ gal}} = 15.6 \text{ acres/tank}$$

Step 5.

$$\frac{1 \text{ lb Lexone}}{0.75 \text{ lb a.i.}} \times \frac{0.3 \text{ lb a.i.}}{\text{acre}} \times \frac{15.6 \text{ acres}}{\text{tank}}$$
$$= 6.25 \text{ lb Lexone 75DF/tank}$$

Granular Application Equipment and Calibration

Granular formulations are used directly from the package and are applied with granular applicators. Calibration of granular applicators consists of calculating the output of the applicator and adjusting the delivery until the desired amount of herbicide is applied.

Gravity flow applicators, in which the granules drop from the applicator (Figure 17.22),

Figure 17.22 Broadcast spinning applicator. Granules drop into spinning plate (arrow), which spreads the granules.

and pneumatic equipment (Figure 17.23), in which the granules are propelled by air, are available for the application of dry fertilizers and granular herbicides.

There are three typical designs of gravity flow granular applicators. Broadcast applicators have openings along the bottom of the hopper and drop the granules directly onto the soil. Broadcast spinning applicators (Figure 17.22) drop the granules onto a spinning plate that spreads the granules in a swath. Other applicators drop the granules into feeder tubers that are attached to spreaders. Applicators with tubes and spreaders are used for band application but also can be adapted for broadcast application when proper spreaders are added. Applicators used for band application have an individual unit (hopper, feeder tube, and spreader) for each crop row attached behind each planter unit.

The parts of a gravity flow granular applicator include an agitator, a feeder gate, a hopper, an on–off lever, and a feeder gate control. Some have feeder tubes and spreaders as well. The on–off lever opens and closes the feeder gate to start and stop the flow of material. The feeder gate control is adjusted to yield different feeder gate openings and thus determine the rate of flow.

Before calibrating a granular applicator, remove rust and corrosion on the feeder plates, agitator, and other working parts. Check nuts, bolts, and other connections for tightness. Check feeder tubes and spreaders for obstructions, kinks, or corrosion that could interfere with delivery. Repair, clean, and lubricate equipment as needed.

Factors affecting the delivery rates of gravity flow granular applicators include size of the feeder gate opening, ground speed of the applicator, speed of the hopper agitator, nature and size of the granules, roughness of the ground,

Figure 17.23 Pneumatic delivery system for dry fertilizers and granular herbicides. The granules are distributed evenly across the entire width of the applicator. (Top) Close-up of fan system and distribution tubes. (Bottom) Full width view of distribution units.

and relative humidity. Granules vary enough in density, size, and flow characteristics that equipment must be calibrated for each granular herbicide formulation used. The variability among granular formulations makes it inadvisable to mix different granular formulations in a single application.

Pneumatic delivery applicators (Figure 17.23) have fans that propel granular particles into a series of dividers, which results in the dispersion of the granules into individual tubes and spread-

ers. The distribution of the granules tends to be more uniform across the total width of the swath than when distributed with a gravity flow applicator with a spinning plate.

Calibrating Granular Applicators for Broadcast Application

1. Determine the required dosage for the formulation from the herbicide label.
2. Set the openings to apply the desired dosage as suggested in the operator's manual. The

granular applicator then should be calibrated under operating conditions to ensure that it is actually delivering the expected rate of herbicide.

3. Mark the hopper at a given level and fill to the mark.
4. Apply the granules to a measured area under the conditions and at the speed to be used during application.
5. Weigh the amount of herbicide granules required to refill the hopper to the mark or determine the amount applied by capturing and measuring the granules as they come out of the machine during the calibration run.
6. Determine the delivery rate of the applicator using the following formulas:

> (17.11)   Width of the application × length of calibration run = area covered

> (17.12) $\dfrac{\text{Amount applied during calibration run}}{\text{Area covered during calibration run}}$
> = Area covered during calibration run

Area covered during calibration run

7. If the applicator has individual boxes, each box must be calibrated to ensure uniform delivery. Variation between boxes is a major source of application error.

Example Problem

A 10G granular formulation is to be applied in a broadcast application at a rate of 30 lb/acre. The granular applicator applies a 10-ft swath. In a calibration run of 1089 ft, 10 lb of 10G are applied. The applicator is adjusted to apply 30 lb/acre according to the manual. What is the actual delivery rate of the applicator and how should it be adjusted if it does not match the expected rate?

Step 1 (equation [17.11]).

$$10 \text{ ft swath} \times \frac{1089 \text{ ft}}{\text{calibration run}} \times \frac{1 \text{ acre}}{43,560 \text{ ft}^2}$$
$$= 0.25 \text{ acres/calibration run}$$

Step 2 (equation [17.12])

$$\frac{10 \text{ lb/calibration run}}{0.25 \text{ acre/calibration run}} = 40 \text{ lb/acre}$$

The actual delivery rate of the granular applicator is 40 lb/acre rather than the desired 30 lb/acre. Therefore, an adjustment must be made to reduce the amount of herbicide applied per acre. This adjustment can be made by closing the feeder gate opening to provide three-fourths of the delivery or, if conditions permit, by increasing the speed of the applicator, for example, from 4 to 5.3 mph. *The applicator must then be recalibrated* to ensure that the adjustments are correct.

In the second run of 1089 ft, 7.5 lb of 10G are applied.

$$\frac{7.5 \text{ lb/calibration run}}{0.25 \text{ acre/calibration run}} = 30 \text{ lb/acre}$$

Calibrating Granular Applicators for Band Application

Equipment for band application of granular materials can be calibrated exactly the same way as for broadcast applications. The band width and length of travel are used to calculate a broadcast dosage.

Example Problem

A planter has six planter units, each with an attached granular applicator unit set up for a 12-in. band application. The effective swath width of each applicator is thus 1 ft. The units are spaced 36 in. apart. Each applicator unit will hold 45 lb of granular herbicide. Each applicator unit applies 1.2 lb of formulated granule in a 1760-ft calibration run. What is the delivery rate of each applicator unit based on (1) the banded area only and (2) the total planted area?

Step 1 (equation [17.11]).

$$1 \text{ ft swath} \times \frac{1760 \text{ ft}}{\text{calibration run}} \times \frac{1 \text{ acre}}{43,560 \text{ ft}^2}$$
$$= 0.04 \text{ acres/calibration run}$$

Step 2 (equation [17.12]).

$$\frac{1.2 \text{ lb/calibration run}}{0.04 \text{ acre/calibration run}}$$
$$= 30 \text{ lb/acre of banded area}$$

Since the swath width is 12/36 or 1/3 of the total row width, the delivery rate per acre planted is

$$1/3 \times 30 \text{ lb} = 10 \text{ lb/acre of planted area}$$

If all six hoppers in the previous problem are filled, how many acres can be planted before the hoppers must be refilled?

$$\frac{45 \text{ lb}}{\text{box}} \times 6 \text{ boxes} = 270 \text{ lbs in all 6 boxes}$$

$$\frac{270 \text{ lb}}{10 \text{ lb/acre}} = 27 \text{ acres}$$

Cleaning the Sprayer

Herbicide residues in sprayers can result in considerable damage when these pieces of equipment are used on sensitive crops. Although not cleaning the sprayer prior to soil application can create problems, injury from low levels of residues is more likely to appear with postemergence treatments. In many cases, flushing residues from the tank followed by a thorough washing is adequate. When auxin-type growth regulator (phenoxy, benzoic, or picolinic acid derivatives) or sulfonylurea herbicides have been used, an extra thorough cleaning should be done before making foliar applications to sensitive crops.

Water-soluble herbicides and wettable powders are usually removed by thorough washing with detergent and water or ammonia and water. Oil-soluble herbicides may require washing with oil or kerosene followed by detergent and water or ammonia and water. Oil-soluble herbicides such as the phenoxy esters are absorbed into rubber hoses and even thorough washing may not completely remove these residues from the sprayer. Changing hoses will solve this problem. Consult directions for sprayer cleanup on individual herbicide labels.

MAJOR CONCEPTS IN THIS CHAPTER

Herbicides must be placed properly in relation to the crop and the weed and also must be accurately and evenly distributed.

Improper application accounts for a large share of problems encountered with herbicides.

Problems with herbicides include injury to off-target vegetation, nonperformance, crop injury, and carryover to subsequent crops.

The most frequently employed device for herbicide application is the sprayer. Sprayers must be properly maintained, calibrated, and operated to avoid herbicide-related problems.

Commonly used pumps include centrifugal, turbine, roller, piston, and diaphragm pumps. Centrifugal, turbine, diaphragm, and roller pumps normally are operated at relatively low pressures, but these pressures are suitable for herbicide applications.

Commonly used nozzles for applying herbicides include flat fans, even flat fans, double-orifice flat fans, extended-range flat fans, deflectors (floods), cone-jets, and swirl (whirl) chambers. Nozzles should be selected to match application needs.

The recirculating sprayer (RCS) and wipe-on devices provide a means of applying nonselective herbicides during periods when crops are present. To date, most use has been for removal of tall weeds in shorter crops. The simplicity of

the somewhat less effective wipe-on devices has made them more popular than the more complicated RCS.

Compressed gas sprayers equipped with regulators can be used to apply herbicide accurately to small areas.

Injury to off-target vegetation occurs either from volatilization and vapor drift or spray drift and frequently is highly visible. Both can and should be minimized in most situations.

Drift reduction depends on eliminating fog- and mist-sized particles from spray patterns.

Factors determining droplet size in a spray pattern include spray pressure, nozzle orifice size and shape, spray viscosity, liquid volume, and spray angle.

Invert emulsions and drift reduction adjuvents (polyvinyl polymers) reduce fine particles by increasing viscosity of the spray solution.

Large droplets are produced by nozzles that deliver spray at low pressures and by specially designed equipment that provides streams of spray through tubes. A range of droplet sizes are produced, but the average size is large. Such large droplets may not provide adequate coverage for foliar applications.

The controlled droplet applicator (CDA) produces uniform droplets and provides them at a maximum number per volume of spray solution. Theoretically, coverage should be very good. Problems with equipment and spray characteristics, however, have limited their success.

Keeping the nozzles as close to the target as possible and avoiding wind favor more herbicide reaching the target.

Using shielded sprayers, selective hand placement, and subsurface injection into soil and water; making applications when sensitive vegetation is not present; and employing other similar practices also help reduce injury off target.

If drift control technology was fully utilized, problems could be reduced substantially.

When soil-applied herbicides cannot be moved dependably into the soil with rainfall or overhead irrigation, performance depends on mechanical mixing.

Suitability for incorporation depends on herbicide characteristics as well as on soil and weed characteristics. Highly volatile or photodegradable herbicides are incorporated routinely. Incorporation also can provide better placement in relation to deep-germinating propagules. Herbicides that are tightly adsorbed to soil colloids or those that are easily leachable are least suitable for mechanical incorporation. When selectivity is based on separating the herbicide from the seed and developing roots by covering the seed with soil and then keeping the chemical on or near the soil surface, incorporation can result in crop injury.

Mechanical incorporation of herbicides can increase trips over the field, cause drying of the soil, and involve special equipment. Mechanical incorporation of herbicides generally is not a viable option in reduced tillage operations.

Equipment used for secondary tillage most frequently is used for herbicide incorporation. Field finishing (tandem) disks, field cultivators, seedbed conditioners, power-driven tillers, rolling cultivators, mulch treaders, and power harrows all have been used successfully.

Wet, cloddy, trashy, and uneven soil surfaces all interfere with herbicide performance and incorporation.

Usually one secondary tillage pass is needed before incorporation is initiated. Whether one or two additional passes will be needed depends on soil conditions, equipment, herbicide, crop, and weeds present.

Accurate herbicide application is critical to good chemical weed control and involves equipment calibration and associated procedures. Accurate application requires the determination of the prescribed dosage, sprayer output (calibration per se), and the amount of herbicide to mix into the spray tank.

TERMS INTRODUCED IN THIS CHAPTER

Airfoil needle boom
Calibration
Centrifugal pump
Compressed gas powered sprayer
Cone-jet nozzle
Controlled droplet applicator (CDA)
Deflector (flood) nozzle
Diaphragm pump
Disc-core nozzle
Double-orifice flat fan nozzle
Drift control nozzle
Even flat fan nozzle
Extended-range flat fan nozzle

Flat fan nozzle
Granular applicator
Herbigation
Hollow cone nozzle
Incorporation
Invert emulsion
Off-target injury
Piston pump
Polyvinyl polymer
Raindrop nozzle
Recirculating sprayer (RCS)
Roller pump
Rope wick applicator
Shielded sprayer
Sponge wick applicator
Sprayer
Subsurface injection
Swirl (whirl) chamber nozzle
Turbine pump
VMD (volume mean diameter)
Wedge wick applicator
Wipe-on applicator

SELECTED REFERENCES ON HERBICIDE APPLICATION

Bode, L. E. et al., Eds. 1992. Pesticide Formulation and Application Systems. Special Technical Series No. 1112, ASTM.

Bohmont, B. L. 1997. The Standard Pesticide User's Guide, 4th ed. Prentice-Hall, Upper Saddle River, New Jersey.

McWhorter, C. G., and M. R. Gebhardt, Eds. 1987. Methods of Applying Herbicides. No. 4, Monograph Series of the Weed Science Society of America, Lawrence, Kansas.

Literature on spraying and other application equipment is available from:

Delavan Corporation
811 Fourth Street
West Des Moines,
IA 50265

Hypro
375 Fifth Avenue N.W.
St. Paul, MN 55112

Micron Corporation
1424 West Belt
Drive North
Houston, TX 77043

Spraying Systems Co.
North Avenue at
Schmale Road
P.O. Box 7900
Wheaton, IL 60189-7900

Greenleaf Technologies
P.O. Box 364
Mandeville, LA
70470-0364

TILLAGE EQUIPMENT

The development of equipment for efficient machine tillage has been synonymous with advances in mechanized agricultural production. Modern tillage implements provide effective and economical weed control for food and fiber crops. The contributions these implements have made and will continue to make should not be overlooked. In this chapter, we describe the equipment used in conventional and conservation tillage systems and discuss how they relate to weed control.

THE CONVENTIONAL TILLAGE SYSTEM

The tillage system historically used for producing annual crops in the United States consists of the initial breaking of the soil with a primary tillage implement such as a moldboard plow, then seedbed preparation followed by planting and selective cultivation (Figure 18.1). This overall system generally is referred to as the *conventional tillage system* and is used successfully over a wide range of soils, climates, and crops.

The initial soil-breaking operation is called primary tillage. The moldboard plow cuts the soil, inverts it, breaks it, and buries growing plants and plant residues. This operation leaves the soil generally loosened, rough, and relatively free of surface residues and green vegetation.

Secondary tillage refers to the additional operations used for seedbed preparation. Tandem disks, field cultivators, bed conditioners, powered tillers, and similar shallow-running implements are used to further level, break up, and firm the soil

as well as to remove any weed growth. The desired result is a weed-free seedbed suitable for successful seeding and crop establishment.

Selective cultivation is used to remove weeds after the crop has been planted or has emerged from the soil. The goal is to remove the weeds selectively without injury to the crop either by using shallow-penetrating broadcast implements over deep-seeded or well-rooted crops or by cutting out all vegetation between the crop rows. Rotary hoes and light harrows are used for the first purpose, and cultivators are used for removing weeds between rows.

PRIMARY TILLAGE EQUIPMENT
Moldboard Plow

The moldboard plow, often called "the plow," is the standard primary tillage tool in conventional crop production systems. This implement was perfected in the early 1800s. The moldboard plow basically functions as a triple wedge (Figure 18.2). It cuts the soil loose from the bottom and one side of the furrow, then lifts, inverts, and shatters the furrow slice. In the process, it covers surface residues. One pass with the moldboard plow should result in maximum coverage of surface residues (Figure 18.3). Coulters (rolling disks) attached ahead of the plow cut through trash and sod, preventing clogging of the plow while providing a clean furrow edge.

Individual plow units (or bottoms) usually are designed to cut furrow slices 14, 16, 18, or

Figure 18.1 Operations for conventional tillage. Upper left: Plowing. Upper right: Seedbed preparation with tandem disk and spike tooth harrow. Lower left: Planting. Lower right: Selective cultivation. (Lower left photo courtesy of International Harvester, Chicago, Ill.; lower right photo courtesy of Chester A. Allen and Son, West Lafayette, IN)

20 in. wide. Several units are hooked together on a common frame to make wider plows (e.g., three, four, six, or more bottoms).

Relatively large energy requirements and exposure of the soil to erosion are disadvantages attributed to this implement.

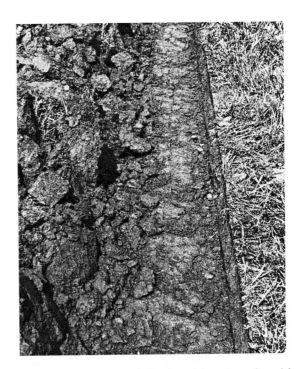

Figure 18.3 Furrow left after passage of moldboard plow. Good coverage of plant residues is obtained.

Figure 18.2 The moldboard plow.

Disk Plow

The disk plow was perfected in the late 1800s. There are two basic types: the standard disk plow and the wheatland or one-way plow.

The standard disk plow has 20- to 30-in. blades tilted backward 15 to 25 degrees (Figure 18.4). Each disk blade is mounted on its own axle and standard and turns independently. Soil inversion is somewhat less complete than with the moldboard plow so more residues remain on the surface. The disk plow works more effectively than the moldboard plow in loose, hard, rocky, or sticky soils (where scrapers are added), or where stumps are a problem.

The wheatland disk plow uses smaller disks (20 to 24 in. in diameter), and the individual disk blades are mounted vertically on a common axle to form a gang (Figure 18.5). Individual disks are spaced 8 to 10 in. apart. This plow generally is operated at a shallow depth (4 to 5 in.). The amount of soil movement and subsequently the power requirements are considerably less than for the moldboard plow. The incomplete soil inversion resulting from use of the wheatland plow leaves nearly half the plant residues on the soil surface.

Chisel Plow

Chisel plows, field cultivators, and small sweep plows are similar implements, and their designations are somewhat confusing. We will designate the sturdier implements used for primary tillage as chisel plows and the light implements used for secondary tillage as field cultivators.

The chisel plow, originally developed in the high plains, has been rapidly adopted as an alternative to the moldboard plow. In the eastern Corn Belt, it is the most frequently used primary

Figure 18.5 Wheatland disk plow.

tillage implement. The modern chisel plow consists of a heavy frame mounted on wheels. Curved spring steel shanks (standards) are attached to the frame, and chisel points are bolted to the bottom ends (Figure 18.6). Chisel points vary in width from 2 to 4 in. and can be either straight or curved. The shanks usually are spaced 1 ft apart and the points are operated at moldboard plow depth or slightly below. A tearing, loosening action rather than inversion results, so the soil is opened up to receive water. As much as 30 to 75% of the plant residues remain on the soil surface, depending on the design and operation of equipment (Figure 18.7, middle). This can be compared to 0 to 5% residue cover after primary tillage with the moldboard plow (Figure 18.7, left) and 90 to 100% cover in no-till (Figure 18.7, right). When operated at equal depths, the chisel plow has a lower power requirement than the moldboard plow.

When used on dry soil, the chisel plow can be used to shatter shallow soil barriers such as plow pans. On wet soil, unmanageable, large clods can result. In fact, where wind erosion is

Figure 18.4 Standard disk plow.

Figure 18.6 Chisel plow.

Figure 18.7 Comparison of amount of corn residue left on soil surface after planting following three different tillage systems. Left: Moldboard plow. Middle: Chisel plow. Right: No-till.

occurring (e.g., on the Great Plains), one of the uses of the chisel is to bring wet clods to the surface as well as to make ridges. Clods and ridges reduce the movement of soil by wind.

Residues can interfere with the action of the chisel plow or with later planting and cultivation if left on the surface. Where large amounts of plant residues are present, chopping or disking prior to plowing will prevent the clogging of some chisel plows. Some chisel plows are equipped with coulters or gangs of coulters or disks to facilitate the cutting and passage of residues, thus eliminating the need for prior stalk preparation. Many chisel plows also have wider and more deeply curved points with special attachments to cover heavy residues more completely.

The type of secondary tillage used after the chisel plow largely determines soil condition at planting time. If followed by sweeps (on a chisel or field cultivator), 50% of the trash can be left on top. If followed by a tandem disk, the seedbed will look like a seedbed prepared by a tandem disk behind a moldboard plow. The chisel followed by one pass with a tandem disk should leave less than 25% of the crop residue on the surface and provide a good seedbed.

Sweep Plow

The sweep plow is used almost exclusively where there is a limited amount of residue and the objective is to leave it on the surface. The two basic types are the large sweep "Noble" plow with sweeps 6 to 8 ft wide (Figure 18.8) and the sweep plow with sweeps 12 to 30 in. wide. The smaller sweeps frequently are interchangeable with

narrow chisel shovels (points). These plows have large, heavy frames on which appropriate standards are mounted for the sweeps. The sweeps generally are operated 4 to 6 in. deep.

The wheatland disk plow, chisel plow, small sweep plow, and large sweep or Noble plow are primary tillage implements that have been used extensively in the Great Plains, where wind erosion and moisture loss are serious problems. These implements control weeds to prevent moisture loss while protecting the soil from erosion by leaving residues on the surface or by leaving the surface rough or ridged.

SECONDARY TILLAGE EQUIPMENT

Secondary tillage implements are used following primary tillage to control weeds; to level, firm, and further break up the soil; and to leave a suitable

Figure 18.8 Noble sweep plow.

seedbed. Some of these implements are used to mix plant residues, fertilizers, lime, and pesticides into the soil. Normally the last secondary tillage operation is done just prior to planting. Secondary tillage implements include tandem disks, field cultivators, spring tooth harrows, mulch treaders, spike tooth harrows, spring tine (vibrating) harrows, coil tine harrows, rollers and packers, rod weeders, pulverizer-mulchers, bed conditioners, and power driven tillers.

Tandem Disk (Tandem Disk Harrow)

The tandem disk probably is the most widely used secondary tillage tool (Figure 18.9). Modern tandem disks consist of two sets of paired gangs. Each gang is made up of concave disks (blades) attached to a single axle. In operation, the entire gang turns as a unit. The standard tandem disk has two front gangs in a V with the blades facing out and moving the soil outward from the center. A second set of gangs in an inverted V behind the first set moves the soil back toward the center of the implement. The action moves the soil laterally, cuts off weeds, chops residues, cuts clods, and inverts the soil. When properly operated, considerable leveling, mixing, and pulverization are achieved.

Tandem disks come in three basic sizes: field finishing disks, heavy-duty disks, and multi-purpose disks. Some specifications for disk harrows from John Deere and Company follow.

The field finishing disk has 18- to 20-in. blades spaced 7 1/4 in. apart and is designed primarily for seedbed preparation following a primary tillage operation. This unit also provides excellent incorporation of herbicides into the soil.

The heavy-duty disk uses 22- to 26-in. blades spaced 11 in. apart. This unit has a heavy frame and its tillage capabilities closely approach those described for disk plows. With one pass, this unit can cut heavy crop residues into the soil. Where residues are light, it can be used as the primary tillage tool. It should not be considered a tool for herbicide incorporation.

Another type of heavy-duty disk is the offset disk, which consists of a single straight gang in front with blades facing in one direction and a second gang behind with blades facing and throwing soil in the opposite direction (Figure 18.10). Most offset disks are relatively unsuited for field finishing and herbicide incorporation.

The multipurpose disk has 20- to 22-in. blades spaced 9 in. apart and is supposed to provide acceptable seedbed finishing as well as serve as a heavy-duty implement. When carefully operated on most soils, it can result in acceptable incorporation of herbicides.

Figure 18.9 Tandem disk. (International Harvester, Chicago, IL)

Figure 18.10 Offset disk being used as a cutting tool to chop residues.

Field Cultivator

Field cultivators consist of curved spring steel shanks attached to a frame (Figures 18.11, 18.12). The frame often is constructed with wheels to control the depth of penetration. Sweeps, duckfeet, or shovels usually are attached to the shanks. The field cultivator works best following a primary tillage tool. It kills weeds and stirs, levels, and breaks up the soil. Incorporation of surface residue is limited. When the field cultivator is used for secondary tillage after the moldboard plow, the resulting seedbed is similar to that obtained with a tandem disk. The field cultivator has some limitations as a herbicide-incorporation tool when residues are present.

Spring Tooth Harrow

The spring tooth harrow usually consists of two or three ranks of spring steel teeth mounted on a frame that rides on the soil surface. It can be used much the same way as a tandem disk or field

Figure 18.12 Field cultivator with Danish tines followed by rolling baskets.

cultivator. The flexing action of the tines tends to bring shallow roots and rhizomes to the soil surface, where they dry out. It is thus a popular implement in regions where quackgrass has been a problem. Some new spring tooth harrows differ slightly in design from the older models, but they yield the same results.

Mulch Treader

The mulch treader is similar in design to the tandem disk except that wheels with curved spikes are used in place of disk blades to make up the gangs (Figure 18.13). The treader runs shallower and moves less soil than the disk. It leaves most of the clods and trash on the soil surface while killing small weeds and firming the seedbed. The major use of the treader has been for secondary tillage in stubble mulch systems in the Great Plains, but it also has been used successfully as a herbicide-incorporation implement.

Figure 18.11 Field cultivator with C-shanks and sweeps.

Figure 18.13 Mulch treader.

Figure 18.14 Left: Spike tooth harrow behind tandem disk. Right: Close-up of spike tooth harrow.

Light Harrows

A number of light harrows generally are used after the tandem disk, the field cultivator, or the spring tooth harrow for final seedbed preparation. They are used most frequently and most efficiently when hitched directly behind or attached to these implements. When used in this manner, light harrows further fragment, firm, and level the soil before it dries and the clods harden. Control of weeds also is increased. These implements are capable of stirring the soil to a depth of no more than 3 or 4 in.

Light harrows can be used by themselves, but combining them with the other implements eliminates a trip over the field, thus saving time and fuel. Combining two implements also eliminates the problem of covering tractor tracks when only the light harrow is used. Common harrows include the spike tooth harrow (Figure 18.14), the spring tine (vibrating) harrow, and the coil tine harrow (Figure 18.15).

Rolling devices (e.g., Figures 18.12 and 18.19) are attached to equipment and serve the same purposes as harrows.

Rollers and Packers

Rollers and packers are used to finish a seedbed beyond the finishing obtained with harrowing. These implements are made up of closely spaced individual wheels that may be corrugated, crowfoot, notched, or smooth. They crush clods, firm the seedbed, and frequently are used for small-seeded crops. The resulting fine, firm seedbed also is ideal for weed seed germination.

Rod Weeder

The rod weeder is used almost entirely for controlling weeds in stubble mulch fallow systems (Figure 18.16). This machine has a rod that slowly revolves under the soil, cutting off weeds and depositing coarse material on the surface while the fine material remains below. The rod is chain driven from the wheel of the implement. The rod weeder is limited to soils that can be penetrated easily and are free of rocks.

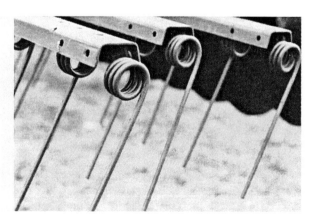

Figure 18.15 Left: Coil tine harrow behind field cultivator. Right: Close-up of coil tine harrow.

Figure 18.16 Rod weeder. Arrow points to revolving rod.

Multicomponent Tillage Implements

Some secondary tillage machines are made up of two or more individual implements. In many instances, one pass of these machines provides a seedbed ready for planting.

The pulverizer-mulcher (cultipacker) consists of clod-crushing front rollers followed by two rows of spring steel teeth similar to those of a spring tooth harrow (Figure 18.17). This unit is followed by another set of crushing, packing rollers. The components are all carried on a single frame.

The do-all type bed conditioner is made up of a field cultivator, sometimes preceded by disks followed by a spiral reel to cut clods and chop residues, then a spike tooth harrow to pulverize and level, and finally a rear smoothing board to finish leveling the surface. A field cultivator type bed conditioner can have a set of disk gangs in front to chop up residues and loosen soil, followed by a field cultivator with five or more tool bars, followed by harrows or rolling baskets (Figure 18.18). The bed conditioner frequently is used for one-pass herbicide incorporation.

POWER-DRIVEN TILLAGE EQUIPMENT

Power-driven tillers can be used to combine a number of tillage operations into one. On small areas they can be used to obtain the equivalent of both primary and secondary tillage; on large areas they are more likely to be used for secondary tillage.

Rototiller

The rototiller (rotovator) is driven from the engine if self-propelled or from the power takeoff if tractor mounted. It consists of a horizontal rotating shaft that contains knives, spikes, or other suitable tools that tear into and pulverize the soil (Figure 18.19). The rototiller is an excellent tool for precise and thorough incorporation of herbicides into the soil and will incorporate to the depth of cut (up to 4 in.). Excess pulverization and fluffing of the soil can be a problem.

Vertical Tine Tiller

The vertical tine tiller (Roterra) has sets of rotating tines that enter the soil vertically (Figure 18.20). Their spinning action prepares the seedbed as the machine moves forward. Two tines are mounted on a T-shaped holder to form a fork. A number of these individual units are mounted in a straight line and rotated in synchronization (much like an eggbeater). This machine is less efficient at incorporating materials deposited on the surface than is the standard rototiller. Herbicides can be incorporated only to a depth of 2 in.

Power harrows are vertical tine implements, but the tines oscillate laterally rather than rotate. A smooth seedbed should result. When used for herbicide incorporation, shallow mixing should be expected.

EQUIPMENT FOR SELECTIVE WEEDING

The use of modern cultivation tools to remove weeds effectively from established crops is a major contributor to efficient crop production. This selective weeding in established crops is called cultivation (in the most limited definition of the word) and also intertillage.

Light harrows (spike tooth or coil tine) and rotary hoes can be used as total coverage implements; row crop cultivators are used to cut weeds between crop rows.

Light Harrows

Spike tooth or coil tine harrows sometimes are used for removing small weeds in well-rooted crops. These implements are pulled at relatively high speeds at an angle to the crop rows to prevent long strips of crops from being torn out.

Rotary Hoe

Rotary hoes are used most commonly in the Corn Belt for removing small weeds in corn and soybeans (Figure 18.21). The rotary hoe is made up of two gangs of wheels with curved spikes

Figure 18.17 Pulverizer-mulcher. (International Harvester, Chicago, IL)

mounted on a single frame. Each wheel turns independently. The hoe is pulled at high speeds parallel with the crop rows. This implement can aid in crop emergence by breaking crusted soil.

Row Crop Cultivators

Row crop cultivators consist of frames to which appropriate cutting and soil-stirring tools are attached by means of shanks and clamps. A reference to "the cultivator" usually means this implement (Figure 18.22).

Most row crop cultivators currently in use are rear mounted. They can be attached easily to the standard three-point hitches on modern farm tractors in a matter of minutes. Cultivators were mounted between the front and rear wheels on old tractors. These units were much more difficult and time consuming to attach to

the tractor but permitted cultivation closer to the row because corrections in steering move a front-mounted cultivator less in a lateral direction than they do one that is rear mounted. Modern herbicides have greatly reduced the need for early-season close cultivation, so the more convenient rear-mounted units provide adequate weed control. Rear-mounted cultivators usually are equipped with guide fins (coulters) to reduce the lateral movement due to changes in steering.

Tools for row cultivators come in various shapes and sizes and serve several purposes (Figure 18.23). Some that are commonly used on the standard row cultivator are described briefly here.

Shovels or *points (narrow shovels)* are flat or slightly curved pieces of steel that penetrate and tear the soil (Figure 18.23A). They frequently penetrate to a depth of 4 to 6 in. Specially designed

Figure 18.18 Cultivator-type bed conditioner.

Figure 18.19 Rototiller.

Figure 18.20 Left: Vertical tine tiller. Right: Arrangement of tines on the vertical tine tiller.

shovels are used for making irrigation furrows or are used as hillers.

Sweeps are shallow-running tools that have a curved shank with two swept-back wings that come to a V in front of the shank (Figure

Figure 18.21 Top: Rotary hoe in a soybean field. Bottom: A modern rotary hoe. (Top photo courtesy of Chester A. Allen and Son, West Lafayette, IN)

18.23B). *Half-sweeps* have only one wing and are used next to the crop row (Figure 18.23C).

Duckfeet are small, flat running sweeps with small wings (Figure 18.23D).

Knives are thin flat pieces of metal that have an upright piece to allow mounting on a shank (Figure 18.23E). The section perpendicular to the upright section forms a swept-back cutting blade. These tools are used most commonly for shallow cutting of small weeds in vegetable crops, sugar beets, and small field crops.

Disks or *disk hillers* are single cup-shaped disks mounted on a single axle attached to a shank. They are set at an angle to cut weeds and trash and to move soil.

The choice of cultivator and selection of tools depend on a number of factors, including the crop, crop size, weed size, amount of surface trash, presence of vines, soil type, soil condition, and to a large extent, personal experience and preference.

A standard setup for a cultivator in the Corn Belt is to align a half-sweep next to the crop row with one, two, or three sweeps staggered behind.

Figure 18.22 Row crop cultivator.

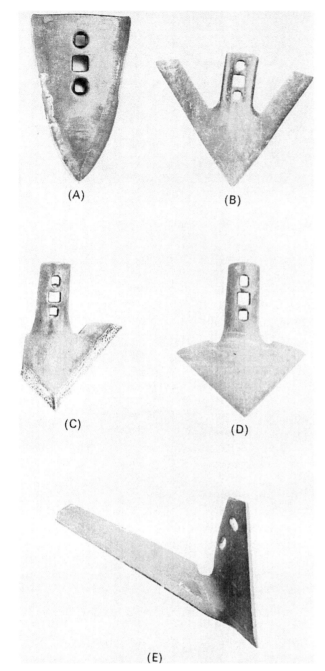

Figure 18.24 Disk hiller mounted on a cultivator. (D. R. Griffith, Purdue University, West Lafayette, IN)

Figure 18.23 Tools for row crop cultivators: (A) shovel, (B) sweep, (C) half-sweep, (D) duckfoot sweep, (E) knife.

The number of sweeps needed per row is determined by row width and the size of the sweeps.

Disks (disk hillers) either alone or in combination with sweeps or shovels can be used where trash or vines are present or where hilling (ridging soil around the crop) is desired (Figure 18.24). The rolling, cutting action of the disk prevents clogging. Knives frequently are used in combination with sweeps or duckfeet when small crops are close cultivated and weeds are small. The knives are placed adjacent to the row and in front of the other tools. Rows

as narrow as 20 in. can be cultivated with this combination.

Shields often are used to prevent soil from covering the crop during cultivation (Figure 18.25).

A coulter placed in front of a sweep permits cultivation through plant residues.

The rolling (rotary) cultivator uses sets of curved, spoked wheels set on a common axle to form gangs (Figure 18.26). The tines (spokes) are twisted and beveled to enhance the cutting action. Individual gangs can be adjusted to move soil either away from or toward the crop and can be used for ridging. One gang is positioned next to one side of a crop row. A four-row cultivator requires eight gangs. This type of cultivator can be operated at high speeds and leaves a smooth finish on the bed. Large weeds frequently escape. The rolling cultivator also is used for shallow herbicide incorporation. In this case, the gangs are adjusted to conform to the contours of the soil surface.

The spring tine (vibrating shank, Danish tine) cultivator uses curved spring steel vibrating shanks with small duckfeet-like points at the ends. These shallow-running, low-draft cultiva-

Figure 18.25 Spring tine cultivator with shields.

Figure 18.26 Tines on a rolling cultivator.

tors are effective against small weeds. They tend to miss large ones.

Multiple shanks spaced closely together for cultivation are subject to wrapping by weeds and clogging of the implement (Figure 18.27). The spread of perennial reproductive structures also can be the end result.

Other cultivator types and tool combinations can be and are used. The ones discussed are used frequently. A summary of the effects of tillage equipment on weeds, surface residues, and soil is presented in Table 18.1.

PURPOSES OF TILLAGE

Weed control is the most cited contribution tillage makes to crop production. Tillage, however, is used for other important purposes. Some purposes can be accomplished with a single operation; others require several sequential operations.

1. Seedbed preparation is an integral function of tillage operations in conventional systems. The desired result is a weed-free, firm yet mellow, relatively level, and trash-free medium that facilitates planting and crop establishment.

2. Tillage that inverts the soil and buries crop residues aids in controlling plant pests such as plant disease organisms, insects, and rodents.

3. Improvement of soil physical conditions can be obtained at least temporarily from tillage operations that loosen and break up the soil. In other cases, a firming action can be provided.

4. Surface conditions can be manipulated to improve reception of rainfall, manage plant residues, and alter surface roughness and configuration. Beds, ridges, and irrigation furrows are established and maintained by tillage.

5. Tillage can be employed to incorporate substances such as plant residues, manure, lime, commercial fertilizer, herbicides, and other pesticides into the soil.

6. Tillage can be used to manipulate soil moisture and temperature. For example, operations that loosen the soil or eliminate surface residues result in drying and rapid warming. The tops of ridged soils also dry out and warm up rapidly, providing an earlier seedbed for establishing row crops.

7. Special tillage techniques can be used to prevent soil erosion and conserve water. These are discussed in the following section.

CONSERVATION TILLAGE SYSTEMS

A number of tillage systems differ significantly from the conventional system. The need to reduce soil erosion has been the primary impetus behind the development of alternative systems. The need to conserve time, energy, and capital outlay for equipment also has contributed to their widespread use. These systems are referred to variously as conservation, minimum, and no-tillage (no-till) systems. The first two names are general and refer to the desired goal of the system rather than designating any tillage implement or sequence of implements. Each of the two can be accomplished with several different tillage procedures.

Minimum tillage is the minimum amount of tillage required for crop production or for meeting the tillage requirement under existing soil and climatic conditions. Minimum tillage refers to eliminating excess tillage; for example, reducing four secondary tillage steps to two.

Conservation tillage is any tillage system that reduces the loss of soil or water when compared to a nonridged clean tillage system (e.g., the conventional tillage system described previously).

Examples of conservation tillage systems include the use of the chisel plow in conjunction

Figure 18.27 Field bindweed wrapped around cultivator shanks can be spread to other areas.

Table 18.1 Effects of Tillage Equipment on Weeds, Surface Residues, and Soil

	Primary Tillage					Secondary Tillage															Selective Tillage			
	Moldboard plow	Disk plow	Chisel plow	Wheatland plow	Sweep plow	Field finishing disk	Multipurpose disk	Heavy-duty disk	Field cultivator (sweeps)	Spring tooth harrow	Field cultivator (Danish tines)	Spike tooth harrow	Spring tine harrow	Rollers	Mulch treaders	Do-alls (bed conditioners)	Pulverizer-mulchers	Rototillers	Vertical tine tillers	Power harrows	Rotary hoe	Rolling cultivators	Sweep cultivators	Danish tine cultivators
IMPACT ON WEEDS																								
Disruption of established perennials	X	X	P	P	X	P	P	P	P	P	P	P	P			P	P	P	P	P		P	P	P
Disruption of large established annuals	X	X	X	X	X	X	X	X	X	X	P	P	P			P	P	X	P	P	P	P	X	P
Destruction of most seedlings and young weeds	X	X	X	X	X	X	X	X	X	X	X	X	X		X	X	X	X	X	X	P	X	X	X
Destruction of small shallow-germinating seedlings	X	X	X	X	X	X	X	X	X	X	X	X	X	X	X	X	X	X	X	X	X	X	X	X
IMPACT ON SURFACE RESIDUES																								
Cutting or chopping action	X	X		X		X	X	X								X		X	X					
Effective burial	X	P		P		P	P	P								P		X						
Leaves residues on surface			X	X	X				X	P	X	X	X	X	X	X	X				P	X	X	X
Subject to clogging			X						X	X	P	X		X	X	X	X				X	X	X	X
TYPE OF ACTION ON SOIL																								
Inversion	X	P	P	P		P	X	P			P					X	X	X	X	X				X
Tearing	X	P	X	P	X	P	X	P	X	X	X	X	X		X	X	X	X	X	X	X	X	X	X
Lifting	X	X	X		X											X								
Crushing									X	X	X	X	X	X	X	X	X				X	X		X
Light stirring	X					X	X	X	X	X	X	X	X	X	X	X	X		X	X	X	X		X
Firming	X								X	X	X	X		X		X							X	
Subsurface cutting	X	X		X	X				X	X						X				X			X	X
Loosening	X	X	X	X	P	X	X	X			X					X	X	X	X	X	X	X	X	X

X = effective; P = partial.X = effective; P = partial.

with tandem disks or chisel sweeps (chisel/disk and chisel/sweep system, respectively; see Chapter 14). The chisel plow is used in place of the moldboard plow for the primary tillage operation. Because the chisel plow leaves some 30 to 80% of the plant residues on the surface, the soil is much better protected from erosion than when the moldboard plow is used. The amount of crop residue (e.g., very little in the case of soybeans) and the design of the chisel plow determine the final cover of residue. This method of conserving soil is particularly effective when long periods of time elapse between the primary and secondary tillage operations; for example, when primary tillage is conducted in the fall and secondary operations are not conducted until the following spring. Secondary tillage can then be accomplished with one or two passes of a tandem disk (chisel/disk system) to incorporate the plant residues. The resulting seedbed actually is similar to the one prepared when the moldboard plow is followed by a tandem disk or field cultivator.

Chisel sweeps can be used in place of tandem disks if more crop residues are wanted on the finished seedbed (chisel/sweep system). Much of the crop residue remains on the surface, yet the soil is loosened and the weeds are cut off underground. Using the chisel plow followed by sweeps probably is the least amount of tillage that should be used when severe stands of perennial weeds are present.

Farm machinery companies have developed multicomponent implements that provide one-pass seedbed preparation; that is, they combine primary tillage, secondary tillage, and seedbed preparation with a single pass. Some of these implements consist of as many as five sets of tools, for example (from front to back), disks, chisel plow sweeps, cultivator sweeps, harrows,

and rolling baskets. The result is a seedbed ready for planting but with substantial amounts of residue remaining.

Another example of conservation tillage is the use of the one-way disk plow in the Great Plains states. This primary tillage tool partially inverts the soil and covers only a portion of the surface residues. Recently, large, heavy-duty tandem disks and offset disks have been employed in the Corn Belt in a manner similar to that of the one-way disk plow. It is possible to eliminate one secondary tillage operation when any of these implements is used in place of the moldboard plow. In the stubble mulch fallow systems of the Great Plains, where mulch is maintained for 18 mo between wheat crops, a combination of the one-way disk and sweep plow or chisel plow is used to keep down weeds and maintain crop residues at the desired levels.

The ridge-till system (see Chapter 14), an example of a conservation tillage system, employs wide sweeps or other devices to remove the tops of ridges from the previous season's crop rows (Figure 18.28). The soil and crop residues are deposited in undisturbed areas between the rows. The ridge-removal device is attached to the planter unit so that seedbed preparation and planting become a one-pass procedure. The ridges are rebuilt by cultivation in the growing crop with a row cultivator, or they may be formed after harvesting in the fall (this buries protective residues).

To date, the ultimate system for maintaining crop residues on the soil surface and reducing the number of operations is the *no-till* plant system in which a 1-in. wide seedbed is prepared with a no-till coulter attached immediately ahead of the planter unit (Figure 18.29). Other than the slit for seed planting, no further disturbance of the soil occurs (Figure 18.30).

Figure 18.28 Ridge-till system. Left: Ridges visible after stalk chopping. Right: Planting into ridges with ridge planter. (D. R. Griffith, Purdue University, West Lafayette, IN)

Figure 18.29 No-till coulter (right) ahead of furrow opener. The planter unit sits above the furrow opener but is not visible in this photo.

Several of the systems described eliminate various tillage operations that normally contribute significantly to weed control. In some cases (for example, when sweeps are used in place of a disk) the alternate tillage operation does provide weed control while leaving the soil protected from erosion by a cover of plant residues. In systems in which one or more tillage operations are eliminated, the increment of weed control obtained from that operation is lost. In these instances, alternate methods of weed control must be used. In fact, the no-till concept of crop production was not a viable possibility until dependable selective chemical control methods were developed.

MAJOR CONCEPTS IN THIS CHAPTER

The conventional tillage system, which consists of the initial breaking of the soil (primary tillage) with the moldboard plow; seedbed

Figure 18.30 Planting with a no-till planter. Note limited amount of disturbed soil in planted row. (Photo courtesy of D. R. Griffith, Purdue University, West Lafayette, IN)

preparation (secondary tillage) with tandem disks, field cultivators, and harrows; planting; and weed removal in the crop (selective cultivation) with cultivators has been used successfully for producing annual crops in situations involving a wide range of soils, climates, and crops. This efficient machine tillage system has come to symbolize modern agriculture.

The conventional tillage system exposes soil to erosive forces (wind and water) and consequently is unsuitable for many situations. Chisel plows, wheatland disk plows, and sweep plows are all primary tillage tools that leave soil less subject to soil erosion because they leave a portion of the crop residue on the soil surface. The chisel plow is rapidly replacing the moldboard plow as a primary tillage tool.

Secondary tillage tools include tandem disks, field cultivators, spring tooth harrows, numerous other harrows, mulch treaders, power-driven rototillers, power-driven vertical tine tillers, and a number of multicomponent implements (e.g., bed conditioners). These implements provide weed control and seedbed preparation and can be used to mix herbicides into the soil.

Once a crop is planted, selective mechanical weeding most frequently is achieved with a cultivator that passes between the crop rows. Numerous types of cultivators are available. Rotary hoes and light harrows have limited utility for broadcast selective weeding shortly after crop emergence.

A major purpose of tillage has been to provide weed control. In addition, tillage provides easy-to-plant seedbeds, a means of mixing crop residues, a means of mixing chemicals into the soil, some disease and insect control, and a means of manipulating soil surface conditions.

Conservation tillage systems have been developed to help save soil and, theoretically at least, to save on time, fuel, and money. Conservation systems range from changing the primary tillage implement from the moldboard plow to the chisel plow to adopting a no-till system in which tillage consists of using a narrow coulter-prepared strip just ahead of the furrow openers on a planter. Some limited-tillage systems include chisel/disk, chisel/sweep, sweep plow, and ridge-till systems.

Many conservation tillage systems depend heavily on herbicides for adequate and dependable weed control.

TERMS INTRODUCED IN THIS CHAPTER

Bed conditioner
Chisel plow
Coil tine harrow
Coulter
Cultivator
Cultivator knives
Cultivator shovels
Danish tine cultivator
Disk hiller
Disk plow
Do-all
Duckfoot
Field cultivator
Field finishing disk (tandem disk)
Furrow-slice
Half-sweeps
Harrow
Heavy-duty tandem disk
Minimum tillage
Moldboard plow
Mulch treader
No-till coulter
Offset disk
Packers
Primary tillage
Pulverizer-mulcher
Rod weeder
Rollers
Rolling cultivator
Rotary hoe
Rototiller
Secondary tillage
Selective cultivation
Shields
Spike tooth harrow
Spring tine harrow
Spring tooth harrow
Stubble mulch fallow system
Sweep plow
Sweeps
Tandem disk
Vertical tine tiller

SELECTED REFERENCES ON TILLAGE EQUIPMENT AND SYSTEMS

ASAE Standards, 40th ed. 1993. See section on Terminology and Definitions for Agricultural Tillage Implements. American Society of Agricultural Engineers, 2950 Niles Road, St. Joseph, MI 49085-9659.

Buckingham, F. 1993. Fundamentals of Machine Operation. John Deere Service Publications, Moline, Illinois. This and other publications are available from John Deere Publishing, Dept. 373, 5440 Corporate Park Drive, Davenport, IA 52807; phone 1-800-522-7448; fax 1-319-355-3690; e-mail jdp@deere.com.

Bull, L. 1993. Residue and Tillage Systems for Field Crops. U.S. Department of Agriculture, Economic Research Service, Resources and Technical Division, Washington, D.C.

Finner, M. F. 1978. Farm Machinery Fundamentals. American Publishing Co., Madison, Wisconsin.

Wicks, G. A., O. C. Burnside, and W. L. Felton. 1994. Weed control in conservation tillage systems. Pages 211–244 in Managing Agricultural Residues, Ed. P. W. Unger. Lewis Publishers, Boca Raton, Florida.

Wiese, A. F., Ed. 1985. Weed Control in Limited Tillage Systems. Weed Science Society of America, Lawrence, Kansas.

For publications on farm equipment, visit your local dealer or check Web sites at

http://www.autodigest.com/automnl/agri_man/agri_man_index/html

http://www.deere.com/aboutus/pub/jdpub/index.htm

TROUBLESHOOTING

Anyone who works as a professional in the weed control area probably will find himself or herself in the position of having to determine the cause of a herbicide performance or injury complaint. The intent of this chapter is to provide an approach for handling such a situation in as professional and scientific a manner as possible.

Weed control problems generally involve two types of requests: those that require information or recommendations to solve a problem and those that are the result of a specific complaint or dissatisfaction with the outcome of a control method or program.

The first situation usually can be solved by obtaining enough information from the person with the weed problem to make a diagnosis and provide a suitable solution. The steps in proper diagnosis have been discussed in Chapter 4. Frequently, the first step is to identify the weed or weeds properly. Identification can be accomplished by referring to plant or weed identification guides that usually are available on a state or regional basis (Appendix A). Sometimes it is necessary to receive or send plant samples for positive identification. Suggestions for collecting, transporting, or preserving plants for later identification are provided in Appendix B.

The major emphasis of this chapter is the solving of problems that arise after the application of a weed control method, usually a herbicide, has been made. The most frequent problems involving herbicides are lack of weed control,

crop injury, and damage to nontarget vegetation. Less frequently, fish kills may be involved.

The major objective for an individual called on to investigate such problem situations is to determine accurately the extent and cause of the problem. Sometimes the objective also will be to minimize or alleviate hard feelings or anger. You may find yourself in this investigative capacity as the representative of the manufacturer of a product, the distributor of a product, the custom applicator, a member of a commercial advisory service, or a mediator or neutral party whose advice is requested (a university extension specialist or county extension agent, for example).

When making such calls, you must be prepared psychologically for just about anything. Fortunately, most complaints generally are easily resolved, and the primary interest of growers is to determine what happened and why, so they can improve their weed control methods and avoid similar problems in the future. Such calls create excellent opportunities to prove yourself as a valuable aide and sincere professional who will assist the grower, dealer, and other individuals in determining the facts. Good service is mandatory in the pesticide industry in order to maintain a clientele of satisfied repeat customers.

On the other hand, you may find that at least one party is unhappy and angry, and you or the organization you represent may end up in court. You also may find that some other "expert" already has been on the scene to provide an opinion that may be contradictory to yours, or you may find that there is no clear-cut solution to the problem. It is important to remember that if two or more disagreeing parties are involved, they may all be important clients or constituents of yours. The key to handling these situations is *to respond quickly and to remain calm, courteous, and analytical.* This approach may keep you from making embarrassing mistakes in handling or diagnosing the situation. The burden on your shoulders is to *improve relations,* if at all possible, while providing a reasonable and correct solution or explanation.

Some specific hints for approaching complaint calls follow.

YOUR CONDUCT

Answer the complaint as soon as possible. A prompt answer demonstrates your concern and gives you the opportunity to observe the problem as soon as possible after it developed. Some situations can change appreciably in a few days, so immediate and initial observations are extremely important. Remember also that if the possibility of replanting a crop is involved, time is money to the grower. Repeat visits may be necessary to determine the long-term effects of a problem.

Be courteous and attentive. You should be prepared to accept some verbal abuse graciously. Listen carefully to the complainant and take notes. Not only show concern but use this initial opportunity to find out as much about the situation as you can. Have the grower repeat the details several times. Walk the field and have the user again explain what happened and listen. Sometimes by the third or fourth review some insight into the problem and its underlying causes can be obtained, or the problem may even begin to seem insignificant. Continue to seek information as long as this procedure appears productive.

Do not be afraid to ask questions. Ask the parties involved what they think caused the problem. One of them may actually provide the exact answer. Seek any additional information that they can or will provide.

Make it clear from the start that the first visit is a fact-finding trip and that you intend to study your findings prior to reaching a conclusion. Frequently you will not be able to reach a firm conclusion based solely on your initial visit. Delaying may leave an initial impression of indecision, but it will help keep you from having to contradict yourself later. This practice does not prevent you from explaining the symptoms, conditions, patterns, and so forth, that you observe during your visit. What it should do is help you to avoid drawing a premature conclusion.

Make a thorough and systematic investigation of your own; that is, investigate as though you know nothing of the situation from the parties involved. As objectively as possible, make your observations and record the data required to make a decision. Keep your eyes and mind open and do not let others pressure or prejudice you into drawing incorrect conclusions. Remember that people are involved, and sometimes personalities are the major source of the problem.

Make a record of your observations. The best way to make a record is to take pictures of the site and immediately write down observations and other pertinent information such as dates, application rates, and weather conditions. All too often information left to memory is forgotten and must be reconstructed from incorrect recollections later. Even if you do not think a piece of information is important, record it. You may later discover some new facts that make a seemingly insignificant detail important. It is particularly important to document the observations of the parties involved because their memories also may wane or change as time goes on.

Take any plant or soil samples needed for analysis of the site. Make absolutely certain that the samples are clearly marked with specific site, date, and other pertinent information.

Graciously exit from the scene without making any incorrect or embarrassing statements.

At the first opportunity, isolate yourself and carefully study your observations and discussions. Fill in any data you might not have taken down while in the field. Study your data and check reliable sources of information. Does there appear to be a clear-cut answer to the problem? What is it? Is it the only possible explanation? Could the problem also have been caused by some factor with which you are less familiar, such as mechanical damage, insects, diseases, seed quality problems, or weather? Can any of these be ruled out?

Seek additional expert help or advice if you are unsure of your conclusions.

Provide the parties with an answer as soon as possible. Be accurate, fair, and clear in your response.

YOUR INVESTIGATION

Your investigation into the causes of a herbicide complaint should be systematic and professional. The information you gather should include pattern of herbicide injury or surviving weeds in the field, type of damage on the plants, soil texture and organic matter content, drainage patterns, application equipment, seed source, other management practices, prevailing wind patterns, and weather records (see Figure 19.1). Two of the most important of these aspects, field injury pattern and plant injury symptoms, are described in the following sections.

Determining Patterns in the Field (or Problem Area)

The pattern of plant injury or uncontrolled weeds in a field can be extremely helpful in determining the cause of a herbicide-related problem (Figure 19.2).

A pattern of injury that starts on one side of an area and diminishes gradually and uniformly away from that area is typical of application drift. The presence of weeds showing similar symptoms immediately upwind from the treated area in an adjacent field, right-of-way, or ditch bank also is indicative of drift. Not all plants will show the same amount of injury. You may need to look for sensitive species in the drift area and resistant ones in the treated zone to find comparable injury symptoms.

A pattern in which injury occurs in irregular patches may be indicative of volatilization of a herbicide and movement of the vapors. In this case the injury usually appears in low-lying areas where air masses settle; such injury probably occurred under totally still atmospheric conditions. Microclimates also can cause localized areas of disease or frost pockets that may appear to be caused by herbicide injury.

Strips of injured areas or surviving weeds at predictable intervals indicate the possibility of a mechanical problem associated with application. Determine the width of the equipment used for application, incorporation, or in some cases tillage. Strips of surviving weeds can be the result of either driving too wide with herbicide application equipment or a plugged or too small nozzle tip on the spray boom. Strips of injured plants occurring at intervals approximately the width of the herbicide application equipment indicate an oversized or worn nozzle that provided considerably greater spray volume than others on the boom or an overlapping spray pattern resulting from driving too close. A shallow-running planter unit or a malfunctioning furrow opener on a planter that leaves the seed uncovered while the other units or openers operate properly can result in increased injury from herbicides to individual rows. Another possible cause may be carryover of a residual herbicide from bands that were applied the year before. In a solid-seeded crop, the carryover injury will be seen at inter-

---

Locate and determine the pattern of crop injury (or herbicide nonperformance) in the field. Make a map of the area.

Assess the crop condition. Take stand counts, record injury symptoms, and take photographs of the general area as well as close-ups of plants. Collect plants with a spectrum of symptoms (from injured to uninjured plants) for later study. Be sure to collect shoots, roots, or other underground structures as well as soil in case analysis will need to be conducted.

Determine whether the herbicide label was followed.

Record all important information.

Dates: of seedbed preparation, herbicide application, planting, etc.

Weather: before, during, and following herbicide application or planting, including air temperatures, rainfall, wind, and hail.

Soil conditions: moisture and condition (roughness, trash, cloddiness) at planting, at herbicide application, and now.

Soil parameters: texture, organic matter, variability, residues, other special problems. If soil samples are taken, label them with site and date of collection. Place any samples (particularly those for chemical analysis) on ice immediately.

Previous crop: herbicides used, lime, yield, weeds present.

Seedbed preparation and tillage: How, when, with what equipment, and under what conditions were these operations conducted?

Herbicide application: equipment, speed, pressure, nozzle type, operator, amount of herbicide per tank, amount of herbicide per treated area, carriers or additives. Ask to examine the sprayer and, if used, the incorporation equipment.

Fertilizer application and dates: equipment, type of fertilizer, lime.

Operations on adjacent areas: herbicide applications at about the same time the problem being investigated occurred, for example.

Weed species: If in doubt, collect plants for positive identification. Dig for perennial structures.

Weed size: at the time of treatment, tillage, planting, etc.

Determine the number of other complaints and settlements the parties have been involved in previously.

Figure 19.1 Information to collect when making an on-site investigation.

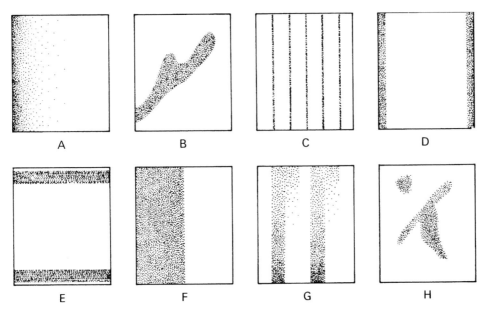

Figure 19.2 Illustrations of various types of field patterns of crop injury or nonperformance: (A) herbicide drift; (B) volatilization and vapor movement of herbicide into low-lying areas; (C) crop injury in strips the width of equipment, indicating overlapping spray pattern or worn nozzles; (D) poor control at edges of field due to light application from outside nozzle (common with normal flood setup); (E) crop injury at ends of field due to double application; (F) definite break between normal and uninjured parts of the field; (G) strips of injury due to poor mixing or inadequate spray tank agitation; and (H) spots of injury (or lack of weed control) due to soil variables.

vals that match the row width from the previous season. If row crops were grown both years, row width patterns may be more difficult to detect.

Poor control at the edges of a field can result from only half coverage by the last nozzle on a boom equipped with flood nozzles, increased weed pressure from weeds encroaching from a fencerow, lack of shading due to a thin stand of the crop, or by a combination of these factors.

Injury limited to the end rows or ends of the field usually is due to double application of the herbicide to the area either the year before or the year the injury is observed. Extra lime applications along the end rows also can cause increased carryover of herbicides that are sensitive to high pH.

A definite break between the normal or uninjured part of the field and the remainder of the field usually indicates some major difference between the two sides of the field. If crop injury is the problem, check tillage methods and dates. For example, if the uninjured side of the field was worked with a moldboard plow and the injured side with a disk or chisel plow, carryover due to lack of herbicide dilution during the disking or chisel plowing should be suspected. Carryover also may be more detectable when the crop variety is changed from a tolerant to a more sensitive one. Different dates of herbicide application and planting can be important if precipitation occurred after one but not the other. A new tankload of herbicide may be responsible

for injury if a mistake was made in mixing or if the sprayer was not adequately cleaned of a herbicide toxic to the crop.

A pattern in which crops and weeds are both killed in a strip along the edge of the field, the crops are injured in the next strip, and (in some cases) weed control is normal in the next strip indicates inadequate or poor agitation of a wettable powder or dispersible granule in the spray tank. This pattern frequently will appear again in the middle of the field or in another field when the tank is refilled. Adding herbicide to an empty tank and not properly mixing it can result in the suction line picking up high concentrations of herbicide and causing an initial overapplication followed by diminishing injury and loss of weed control. Inadequate agitation of other herbicide formulations may result in their separating into a top layer in the spray tank. Separation theoretically could cause the opposite pattern of that expected from a suspended solid that has settled in the tank.

A pattern in which individual spots in the field show crop injury most frequently is observed with soil-applied herbicides. Soil texture, organic matter, soil pH, and moisture should be determined for these areas and compared with the rest of the field. This type of injury frequently occurs on coarse-textured, low organic matter soils (see Chapter 7). Areas with high pH soils can increase the availability and herbicide toxicity of triazine herbicides. High pH extends the persistence of sulfonylurea herbicides. Other causes of spot injuries include herbicides carried into the field with crop residues, blowing soils, or water movement; allowing the sprayer to run while nozzles are being checked; water ponding; and use of highly persistent herbicides as spot treatments in previous years.

Individual spots in the field that show poor weed control may be the result of a nonsusceptible weed species, a patch of a perennial weed, a new infestation of a resistant species, or a resistant population in a formerly susceptible species. Other potential causes could be a wet soil surface when a volatile herbicide was applied, presence of a fine-textured, high organic matter area in the field, a skip in the spraying, or low pH spots (triazine herbicides). Reduced crop stand and vigor from localized flooding, poor crop emergence, disease problems, or insects also will allow weeds to grow due to lack of a competitive crop.

Identifying Injury Symptoms on Individual Plants

Symptoms that develop following the application of herbicides differ depending on herbicide mode of action. Herbicide symptoms provide valuable information for diagnosis and will be reviewed here. As the symptoms of the herbicide groups are described, you should refer to the photos in Figures 19.3 and 19.4. Additional figures can be found in Chapters 9 through 12.

Foliar-Active Herbicides

The three groups of foliar-active herbicides are those that show symptoms on new growth first (symplastically translocated herbicides), those that show symptoms on old growth first (apoplastically translocated *only* herbicides), and herbicides that cause rapid localized damage (contact herbicides that have little or no translocation).

Herbicide mode-of-action groups showing injury symptoms on new growth (smplastically translocated herbicides) are the auxin-type growth regulators, aromatic amino acid inhibitors, branched-chain amino acid inhibitors, carotenoid pigment inhibitors, and lipid synthesis inhibitors. Whether taken up by the roots or the shoots, these compounds initially move to the meristems and typically injure the youngest tissues of the plant. The young tissue will be discolored and start to die, or new growth may cease. Leaves formed prior to herbicide application will show minimal damage until the plant is near death. If plants

survive and additional growth occurs, new growth will show discoloration or malformation, and the injury will persist for several sets of leaves. Plants respond slowly to these herbicides and generally require a week or more to develop symptoms and even longer to die. Some symplastically translocated herbicides can produce symptoms both above and below the ground.

Auxin-type growth regulator herbicides such as 2,4-D differ from other symplastically translocated herbicides since they cause twisting and bending of leaf and stem tissue almost immediately after application (Figure 19.3A). Later-developing auxin-type growth regulator effects include malformed leaves and shoots, proliferation of cortex tissue in the stem, malformed and adventitious roots, and fusion of plant parts.

Aromatic amino acid (EPSPS) inhibitors (glyphosate) show yellowing of shoot tips (Figure 19.3B) a few days to a week or more after application depending on dose and sensitivity of the treated plant. Injury progresses from new to old growth.

Branched-chain amino acid (ALS) inhibitors include the imidazolinone (e.g., imazapyr), sulfonylurea (e.g., chlorsulfuron), and triazolopyrimidine sulfonanilide (e.g., flumetsulam) herbicides. New shoot growth ceases, and yellow streaks develop between veins in grass leaves (Figure 19.3C). In broadleaves the blades sometimes yellow and the veins become red to black (Figure 19.3D). Roots develop a bottlebrush appearance (Figure 19.4; see also Figure 9.9). Symptom development takes a week or more.

Carotenoid pigment inhibitors include norflurazon, amitrole, clomazone, and fluridone. White shoot tips are the first symptoms to develop (Figure 19.3E). Plants continue to generate bleached leaves, and plant death usually is extended over several weeks. Plants not killed eventually recover and show a group of leaves with pigment loss. Leaves above this point show gradually increasing amounts of green until all leaves appear normal.

Lipid biosynthesis (ACCase) inhibitors (aryloxyphenoxy propionate and cyclohexonedione herbicides) destroy grass meristems. Mushy sheaths develop on the newest leaves, and the growing points just above the nodes disintegrate. It requires about a week to kill susceptible annuals and somewhat longer for established perennials. Plants other than grasses rarely respond to these herbicides.

Herbicides that show symptoms on old growth first (apoplastically translocated only herbicides) consist of the photosynthesis inhibitors such as the triazine, urea, and uracil herbicides. Their movement is up

Figure 19.3 Injury symptoms caused by foliar active herbicides: (A) Auxin-type growth regulator; (B) Aromatic amino acid (EPSPS) inhibitor; (C) Branched chain amino acid (ALS) inhibitor—grass; (D) Branched-chain amino acid (ALS) inhibitor—broadleaf; (E) Carotenoid pigment inhibitor; (F) Photosynthesis inhibitor; (G) Rapidly acting cell membrane destroyer.

Figure 19.4 Injury symptoms caused by herbicides applied to the soil only. From left to right: Untreated shattercane seedling (10 days old); shoot inhibitor at normal dose; shoot inhibitor at high dose; root inhibitor at normal dose.

and out of the plant with the transpiration stream; thus, symptoms are observed primarily in the shoots. The older leaves are affected first; very young leaves remain normal as long as the plants survive (Figure 19.3F). Thus, the plant shows injury progressing from the bottom to the top of the shoot. Individual leaves show most injury at their tips and margins or between the veins. Usually only the true leaves, not the cotyledons of injured plants, show symptoms. No obvious injury to underground portions of the plant occurs except for eventual stunting due to poor shoot growth. If excessive doses are applied or a species is very sensitive, symptom development will occur so rapidly that injury will appear much the same as that for cell membrane destroyers. The only difference is that injury symptoms caused by photosynthesis inhibitors do not appear until 2 to 5 days after treatment, whereas injury from cell membrane destroyers occurs within a day or so of application. Once symptoms appear, the plants die very rapidly.

Rapidly acting cell membrane destroyers *(contact herbicides)* include the photosystem I energized bipyridilium herbicides, herbicides that inhibit protophyrinogen oxidase, herbicides that inhibit glutamate synthesis, and herbicides that cause the direct destruction of membranes. Although several modes of action are involved, the symptoms look the same. Plant tissue is killed on the parts of the plant contacted by the spray (Figure 19.3G). Movement is limited or nonexistent, so incomplete coverage results in spotting or kill in localized areas. Surviving tissue continues to grow. Thus, if buds are not killed, the new growth will be normal. Most of the damage from such treatments is evident within a few days of the initial application. Recovery, if any, will commence within a week. Top kill is about all that can be anticipated with these herbicides.

Soil-Active Herbicides

Herbicides applied almost exclusively to the soil act primarily on meristems of germinating seedlings. Young plants will exhibit the most symptoms. We distinguish three groups of herbicide symptoms: herbicide symptoms that show up first on roots by inhibiting microtubule and spindle fiber formation and function and cause club-shaped roots; those that show up first as malformed, twisted, and stunted shoots; and those that are less specific in their effects.

Microtubule-spindle apparatus inhibitors (root inhibitors) are characterized by poor root development and swollen root tips. The shoots remain relatively normal, although at high doses the shoots also can be stunted. The dinitroanilines and DCPA are the most common examples of herbicides that cause this type of injury. The plants must be dug up to detect symptoms on the roots. The shoots may show water stress and an accumulation of reddish pigments.

Shoot inhibitors cause the tops of the plants to be malformed and twisted. Chloroacetamides and thiocarbamates cause these symptoms on grasses. The tips and edges of the leaves are most affected. Where injury is moderate, looping of the leaves often occurs because the base of the leaf continues to grow while the leaf tips remain stuck together. A dark green color often can be observed when injury occurs. Chloroacetamides produce less obvious effects on the leaves of broadleaves than thiocarbamates so that grasses are less likely to recover from injury than broadleaves.

Figure 19.4 shows a side-by-side comparison of shattercane seedlings that have been treated with soil-applied herbicides. Compared to the untreated seedling, the shoot inhibitors at normal doses cause stunted and twisted shoots but relatively normal roots. When the dose of shoot inhibitor is increased, both shoots and roots are inhibited. The root inhibitors result in seedlings with sparse, club-shaped roots and normal shoots.

The imidazolinone and sulfonylurea herbicides, which can be either foliar or soil applied, affect newly developing tissues by inhibiting branched-chain amino acid synthesis. Although their mode of action differs from those of the root and shoot inhibitors, their symptoms on seedlings result in both stunted (bottlebrush) roots and stunted shoots (Figure 19.5).

Nematodes, soil compaction, and cold weather can cause injury symptoms that can be confused with those caused by the root inhibitor herbicides. Interactions between herbicides and insecticides also can cause crop injury. An example is the organophosphate insecticides, which increase injury from the sulfonylurea herbicides on corn and injury from metribuzin on soybeans. Poor performance sometimes can be attributed to antagonistic effects, such as when the lipid biosynthesis (ACCase) inhibitors are mixed with other postemergence herbicides (see Chapter 16).

INTERPRETING YOUR FINDINGS

Once you have collected the information from your investigation, you will need to study it carefully and thoroughly to determine what it

Figure 19.5 Injury symptoms on shattercane seedlings caused by branched-chain amino acid synthesis inhibitors. Left: untreated seedling; Right: treated seedling showing bottlebrush root and stunted shoot.

says. You should ask yourself several questions. Do you have enough data available to reach a clear-cut conclusion? Is there some element of the story that is inconsistent with the other data or with what is generally known about this type of problem? Are herbicides the only possible cause for the problem? (A major problem is having enough experience and breadth of knowledge to recognize all possible causes.) Does another expert need to be consulted? Are any other data available? If so, how long will it take you to obtain them and from what sources are they available?

Always consider the possibility of a combination of factors such as cold, wet soil and a marginally high rate of herbicide, which together lead to crop injury. This combination may be the cause of the problem for a grower who, until this year, had been getting away with using marginal or even unlabeled rates of the herbicide.

Eventually, you must reach a conclusion as to what did or did not cause the problem. Providing an answer within a few hours or days of the initial visit often is expected and may be necessary. If no clear-cut solution is available, then a tentative conclusion (if warranted) or at least a report of the findings to date is required. Avoid giving conclusions, statements, and recommendations that you cannot substantiate. Figure 19.6 presents a starting point for considering possible causes for the two major herbicide-related problems, lack of performance and crop injury.

The results of an investigation can be used for a variety of purposes ranging from simply satisfying someone's curiosity about what went wrong to providing expert witness data for a court case. In other instances, the results may be used to change practices in order to avoid the problem in the future, to make an immediate decision that will minimize the impact of the current problem, or to project a probable yield loss for a claim settlement.

When quick action is needed to minimize the impact of a problem, decisions as to cause and a course of action must be made as soon as possible. In the case of lack of performance, the decision is one of selecting an appropriate control method still available under the circumstances. The decisions related to preventing further herbicide injury tend to be more involved and difficult to make.

One set of conditions requiring rapid action arises when relatively small areas have received an unwanted soil residual herbicide. The problem may have arisen from application of a herbicide to the wrong place, application of the wrong herbicide, a spill from a container or sprayer, or

Lack of herbicide performance
Resistant weed species
Weed at incorrect stage when sprayed
Weed under stress at time herbicide applied
Environmental conditions affected herbicide performance (cold temperatures prevent activity; low humidity with foliar applications)
No rain after application of soil-applied herbicide
Excessive rain caused herbicide to leach out of weed seed zone
Rain washed off foliar-applied herbicides
No wetting agent added to foliar-applied herbicide solution
Poor coverage with a contact herbicide
Rates improper for soil type
Used a lower rate than suggested
Improper application or poorly calibrated equipment
Picked a herbicide inconsistent in performance for the weed species present
Loss of activity due to antagonism with another chemical
Weeds had time to recover after application of short-lived herbicide
Late weed germination after herbicide effectiveness dissipated
Used slow-acting herbicide for which symptoms are not yet obvious
Given evidence available, reason cannot be determined

Crop injury
Drift during application (or volatilization and then drift of vapors)
Too high a rate applied
Herbicide applied at improper stage of crop growth
Foliar applications made in liquid fertilizer, crop oil, or some other chemical when they should not have been
Weather resulted in higher than normal injury to crop (high temperatures and humidity with foliar-applied herbicides; cool, cloudy, wet conditions with soil-applied herbicides)
Another problem such as insect, insecticide, or disease damage that, when added to the herbicide injury, resulted in considerably greater crop damage than any single factor alone
Crop variety susceptible to herbicide
Herbicide dependent on depth placement for selectivity placed too near or on the seed
Excessive rain on a normally immobile herbicide

Figure 19.6 Possible causes of lack of performance and crop injury associated with herbicides.

vandalism. The goal should be to remove the herbicide from the area and to reduce as much as possible the potential for injury beyond that already encountered. The first step is to contain the affected area by building a dike or to remove the contaminated soil and dispose of it in a manner that will not cause additional problems or violate disposal of hazardous waste regulations. Soil itself is a reasonably good absorbent, so removal before the herbicide has had a chance to leach through the soil or move with runoff water can serve to reduce greatly the amount of chemical present. A second step should be to use a strong substance such as activated charcoal to help adsorb any remaining herbicide. Most triazine and urea herbicides respond well to such treatments. Approximately 100 to 400 times (on a pound/pound basis) as much charcoal as remaining herbicide should be used. The last step is to replace the soil with uncontaminated soil and to manage the vegetation as appropriate.

A second set of conditions requiring a quick decision involves the replanting of crops. Herbicide injury may have resulted in stand loss or in severe stunting, and an immediate decision must be made on whether to leave the crop in the field or to tear it up and replant. Three important factors should be considered when making this decision.

The first is the current crop stand. What will be the probable yield of the surviving crop if left until harvest? This decision requires a judgment as to how many plants can be expected to recover. Plants that show no injury can be counted as providing full yields. The difficult part is to determine the response of the injured plants. If the herbicide causes short-term injury (e.g., a contact herbicide or a herbicide having only short-term soil activity), the plants probably will develop normally as soon as normal growth reappears. When effects tend to continue for long periods of time (e.g., with symplastically translocated or residual soil-applied herbicides), estimating damage is much more difficult. As long as there is life there usually is hope, but whether that life can be sustained depends on the nature of the herbicide and the condition of the growing point of the crop plant involved.

The second factor to be considered in replanting is the crop yield that can be obtained with replanting. A 3- to 6-wk delay from planting may have taken place by the time the injury is noted and the investigation conducted. Thus, yields may be considerably reduced even if the crop is replanted. Options are to leave the stand as it is, to replant and take the reduced yield, or to switch to another crop that can tolerate the levels of herbicide remaining. If severe loss in stand appears likely or has occurred, the decision may be to take the loss rather than incur further expense by replanting.

The third factor relates to whether the replanting actually will help to circumvent the problem. If a residual herbicide at excessive dose is involved, replanting will result in another crop of injured plants (unless the herbicide can be diluted with tillage or the plants can be protected with adsorbent).

In cases in which a damage claim is involved (or the potential exists for such a claim), it is absolutely essential that good notes and photographs be taken, dated, and kept. Such claims frequently are settled by an insurance company and an assessment of probable yield loss usually is required. The most precise way of determining loss is to take yields at harvesttime on plants not under stress but otherwise under identical conditions to the injured plants. Finding an appropriate uninjured area can be a problem.

Some claims invariably end up in court. When this happens, something has gone wrong: either one or both sides are unrealistic, someone is trying to take unfair advantage of a difficult situation, or an untenable or inflexible stand is taken by one or both of the parties involved. Such a situation is perhaps the greatest test of the person who works in the herbicide area and requires some experience with people and an intimate knowledge and understanding of both specific information and basic principles related to herbicides and weed control.

MAJOR CONCEPTS IN THIS CHAPTER

Complaints involving herbicide use most frequently involve poor weed control, crop injury, and off-target vegetation injury.

As an expert called into a problem situation, your goal is to determine accurately the extent and cause of the problem and to avoid antagonizing or alienating the parties involved. If more than one party is involved, chances are good that all are clients or constituents of yours.

Be prepared psychologically to encounter an unhappy constituent. In doing so you can keep a cool head and avoid embarrassing mistakes and even improve relations.

Answer the complaint as soon as possible and make it clear from the start that your initial visit is a fact-finding trip.

Ask questions and be attentive to answers. Have the grower repeat the details several times, then inspect the problem area and have the grower explain the situation again. Ask the parties involved what they think caused the problem.

Make a thorough investigation on your own; that is, investigate as though you know nothing about the situation. Keep your mind and eyes open. Inspect appropriate application, incorporation, and planting equipment if possible.

Record your observations and the information provided by the involved parties. Take any data, samples, or pictures needed to document or solve the problem. Separate opinion from facts.

Leave the scene without making any incorrect or embarrassing statements.

At the first opportunity, isolate yourself and carefully study your data, observations, and discussions.

Seek additional expert help or advice if you are unsure of your conclusions.

Provide the parties with an answer as soon as possible. Urgency may be the key to actions that can minimize the impact of the problem. Be accurate, fair, and clear in your response.

Determine the pattern of injury in the field or problem area. Look for a gradation of injury, spots, sharp breaks, patterns corresponding to equipment widths, areas limited to end rows, differences in soil type, low-lying areas, and so on.

Remember to consider causes of injury other than herbicides: other pesticides, frost, drought, flooding, fertilizer burn, soil compaction, pest damage. Consider the possibility of herbicide interaction with some other problem.

Examine individual plants closely to determine symptoms and injury patterns. Note the condition of new growth and buds. Observe species differences.

If some type of action is needed to minimize the impact of the problem, report as soon as you can reach a sound conclusion. Potential actions include controlling weeds, replanting a crop, and removing a herbicide residue (usually from a limited area).

Try to make the interaction with the grower or complainant helpful and educational so the problem can be avoided in the future.

SELECTED REFERENCES ON TROUBLESHOOTING

Agri Growth Research, Inc. 1990. Guide to Herbicide Injury Symptoms in Corn. Agri Growth Research, Inc., 4040 Vincennes Circle, Suite 500, Indianapolis, IN 46268.

Agri Growth Research, Inc. 1990. Guide to Herbicide Injury Symptoms in Soybeans. Agri Growth Research, Inc., 4040 Vincennes Circle, Suite 500, Indianapolis, IN 46268.

Tickes, B., D. Cudney, and C. Elmore. 1996. Herbicide Injury Symptoms. University of Arizona, Cooperative Extension, Publication No. 195021.

Web Sites for Herbicide Injury Symptoms:

Crop Injury Symptoms, University of Illinois:

http://w3.ag.uiuc.edu/maize-coop/crop-inj.html

Guide to Herbicide Injury Symptomology, University of Missouri:

http://psu.missouri.edu/agronx/weeds/muhrbinj.html

Herbicide Injury Symptoms on Corn and Soybeans, Purdue University:

http://www.btny.purdue.edu/Extension/Weeds/HerbInj/InjuryHerb1.html

Herbicide Symptomology for Southern Crops, University of Georgia:

http://weed1.cropsoil.uga.edu/css434/symp.html

WEED CONTROL IN AN AGE OF RAPIDLY CHANGING TECHNOLOGY

Even to the most casual observer, advances in science and technology must seem to be occurring at an extremely rapid pace as we approach and enter the new millenium. Along with these new technologies, pressures to save the environment in the face of an everincreasing world population and the need to produce food, feed, and fiber profitably in a global economy present numerous challenges as well as opportunities for the agricultural sector. Such challenges and opportunities also await the science of weed management. They include the application of genetic engineering and other new technologies, some of them space age in nature, to weed control and many other aspects of agriculture and everyday life. Changes in how agribusiness discovers, markets, and targets new products, and how agribusiness responds to increasing laws and regulations all will have an impact on how weeds are controlled in the future.

Let us take a look at some of the new technologies, approaches, and limitations that will govern how agriculture and weed management are conducted in the future. As we do this, however, keep in mind that not all new equipment, ideas, and products will pan out. New technology, no matter what it is, often comes with a level of promise and promotion that results in unrealistic expectations. In addition, when technology is hardware dependent, the initial equipment may be cumbersome, difficult to operate, poorly fitted to the intended use, and fragile (just ask people who first started working with

computers). The early adopter of such equipment may spend far more time and effort learning how to use it or adapt it to the situation than is justified by the results.

If you are considering adopting new technology, approach it in an analytical manner. Who is promoting the technology and why? What are the initial and continuing costs for equipment? How much effort on your part will be required to make the equipment work? Are you willing to make or capable of making the commitment needed? If a service provider is involved, does this lock you into an elaborate management program? How user-friendly and dependable is the technology? Is good support service available? Can you accomplish the same objective more economically some other way?

Even with these precautions, however, the way in which agriculture and weed control will be practiced in the future is changing. The major drivers of these changes are economics (maintaining competitiveness in a global economy) and environmental awareness and its accompanying regulations. To achieve these goals, major efforts are being made to reduce external inputs (e.g., fertilizers, pesticides) and to reduce the negative impacts of agricultural practices on the soil, water, air, and surrounding urban environment (e.g., the prevention of erosion). The challenge of agribusiness is to provide environmentally friendly products at a price that will be competitive in the marketplace but profitable to the manufacturer or provider. We will discuss the following technologies, trends, and approaches: the application of space age technology (GPS/GIS and remote sensing); the trends in agribusiness discovery, marketing, and direction; weed modeling; and management approaches to minimize environmental concerns.

APPLICATIONS OF SPACE AGE TECHNOLOGY

Many of the ongoing changes in agriculture involve the use of space age technology to monitor, map, and rapidly process the information needed to make decisions that impact crop and weed development and management inputs. This technology can be used for specific locations, including individual fields and portions of individual fields. It allows the application of management practices, such as the application of fertilizers or herbicides, to only those portions of the field that require them. In this way, the grower saves money, and fewer inputs are added to the environment. Crop production using this technology is called Site Specific Farming or Precision Agriculture. Such an approach requires the accurate location of data collection points in the field, the mapping of such points, and the manipulation and use of the data for decision making and field operations.

These technologies fall into two major categories: those obtained through electronic signals (GPS) and those obtained through visual imaging (remote sensing).

Global Positioning Systems (GPS) and Geographic Information Systems (GIS)

GPS and GIS technology is associated with satellite systems initially deployed for use by space programs and the military. GPS is based on the use of a military network of two dozen satellites circling the earth in six different orbits to provide a potential of eight satellites electronically visible from any spot on earth at any one time. Obstacles on the ground (mountains, buildings, trees) can block access of some satellites, others may not function properly, and some may be in less than ideal positions to provide optimal data. As a result, good signals may be available from fewer than eight satellites.

A receiver on the ground communicates with "visible" satellites via electronic signals, which in turn allows the receiver to record its exact location on the earth's surface as well as receive information about its location. Receivers can be handheld, mounted in an automobile, or for agricultural purposes, mounted on a combine or soil-sampling vehicle. Three satellites are needed to provide horizontal location and four are needed to provide both horizontal and vertical location. Fairly simple and low-cost receivers provide accuracy to just over the length of a football field (100 m). One way to increase accuracy to smaller distances is to use the receiver in conjunction with land-based transmitters, such as a Coast Guard beacon, tied in to the satellite systems. With this technology, measurements can be processed and corrected to pinpoint a location within 1 to 5 m.

In addition to the GPS location information, features of the location are recorded by on-site sampling for crop yield, soil type, pH, organic matter, fertility, moisture level, and other pertinent parameters. Additional data sometimes are obtained from remote sensing (see the following section). All of this information is fed into GIS software to store, arrange, and apply the data for mapping, interpretation, and other forms of utilization.

Remote Sensing

Remote sensing is a term for photographs or digitized images taken from satellites or aircraft to show plant condition or soils. Light reflected by the soil or plant surface is captured either as digitized images or by certain kinds of film, such as color infrared film, for photographs. Captured red wavelengths show chlorophyll amounts in plants, and near infrared wavelengths are highly correlated with plant density. Parameters such as the health, distribution, and density of vegetation can be estimated. Remote sensing coupled with on-site inspection and GPS/GIS-generated maps can zero in on problems such as those caused by disease, moisture stress, soil compaction, insect infestations, and invading weeds, as well as other problems limiting crop development. Remote sensing satellites for agriculture and business are being placed in orbit through the cooperative efforts of NASA, the USDA, and private companies.

Spatial resolution (the diameter of the smallest detectable area that makes up a dot [pixel] on the image produced) limits the utility of remote sensing for weed detection. State-of-the-art satellite equipment available for non-military use is limited to 4 to 15 m at best. Aircraft can get resolution to just under 1 ft.

Using GPS/GIS and Remote Sensing Technologies

Although large weed patches and blankets of weeds should be visible using remote sensing, emerging weeds may not be detectable soon enough to determine optimal removal times. The technology can be used to distinguish certain weed species with spectral qualities that allow them to be detected from among a rather uniform plant population, such as a grassland. For example, invasive rangeland weeds such as leafy spurge, St. Johnswort, and yellow star thistle have been accurately mapped with the combination of GPS/GIS and remote sensing technologies (Everitt et al. 1995, Lass et al. 1996). Such maps provide important baseline data with which to monitor spread of the weed species or the effect of weed control efforts such as the release of biocontrol agents. Unfortunately, the technology to detect individuals in a mixed stand composed of many weed species, such as in a typical cropping situation, is still in the development stage.

The major uses for GPS/GIS and remote sensing technologies in agriculture are to monitor crop yields and to apply variable amounts of seed, fertilizers, or pesticides to fields with variable soil types or pest populations. An example of how GPS/GIS and remote sensing are used in these ways is as follows.

Soil samples are taken from a field (using a grid or modified grid sampling technique or by soil type) and analyzed for parameters such as pH, phosphorus, potassium, organic matter, and soil texture. Each sample location is determined using a GPS receiver. A map then is generated that shows the outline of the field and the sample locations. Overlays for the map show the distribution of each parameter. For example, one overlay may show the amounts of organic matter at each site in the field and another overlay may show the soil pH at each site (Figure 20.1). Recommendations of suggested doses of chemical (e.g., lime, fertilizer, or soil-applied herbicides that are affected by soil parameters such as organic matter or pH) are made to match each sample site. An application map is made based on the recommended chemical doses and sample locations. Chemical is then applied to the field with a GPS-assisted variable-rate applicator that adjusts the amount of the chemical applied according to its location in the field. (Equipment adapted for GPS-assisted variable-rate applications can be purchased from a number of equipment companies.) At harvesttime, the combine is equipped with a yield monitor to provide a final map overlay of crop yield at each location in the field. Problems are noted and recorded.

During the growing season, fields should be monitored for escaped weeds, disease or insect problems, flooded areas, and other problems. Monitoring is usually done with scouting (people trained to identify and count weeds and insects, or assess disease severity). The impact of these adverse conditions on the health of the crop can be detected using remote sensing. Remedial action is taken as needed during the growing season. It is always useful, however, for the combine operator to compare exact locations of weeds at harvest with the crop yield data to determine if weed growth might have been the cause of any yield loss.

Weeds present at harvest should be recorded and mapped. Caution, however, is needed when associating the presence of weeds at harvest with yield. The logical explanation (and frequently the correct one) is that the weeds interfered directly with crop development. On the other hand, the weeds may be there simply because the crop lacked enough stand or vigor to provide shading, and the weeds came in too late to cause a yield loss. Poor crop stands, disease problems, standing water, or dry spots could be the culprits rather than poor weed control. The information on weed location and yield coupled with the data on all the other pest and cultural variables can be used to help determine actions to be taken the next season.

AGRIBUSINESS DISCOVERY, MARKETING, AND DIRECTION

The introduction of genetically altered crop cultivars, the introduction of herbicide combinations as tank mixes and prepackaged multiple component mixtures, and the impact of mandates regarding pesticide registration and reregistration on profitability are all factors that are forcing changes in the herbicide industry. The possible impact of each of these trends is discussed in the following sections.

Genetically Altered Crop Cultivars

The number of genetically altered crop cultivars with tolerance to herbicides is increasing rapidly (the technology and benefits and risks associated with genetically modified crops are discussed in Chapter 8). The ability to insert herbicide resistance genes into formerly susceptible crops holds exciting potential and will continue to improve the way weeds are controlled.

With any new technology, however, there are also concerns. Some of these concerns are due to the major changes that genetic engineering is forcing in the herbicide industry. The introduction of herbicide-resistant cultivars requires cooperation between chemical manufacturers, the source

Field 203 – May – Modified Grid – Fall 1997

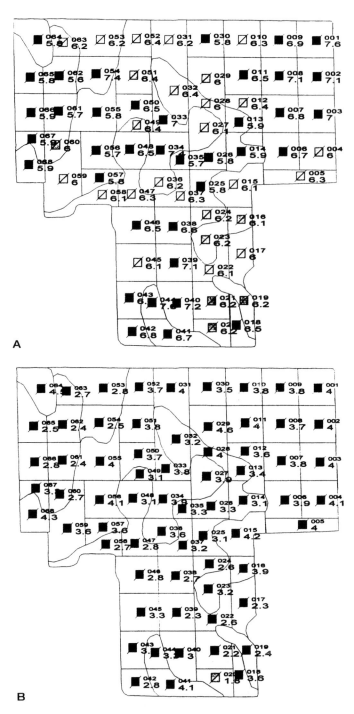

Figure 20.1 Overlays of (A) pH and (B) organic matter values for a single field. Soil type boundaries also are visible. Sample locations, maps, values, and soil type boundaries provided by a GPS/GIS system.

of most resistance genes to this point, and seed producing companies. Several herbicide companies have acquired formerly independent seed companies, have a major shareholding interest in such companies, or have written agreements with them. The potential for a single company to control both the herbicide technology and the seed production components of crop production is very real. Should these changes cause concern? They will if they result in monopolies, in other words, a few companies that control development of resistant varieties and reduce competition within the industry. In addition, there is a potential to end up with restricted germ plasms that may not appropriately meet new problems, or to lose the incentive to develop new products and strategies for weed control. Another concern about genetic engineering technology is the extra workload it places on plant breeders. The effort to introduce a herbicide resistance gene into a crop competes with breeding for higher yield, disease resistance, increased crop quality, and the insertion of genes for other pests (e.g., the BT gene for insect control). The seed producer will need to recover the cost of extra breeding, probably from the grower.

We hope that industry pays heed to these concerns and that sound decisions are made as genetically altered crops are introduced to the market.

Herbicide Combinations

Until now we have purposely limited our discussion in this book to individual herbicides. In fact, more individual herbicides are available than we can cover adequately in an introductory textbook. Discussing herbicide combinations does little to illustrate most of the principles of weed management other than antagonism and synergism and possibly the complexity involved when selecting herbicides for solving weed problems. Herbicides, however, have been tank mixed or sequenced ever since herbicide use began, and the mixing of herbicides is a major strategy.

Prepackaged multiple component mixtures, consisting of two or more basic chemicals, are becoming more and more common in the marketplace. Some of the available prepackaged mixtures for corn and soybean use are listed in Chapter 14.

The most compelling reason to mix herbicides is to broaden the spectrum of weed species controlled. Nearly all mixtures do this. Several other performance benefits can be realized as well. The negative impacts of a herbicide can be reduced when its dose is lowered (while retaining some of its weed control benefits). The loss in weed control can be offset by the addition of low doses of the complementary herbicide. For example, lower doses of two herbicides in a mixture may result in less carryover (one herbicide is more persistent at normal doses), greater crop safety (one herbicide is more likely to cause injury at the normal dose), or reduced inconsistency due to weather (one herbicide is highly mobile and could move out of the root zone with excessive rainfall). A more environmentally sound product can be the result of proper mixing. More research has been conducted on the performance of prepackaged mixtures than on tank mixes of herbicides.

Industry has additional incentives for marketing prepackaged herbicide mixes. One is to match a high-priced product with a low-priced product to provide a less costly but effective combination. Two or more herbicides with limited utility can make a single broader spectrum product that is competitive in the marketplace with single herbicides. Prepackaged mixing is a particularly good strategy for a company that has proprietary rights to all of the chemicals in a mixture. When sold individually, the products may not gain nearly the market share that they do when combined. Recently the trend has been to combine a herbicide that has no stand-alone market with two or three other herbicides.

Prepackaged mixing has many advantages. Probably its major disadvantage is that it is resulting in a proliferation of packaged products that makes it difficult for the grower or dealer to keep track of and store a large inventory.

Mandates from Regulatory Agencies

The restrictions mandated by the EPA and other regulatory agencies regarding pesticide registration and reregistration will dictate which new herbicides will make it to the market and which old herbicides will remain on the market. Clearly this trend will result in products that are more environmentally safe and more benign to people, domestic animals, and wildlife. Another positive result of environmental awareness on the part of industry has been the development of returnable herbicide containers or disposable containers (e.g., water-soluble packets in noncontaminated bags, boxes, or jars; see Chapter 16). This technology reduces not only environmental contamination but human exposure as well.

Such environmentally sound approaches, whether regulated or voluntary, come at a financial cost. The cost of registration and reregistration continue to increase due to the increasing demands for more data or new kinds of data on residues in food and environmental effects. As far as industry is concerned, profits from sales must exceed the cost of development. At present, the estimated cost to develop a new herbicide for market is $50 million. Patents protect the right of the developer to market the product for less than two decades, but it may take one third of that time just to reach the market with the new product. Keeping a herbicide on the market requires reregistration, and this can be costly as well. An unfortunate consequence of high costs is the loss from the label of otherwise good and environmentally sound uses that generate limited sales, even if the product itself survives.

WEED MODELING

Weed modeling is the mathematical simulation of weed development based on a number of weed-related parameters and their effects on crop development. Examples of weed parameters might include weed density, weed seed numbers in the soil, weed seedling emergence from the soil, or plant height or leaf area at different stages of development or in response to some environmental variable. These parameters usually are evaluated for their effect on crop yield or quality. Measurements are taken of weed and crop development, usually over a period of several years; numerical values are assigned to these parameters; and a mathematical model is developed from them. Modeling can be used to predict the impact of a weed on a crop based on initial weed seed densities in the soil, or to predict how a management strategy might affect weed density, which in turn will affect crop yield. Often economic values are added to the model so that the user can predict the amount of money gained or lost through various management scenarios. Models can be developed and used to predict interactions in systems other than weed–crop interactions. An example includes the surface movement of a herbicide in response to rainfall, slope, soil, management practices, and time of herbicide application.

The major practical application of models is that it allows users to have access to a greater database than they would otherwise have. Ideally, the user might take weed seed counts prior to the beginning of the growing season and enter them into a personal computer-based predictive model. The model may then allow the user to determine how many seeds will germinate, what the impact of the germinating seedlings will be on crop yield, and the alternative control practices.

Such elaborate models have been slow to reach a stage at which growers can use them. The problem is that numerous measurements have to be taken on many weed species, crops, and environmental parameters in order to generate the data on which to base the model. The model then must be validated over a period of years in the field. Considerable research is being conducted to develop more complex models with a precision that will be valid under most circumstances.

The first models available are relatively simple and assist the user with herbicide selection. Several states provide herbicide selection programs based on models. The user enters weed species, rank of abundance or severity of weed species in the field, soil parameters if the choice of herbicide is to be soil applied, management system (e.g., no-till, conventional, chisel, etc.), crop, and sometimes expected yield and price. The computer model will then provide a list of herbicide choices or other management methods best suited for that situation.

We anticipate that as models continue to be developed and improved, they will become an important part of the decision-making process in weed management.

MANAGEMENT APPROACHES TO MINIMIZE ENVIRONMENTAL CONCERNS

Political and social pressures to protect the environment have resulted in increasing numbers of laws and regulations that pertain to agriculture. The Environmental Protection Agency is the major regulatory agency responsible for protecting the environment. One of the EPA's missions is to prevent environmental damage from pollutants. Targeted pollutants with an impact on weed management include eroding soil, herbicides, and organic solvents.

Efforts to promote management practices considered to be environmentally friendly have resulted in terms such as *integrated pest management* (IPM), *integrated weed management* (IWM), *sustainable agriculture, environmental stewardship,* and *best management practices.*

Integrated pest management originally was intended to reduce the inputs of environmentally damaging insecticides. The approach combines information on the biology and ecology of the pest with all available control technologies in an effort to develop an integrated management program. The implication is that if all applicable methods, including nonchemical methods, are used, then the emphasis on chemical approaches can be minimized, and perhaps even eliminated.

Integrated pest management has since been expanded to include the management of other pests, including weeds (IWM), and even total crop management. What is now called integrated crop management (ICM) takes into account additional agricultural decision-making tasks such as fertilizer and soil water management. This broader view of crop management seems more in tune with the goals of best management practices, sustainable agriculture, and environmental stewardship. When applied to vegetation management, all of these goals are committed to efficient production, judicious use of resources, and maintaining or improving environmental quality.

The term *sustainable agriculture* is used to define management systems in which minimal inputs are added to the system. The goal of sustainable agriculture is that the system be self-sustaining—for example, animal manure is recycled for fertilizer, pesticides are eliminated or limited, and animals are fed from crops grown on site. From the standpoint of weed control, the elimination of herbicides means that the grower is dependent on tillage and hand weeding. Sustainable agriculture is likely to be labor intensive; therefore, the cost of production under this system tends to be high and may limit its utility to high-value crops. Although sustainable agriculture has not been adopted rapidly in this country, its popularity, and that of organic gardening, may well increase as the public accepts the fact that protecting the environment is going to be expensive. Sustainable agriculture is most suitable for high-value, low-acreage crops and for developing countries where hand labor is cheap and plentiful and external inputs, such as fertilizers and pesticides, are relatively expensive.

Implications for Weed Control

As we have noted in other parts of this textbook, the most economical, broad spectrum, dependable, and effective control practices for weeds are tillage, herbicides, and sometimes mowing. There are no likely replacements for these methods in the first decade of the millenium.

A continuing problem will be the general perception that synthetic chemicals are bad (especially any used to control weeds or other pests) and that any alternative to pesticides is good. We maintain that modern herbicides are efficient, dependable, and environmentally safe, and that with special efforts they can be even more so.

This is not to say, however, that the goals of IPM or IWM, or even sustainable agriculture, cannot or should not be achieved. Most herbicide applications today are planned ahead of the growing season based on the weed species present, soil types, and crop. The intent is to make the planned herbicide application(s) at a high enough dose to yield a crop relatively free of weeds at harvest in spite of less than ideal timing and weather conditions. Additional control may be required in the occasional season when very unusual conditions arise. The alternative approach, and one more in tune with IWM, is to

carefully determine weed distribution, density, and stage of growth and apply the correct amount of herbicide needed to obtain control at the time of initial action. Monitoring is continued throughout the growing season to identify escapes or later-emerging populations, which will be treated only when warranted. This approach requires a certain level of sophistication. Scouts must monitor the field, which increases the expense, although the reduction in chemical use should offset scouting, mapping, and respraying costs. This sort of approach has been successful in the fruit and vegetable industry in California and for insect control in cotton when applying the pesticide only when needed can greatly offset the expense of traditional multiple pesticide applications. The approach has not been as widely adopted in low-value crops such as corn and soybeans. In the future, the goals of IPM and IWM can be enhanced by the integration of GPS/GIS or remote sensing and computer models into the overall management plan.

MAJOR CONCEPTS IN THIS CHAPTER

Rapidly changing technology; pressures to save the environment in the face of increasing populations; and the need to produce food, feed, and fiber in a global economy present numerous challenges and opportunities for agriculture and weed management.

New technologies must be approached with caution and in an analytical manner. Sometimes new technology promises more than it can deliver. In addition, the user must determine whether the technology is suited for his or her needs.

The future for weed management will be dictated in part by the application of space age technologies; new trends in agribusiness discovery, marketing, and direction; weed modeling; management approaches to minimize environmental concerns; and new equipment.

Space age technologies include GPS/GIS systems and remote sensing. GPS and remote sensing use satellite or airplane imaging to map fields. On-site sampling provides data on such parameters as soil pH, organic matter, fertility, and pest species and populations. These data are plotted on the maps using GIS technology.

The major uses for GPS/GIS and remote sensing technologies are to monitor crop yields and to apply variable amounts of seed, fertilizer, or pesticides to fields with variable soil types or pest populations.

Changes that are occurring in agribusiness include the introduction of genetically altered crop cultivars, the introduction of numerous herbicide combinations as prepackaged mixtures, and the impact of mandates regarding pesticide registration and reregistration on profitability.

Weed modeling, or the mathematical simulation of weed and crop development and their interaction, allows the user to predict the impact of weeds on the crop and the impact of management methods on weeds and thus on the crop. Initial grower use of computer models has been to aid in herbicide-selection.

Efforts to protect the environment have resulted in crop management systems that minimize external chemical inputs, particularly pesticides. Such systems include integrated pest management (IPM) and sustainable agriculture. IPM can be accomplished by monitoring fields during the growing season and recommending pesticide application only when needed.

TERMS INTRODUCED IN THIS CHAPTER

Geographic Information Systems (GIS)
Global Positioning Systems (GPS)
Integrated Crop Management (ICM)
Integrated Pest Management (IPM)
Integrated Weed Management (IWM)
Precision Agriculture
Remote sensing
Site Specific Farming
Sustainable agriculture
Weed model

SELECTED REFERENCES ON NEWLY EMERGING TECHNOLOGIES

Buhler, D. D., R. P. King, S. M. Swinton, J. L. Gunsolus, and F. Forcella. 1997. Field evaluation of a bioeconomic model for weed management in soybean (*Glycine max*). Weed Sci. 45:158–165.

Mitchell, K. M., D. R. Pike, and H. Mitasova. 1996. Using a geographic information system (GIS) for herbicide management. Weed Technol. 10:856–864.

Oderwald, R. G., and B. A. Boucher. 1997. An Introduction to Global Positioning Systems. Kendall/Hunt, Dubuque, Iowa.

Radosevich, S., J. Holt, and C. Ghersa. 1997. Weed Ecology, Implications for Management, 2nd ed., John Wiley & Sons, New York.

Wilkerson, G. G., S. A. Modena, and H. D. Coble. 1991. HERB: Decision model for postemergence weed control in soybean. Agron. J. 83:413–417.

Addresses or Web sites for models:

CORNHERB, herbicide options for corn weed control: Michigan State University Bulletin Office, 10-B Agriculture Hall, Michigan State University, East Lansing, MI 48824-1039.

WeedCast, software to forecast weed seedling emergence and growth in crop environments, North Central Soil Conservation Research Laboratory, Morris, MN: http://www.infolink.morris.mn.us/~lwink/products/weedcast.htm

Indiana Herbicide Selector: http://www.btny.purdue.edu/software/IN_Herbicide_Selector/

NebraskaHERB, for postemergence weed control decisions: Dr. John McNamara, Department of Agronomy, University of Nebraska, Lincoln, NE 68583-0910; phone 1-402-472-1544.

SOYHERB, herbicide options for soybean weed control: Michigan State University Bulletin Office, 10-B Agriculture Hall, Michigan State University, East Lansing, MI 48824-1039.

Web sites for sustainable agriculture or Integrated Pest Management programs:

Ohio Pest Management and Survey Program: http://www.ag.ohio-state.edu/~ohioline/icm-fact/fc-01.html

National Integrated Pest Management (IPM) Network: http://www.reeusda.gov/nipmn/

University of California Statewide Integrated Pest Management Project: http://www.ipm.ucdavis.edu/

University of California Sustainable Agriculture Research: http://www.sarep.ucdavis.edu/

University of Connecticut Integrated Pest Management Program: http://www.canr.uconn.edu/ces/ipm/index.html

WEED IDENTIFICATION GUIDES

Many state and regional publications on weed identification are available. The following list represents only a few of the commonly used guides.

TERRESTRIAL WEEDS

Delorit, R. 1970. Illustrated Taxonomy Manual of Weed Seeds. Available from the North Central Weed Science Society, 1508 W. University Ave., Champaign, IL 61821-3133; phone 1-217-352-4212.

Elmore, F. H. 1976. Shrubs and Trees of the Southwest Uplands. Southwest Parks and Monuments Association, 221 North Court, Tucson, AZ 85701.

Haselwood, E. L., and G. G. Motter. 1991. Handbook of Hawaiian Weeds. University of Hawaii Press, Order Department, 2840 Kolowalu St., Honolulu, HI 96822.

Healy, E. A., J. M. DeTomaso, G. F. Hrusa, and E. Dean. Expected release date 2001. Weeds of California. Division of Agriculture and Natural Resources, University of California. Contact Dr. DeTomaso by phone 1-530-754-8715 or e-mail ditomaso@vegmail.ucdavis.edu for details.

Lym, R., and K. Christianson. 1996. The Thistles of North Dakota. North Dakota State University Extension Distribution Center. Phone 1-701-231-7882.

Ontario Weeds. 1992. Publications Ontario, 50 Grosvenor St., Toronto, Ontario, M7A 1N8 Canada.

Parker, K. F. 1986. An Illustrated Guide to Arizona Weeds. University of Arizona Press, Tucson, AZ.

Stubbendieck, J., G. Y. Friisoe, and M. R. Bolick. 1994. Weeds of Nebraska and the Great Plains. Nebraska Department of Agriculture, Bureau of Plant Industry, P.O. Box 94756, Lincoln, NE 68509-4756.

Uva, R. H., J. C. Neal, and J. M. DiTomaso. 1997. Weeds of the Northeast. Cornell University Press, Ithaca, NY.

Weeds of the North Central States. 1981. North Central Regional Publication 281. College of Agriculture, University of Illinois, Agricultural Publications Office, 123 Munsford Hall, Urbana, IL 61801.

Whitson, T. D., Ed. 1996. Weeds of the West, 5th ed. Western Society of Weed Science, P.O. Box 963, Newark, CA 94560 (also available on diskette in a diagnostic format; entitled W.E.E.D.S., Western Expert Educational Diagnostic System).

WSSA. 1989. Composite List of Weeds. Weed Science Society of America, P.O. Box 1879, 810 E. 10th St., Lawrence, KS 66044-8897; phone 1-800-627-0629 (also on computer disk).

Available on CD-ROM:

Broadleaf Weed Seedling Identification. 1998. Purdue University Media Distribution Center, 301 South 2nd Street, Lafayette, IN 47901-1232; phone 1-888-398-4636 (Grass Weed Seedling Identification will be available in 1999).

Weeds of the United States. 1995. Southern Weed Science Society, 1508 W. University Ave., Champaign, IL 61821-3133; phone 1-217-352-4212.

Web sites for terrestrial weed identification:

Common Weeds of No-Till Cropping Systems, Purdue University: http://www.btny.purdue.edu/Extension/Weeds/NoTillID/NoTillWeedID1.html

Common Weed Seedlings of Michigan, Michigan State University: http://www.msue.msu.edu/msue/iac/e1363/31363.htm

Federal Interagency Committee for the Management of Noxious and Exotic Weeds: http://bluegoose.arw.r9.fws.gov/FICMNEWFiles/FICMNEWHomePage.html

Noxious Weeds, and Exotic and Invasive Plant Management Resources, The Nature Conservancy: http://refuges.fws.gov/NWRSFiles/InternetResources/Weeds.html

Noxious Weeds in Kansas, Plant Health Division, Kansas Department of Agriculture: http://www.ink.org/public/kda/phealth/phprot/weeds.html

PhotoHerbarium of Weeds, Weed Science Society of America: http://piked2.agn.uiuc.edu/wssa/subpages/weed/weedid.html

Weed Guide, American Cyanamid Company: http://www.cyanamid.com/tools/weedguide/index.html

Weed Photo Gallery for Small Grains, University of California, Davis: http://www.ipm.ucdavis.edu/PMG/r730700999.html

Weed Seedling Identification, Iowa State University: http://extension.agron.iastate.edu/extweeds/weed-id/weedid.htm

Weeds of the World: http://ifs.plants.ox.ac.uk/wwd/wwd.html

AQUATIC WEEDS

Aulbach-Smith, C. A., and S. J. deKozlowski. 1996. Aquatic and Wetland Plants of South Carolina. Available from K. Horan, SCDNR, Water Resources Division, 1201 Main St., Suite 1100, Columbia, SC 29201; phone 1-803-737-0800.

Beal, E. O., and J. W. Thieret. 1996. Aquatic and Wetland Plants of Kentucky. Available from Kentucky State Nature Preserves Commission, 801 Schenkel Lane, Frankfort, KY 40601; phone 1-502-573-2886.

Fasset, N. C. 1957. A Manual of Aquatic Plants. University of Wisconsin Press, Madison, WI 53706.

Fink, D. F. 1997. A Guide to Aquatic Plants. Ecological Services Section, Minnesota Department of Natural Resources, 500 Lafayette Road, St. Paul, MN 55155-4025.

Hoyer, M. V., D. E. Canfield, Jr., C. A. Horsburgh, and K. Brown. 1996. Florida Freshwater Plants, A Handbook of Common Aquatic Plants in Florida Lakes. Publications Center, University of Florida, P.O. Box 11011, Gainesville, FL 32611-0011; phone 1-800-226-1764.

Lembi, C. A. 1997. Aquatic Pest Control Manual. Available from Purdue Pesticides Program, Department of Botany and Plant Pathology, Purdue University, West Lafayette, IN 47907; Phone 1-765-494-4566.

Winterringer, G. S., and A. C. Lopinot. 1977. Aquatic Plants of Illinois. Illinois State Museum, Springfield, IL 62706.

Web site for aquatic weed identification and information:

Center for Aquatic Plants, University of Florida: http://aquat1.ifas.ufl.edu/

POISONOUS PLANTS

Fuller, T. C., and E. McClintock. 1986. Poisonous Plants of California (California Natural History Guide 53). University of California Press, Berkeley.

Kingsbury, J. M. 1964. Poisonous Plants of the United States. Prentice-Hall, Englewood Cliffs, New Jersey.

Lampke, K. F., and M. A. McCann. 1985. AMA Handbook of Poisonous and Injurious Plants. American Medical Association, Chicago Review Press, Chicago.

Stephens, H. A. 1980. Poisonous Plants of the Central United States. University Press of Kansas, 2501 W. 15th St., Lawrence, KS 66044.

Westbrooks, R. G., and J. W. Preacher. 1986. Poisonous Plants of Eastern North America. University of South Carolina Press, Columbia.

Web sites for poisonous plant identification:

California Poisonous Plants (list only): http://veghome.ucdavis.edu/weedsci/WWW/Poisontbl.html

Canadian Poisonous Plants: http://res.agr.ca/brd/poisonpl/illust.html

Indiana Plants Poisonous to Livestock and Pests, Purdue University: http://vet.purdue.edu/depts/addl/toxic/cover1.html

Poisonous Plants Web Pages, Cornell University: http://www.ansci.cornell.edu/plants.html

Plants Toxic to Animals, University of Illinois: http://www.library.uiuc.edu/vex/toxic/intro/html

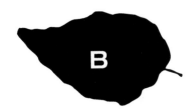

COLLECTING PLANTS FOR IDENTIFICATION

Plants can be identified (1) immediately, (2) within 2 or 3 days after collection, or (3) after permanent preservation.

1. The best time to identify a plant is immediately upon collection. To do so, you need equipment (handbooks, hand lens, knife, shovel, or digger) *and* experience and a knowledge of what characteristics to look for.

2. If you must wait 2 or 3 days to identify the plant, following these suggestions will help:

    a. The best procedure is to collect the root ball as well as the shoot portion, place the whole plant in a plastic bag, seal it, and store it in a cool place.

    b. The next best alternative is to take as much of the shoot and root as practical, place it in a plastic bag immediately, seal the bag, and place it in a cool area.

    c. *Do not* throw plants in the back seat of your car or pickup and let warm air pass over them. They will dry out rapidly and become withered and eventually brittle. They will be very difficult to handle in this condition.

    d. *Do not* put plants in plastic bags with excess water and then leave them in a warm place or send them through the mail. They will rot rapidly and become unidentifiable. The only time you might consider adding water to a plant is if the root ball is extremely dry. In that case, wet the root ball lightly, but do not allow the shoot portion of the plant to come in contact with the moistened areas.

3. If a plant is to be preserved permanently, it is best to identify the plant first. Better yet, collect duplicates of each species; press one and save the other for identification while it is still in good condition. The most common method of preservation is to press the plant. For pressing you will need:

    plant press (two wood frames with straps)
    blotters
    cardboard separators
    newspaper (but not the comics section, which is not absorptive enough)

    a. Procedures for pressing a plant:

    (1) Select plants that are in good condition and that show as many descriptive features as possible.

    (2) Slit any thickened parts (such as roots, stem bases) with a knife. You also can trim away any leaves or branches that clutter the specimen, but make sure you leave enough to show the characteristics.

    (3) Arrange the plant between the folds of a newspaper. If the plant is long, bend or twist it so that it all fits in the newspaper and in the size of the mount to be used.

    (4) Place a label with the plant noting the name if known, date of collection, and site of collection.

(5) Place the newspaper between two blotters.

(6) Alternate blotters, newspapers, and cardboard separators. If separators are not plentiful, use them after every four or five plants.

(7) Place in press. Tighten straps by standing on the press.

(8) Change both the newspapers and blotters once a day until the plants are dry to the touch.

(9) Place the presses in a place where it is dry and warm and, if possible, where the air circulates.

(10) Once the plant has dried, you can remove it from the newspaper and mount it on herbarium paper using an adhesive. Elmer's Glue works well.

(11) If the mounted specimens are to be handled often or moved frequently, cover them after they are *completely* dry. You can sandwich the mounted specimen between a stiff backing and a sheet of clear plastic.

b. Attach a label to each specimen with the following information:

Common name of plant	Johnsongrass
Scientific name of plant	*Sorghum halepense* L.
Habitat from which collected	Soybean field
Geographical location where collected	Jefferson Co., Ind.
Life span of plant	Creeping perennial
Name of person collecting plant	William B. Smith
Date collected	September 4, 1998

C

CONVERSION FACTORS USEFUL IN FIELD PLOT WORK

AREA EQUIVALENTS

township = 36 sections = 23,040 acres

square mile = 1 section = 640 acres

acre = 43,560 ft^2

    = 160 square rods (rd)

    = 0.405 hectare (ha)

    = 4840 yd^2

    = an area 16 1/2 ft wide and 1/2 mile long

are = 100 m^2

hectare = 100 are = 2.471 acres

LIQUID EQUIVALENTS

U.S. gal = 4 qt = 8 pt = 16 cups

    = 128 fluid oz

    = 256 tablespoons (Tbs)

    = 3.785 l = 3785 ml

    = 231 cu in. = 0.1337 cu ft

    = 8.3370 lb water

    = 3785.4 cu cm

qt = 0.9463 l = 2 pt = 32 fl oz

    = 4 cups = 64 Tbs

l = 1000 ml = 1.0567 liquid qt (U.S.)

tablespoon = 14.8 ml = 3 teaspoons

    = 0.5 fl oz

oz (U.S. fluid) = 29.57 ml = 2 tablespoons

oz (British fluid) = 28.41 ml

cu ft water = 62.43 lb = 7.48 gal

TEMPERATURE EQUIVALENTS

(°F − 32) × 5/9 = °C

(°C × 9/5) + 32 = °F

LENGTH EQUIVALENTS

centimeter = 0.394 in.

meter = 3.28 ft = 39.4 in.

kilometer = 0.621 statute mile

inch = 2.54 cm

foot = 30.48 cm

yard = 0.914 m

rod (16.5 feet) = 5.029 m

statute mile (1,760 yards) = 1.61 km = 5280 ft

    = 1760 yd = 160 rods

PRESSURE EQUIVALENTS

1 psi = 6.9 kPa

1 atmos = 760 ml mercury

1 ft lift of water = 0.433 lb pressure per sq in. (psi)

WEIGHT EQUIVALENTS

pound (avdp) (16 oz) = 453.6 g

ton, gross or long (2,240 lb) = 1.016 metric tons

ton, short or net (2,000 lb) = 0.907 metric ton

milligram (mg) = 10^{-3} g   1 mg/g = 1000 ppm

microgram (µg) = 10^{-6} g   1 µg/g = 1 ppm

nanogram = 10^{-9} g   1 nanogram/g = 1 ppb

picogram = 10^{-12} g   1 picogram/g = 1 ppt

1 mg/kg or 1 mg/L

    = 1 ppm

1 µg/kg or 1 µg/L

    = 1 ppb

USEFUL CONVERSIONS

Multiply	by	To obtain
ft	0.3048	m
gal (U.S.)	3785	cubic centimeters
gal (U.S.)	3.785	liters
gal (U.S.)	0.83	gal (Imperial)
gal	128	fluid oz
gal/min	2.228×10^{-3}	cu ft/sec
gal/acre	9.354	L/ha
hectares	2.471	acres (U.S.)
kilograms	2.205	lb (advp)
kg/ha	0.892	lb/acre
liters	0.0353	cu ft
liters	1.05	qt (U.S.)
liters	0.2642	gal (U.S.)
liters/ha	0.107	gal/acre
meters	3.281	ft
miles/hr	88	ft/min
miles/hr	1.61	km/hr
oz (fluid)	29.573	milliliters
oz	28.35	grams
lb	453.59	grams
psi	6.9	kilopascals
lb/gal	0.12	kg/L
lb/square in.	0.070	1 kg/cm2 (atm)
lb/1000 square ft	0.489	kg/acre
lb/acre	1.12	kg/ha
square in.	6.452	cm²
yards	0.9144	meters
parts per million	2.719	lb a.i./acre-foot of water

Adapted from the Herbicide Handbook of the Weed Science Society of America (1994) and Klingman and Ashton (1982).

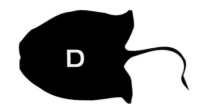

WEED SCIENCE JOURNALS, PROCEEDINGS, AND SOCIETIES

WEED SCIENCE JOURNALS AND PROCEEDINGS

Weed Science
Weed Research
Weed Technology
Reviews of Weed Science
Aquatics
Aquatic Botany
Journal of the Aquatic Plant Management Society
Proceedings of the Northeastern Weed Science Society
Proceedings and Research Reports of the North Central Weed Science Society
Proceedings of the Southern Weed Science Society
Proceedings and Research Reports of the Western Society of Weed Science

WEED SCIENCE SOCIETIES

Weed Science Society of America, P.O. Box 1879, 810 E. 10th St., Lawrence, KS 66044-8897; phone 1-800-627-0629; fax 1-785-843-1274; e-mail wssa@allen-press.com. Web site for WSSA is at http://piked2.agn.uiuc.edu/wssa

International Weed Science Society. Web site for IWSS is at http://www.orst.edu/Dept/IPPC/IWSS

European Weed Research Society. Web site for EWRS is at http://www.res.bbsrc.ac.uk/ewrs/

Northeastern Weed Science Society, Dr. Andrew Senesac, Secretary/Treasurer, 3059 Sound Avenue, Riverhead, NY 11901-1098; phone 1-516-727-3595; fax: 1-516-727-3611; e-mail senesac@cce.cornell.edu

North Central Weed Control Conference, 1508 West University Avenue, Champaign, IL 61821-3133; phone 217-352-4212.

Southern Weed Science Society, 1508 West University Avenue, Champaign, IL 61821-3133; phone 1-217-352-4212.

Western Society of Weed Science, 35806 Ruschin Drive, Newark, CA 94560. Web site for WSWS is at http://www.uidaho.edu/ag/wsws/

Aquatic Plant Management Society, P.O. Box 1477, Lehigh, FL 33970-1477; phone 1-941-694-1129.

Florida Aquatic Plant Management Society (publishes *Aquatics*), Orange County Environmental Protection Department, 2002 E. Michigan St., Orlando, FL 32806; phone 1-407-836-7400.

North American Lake Management Society, P.O. Box 5443, Madison, WI 53705-5443; phone 1-608-233-2836; fax 608-233-3186; Web site for NALMS is at http://www.nalms.org/abtnalms/abtnalms.htm

GLOSSARY

Absorption. The process by which a compound passes from one system into another (e.g., a herbicide passing from the soil solution into the plant root or from the leaf surface into the leaf tissue).

Acid-equivalent (a.e.). the theoretical yield of parent acid from the active ingredient content of a formulation.

Acre-foot. A volume measurement frequently used for bodies of water. A surface acre 1 ft deep.

Active ingredient (a.i.). The chemical in a herbicide formulation that is primarily responsible for its phytotoxicity and that is identified as the active ingredient on the product label.

Acute toxicity. The ability of a substance to cause injury or illness shortly after exposure to a relatively large dose (see also *chronic toxicity*).

Additive effect. Occurs when the effect of two chemicals is equal to the predicted effect based on the activity of each chemical alone.

Adjuvant. Any additive used with a herbicide that enhances the performance or handling of the herbicide.

Adsorption. The association of an extremely thin layer of molecules or ions with the surface of a solid.

After-ripening. The period of time during which seeds become ready to germinate.

Aliphatic compounds. Compounds that primarily consist of carbon chains.

Allelopathy. Production by plants of chemical compounds that inhibit the growth of other nearby plants.

Amino acid. The chemical compounds that make up proteins. Characterized by the presence of an amino group and a carboxyl group. Inhibition of amino acid synthesis reduces protein synthesis and growth.

Anion. An atom with a negative (−) charge.

Annual weed. A weed that completes its life cycle in one growing season. These plants germinate, grow vegetatively, flower, produce seed, and die in 12 mo or less.

Antagonism. The interaction of two or more chemicals such that the effect when combined is less than the predicted effect based on the activity of each chemical alone.

Apical dominance. The ability of a terminal bud to inhibit the sprouting and growth of adjacent lateral buds.

Apoplast. The nonliving, more or less continuous, system of cell walls, intercellular spaces, and interconnecting cells of xylem tissue. The major pathway for movement of substances upward in the plant.

Apoplastic translocation. The movement of water and solutes through the apoplast. The driving force for apoplastic movement is transpiration.

Aquatic plant. A plant that is structurally adapted for life on, in, or around a body of water.

Aromatic compounds. Compounds that contain an unsaturated ring structure (e.g., benzene).

Auxin. Any of a group of growth-regulating substances that promote cell elongation in plants.

Band application. A treatment applied to a restricted linear strip on or along a crop row rather than applied continuously over the entire area (broadcast).

Basal bulb. The bulblike structure at the base of stems of yellow nutsedge and purple nutsedge.

Basal rosette. A plant growth form in which the stem tip is located close to or below ground level. The plant is low growing, and the stem does not elongate until flowering.

Basal spray. A herbicide treatment applied to encircle the stem above or at the ground line such that foliage contact is minimal. A term usually restricted to treatment of woody plants.

Biennial weed. A weed that completes its life cycle over two growing seasons. It germinates and forms a rosette the 1st yr. The overwintering rosette is vernalized, then bolts, flowers, seeds, and dies the 2nd yr.

Bioassay. Determination of herbicide concentration by use of sensitive indicator plants or other organisms. The whole plant or organism or some portion thereof may be used.

Bioherbicide. A product containing a disease-causing organism that is used like a herbicide.

Biological weed control (biocontrol of weeds). The control or suppression of weeds by the action of one or more organisms through natural means or by manipulation of the weed, organism, or environment.

Biotransformation. A biologically driven process that leads to the alteration and inactivation of herbicides.

Biotype. A population within a species that has distinct genetic variation, e.g. resistance to a herbicide.

Broadcast application. A treatment applied as a continuous sheet over the entire area.

Bud. Tissue located in the axils of leaves that will produce new shoots.

Bulb. Underground leaf tissue modified for food storage and vegetative reproduction.

Burn-down herbicides. Herbicides that kill emerged vegetation on no-till. Typically used ahead of, at, or just after planting but before the crop emerges.

Carcinogenic. Capable of causing cancer in animals.

Carotenoid. A yellow to red pigment that protects chlorophyll from damage by excess energy.

Carrier. A gas, liquid, or solid substance used to dilute or suspend a herbicide during its application.

Cation. An atom with a positive (+) charge.

Chemical name. The name applied to a herbicide active ingredient that describes its chemical structure according to rules prescribed by the American Chemical Society and published in the *Chemical Abstracts Indexes.*

Chemical weed control. The control or suppression of weeds using herbicides.

Chlorophyll. The green pigment that captures light energy for photosynthesis.

Chloroplast. The cellular site of photosynthesis.

Chlorosis. Loss of green color (chlorophyll) from foliage.

Chronic toxicity. The ability of a substance to cause injury or illness after repeated exposure to small doses over an extended period of time (see also *acute toxicity*).

Classical biocontrol. An approach by which a biocontrol organism, usually an insect, is introduced and allowed to spread on its own through the weed-infested area.

Climax community. A stable vegetation community. Left undisturbed, it can maintain itself indefinitely.

Closed handling system. A system that allows the transfer of the herbicide from the container to the spray tank without operator exposure.

Coleoptile. A leaflike sheath that protects the shoot tip and leaves of a seedling grass as it emerges from the soil.

Colloids. Particles of extremely small size (1.0 to 0.001 µm) that result in a high surface-to-volume ratio, thus providing a large amount of reactive surface. Colloidal particles in soils are clays and organic matter.

Common name. A name applied to a herbicide active ingredient usually agreed upon by the American National Standards Institute and the International Organization for Standardization.

Companion cells. Cells associated with sieve tube members. Such cells are not conductive but contain the full complement of organelles. Their role in phloem transport is as yet unknown.

Compatibility. The quality of being mixable in the formulation or in the spray tank for application in the same carrier without causing an alteration in the characteristics or effects of the individual components.

Competition. The process by which plants interact with one another to obtain limited supplies of light, water, and nutrients.

Conifer release. The use of herbicides to control unwanted deciduous trees, thus stimulating the growth of coniferous trees.

Conjugate. A product produced when a herbicide binds to another molecule. Conjugation usually detoxifies the herbicide.

Conservation tillage. Any tillage system that reduces the loss of soil or water when compared to a clean tillage system such as the conventional tillage system.

Contact herbicide. A herbicide that causes localized injury to plant tissue where contact occurs.

Control. The suppression of a weed to the point that its economic impact is minimized.

Controlled droplet applicator. A low-volume spinning applicator in which droplets produced are of a relatively uniform size.

Conventional tillage system. A tillage system that consists of breaking and inverting the soil, seedbed preparation, and selective cultivation. The soil surface is left relatively residue free.

Cotyledon. The seed leaf.

Cracking (at emergence). The time when the soil above the emerging crop seedlings begins to crack open.

Creeping perennial weed. A weed that can both overwinter and produce new independent plants from vegetative reproductive structures.

Creeping root. A root modified for food storage and vegetative reproduction.

Critical period. The period of time when weeds have a negative impact on the crop yield.

Crop oil concentrate. Nonphytotoxic light oil that also contains surfactants. It is used to enhance the effectiveness of aqueous herbicide solutions applied to foliage.

Cross-resistance. A property of weed populations that are resistant to two or more herbicides that have the same mode of action.

Cultivation. Mechanical soil agitation after crop emergence for the purpose of killing weeds.

Cultural weed control. The control or suppression of weeds by utilizing the weed control properties of a vigorously growing smother crop, competitive crops, or crops that are allelopathic.

Cut and spray. Spraying cut stumps of woody plants with herbicide to prevent sprouting.

Cuticle. The waxy outer layer of a plant shoot.

Cutin. A component of the cuticle. More water soluble than wax, it can become hydrated in the presence of water.

Degradation. The breakdown of materials, such as plant and animal residues, organic matter, or pesticides.

Depth protection. Selectivity based on mobility of the herbicide in the soil, differential root growth between weed and crop, and the ability to position a herbicide in the soil.

Desiccant. A substance used to accelerate the drying of plant tissue.

Dicot. A "broadleaved" plant; member of the dicotyledonous (two cotyledon) plants.

Diluent. Any gas, liquid, or solid material used to reduce the concentration of an active ingredient in a formulation.

Directed application. The precise application to a specific area such as a row or bed or a plant organ such as the leaves or stems.

Directed postemergence. Application under or on the lower portions of an established crop and onto the soil or over the top of smaller weeds.

Dormancy. The condition in which seeds or buds fail to germinate or bud even when the proper amounts of moisture, air, light, and temperature are available.

Drawdown. The draining of a body of water to expose aquatic weeds to adverse conditions such as the freezing and thawing conditions of winter.

Drift. The movement of herbicides off target during application.

Early postemergence. Herbicide applications made when weeds and crops are small, usually when weeds are less than 2 to 3 in. tall.

Economic threshold. The point at which the loss due to pest damage equals the cost of control.

Ecotype. A population within a species that has developed a distinct morphological or physiological characteristic as an adaptation to a specific environment and which persists when individuals are moved to a different environment.

Emergence. The event in seedling or perennial growth when a shoot becomes visible by pushing through the soil surface.

Emergent plants. Aquatic plants that extend above the water surface and are rooted in sediments at water depths less than 2 to 3 ft.

Emulsifiable concentrate. A single-phase liquid formulation that forms an emulsion when added to water.

Emulsifier. A substance that promotes the suspension of one liquid in another (e.g., oil in water).

Endodermis. A specialized layer of cells surrounding the vascular cylinder in the plant root. Because of waxy cell walls (Casparian strip), water solutions must pass through the cytoplasm of the endodermal cells to reach the vascular cylinder.

Epicotyl. The portion of the stem of a seedling above the cotyledon(s).

Epinasty. The downward bending or curvature of a plant caused by more rapid cell elongation on one side of a plant organ or part. Most often caused by auxin-type compounds.

Eradication. The elimination of all plants and plant parts of a weed species from an area.

Experimental use permit (EUP). A permit obtained from the Environmental Protection Agency that allows the company to test a pesticide under actual user conditions.

Exposure. The probability of encountering a harmful dose of a chemical.

Fertilization. The union of sperm and egg.

Flowable. A two-phase formulation that contains solid herbicide suspended in liquid and that forms a suspension when added to water.

Formulation. A herbicide preparation supplied by a manufacturer for practical use, or the process, carried out by the manufacturer, of preparing herbicides for practical use.

Free-floating plants. Aquatic plants that float freely on or beneath the water surface.

Frill injection. The application of herbicide to cuts in the trunk of a tree.

Gel. Thickened emulsifiable concentrates packed in water-soluble bags.

General use herbicide. A herbicide regarded by the EPA as not causing unreasonable adverse effects on the environment and safe for application by the general public without special training.

Genetic engineering. The introduction of genes from virtually any organism into other organisms.

Geographic information system (GIS). The storage of information determined from globally positioned systems that can be used for mapping.

Germination. The process of initiating growth in seeds.

Global positioning system (GPS). The capture of signals from satellites to determine coordinates on the earth's surface. Used in conjunction with geographic information systems.

Granular. A dry formulation consisting of discrete particles generally less than 10 mm^3 and designed to be applied without a liquid carrier.

Grass carp. A herbivorous fish used as a biological control for submersed aquatic weeds.

Groundwater. Subsurface supplies of water.

Growth regulator. Frequently applied to herbicides that produce auxin-type symptoms (e.g., 2, 4-D).

Growth stages for cereal crops.
1. **Tiller** or **tillering:** when additional shoots are developing from the crown.
2. **Joint** or **jointing:** when stem internodes begin elongating.
3. **Boot** or **booting:** when developing seed heads can be detected in the top leaf sheath.
4. **Head** or **heading:** when the seed head is emerging or has emerged from the sheath.

Hazard. The probability of encountering a harmful dose of a compound. Toxicity, dose, time of exposure, frequency of exposure, formulation, persistence, etc., combined determine hazard.

Herbaceous plant. A vascular plant that does not develop persistent woody tissue above ground.

Herbicide. A chemical used to control, suppress, or kill plants, or to severely interrupt their normal growth processes.

Herbicide metabolism. The process that leads to the alteration and inactivation of herbicides.

Heterotrophic organism. An organism that cannot manufacture its own food; that is, an organism that requires an external source of organic carbon. Includes fungi, most bacteria, and animals.

Highly persistent herbicide. A herbicide that harms susceptible plants during the second season and sometimes for longer periods of time. Formerly called a soil sterilant or a long residual herbicide.

Hydration. To combine with water; to become wetted.

Hydrophilic. Water-loving; attracted to water. Characteristic of polar compounds.

Hypocotyl. The portion of the stem of a seedling between the cotyledon(s) and roots.

Incorporation. The mixing or blending of a herbicide into the soil.

Inflorescence. A cluster of flowers.

Integrated pest management. A crop management approach that combines information on the biology and ecology of the pest with all available control technologies to develop an integrated management program.

Intercalary meristem. A meristematic region at the base of each internode that accounts for stem elongation in grasses.

Interference. The combined effect of plant competition and allelopathy. The total impact of one plant on another.

Internode. The region of the stem between two successive nodes.

Invert emulsion. A high-viscosity mixture of oil, water, and emulsifier.

Ion trapping. The phenomenon by which herbicides that are weak acids (e.g., 2,4-D) enter into and are trapped inside the cytoplasm of the cell.

Label. The directions for using a pesticide approved as a result of the registration process and attached to the herbicide container.

Lag phase. A period of low or no activity that precedes a more rapid response.

Late bud stage. The period in seasonal development of a plant just before flower opening.

Late postemergence. Postemergence applications made after the crop or weeds are well established.

Lateral movement. The movement of a herbicide through soil in a horizontal plane from the original site of application.

Layby application. Applications made with or just after the last cultivation of a crop.

LC$_{50}$. The concentration of a chemical in air (inhalation toxicity) or water (aquatic toxicity) that will kill 50% of the organisms in a specific test situation.

LD$_{50}$. The dose (quantity) of a chemical calculated to be lethal to 50% of the organisms in a specific test situation. It is expressed as weight of the chemical per unit of body weight (mg/kg). The toxicant may be fed (oral LD$_{50}$), applied to the skin (dermal LD$_{50}$), or administered in the form of vapors (inhalation LD$_{50}$).

Leaching. The downward movement of a water solution through the soil.

Lipid. A fatty substance. Lipids are major components of phospholipids, which along with proteins, make up cell membranes.

Lipid peroxidation. The oxidative destruction of the lipid components of cell membranes.

Ligule. In grass leaves, an outgrowth from the upper and inner side of the leaf blade where it joins the sheath.

Lipophilic. Fat-loving; attracted to fatty or waxy substances. Characteristic of nonpolar compounds.

Manipulated biocontrol agents. Disease organisms or insects that are grown, packaged, and introduced repeatedly into the environment to control target weed species.

Material safety data sheet (MSDS). Literature accompanying hazardous materials that provides information on safety considerations.

Mechanical weed control. The control or suppression of weeds by mechanical methods that range from hand hoeing to tillage.

Meristem. Undifferentiated tissue, the cells of which are capable of active cell division and differentiation into specialized tissues.

Metabolite. A compound derived from metabolic transformation of a herbicide by plants or other organisms.

Microencapsulated formulation. Small particles that consist of a herbicide core surrounded by a barrier layer, usually a polymer shell.

Microtubule. Tubular protein structures that make up the spindle apparatus in dividing cells. Disruption of microtubule synthesis or function prevents cell division and growth.

Mineral soil. Soil low in organic matter, usually containing between 1/2 and 6% organic matter.

Minimum tillage. The minimum amount of tillage required for crop production or for meeting the tillage requirement under existing soil and climatic conditions. Elimination of excess tillage.

Mode of action. The way in which a herbicide affects a plant at the tissue or cellular level.

Monocot. A "grasslike" plant; member of the monocotyledonous (one cotyledon) plants.

Mulch. Material spread on the ground to inhibit growth of weeds (e.g., straw, wood chips, black plastic).

Multiple resistance. Weed populations that are resistant to two or more herbicides from different chemical families and with different modes of action.

Mutagenic. Capable of causing genetic damage.

Necrosis. The localized death of tissue usually characterized by browning and desiccation.

Node. The slightly enlarged portion of the stem where leaves and buds arise and where branches originate.

Nonselective herbicide. A herbicide that is generally toxic to all plants. Selective herbicides may become nonselective if used at high rates.

Nontarget species. Species not intentionally affected by a pesticide.

No-observable adverse effect level (NOAEL). The highest pesticide dose that does not cause observable harm or side effects to experimental animals.

No-till. Planting crop seed directly into stubble or sod with no more soil disturbance than is necessary to get the seed into the soil.

Noxious weed. A weed specified by law (usually a seed certification law) as being especially undesirable, troublesome, and difficult to control. Precise definition varies from state to state.

1. Primary or prohibited noxious weed: none allowed in crop seed for planting.

2. Secondary or restricted noxious weed: few allowed in crop seed for planting.

Nurse crop. A temporary crop planted to protect the crop of value during development.

Organic matter. The stable fraction of decaying plant and animal residues in soils (humus).

Organic soil. Soil high in organic matter; may contain between 10 and 80% organic matter.

Over-the-top application. Applied over the top of transplanted or growing plants (e.g., by airplane or a raised spray boom on a ground rig). Any broadcast or banded application above the plant canopy.

Partitioning. The movement of one substance into another, e.g., oil into wax.

Pectin. A carbohydrate component of cell walls and cuticle. Pectin is the most hydrophilic of the cuticle components, and may provide pathways for the uptake of water-soluble herbicides.

Pelleted formulation. A dry formulation consisting of discrete particles usually larger than 10 mm³ and designed to be applied without a liquid carrier.

Perennial weed. A weed that produces vegetative structures that allow it to live for more than 2 yr.

Pericycle. A specialized meristematic layer of cells in the root responsible for the formation of lateral roots.

Persistent herbicide. A herbicide that harms susceptible plants during the first and sometimes the second season after application.

Phloem. The living tissue in plants that functions primarily to transport sucrose from the site of synthesis or storage to the site of utilization. The component transport cells are called sieve tube members.

Photosynthesis. The process by which plants convert carbon dioxide and water into sugar and oxygen. Occurs in the chloroplast.

Phreatophyte. Plants, usually woody, that draw water from underground water supplies as well as from soil water.

Physical weed control. The control or suppression of weeds by physical methods, such as mulching, flooding, and fire.

Phytochrome. A light-sensitive pigment in seeds that regulates germination in some species.

Phytoplankton. Microscopic algae suspended in the open waters of lakes and ponds. Heavy growths result in algal "blooms."

Phytotoxic. Injurious or lethal to plants.

Plasma membrane. The membrane enclosing the cell cytoplasm. Is a living structure, in contrast to the cell wall, which is a nonliving structure.

Plasmodesmata. Protoplasmic channels that link the protoplasts of adjacent cells.

Pollination. The process by which pollen, which contains sperm, is delivered to the vicinity of the egg.

Postemergence herbicide (post). A herbicide applied after the emergence of the specified weed or crop.

Preemergence herbicide (pre). A herbicide applied to the soil prior to the emergence of the specified weed or crop.

Preplant herbicide. A herbicide applied before planting or transplanting a crop either as a foliar application to control existing vegetation or as a soil application.

Preplant incorporated herbicide (ppi). A herbicide applied and tilled into the soil before seeding or transplanting.

Prevention. Keeping a weed from being introduced into an uninfested area or reproducing.

Primary tillage. The initial soil breaking operation; plowing.

Protectant. See *safener.*

Protein. A complex molecule made up of amino acids. Proteins make up enzymes and also are major components of cell membranes, along with lipids.

Quarantine. A regulation imposed to isolate and prevent the spread of noxious weeds.

Quiescent. A condition in which seeds or buds fail to germinate or sprout because some growth factor such as water, oxygen, or temperature is lacking; a period of inactivity.

Rate. Frequently used as the equivalent of dose: The amount of herbicide applied per unit area or other treatment unit.

Registration. The process designated by FIFRA and carried out by the EPA by which a pesticide is legally approved for use.

Reregistration. The process by which pesticides originally registered prior to November 1, 1984, are reregistered according to current standards and test procedures.

Residual herbicide. A herbicide that persists in the soil and injures or kills germinating weed seedlings over a relatively short period of time (in contrast to a highly persistent herbicide).

Residue. The undecayed plant material on the soil surface.

Resistance. The ability of a formerly susceptible plant population to survive herbicide doses above those that were once used to control the original plant population.

Restricted-use herbicide. A herbicide that can be purchased and used only by trained and certified applicators to avoid possible adverse health or environmental effects.

Returnable container. A herbicide container designed to be returned to the dealer and reused. Also called a two-way container.

Rhizome. An underground, horizontal stem. Usually contains stored food and numerous buds.

Root. Underground plant organ that serves to anchor the plant and to absorb and conduct water and mineral nutrients.

Rooted floating plants. Aquatic plants rooted in the bottom sediments at water depths of roughly 1 to 5 ft. Floating leaves originate from underground rhizomes.

Rootstock. A loosely used term referring to underground vegetative reproductive structures. Properly used, it refers to a rhizome.

Rosette. A cluster of basal leaves that originate from a crown due to lack of stem elongation.

Safener. A substance that reduces the toxicity of a specific herbicide to the crop.

Secondary tillage. Tillage operations used for seedbed preparation.

Seed bank. The accumulated deposit of seeds present in the soil.

Seed protectant. A substance applied to seed before planting to control pests.

Selection pressure. Occurs when continued use of a repeated practice or herbicide kills susceptible plants but leaves resistant plants, which can then survive and increase in number.

Selective cultivation. The removal of weeds without injury to the crop either by using shallow-penetrating broadcast implements over deep-seeded or well-rooted crops or by cutting out all vegetation between the crop rows.

Selective herbicide. A chemical treatment that is toxic to some plant species but does not damage others.

Shoot. A general term for the aboveground portion of a plant.

Sieve tube member. The cell type that makes up the food-transporting (phloem) tissue in the plant.

Simple perennial weed. A weed that can live more than 2 yr by means of vegetative structures but, unlike a creeping perennial weed, is unable to produce new individual plants from these structures.

Smother crop. Crop which, when maintained in dense stands, can inhibit the growth of weeds.

Soil injection. The placement of a herbicide beneath the soil surface with a minimum of mixing or stirring of the soil (e.g., with an injection blade, knife, or tine).

Soil layering. The placement of a herbicide beneath the soil surface in a continuous layer with a minimum of mixing.

Soluble liquid. A liquid formulation that forms a true solution (completely dissolves) when added to water.

Soluble packet. Water-soluble packaging that usually contains dry formulations such as wettable powders and water-dispersible granules. The package dissolves in the spray tank.

Soluble powder. A dry formulation that forms a true solution (completely dissolves) when added to water.

Solution. A homogeneous or single-phase mixture of two or more substances.

Sorption. The retention of a substance on or in a solid phase.

Source to sink. A concept describing the mass flow of solutes from regions of high solute concentration (e.g., photosynthetically active leaves) to regions of low solute concentration (e.g., actively growing plant parts).

Spot treatment. Treatment to a restricted area of a larger unit (e.g., herbicide treatment of spots or patches of weeds in a large field).

Stale seedbed. A tillage system in which the last secondary tillage operation is completed weeks or months prior to planting so that the soil is not disturbed at planting.

Stele. The central cylinder of conducting tissue in a plant.

Stem. The plant organ specialized for support, leaf production, some photosynthesis (aboveground stems), and food storage.

Stolon. An aboveground, horizontal stem.

Stratification. The process of exposing seeds to a cold period to break dormancy.

Submersed plant. An aquatic plant that grows with all or most of its vegetative tissue below the water surface.

Succession. The long-term transition of vegetation types to a condition of stability.

Summer annual. A plant that germinates from seed in the spring, flowers and produces seed in mid- to late summer, and dies in the fall.

Surface runoff. The lateral movement of herbicide molecules (and other substances) with water on the soil surface to areas outside of the target area.

Surfactant. A material that improves the emulsifying, dispersing, spreading, wetting, or other surface-modifying properties of liquids.

Susceptibility. A condition in which a plant population dies at herbicide doses that normally do not injure other plant species.

Suspension. A mixture containing finely divided particles evenly dispersed in a solid, liquid, or gas.

Suspension concentrate. Finely divided solids suspended in a liquid such as water or oil.

Sustainable agriculture. A management system that can be maintained indefinitely with minimal extraneous inputs added to the system.

Symplast. The living, more or less continuous system of protoplasts of plant cells (connected by plasmodesmata) that includes the interconnected protoplasts of the phloem tissue. The major pathway for movement of substances from the leaves to sites of sugar storage or utilization.

Symplastic translocation. The movement of sugars through the symplast. The driving force for symplastic movement is the mass flow of sugars from a source to a sink.

Synergism. The interaction of two or more chemicals such that the effect when combined is greater than the predicted effect based on the activity of each chemical alone.

Systemic herbicide. Synonymous with translocated herbicide but more often used to describe the action of insecticides and fungicides. This term does not differentiate between apoplastic and symplastic mobility.

Tank mix combination. The mixing of two or more pesticides or agricultural chemicals in the spray tank at the time of application.

Teratogenic. Capable of producing birth defects.

Terrestrial plant. A plant that lives on land.

Tillage. Preparation of a seedbed involving soil disturbance or displacement.

Tolerance. The ability of a plant population to remain uninjured by herbicide doses normally used to control other plant species. Also refers to the concentration of herbicide residue allowed in or on agricultural products.

Toxicity. The quality or potential of a substance to cause injury or illness.

Toxicology. The study of the principles or mechanisms of toxicity.

Trade name. A trademark applied to a herbicide formulation by its manufacturer.

Transformation. The introduction and successful expression of a foreign gene into a plant.

Transgenic plant. A plant that has been genetically altered by genetic engineering.

Translocated herbicide. A herbicide that is moved within the plant. Translocated herbicides may be either phloem mobile or xylem mobile, but the term frequently is used in a more restrictive sense to refer to herbicides that are applied to the foliage and move downward through the phloem to underground parts.

Transpiration. The giving off of water vapor from the surface of leaves. Acts as a driving force for water movement upward in the plant.

Tuber. The thickened underground stem borne on a rhizome.

Vapor drift. The movement of chemical vapors from the area of application. Some herbicides, when applied at normal rates and normal temperatures, have a sufficiently high vapor pressure to change into vapor form, which may cause injury to susceptible plants some distance from the application site.

Vascular cambium. An internal meristematic region that produces new xylem and phloem during secondary growth. Characteristic of woody stems.

Vegetative reproductive structure. The nonflowering part of the plant (usually a modified stem or root) that possesses buds capable of generating new plants.

Vessel member. One of two cell types (the other is the tracheid) that make up the water-conducting (xylem) tissue in the plant.

VMD (volume mean diameter). Spray droplet size at which one half of the spray volume consists of larger droplets and one half of the spray volume consists of smaller droplets.

Volatility. The evaporation potential of a herbicide. The tendency to volatilize is determined by the vapor pressure.

Water-dispersible granule (dry flowable). A dry formulation of granular dimensions that disperses completely in water to form a suspension.

Water-dispersible liquid (flowable or liquid). A formulation in which finely ground solids are suspended in a liquid system. Readily disperses in water to form a suspension.

Water hardness. A measure of hardness ions (primarily Ca^{++} and Mg^{++}) in water.

Wax. A component of the cuticle. A very water-repellent layer of hydrocarbons.

Weed. A plant that interferes with the growth of desirable plants and is unusually persistent and pernicious. Weeds negatively impact human activities and as a result are undesirable.

Weed modeling. The mathematical simulation of weed development.

Wettable powder. A finely divided dry formulation that can be suspended readily in water.

Wetting agent. A substance that serves to reduce interfacial tensions and causes spray solutions or suspensions to make better contact with treated surfaces. Also refers to a substance in a wettable powder formulation that causes it to wet readily when added to water.

Winter annual. Plant that germinates from seed from late summer to early spring, flowers and produces seed in mid- to late spring, and dies in the summer.

Wipe-on treatment. Herbicide treatment to a plant using an applicator (e.g., sponge, rope) that makes contact with the plant directly.

Woody perennial plant. A vascular plant that develops persistent, woody tissue above ground.

Xylem. The nonliving tissue in plants that functions primarily to conduct water and mineral nutrients from roots to the shoot. The component transport cells are called vessels and tracheids.

LITERATURE CITED

Abbas, H. K., and C. D. Boyette. 1992. Phytotoxicity of fumonisin B_1 on weed and crop species. Weed Technol. 6:548–552.

Abbas, H. K., T. Tanaka, S. O. Duke, and C. D. Boyette. 1995. Susceptibility of various crop and weed species to AAL-toxin, a natural herbicide. Weed Technol. 9:125–130.

Altom, J. D., and J. F. Strizke. 1973. Degradation of dicamba, picloram, and four phenoxy herbicides in soils. Weed Science 21:556–560.

Anderson, D. M., C. J. Swanton, J. C. Hall, and B. G. Mersey. 1993. The influence of temperature and relative humidity on the efficacy of glufosinate-ammonium. Weed Res. 33:139–147.

Anderson, L. W. J. 1989. Aquatic weed problems and management in the western United States and Canada. In Aquatic Weeds, The Ecology and Management of Nuisance Aquatic Vegetation. A. H. Pieterse and K. J. Murphy, Eds. Oxford University Press, pp. 371–391.

Armstrong, T. F. 1975. The problem: Yellow nutsedge. Proceedings of North Central Weed Control Conference. 30:120–121.

Bailey, G. W., and J. L. White. 1964. Soil-pesticide relationships. Review of adsorption and desorption of organic pesticides by soil colloids, with implications concerning pesticide bioactivity. J. Agric. Food Chem. 12:324–332.

Balyan, R. S., and R. K. Malik. 1989. Influence of nitrogen on competition of wild canary grass. Pestology 13f:5–6

Barko, J. W., and R. M. Smart. 1980. Mobilization of sediment phosphorus by submersed freshwater macrophytes. Freshwater Biology 10:229–238.

Barton, L. V. 1965. Seed dormancy: General survey of dormancy types in seeds and dormancy imposed by external agents. Encycl. Pl. Physiol. 15:699–720.

Bauman, T. T., and M. A. Ross. 1983. Effect of three tillage systems on the persistence of atrazine. Weed Sci. 31:423–426.

Baysinger, J. A., and B. D. Sims. 1991. Giant ragweed (Ambrosia trifida L.) interference in soybeans (Glycine max). Weed Sci. 39:358–362.

Baziramakenga, R., and G. D. Leroux. 1994. Critical period of quackgrass (Elytrigia repens) removal in potatoes (Solanum tuberosum). Weed Sci. 42:528–533.

Blackshaw, R. E. 1993. Downy brome (Bromus tectorum) density and relative time of emergence affects interference in winter wheat (Triticum aestivum). Weed Sci. 41:551–556.

Bodle, M. J., A. P. Ferriter, and D. D. Thayer. 1994. The biology, distribution, and ecological consequences of Melaleuca quinquenervia in the Everglades. In Everglades: The Ecosystem and Its Restoration. S. M. Davis and J. C. Ogden, Eds. St. Lucie Press, Delray Beach, FL, pp. 341–355.

Böger, P. H., and G. Sandman, Eds. 1989. Target Sites of Herbicide Action. CRC Press, Boca Raton, FL.

Boyd, C. E., E. E. Prather, and R. W. Parks. 1975. Sudden mortality of a massive phytoplankton bloom. Weed Sci. 23:61–67.

Bozsa, R. C., L. R. Oliver, and T. L. Driver. 1989. Intraspecific and interspecific sicklepod (Cassia obtusifolia) interference. Weed Sci. 37:670–673.

Bridges, D. C. 1994. Impact of weeds on human endeavors. Weed Tech. 8:392–395.

Bridges, D. C., and R. L. Anderson. 1992. Crop losses due to weeds in the United States. In Crop Losses Due to Weeds in the United States. D. C. Bridges, Ed. Weed Science Society of America, P.O. Box 1879, 810 E. 10th St., Lawrence, KS 66044–8897. pp. 1–60.

Buchanan, G. A., and E. R. Burns. 1971a. Weed competition in cotton. I. Sicklepod and tall morningglory. Weed Sci. 19:576–579.

Buchanan, G. A., R. H. Crowley, J. E. Street, and J. A. McGuire. 1980. Competition of sicklepod (*Cassia obtusifolia*) and redroot pigweed (*Amaranthus retroflexus*) with cotton (*Gossypium hirsutum*). Weed Sci. 28:258–262.

Buchanan, G. A., and E. R. Burns. 1971b. Weed competition in cotton. II. Cocklebur and redroot pigweed. Weed Sci. 19:580–582.

Buchanan, G. A., and R. D. McLaughlin. 1975. Influence of nitrogen on weed composition in cotton. Weed Sci. 23:324–328.

Buckovac, M. J. 1976. Herbicide entry into plants. *In* Herbicides, Physiology, Biochemistry, Ecology. L. J. Audus, Ed. Vol. 1, Chap. 11. Academic Press, New York.

Buick, R. D., B. Robson, and R. J. Field. 1992. A mechanistic model to describe organosilicone surfactant promotion of triclopyr uptake. Pestic. Sci. 36:127–133.

Burnet, M. W. M., Q. Hart, J. A. M. Holtum, and S. B. Powles. 1994. Resistance to nine herbicide classes in a population of rigid ryegrass (*Lolium rigidum*). Weed Sci. 42:369–377.

Burnside, O. C. 1973. Shattercane—a serious weed throughout the central United States. Weeds Today 4(2):21.

Burnside, O. C., and G. A. Wicks. 1967. The effect of weed removal treatments on sorghum growth. Weeds 15:204–207.

Carignan, R., and J. Kalff. 1980. Phosphorus sources for aquatic weeds: Water or sediments? Science 207:987–989.

Cavers, P. B., and J. L. Harper. 1964. Biological flora of the British Isles, *Rumex obtusifolius* L. and *Rumex crispus* L. J. Ecol. 52:737–766.

Chandler, J. M., and J. E Dale. 1974. Comparative growth of four Malvaceous species. Proceedings of Southern Weed Science Society. 27:116–117.

Chikoye, D., S. F. Weise, and C. J. Swanton. 1995. Influence of common ragweed (*Ambrosia artemisiifolia*) time of emergence and density on white bean (*Phaseolus vulgaris*). Weed Sci. 43:375–380.

Crafts, A. S. 1975. Modern Weed Control. University of California Press, Berkeley.

Cruz, M., J. L. White, and J. D. Russell. 1968. Montmorillonite-*s*-triazine interactions. Israel J. Chemistry 6: 315–323.

Dawson, J. H. 1965a. Competition between irrigated field beans and annual weeds. Weeds 12:206–208.

Dawson, J. H. 1965b. Competition between irrigated sugarbeets and annual weeds. Weeds 13:245–249.

DeGennaro, F. P., and S. C. Weller. 1984. Differential susceptibility of field bindweed (*Convolvus arvensis*) biotypes to glyphosate. Weed Sci. 32:472–476.

Delvo, H. W., compiler, 1990. Agricultural Resources—Inputs Situation and Outlook Report. Resource and Technology Division, Economics Research Service, U. S. Department of Agriculture, Washington, D.C. AR-17.

Derscheid, L. A. 1978. Controlling field bindweed while growing adapted crops. Proceedings of North Central Weed Control conference. 33:144–150.

Derscheid, L. A., and K. Wallace. 1959. Control and elimination of thistles. Circular 147. South Dakota Agricultural Experiment Station, Brookings.

Devine, M. D., S. O. Duke, and C. Fedtke. 1993. Physiology of Herbicide Action. Prentice Hall, Englewood Cliffs, NJ.

Dickerson, C. T., Jr., and R. D. Sweet. 1971. Common ragweed ecotypes. Weed Sci. 19:64–66.

DiTomaso, A., A. K. Watson, and S. G. Hallett. 1996. Infection by the fungal pathogen *Colletotrichum coccodes* affects velvetleaf (*Abutilon theophrasti*)–soybean competition in the field. Weed Sci. 44:924–933.

Duke, S. O., Ed. 1985. Weed Physiology, Vol. II. Herbicide Physiology. CRC Press, Boca Raton, FL.

Edmundson, W. T. 1970. Phosphorus, nitrogen, and algae in Lake Washington after diversion of sewage. Science 169:690–691.

Epstein, A. H., J. H. Hill, and F. W. Nutter, Jr. 1997. Augmentation of rose rosette disease for biocontrol of multiflora rose (*Rosa multiflora*). Weed Sci. 45:172–178.

Evans, R. A., and J. A. Young. 1974. Today's weed, Russian thistle. Weeds Today 5(4):18.

Everitt, J. H., G. L. Anderson, D. E. Escobar, M. R. Davis, N. R. Spencer, and R. J. Andrascik. 1995. Use of remote sensing for detecting and mapping leafy spurge (*Euphorbia esula*). Weed Technol. 9:599–609.

Field, R. J., and N. G. Bishop. 1988. Promotion of stomatal infiltration of glyphosate by an organosilicone surfactant reduces the critical rainfall period. Pestic. Sci. 24:55–62.

Forcella, F., K. E. Oskoui, and S. W. Wagner. 1993. Application of weed seedbank ecology to low-input crop management. Ecol. Applications 3:74–83.

Forcella, F., R. G. Wilson, K. A. Renner, J. Dekker, R. G. Harvey, D. A. Alm, D. D. Buhler, and J. Cardina. 1992. Weed seedbanks of the U.S. Corn Belt: Magnitude, variation, emergence, and application. Weed Sci. 40:636–644.

Foy, C. L., and D. W. Pritchard, Eds. 1996. Pesticide Formulation and Adjuvant Technology. CRC Press, Boca Raton, FL.

Frear, D. S., and H. R. Swanson. 1970. Biosynthesis of S-(4-ethylamino-6-isopropyl amino-2-s-triazino) glutathione: partial purification and properties of a glutathione S-transferase from corn. Phytochemistry 9:2123–2132.

Gangstad, E. O., and N. F. Cardarelli. 1989. The relation between aquatic weeds and public health. *In* Aquatic Weeds, The Ecology and Management of Nuisance Aquatic Vegetation. A. H. Pieterse and K. J. Murphy, Eds. Oxford University Press, pp. 85–90.

Ghosheh, H. Z., D. L. Holshouser, and J. M. Chandler. 1996. The critical period of johnsongrass (*Sorghum halepense*) control in field corn (*Zea mays*). Weed Sci. 44:944–947.

Gianessi, L. P. 1992. U.S. Pesticide Use Trends: 1966–1989. Quality of the Environment Division, Resources for the Future, 16168 P Street, N.W., Washington, D.C. 20036.

Gianessi, L. P., and C. Puffer. 1991. Herbicide Use in the United States, National Summary Report. Quality of the Environment Division, Resources for the Future, 16168 P Street, N.W., Washington, D.C. 20036.

Gill, N. T. 1938. The viability of weed seeds at various stages of maturity. Ann. Appl. Biol. 25:447–456.

Glaze, N. C. 1987. Cultural and mechanical manipulation of *Cyperus* spp. Weed Technol. 1:82–83.

Goeden, R. D. 1975. How biological control is working. Weeds Today 6(4):4–9.

Gressel, J. 1991. Why get resistance? It can be prevented or delayed. *In* Herbicide Resistance in Weeds and Crops. Caseley, J. C., G. W. Cussans, and R. K. Atkin, Eds. Butterworth-Heinemann, Oxford, pp.1–25.

Griffith, D. R., and S. D. Parsons. 1980. Energy Requirements for Various Tillage-Planting Systems. Tillage ID-141, Cooperative Extension Service, Purdue University, West Lafayette, IN.

Grover, R. Ed. 1988. Environmental Chemistry of Herbicides, Vol. I. CRC Press, Boca Raton, FL.

Grover, R., and A. J. Cessna, Eds. 1991. Environmental Chemistry of Herbicides, Vol. II. CRC Press, Boca Raton, FL.

Guttieri, M. J., C. V. Eberlein, C. A. Mallory-Smith, D. C. Thill, and D. L. Hoffman. 1992. DNA sequence variation in Domain A of the acetolactase synthase genes of herbicide-resistant and -susceptible weed biotypes. Weed Sci. 40:670–676.

Hagood, E. S., Jr., T. T. Bauman, J. L. Williams, Jr., and M. M. Schreiber. 1980. Growth analysis of soybeans (*Glycine max*) in competition with velvetleaf (*Abutilon theophrasti*). Weed Sci. 28:729–734.

Haigler, W. E., B. J. Gossett, J. R. Harris, and J. E. Toler. 1994. Growth and development of organic arsenical-susceptible and -resistant common cocklebur (*Xanthium strumarium*) biotypes under noncompetitive conditions. Weed Technol. 8:154–158.

Hall, M. R., C. J. Swanton, and G. W. Anderson. 1992. The critical period of weed control in grain corn (*Zea mays*). Weed Sci. 40:441–447.

Hance, R. J., Ed. 1980. Interactions Between Herbicides and the Soil. Academic Press, New York.

Hanson, C. L., C. E. Rieck, J. W. Herron, and W. W. Witt. 1976. The johnsongrass problem in Kentucky. Agronomy Notes, University of Kentucky Cooperative Extension Service, Vol. 9, No. 4.

Harper, J. L., Ed. 1960. The Biology of Weeds. Blackwell Scientific Publications, Oxford, England.

Harvey, R. G. 1976. Quackgrass control in forage. Proceedings of North Central Weed Control Conference. 31:154–155.

Hatzios, K. K. 1991. Enhancement of crop tolerance to herbicides with chemical safeners. *In* Herbicide Resistance in Weeds and Crops. J. C. Caseley, G. W. Cussans, and R. K. Atkin, Eds. Butterworth-Heinemann, Oxford, pp. 293–303.

Hatzios, K. K., and D. Penner. 1982. Metabolism of Herbicides in Higher Plants. Burgess Publishing, Minneapolis.

Hay, W. 1937. Canada thistle seed production and its occurrence in Montana seeds. Seed World 41:6–7.

Helling, C. S., and B. C. Turner. 1968. Pesticide mobility: Determination by soil thin-layer chromatography. Science 162:562–563.

Herbicide Handbook of the Weed Science Society of America. 1994. Weed Science Society of America, Champaign IL.

Hill, G. D. 1982. Impact of weed science and agricultural chemicals on farm productivity in the 1980's. Weed Sci. 30:426–429.

Hill, J. V., and P. W. Santelmann. 1969. Competitive effects of annual weeds on Spanish peanuts. Weed Sci. 17:1–2.

Hoefer, R. H. 1981. Growth and development of Canada thistle. Proceedings of North Central Weed Control Conference. 36:153–157.

Holm, L. G. 1971. The role of weeds in human affairs. Weed Sci. 19:485–490.

Holm, L. G., J. Doll, J. Pancho, and J. Herberger. 1997. World Weeds, Natural Histories and Distribution. John Wiley and Sons, New York.

Holm, L. G., J. V. Pancho, J. P. Herberger, and D. L. Plucknett. 1979. A Geographical Atlas of World Weeds. John Wiley and Sons, New York.

Holm, L. G., D. L. Plucknett, J. V. Pancho, and J. P. Herberger. 1977. The World's Worst Weeds—Distribution and Biology. The University Press of Hawaii, Honolulu.

Holm, L. G., L. W. Weldon, and R. D. Blackburn. 1969. Aquatic weeds. Science 166:699–709.

Holm, L. G., and R. Yeo. 1981. The biology, control, and utilization of aquatic weeds, Part III. Weeds Today 12(1):7–10

Holt, J. S., S. B. Powles, and J. A. M. Holtum. 1993. Mechanisms and agronomic aspects of herbicide resistance. Annu. Rev. Plant Physiol. Plant Mol. Biol. 44:203–229.

Jackson, M. A., D. A. Schisler, P. J. Slininger, C. D. Boyette, R. W. Silman, and R. J. Bothast. 1996. Fermentation strategies for improving the fitness of a bioherbicide. Weed Technol. 10:645–650.

Jasieniuk, M., A. L. Brulé-Babel, and I. N. Morrison. 1996. The evolution and genetics of herbicide resistance in weeds. Weed Sci. 44:176–193.

Jasieniuk, M., and B. D. Maxwell. 1994. Population genetics and the evolution of herbicide resistance in weeds. Phytoprotection 75 (Suppl.): 25–35.

Johnson, B. J. 1971. Effect of weed competition on sunflowers. Weed Sci. 19:378–380.

Johnson, D. W., J. M. Krall, R. H. Delaney, and D. S. Thiel. 1990. Response of seed of 10 weed species to Fresnel-lens-concentrated solar radiation. Weed Technol. 4:109–114.

Kearney, P. C., and D. D. Kaufman, Eds. 1975 and 1976. Herbicides, Chemistry, Degradation, and Mode of Action. Vols. I and II. Marcel Dekker, New York.

Kelley, A. D., and V. F. Bruns. 1975. Dissemination of weed seeds by irrigation water. Weed Sci. 23:486–493.

Kerpez, T. A., and N. S. Smith. 1987. Saltcedar control for wildlife habitat improvement in the southwest United States. U.S. Department of Interior, Fish and Wildlife Service, Resource Publication 169.

Khedir, K. D., and F. W. Roeth. 1981. Velvetleaf (*Abutilon theophrasti* Medic.) seed populations in six corn fields. Abstr. Weed Sci. Soc. Amer., p. 81.

King, J. 1991. A matter of public confidence—consumers' concerns about pesticide residues unjustified. Agric. Eng. 72 (14): 16–18.

King, L. J. 1966. Weeds of the World—Biology and Control. Interscience Publishers, New York.

Kivilaan, A., and R. S. Bandurski. 1981. The one hundred-year period for Dr. Beal's seed viability experiment. Amer. J. Bot. 68:1290–1292.

Klingman, G. C., and F. M. Ashton. 1982. Weed Science Principles and Practices. John Wiley and Sons, New York.

Klingman, T. E., and L. R. Oliver. 1994. Palmer amaranth (Amaranthus palmeri) interference in soybeans (Glycine max). Weed Sci. 42:523–527.

Knake, E. L., and F. W. Slife. 1965. Giant foxtail seeded at various times in corn and soybeans. Weeds 13:331–334.

Lamoureux, G. L., and D. S. Frear. 1979. Pesticide metabolism in higher plants, in vitro enzyme studies. In Xenobiotic Metabolism, In Vitro Methods. G. D. Paulson, D. S. Frear, and E. P. Marks, eds. ACS Symp. Ser. 97, American Chemical Society, Washington, D.C., pp. 77–128.

Lass, L. W., H. W. Carson, and R. H. Callihan. 1996. Detection of yellow starthistle (Centaurea solstitialis) and common St. Johnswort (Hypericum perforatum) with multispectral digital imagery. Weed Technol. 10:466–474.

LeBaron, H. M. 1991. Distribution and seriousness of herbicide-resistant weed infestations worldwide. In Herbicide Resistance in Weeds and Crops. J. C. Caseley, G. W. Cussans, and R. K. Atkin, Eds. Butterworth-Heinemann, Oxford, pp. 27–43.

Locke, M. A., and C. T. Bryson. 1997. Herbicide-soil interactions in reduced tillage and plant residue management systems. Weed Sci. 45:307–320.

Louda, S. M., D. Kendall, J. Connor, and D. Simberloff. 1997. Ecological effects of an insect introduced for the biological control of weeds. Science 277:1088–1090.

Lueschen, W. E., and R. N. Andersen. 1980. Velvetleaf. Hard to wake up! Weeds Today 11(2):21.

Malecki, R. A., B. Blossey, S. D. Hight, D. Schroeder, L. T. Kok, and J. R. Coulson. 1993. Biological control of purple loosestrife. BioScience 43:680–686.

Malik, R. K., and S. Singh. 1993. Evolving strategies for herbicide use in wheat: resistance and integrated weed management. Proc. Indian Soc. Weed Sci. Int. Symp., Hisar, India 1:225–238.

Martin, A. R., and O. C. Burnside. 1980. Common milkweed—weed on the increase. Weeds Today 11(1):19–20.

Marwat, K. B., and E. D. Nafziger. 1990. Cocklebur and velvetleaf interference with soybean grown at different densities and planting patterns. Agron. J. 82:531–534.

Matthews, G. A. 1979. Pesticide Application Methods. Longman Group, London.

McEvoy, P., C. Cox, and E. Coombs. 1991. Successful biological control of ragwort, Senecio jacobaea, by introduced insects in Oregon. Ecol. Applications 1:430–442.

McWhorter, C. G. 1961. Morphology and development of johnsongrass plants from seeds and rhizomes. Weeds 9:558–562.

McWhorter, C. G. 1973, 1981. Johnsongrass as a weed. USDA Farmers Bulletin 1537, Washington D.C.

McWhorter, C. G. 1989. History, biology, and control of johnsongrass. Rev. Weed Sci. 4:85–121.

McWhorter, C. G., T. N. Jordan, and G. D. Wills. 1980. Translocation of 14-C-glyphosate in soybeans (Glycine max) and johnsongrass (Sorghum halepense). Weed Sci. 28:113–118.

McWhorter, C. G., and D. T. Patterson. 1980. Ecological factors affecting weed competition in soybeans. In Proceedings of World Soybean Research Conference II. F. T. Corbin, Ed., pp. 371–392. Westview Press, Boulder, CO.

Mesbah, A., S. D. Miller, K. J. Fornstrom, and D. E. Legg. 1995. Wild mustard (Brassica kaber) and wild oat (Avena fatua) interference in sugarbeets (Beta vulgaris L.). Weed Technol. 9:49–52.

Mitich, L. W. 1987. Colonel Johnson's grass: Johnsongrass. Weed Technol. 1:112–113.

Mitich, L. W. 1995. Poison-ivy/poison-oak/poison-sumac—the virulent weeds. Weed Tech. 9:653–656.

Muenscher, W. C. 1955, 1980. Weeds. Macmillan, New York.

Mulugeta, D., P. K. Fay, and W. E. Dyer. 1992. The role of pollen in the spread of sulfonylurea reistant Kochia scoparia L. (Schrad.). Weed Sci. Soc. Amer. Abstr. 32:16.

Murphy, T. R., B. J. Gossett, and J. E. Toler. 1986. Growth and development of dinitroaniline-susceptible and -resistant goosegrass (Eleusine indica) biotypes under noncompetitive conditions. Weed Sci. 34:704–710.

National Research Council. Committee on Plant and Animal Pests. 1968. Principles of Plant and Animal Pest Control. Vol. 2, Weed Control. National Academy of Science, Washington, D.C.

Nave, W. R., and L. M. Wax. 1971. Effect of weeds on soybean yield and harvesting efficiency. Weed Sci. 19:533–535.

Nelson, D. C., and R. E. Nylund. 1962. Competition between peas grown for processing and weeds. Weeds 10:224–229.

Nieto, J. H., M. A. Brando, and J. T. Gonzales. 1968. Critical periods of the crop growth cycle for competition from weeds. Pest. Artic. News Summ. 14: 159.

Norris, R. F. 1996. Water use efficiency as a method for predicting water use by weeds. Weed Technol. 10:153–155.

Palmer, J., and G. Sagar. 1963. Biological flora of the British Isles: Agropyron repens (L.) Beauv. J. Ecology 51:783–794.

Parker, C. 1991. Protection of crops against parasitic weeds. Crop. Prot. 10:6–22.

Pemberton, R. W., and C. E. Turner. 1990. Biological control of Senecio jacobaea in northern California, an enduring success. Entomophaga 35:71–77.

Peters, E. J., J. F. Stritzke, and F. S. Davis. 1965. Wild Garlic, Its Characteristics and Control. USDA Agriculture Handbook 298, Washington, D.C.

Phillips, W. M. 1978. Field bindweed: The weed and the problem. Proceedings of North Central Weed Control Conference 33:140–141.

Purcell, M. F., and J. K. Balciunas. 1994. Life history and distribution of the Australian weevil Oxyops vitiosa, a potential biological control agent for Melaleuca quinquenervia. Arthropod Biol. 87:866–873.

Putnam, A. R., and W. B. Duke. 1974. Biological suppression of weeds: Evidence of allelopathy in accessions of cucumber. Science 185:370–372.

Putnam, A. R., and L. A. Weston. 1986. The Science of Allelopathy. John Wiley and Sons, New York.

Raleigh, S. T., T. Flanagan, and C. Veatch. 1962. Life History Studies as Related to Weed Control. 4. Quackgrass. Bulletin 365. Rhode Island Agricultural Experiment Station, Kingston.

Randall, J. M. 1996. Weed control for the preservation of biological diversity. Weed Tech. 10:370–383.

Raynal-Roques, A. 1991. Diversification in the genus *Striga. In* Proceedings, 5th International Symposium on Parasitic Weeds, Nairobi 1991. J. K. Ransom, L. J. Musselman, A. D. Worsham, and C. Parker, Eds. CIMMYT, Nairobi, pp. 251–261.

Regnier, E. E., and E. W. Stoller. 1989. The effect of soybean (*Glycine max*) interference on the canopy architecture of common cocklebur (*Xanthium strumarium*), jimsonweed (*Datura stramonium*), and velvetleaf (*Abutilon theophrasti*). Weed Sci. 37:187–195.

Robison, L. R., and L. S. Jeffery. 1972. Hemp dogbane growth and control. Weed Sci. 20:156–159.

Rogers, J. B., D. S. Murray, L. M. Verhalen, and P. L. Claypool. 1996. Ivyleaf morningglory (*Ipomoea hederacea*) interference with cotton (*Gossypium hirsutum*). Weed Tech. 10:107–114.

Ryan, G. F. 1970. Resistance of common groundsel to simazine and atrazine. Weed Sci. 18: 614–616.

Rydrych, D. J. 1974. Competition between winter wheat and downy brome. Weed Sci. 22:211–214.

Salisbury, F. B., and C. W. Ross. 1992. Plant Physiology, 4th ed. Wadsworth Publishing Co., Belmont, CA.

Sauerborn, J. 1991. The economic importance of the phytoparasites *Orobanche* and *Striga. In* Proceedings of the 5th International Symposium on Parasitic Weeds. J. K. Ransom, L. J. Musselman, A. D. Worsham, and C. Parker, Eds. CIMMYT, Nairobi, pp. 137–143.

Schafer, D. E., and D. O. Chilcote. 1970. Selectivity of bromoxynil in a resistant and susceptible species. Weed Sci. 18:725–729.

Schreiber, M. M. 1965. Effect of date of planting and stage of cutting on seed production of giant foxtail. Weeds 13:60–62.

Schreiber, M. M. 1967. Effect of density and control of Canada thistle on production and utilization of alfalfa pasture. Weeds 15:138–142.

Schreiber, M. M. 1992. Influence of tillage, crop rotation, and weed management on giant foxtail (*Setaria faberi*) population dynamics and corn yield. Weed Sci. 40:645–653.

Schroeder, J. 1993. Late-season interference of spurred anoda in chile peppers. Weed Sci. 41:172–179

Schweizer, E. E., and R. L. Zimdahl. 1984. Weed seed decline in irrigated soil after rotation of crops and herbicides. Weed Sci. 32:84–89.

Seifert, S., and C. E. Snipes. 1996. Influence of flame cultivation on mortality of cotton (*Gossypium hirsutum*) pests and beneficial insects. Weed Technol. 10:544–549.

Selleck, G. W. 1980. Fall panicum: Weed on the increase. Weeds Today 11(2):15.

Sharma, M. P. 1979. Wild oat—A billion dollar problem. Weeds Today 10(3):5–6.

Shaw, W. C., D. R. Sheperd, E. L. Robinson, and P. F. Sand. 1962. Advances in witchweed control. Weeds 10:182–192.

Shimabukuro, R. H., G. L. Lamoureux, and D. S. Frear. 1978. Glutathione conjugation: A mechanism for herbicide detoxification and selectivity in plants. *In* Chemistry and Action of Herbicide Antidotes. F. M. Pallow and J. E. Casida, Eds. Academic Press, New York, pp. 133–149.

Shuttleworth, C. L. 1973. The case for reducing wild oats in commercial grain. *In* Let's Clean Up on Wild Oats, Proceedings, Action Proposals and Programs. Agriculture Canada and United Grain Growers Limited Special Seminar.

Simkins, G. S., and J. D. Doll. 1981. Effects of crop rotations and weeding systems on yellow nutsedge (*Cyperus esculentus* L.) tuber populations and control. Abstr. Weed Sci. Soc. Amer., p. 78.

Smeda, R. J., W. L. Barrentine, and C. E. Snipes. 1993. Johnsongrass (*Sorghum halepense* (L.) Pers.) resistance to postemergence grass herbicides. Weed Science Society of America, Abstract No. 53.

Smith, A. E., and D. M. Secoy. 1976. Early chemical control of weeds in Europe. Weed Sci. 24:594–597.

Stannard, M. E., and P. K. Fay. 1987. Selection of alfalfa seedlings for tolerance to chlorsulfuron. Weed Science Society of America Abstracts, No. 61.

Stevens, O. A. 1932. The number and weight of seeds produced by weeds. Amer. J. Botany 19:784–794.

Steward, K. K. 1989. Aquatic weed problems and management in the eastern United States. *In* Aquatic Weeds, The Ecology and Management of Nuisance Aquatic Vegetation. A. H. Pieterse and K. J. Murphy, Eds. Oxford University Press, New York, pp. 391–405.

Stobbe, E. H. 1976. Biology of quackgrass. Proceedings of North Central Weed Control Conference 31:151–152.

Stoller, E. W., and R. D. Sweet. 1987. Biology and life cycle of purple and yellow nutsedges (*Cyperus rotundus* and *C. esculentus*). Weed Technol. 1:66–73.

Stoller, E. W., L. M. Wax, and F. W. Slife. 1979. Yellow nutsedge (*Cyperus esculentus*) competition and control in corn (*Zea mays*). Weed Sci. 27:32–37.

Sturkie, D. G. 1930. The influence of various topcutting treatments on rootstocks of johnsongrass. J. Amer. Soc. Agron. 22:82–93.

Sweet, R. D. 1975. Control of nutsedge in horticultural crops. Proceedings of North Central Weed Control Conference 30:129–130.

Templeton, G. E. 1986. Mycoherbicide research at the University of Arkansas—past, present and future. Weed Sci. 34 (Suppl. 1):35–37.

Thompson, C. R., D. C. Thill, and B. Shafii. 1994. Growth and competitiveness of sulfonylurea-resistant and -susceptible kochia (*Kochia scoparia*). Weed Sci. 42:50–56.

Thompson, L., Jr., J. M. Houghton, F. W. Slife, and H. J. Butler. 1971. Metabolism of atrazine by fall panicum and large crabgrass. Weed Sci. 19:409–412.

Timmons, F. L. 1960. Weed control in western irrigation and drainage systems: Losses caused by weeds, costs and benefits of control. USDA and Bureau of Reclamation, Publication ARS 3414, Washington, D.C.

Tinney, J. R. 1942. Eradication of wild onions. J. Minn. Agr. 44:155–157.

Tumbleson, M., and R. Kommedahl. 1961. Reproductive potential of *Cyperus esculentus* by tubers. Weeds 9:646–653.

Tumbleson, M., and R. Kommedahl. 1962. Factors affecting dormancy in tubers of *Cyperus esculentus.* Bot. Gaz. 123:186–190.

U. S. Dept. Interior, Bureau of Land Management. 1995. Partners Against Weeds—An Action Plan for the Bureau of Land Management Fiscal Year 1995 and Beyond. BLM State Office, Billings, Montana. 39 pp.

Vengris, J., and M. Stacewicz-Sapuncakis. 1971. Common purslane competition in table beets and snap beans. Weed Sci. 19:4–6.

Vigneault, C., D. L. Benoit, and N. B. McLaughlin. 1990. Energy aspects of weed electrocution. Rev. Weed Sci. 5:15–26.

Vogler, R. K., G. Ejeta, and L. G. Butler. 1995. Integrating biotechnological approaches for the control of *Striga.* Afr. Crop Sci. J. 3:217–222.

Waldrep, T. W., and R. D. McLaughlin. 1969. Cocklebur competition in soybeans. The Soybean Farmer 3:26–30.

Warren, L. E. 1976. Controlling drift of herbicides. World of Agricultural Aviation. March, April, May, and June issues.

Warwick, S. I., and L. D. Black. 1994. Relative fitness of herbicide-resistant and -susceptible biotypes of weeds. Phytoprotection 75:37–49.

Weaver, S. E., and C. S. Tan. 1983. Critical period of weed interference in transplanted tomatoes (*Lycopersicon esculentum*): Growth analysis. Weed Sci. 31:476–481.

Weber, J. B., T. J. Monaco, and A. D. Worsham. 1973. What happens to herbicides in the environment? Weeds Today 4(1):16–17.

Webster, T. M., M. M. Loux, E. E. Regnier, and S. K. Harrison. 1994. Giant ragweed (*Ambrosia trifida*) canopy architecture and interference studies in soybean (*Glycine max*). Weed Technol. 8:559–564.

Wicks, G. A., and L. A. Derscheid. 1964. Leafy spurge seed maturation. Weeds 12:175–176.

Wicks, G. A., D. N. Johnston, D. S. Nuland, and E. J. Kinbacher. 1973. Competition between annual weeds and sweet Spanish onions. Weed Sci. 21:436–439.

Wiederholt, R. J., and D. E. Stoltenberg. 1996. Absence of differential fitness between giant foxtail (*Setaria faberi*) accessions resistant and susceptible to acetyl-coenzyme A carboxylase inhibitors. Weed Sci. 44:18–24.

Wiese, A. F., and W. M. Phillips. 1976. Field bindweed. Weeds Today 7(1):22–23.

Wills, G. D., J. D. Byrd, Jr., and H. R. Hurst. 1992. Herbicide-resistant and -tolerant weeds. Proceedings of the Southern Weed Science Society, 45:43.

Wills, G. D., and C. G. McWhorter. 1981. Effect of environment on the translocation and toxicity of acifluorfen to showy crotalaria (*Crotolaria spectabilis*). Weed Sci. 29:397–401.

Wilson, H. P., and R. H. Cole. 1966. Morningglory competition in soybeans. Weeds 14:49–51

Yenish, J., J. Doll, and D. Buhler. 1992. Effects of tillage on vertical distribution and viability of weed seed in soil. Weed Sci. 40:429–433.

York, A. C., and H. D. Coble. 1977. Fall panicum interference in peanuts. Weed Sci. 25:43–47.

Zimdahl, R. L. 1988. The concept and application of the critical weed-free period. *In* Weed Management in Agroecosystems: Ecological Approaches. M. Altieri and M. Liebman, Eds. CRC Press, Boca Raton, FL, pp. 145–155.

Identify the Weed Leaves?

1. Field bindweed. 2. Velvetleaf. 3. Ground Ivy. 4. Prickly sida. 5. Lambsquarters. 6. Tall morningglory. 7. Honeyvine milkweed. 8. Pokeweed. 9. Canada thistle. 10. Ivyleaf morningglory. 11. Redroot pigweed. 12. Common sunflower. 13. Field buttercup. 14. Henbit. 15. Dead nettle. 16. Chickweed. 17. Jimsonweed. 18. Eastern black nightshade. 19. Wild buckwheat. 20. Giant ragweed.

INDEX

NOTE: Bold page numbers indicate tables. Figures are indicated by *f* following page number.